Design of Prestressed Concrete to Eurocode 2

Second Edition

Design of Prestressed Concrete to Eurocode 2

Second Edition

Raymond Ian Gilbert
Neil Colin Mickleborough
Gianluca Ranzi

CRC Press
Taylor & Francis Group
Boca Raton London New York

CRC Press is an imprint of the
Taylor & Francis Group, an **informa** business

CRC Press
Taylor & Francis Group
6000 Broken Sound Parkway NW, Suite 300
Boca Raton, FL 33487-2742

First issued in paperback 2018

© 2017 by Raymond Ian Gilbert, Neil Colin Mickleborough, and Gianluca Ranzi
CRC Press is an imprint of Taylor & Francis Group, an Informa business

No claim to original U.S. Government works

ISBN-13: 978-1-4665-7310-9 (hbk)
ISBN-13: 978-0-367-02791-9 (pbk)

Library of Congress Cataloging-in-Publication Data

Visit the Taylor & Francis Web site at
http://www.taylorandfrancis.com

and the CRC Press Web site at
http://www.crcpress.com

Contents

Preface

For the design of prestressed concrete structures, a sound understanding of structural behaviour at all stages of loading is essential. Also essential is a thorough knowledge of the design criteria specified in the relevant design standard, including the rules and requirements and the background to them. The aim of this book is to present a detailed description and explanation of the behaviour of prestressed concrete members and structures both at service loads and at ultimate loads and, in doing so, provide a comprehensive guide to structural design. Much of the text is based on first principles and relies only on the principles of mechanics and the properties of concrete and steel, with numerous worked examples. Where the design requirements are code specific, this book refers to the provisions of Eurocode 2 (EN 1992–1–1:2004) and other relevant EN Standards, and, where possible, the notation is the same as in the Eurocode. A companion edition in accordance with the requirements of the Australian Standard for Concrete Structures AS 3600–2009 is also available, with the same notation as in the Australian Standard.

The first edition of the book was published over 25 years ago, so a comprehensive update and revision is long overdue. This edition contains the most up-to-date and recent advances in the design of modern prestressed concrete structures, as well as the fundamental aspects of prestressed concrete behaviour and design that were well received in the first edition. The text is written for senior undergraduate and postgraduate students of civil and structural engineering and also for practising structural engineers. It retains the clear and concise explanations and the easy-to-read style of the first edition.

Between them, the authors have almost 100 years of experience in the teaching, research and design of prestressed concrete structures, and this book reflects this wealth of experience.

The scope of the work ranges from an introduction to the fundamentals of prestressed concrete to in-depth treatments of the more advanced topics in modern prestressed concrete structures. The basic concepts of prestressed concrete are introduced in Chapter 1, and the limit states design philosophies used in European practice are outlined in Chapter 2. The hardware required to pretension and post-tension concrete structures is introduced in Chapter 3, including some construction considerations. Material properties relevant to design are presented and discussed in Chapter 4. A comprehensive treatment of the design of prestressed concrete beams for serviceability is provided in Chapter 5. The instantaneous and time-dependent behaviour of cross-sections under service loads are discussed in considerable detail, and methods for the analysis of both uncracked and cracked cross-sections are considered. Techniques for determining the section size, the magnitude and eccentricity of prestress, the losses of prestress and the deflection of members are outlined. Each aspect of design is illustrated by numerical examples.

Chapters 6 and 7 deal with the design of members for strength in bending, shear and torsion, and Chapter 8 covers the design of the anchorage zones in both pretensioned and post-tensioned members. A guide to the design of composite prestressed concrete beams is provided in Chapter 9 and includes a detailed worked example of the analysis of a composite through girder footbridge. Chapter 10 discusses design procedures for statically determinate beams. Comprehensive self-contained design examples are provided for fully-prestressed and partially prestressed, post-tensioned and pretensioned concrete members.

The analysis and design of statically indeterminate beams and frames is covered in Chapter 11 and provides guidance on the treatment of secondary effects at all stages of loading. Chapter 12 provides a detailed discussion of the analysis and design of two-way slab systems, including aspects related to both strength and serviceability. Complete design examples are provided for panels of an edge-supported slab and a flat slab. The behaviour of axially loaded members is dealt with in Chapter 13. Compression members, members subjected to combined bending and compression, and prestressed concrete tension members are discussed, and design aspects are illustrated by examples. Guidelines for successful detailing of the structural elements and connections in prestressed concrete structures are outlined in Chapter 14.

As in the first edition, the book provides a unique focus on the treatment of serviceability aspects of design. Concrete structures are prestressed to improve behaviour at service loads and thereby increase the economical range of concrete as a construction material. In conventional prestressed structures, the level of prestress and the position of the tendons are usually based on considerations of serviceability. Practical methods for accounting for the non-linear and time-dependent effects of cracking, creep, shrinkage and relaxation are presented in a clear and easy-to-follow format.

The authors hope that *Design* of *Prestressed Concrete to Eurocode 2* will be a valuable source of information and a useful guide for students and practitioners of structural design.

Ian Gilbert
Neil Mickleborough
Gianluca Ranzi

Authors

Raymond Ian Gilbert is emeritus professor of civil engineering at the University of New South Wales (UNSW) and deputy director of the UNSW Centre for Infrastructure Engineering and Safety. He has more than 40 years of experience in structural design and is a specialist in the analysis and design of reinforced and prestressed concrete structures. Professor Gilbert has taught successive generations of civil engineering students in Australia on subjects related to structural engineering, ranging from statics and structural analysis to the design of reinforced and prestressed concrete structures. His research activities are in the field of concrete structures, with a particular interest in serviceability. Professor Gilbert has published six books, including *Structural Analysis: Principles, Methods and Modelling* and *Time-Dependent Behaviour of Concrete Structures* which are also published by CRC Press, and more than 350 technical papers and reports. He was awarded Honorary Life Membership of the Concrete Institute of Australia in 2011.

Neil Colin Mickleborough is professor of civil engineering and the director of the Center for Engineering Education Innovation at Hong Kong University of Science and Technology. He has been actively involved in the research, development and teaching of prestressed and reinforced concrete, structural analysis and tall building and bridge design in Australia, Asia and the Middle East for the past 30 years. He has acted as an expert design consultant on tall buildings and long-span bridge projects in both Dubai and Hong Kong. In addition, he is a chartered structural engineer and a Fellow of the Hong Kong Institution of Engineers.

Gianluca Ranzi is professor of civil engineering, ARC Future Fellow and director of the Centre for Advanced Structural Engineering at the University of Sydney. Gianluca's research interests range from the field of structural engineering, with focus on computational mechanics and the service behaviour of composite steel–concrete and concrete structures, to architectural science.

Acknowledgements

The authors acknowledge the support given by their respective institutions and by the following individuals and organisations for their assistance:

Mr Brian Lim (VSL International Limited, Hong Kong)
Mr Brett Gibbons (VSL Australia, Sydney)

Notation and sign convention

All symbols are also defined in the text where they first appear. Throughout the book we have assumed that tension is positive and compression is negative and that positive bending about a horizontal axis causes tension in the bottom fibres of a cross-section.

Latin upper-case letters

A	Cross-sectional area or accidental action
A_c	Cross-sectional area of concrete
$A_{c,eff}$	Effective area of concrete in tension surrounding the tendons with depth $h_{c,ef}$ equal to the lesser of $2.5(h-d)$, $(h-x)/3$ or $h/2$
A_{ct}	Area of the concrete in the tensile zone just before cracking
A_{c0}	Bearing area
A_{c1}	Largest area of the concrete supporting surface that is geometrically similar to and concentric with A_{c0}
A_g	Gross cross-sectional area
$\bar{A}_k$	Area of the age-adjusted transformed section at time t_k
A_p	Cross-sectional area of prestressing steel
$A_{p(i)}$	Cross-sectional area of the prestressing steel at the i-th level
A_{pc}	Cross-sectional area of the precast member
A_s	Cross-sectional area of non-prestressed steel reinforcement or cross-sectional area of a single bar being anchored
$A_{s(i)}$	Cross-sectional areas of non-prestressed steel reinforcement at the i-th level
A_{sb}	Area of transverse reinforcement in the end zone of a pretensioned member (Equation 8.6)
A_{sc}	Cross-sectional area of non-prestressed steel reinforcement in the compressive zone

A_{st}	Cross-sectional area of non-prestressed transverse steel reinforcement *or* cross-sectional area of non-prestressed reinforcement in the tension zone
$A_{s,min}$	The minimum area of bonded longitudinal reinforcement in the tensile zone (Equation 14.9) *or* minimum area of longitudinal reinforcement in a column (Equation 14.16)
A_{sw}	Cross-sectional area of the vertical legs of each stirrup *or* area of the single leg of transverse steel in each wall of the idealised thin-walled section in torsion
$A_{sw,max}$	Maximum cross-sectional area of shear reinforcement (Equations 7.13 and 7.14)
$A_{sw,min}$	Minimum cross-sectional area of shear reinforcement (Equation 7.17)
A_0	Area of the transformed section at time t_0
B_c	First moment of the concrete part of the cross-section about the reference axis
$\bar{B}_k$	First moment of the age-adjusted transformed section at time t_k
B_0	First moment of area of the transformed section about the reference axis at time t_0
C	Strength class of concrete *or* carry-over factor *or* Celsius
D_0	Matrix of cross-sectional rigidities at time t_0 (Equation 5.42)
E (subscript)	Effect of actions
$E_{c,eff}(t, t_0)$, $E_{c,eff}$	Effective modulus of concrete at time t for concrete first loaded at t_0 (Equations 4.23 and 5.56)
$\bar{E}_{c,eff}(t, t_0)$, $\bar{E}_{c,eff}$	Age-adjusted effective modulus of concrete at time t for concrete first loaded at t_0 (Equations 4.25 and 5.57)
E_{cm}	Secant modulus of elasticity of concrete
$E_{cm,0}$	Secant modulus of elasticity of concrete at time t_0
E_p	Design value of modulus of elasticity of prestressing steel
$E_{p(i)}$	Design value of modulus of elasticity of the i-th level of prestressed steel
E_s	Design value of modulus of elasticity of reinforcing steel
$E_{s(i)}$	Design value of modulus of elasticity of the i-th level of non-prestressed steel
F_{bc}	Transverse compressive force due to bursting moment in a post-tensioned end block
F_{bt}	Transverse tensile force due to bursting in a post-tensioned end block (the bursting force)
F_c	Force carried by the concrete
F_{cc}	Compressive force carried by the concrete
F_{cd}	Design force carried by the concrete *or* design compressive force in a strut

F_{cdf}	Design force carried by the concrete flange (Equation 6.36)
F_{cdw}	Design force carried by the concrete web of a flanged beam (Equation 6.37)
$\overline{F}_{e,0}$	Age-adjusted creep factor (Equation 5.60)
F_{pt}	Tensile force carried by the prestressing steel
F_{ptd}	Design tensile force carried by the prestressing steel at the ultimate limit state
F_s	Force carried by non-prestressed steel reinforcement
F_{sc}	Force carried by non-prestressed compressive steel reinforcement
F_{sd}	Design force carried by non-prestressed steel reinforcement
F_{st}	Force carried by non-prestressed tensile steel reinforcement
F_t	Resultant tensile force carried by the steel reinforcement and tendons
$\mathbf{F}_k$	Matrix relating applied actions to strain at time t_k (Equation 5.102)
$\mathbf{F}_0$	Matrix relating applied actions to strain at time t_0 (Equation 5.46)
G	Permanent action
G_k	Characteristic permanent action
I	Second moment of area (moment of inertia) of the cross-section
I_{av}	Average second moment of area after cracking
I_c	Second moment of area of the concrete part of the cross-section about the reference axis
I_{cr}	Second moment of area of a cracked cross-section
I_{ef}	Effective second moment of area after cracking
I_g	Second moment of area of the gross cross-section
$\overline{I}_k$	Second moment of area of the age-adjusted transformed section at time t_k
I_{uncr}	Second moment of area of the uncracked cross-section
I_0	Second moment of area of the transformed section about reference axis at time t_0
$J(t,t_0)$	Creep function at time t due to a sustained unit stress first applied at t_0
K	Slab system factor or factor that accounts for the position of the bars being anchored with respect to the transverse reinforcement (Figure 14.14)
L_{di}	Length of draw-in line adjacent to a live-end anchorage (Equation 5.150)
M	Bending moment
$\overline{M}$	Virtual moment
M_b	Bursting moment in a post-tensioned anchorage zone

$M_{c,0}$	Moment resisted by the concrete at time t_0
M_{cr}	Cracking moment
M_{Ed}	Design value of the applied internal bending moment
$M_{Ed.x}$, $M_{Ed.y}$	Design moments in a two-way slab spanning in the x- and y-directions, respectively
$M_{ext,0}$	Externally applied moment about reference axis at time t_0
$M_{ext,k}$	Externally applied moment about reference axis at time t_k
M_G	Moment caused by the permanent loads
M_{int}	Internal moment about reference axis
$M_{int,k}$	Internal moment about reference axis at time t_k
$M_{int,0}$	Internal moment about reference axis at time t_0
M_o	Total static moment in a two-way flat slab *or* decompression moment
M_Q	Moment caused by the live loads
M_{Rd}	Design moment resistance
M_s	Spalling moment in a post-tensioned anchorage zone
M_{sus}	Moment caused by the sustained loads
M_{sw}	Moment caused by the self-weight of a member
M_T	Moment caused by total service loads
M_u	Ultimate moment capacity
M_{var}	Moment caused by variable loads
M_0	Moment at a cross-section at transfer
N	Axial force
$N_{c,k}$	Axial forces resisted by the concrete at time t_k
$N_{c,0}$	Axial forces resisted by the concrete at time t_0
N_{Ed}	Design value of the applied axial force (tension or compression)
N_{ext}	Externally applied axial force
$N_{ext,k}$	Externally applied axial force at time t_k
$N_{ext,0}$	Externally applied axial force at time t_0
N_{int}	Internal axial force
$N_{int,k}$	Internal axial force at time t_k
$N_{int,0}$	Internal axial force at time t_0
$N_{p,k}$	Axial force resisted by the prestressing steel at time t_k
$N_{p,0}$	Axial force resisted by the prestressing steel at time t_0
N_{Rd}	Design axial resistance of a column
$N_{Rd,t}$	Design axial resistance of a tension member
$N_{s,k}$	Axial force resisted by the non-prestressed reinforcement at time t_k
$N_{s,0}$	Axial force resisted by the non-prestressed reinforcement at time t_0
P	Prestressing force; applied axial load in a column
P_h	Horizontal component of prestressing force
$P_{init\,(i)}$	Initial prestressing force at the i-th level of prestressing steel

P_j	Prestressing force during jacking (the jacking force)
$P_{m,t}$	Effective force in the tendon at time t after the long-term losses
P_{m0}	Initial force in the tendon immediately after transfer after the short-term losses
P_x, P_y	Prestressing forces in a slab in the x- and y-directions, respectively
P_v	Vertical component of prestressing force
P_0	Initial force at the active end of the tendon immediately after stressing
Q	Variable action
Q_k	Characteristic variable action
$R_{A,k}, R_{B,k}, R_{I,k}$	Cross-sectional rigidities at time t_k (Equations 5.84, 5.85 and 5.89)
$R_{A,p}, R_{B,p}, R_{I,p}$	Contribution to section rigidities provided by the bonded tendons (Equations 5.125 through 5.127)
$R_{A,s}, R_{B,s}, R_{I,s}$	Contribution to section rigidities provided by the steel reinforcement (Equations 5.122 through 5.124)
$R_{A,0}, R_{B,0}, R_{I,0}$	Cross-sectional rigidities at time t_0 (Equations 5.35, 5.36 and 5.39)
S	First moment of area
T	Torsional moment
T_{Ed}	Design value of the applied torsional moment
$T_{Rd,c}$	Torsion required to cause first cracking in an otherwise unloaded beam
$T_{Rd,max}$	Maximum design torsional resistance (Equation 7.34)
U	Internal work
V	Shear force
V_{ccd}	Shear component of compressive force in an inclined compression chord
V_{Ed}	Nett design shear force
V_{Rd}	Design strength in shear
$V_{Rd,c}$	Design shear strength of a beam without shear reinforcement (Equations 7.2 and 7.4)
$V_{Rd,max}$	Maximum design shear strength for a beam with shear reinforcement (Equation 7.8)
$V_{Rd,s}$	Design strength provided by the yielding shear reinforcement (Equation 7.7)
W	External work
W_1	Elastic energy (Figure 6.20)
W_2	Plastic energy (Figure 6.20)
Z	Section modulus of uncracked cross-section
Z_{btm}	Bottom fibre section modulus (I/y_{btm})
Z_{top}	Top fibre section modulus (I/y_{top})

Latin lower-case letters

a, a	Distance
b	Overall width of a cross-section or actual flange width in a T or L beam
b_{eff}	Effective width of the flange of a flanged cross-section
b_w	Width of the web on T, I or L beams
c	Concrete cover
d	Effective depth of a cross-section, that is the depth from the extreme compressive fibre to the resultant tensile force in the reinforcement and tendons at the ultimate limit state
d_n	Depth to neutral axis
$d_{n,0}$	Depth to neutral axis at time t_0
d_o	Depth from the extreme compressive fibre to the centroid of the outermost layer of tensile reinforcement
d_p	Depth from the top fibre of a cross-section to the prestressing steel
$d_{p(i)}$	Depth from the top fibre of a cross-section to the i-th level of prestressing steel
d_{ref}	Depth from the top fibre of a cross-section to the reference axis
d_s	Depth from the top fibre of a cross-section to the non-prestressed steel reinforcement
$d_{s(i)}$	Depth from the top fibre of a cross-section to the i-th level of non-prestressed steel reinforcement
d_x, d_y	Effective depths to the tendons in the orthogonal x- and y-directions, respectively
e	Eccentricity of prestress; eccentricity of axial load in a column; axial deformation
f_{bd}	Design value of the average ultimate bond stress (Equation 14.5)
f_c	Compressive strength of concrete
$f_{cc,t}$	Compressive stress limits for concrete under full load
$f_{cc,0}$	Compressive stress limits for concrete immediately after transfer
f_{cd}	Design value of the compressive strength of concrete
f_{ck}	Characteristic compressive cylinder strength of concrete at 28 days
f_{cm}	Mean value of concrete cylinder compressive strength
f_{ct}	Uniaxial tensile strength of concrete
f_{ctd}	Design value of the tensile strength of concrete
$f_{ctm.fl}$	Mean flexural tensile strength of concrete
$f_{ctk,0.05}$	Lower characteristic axial tensile strength of concrete
$f_{ctk,0.95}$	Upper characteristic axial tensile strength of concrete

f_{ctm}	Mean value of axial tensile strength of concrete
$f_{ct,t}$	Tensile stress limits for concrete under full load
$f_{ct,0}$	Tensile stress limits for concrete immediately after transfer
f_p	Tensile strength of prestressing steel
f_{pk}	Characteristic tensile strength of prestressing steel
$f_{p0,1k}$	Characteristic 0.1% proof-stress of prestressing steel
f_t	Tensile strength of reinforcement
f_y	Yield strength of reinforcement
f_{yd}	Design yield strength of reinforcement
f_{yk}	Characteristic yield strength of reinforcement
f_{ywd}	Design yield strength of shear reinforcement
$\mathbf{f}_{cp,0}$	Vector of actions to account for unbonded tendons at time t_0 (Equation 5.100)
$\mathbf{f}_{cr,k}$	Vector of actions at time t_k that accounts for creep during previous time period (Equation 5.96)
$\mathbf{f}_{cs,k}$	Vector of actions at time t_k that accounts for shrinkage during previous time period (Equation 5.97)
$\mathbf{f}_{p,init}$	Vector of initial prestressing forces (Equation 5.44)
$\mathbf{f}_{p.rel,k}$	Vector of relaxation forces at time t_k (Equation 5.99)
h	Overall depth of a cross-section
h_e	Depth of the symmetric prism
h_p	Dimension of a post-tensioning anchorage plate
h_0	Notional size or hypothetical thickness
i	Radius of gyration
i, j, k	Integers
k	Coefficient or factor or angular deviation (in radians/m) or stiffness coefficient
k_r	Shrinkage curvature coefficient
l	Length or span
l_{bpd}	Anchorage length required to develop the design stress in a tendon at the ultimate limit state
$l_{b,rqd}$	Required anchorage length (Equation 14.4)
l_{eff}	Effective span of a slab strip; longer of the two effective spans on either side of a column
l_h	Length of plastic hinge
l_n	Clear span as defined in Figure 2.1
l_{pt}	Transmission length
l_t	Transverse span
l_x, l_y	Longer and shorter orthogonal span lengths, respectively, in two-way slabs
l_0	Distance along a beam between the points of zero moment or effective length of a column or design lap length
m_p	Number of layers of prestressed steel
m_s	Number of layers of non-prestressed reinforcement

q_k	Characteristic uniformly distributed variable action
$r_{ext,k}$	Vector of applied actions at time t_k (Equation 5.75)
$r_{ext,0}$	Vector of applied actions at time t_0 (Equation 5.41)
$r_{int,k}$	Vector of internal actions at time t_k (Equation 5.76)
s	Spacing between fitments
s_f	Spacing between transverse reinforcement in a flange (Figure 7.10)
$s_{l,max}$	Maximum spacing between stirrups (or stirrup assemblies) measured along the longitudinal axis of the member (Equation 14.13)
s_t	Stirrup spacing required for torsion
s_v	Stirrup spacing required for shear
$s_{r,max}$	Maximum crack spacing (Equation 5.201)
$s_{t,max}$	Maximum transverse spacing of the legs of a stirrup (Equation 14.14)
t	Thickness or time
t_0	The age of concrete at the time of loading
u	Perimeter of concrete cross-section
v	Deflection or shear stress
v_{cc}	Deflection due to creep
v_{cs}	Deflection due to shrinkage
v_{cx}, v_{mx}	Deflection of the column strip and the middle strip in the x-direction
v_{cy}, v_{my}	Deflection of the column strip and the middle strip in the y-direction
v_0	Deflection immediately after transfer
v_{max}	Maximum permissible total deflection or maximum final total deflection
v_{min}	Minimum shear stress
$v_{sus,0}$	Short-term deflection at transfer caused by the sustained loads
v_{tot}	Total deflection
w	Uniformly distributed load or crack width
w_{bal}	Uniformly distributed balanced load
w_{Ed}	Factored design load at the ultimate limit state
w_G	Uniformly distributed permanent load
w_k	Calculated crack width; design crack width
w_p	Distributed transverse load exerted on a member by a draped tendon profile
w_{px}, w_{py}	Transverse loads exerted by tendons in the x- and y-directions, respectively
w_Q	Uniformly distributed live load
w_s	Uniformly distributed service load
w_{sw}	Uniformly distributed load due to self-weight of the member

w_u	Collapse load
w_{unbal}	Uniformly distributed unbalanced load
$w_{unbal.sus}$	Sustained part of the uniformly distributed unbalanced load
x	Neutral axis depth at the ultimate limit state
x,y,z	Coordinates
y_{btm}	Distance from the centroidal axis to the bottom fibre of a cross-section
y_c	Distance from reference axis to centroid of the concrete cross-section
$y_{n,0}$	Distance from the reference axis to the neutral axis at time t_0
$y_{p(i)}$	y-coordinate of i-th level of prestressed steel
$y_{s(i)}$	y-coordinate of i-th level of non-prestressed reinforcement
y_{top}	Distance from the centroidal axis to the top fibre of a cross-section
z	Lever arm between internal forces
z_d	Sag (or drape) of a parabolic tendon in a span

Greek lower-case letters

α	Angle or ratio or index or factor
α_c	Modular ratio (E_{cm2}/E_{cm1}) in a composite member
$\alpha_{ep(i),0}$	Effective modular ratio ($E_{p(i)}/E_{c,eff}$) of the i-th layer of prestressing steel at time t_0
$\alpha_{es(i),0}$	Effective modular ratio ($E_{s(i)}/E_{c,eff}$) of the i-th layer of non-prestressed steel at time t_0
$\bar{\alpha}_{ep(i),k}$	Age-adjusted effective modular ratio ($E_{p(i)}/\bar{E}_{c,eff}$) of the i-th layer of prestressing steel at time t_k
$\bar{\alpha}_{es(i),k}$	Age-adjusted effective modular ratio ($E_{s(i)}/\bar{E}_{c,eff}$) of the i-th layer of non-prestressed steel at time t_k
$\alpha_{p(i),0}$	Modular ratio ($E_{p(i)}/E_{cm,0}$) of the i-th layer of prestressing steel at time t_0
$\alpha_{s(i),0}$	Modular ratio ($E_{s(i)}/E_{cm,0}$) of the i-th layer of non-prestressed steel at time t_0
α_1, α_2	Creep modification factors for cracked and uncracked cross-section (Equations 5.184 and 5.185), respectively, or fractions of the span l shown in Figure 11.14
β	Angle or ratio or coefficient or slope
$\beta_{cc}(t)$	Function describing the development of concrete strength with time (Equation 4.2)
β_x, β_y	Moment coefficients (Table 12.3)
$\chi(t,t_0)$	Aging coefficient for concrete at time t due to a stress first applied at t_0

γ	Partial factor
γ_C	Partial factor for concrete
γ_G	Partial factor for permanent actions, G
γ_P	Partial factor for actions associated with prestressing, P
γ_Q	Partial factor for variable actions, Q
γ_S	Partial factor for reinforcing or prestressing steel
Δ	Increment or change
Δ_{slip}	Slip of the tendon at an anchorage (Equation 5.148)
ΔP_{c+s+r}	Time-dependent loss of prestress due to creep, shrinkage and relaxation
ΔP_{el}	Loss of prestress due to elastic shortening of the member (Equation 5.146)
ΔP_{di}	Loss of prestress due to draw-in at the anchorage (Equation 5.151)
ΔP_μ	Loss of prestress due to friction along the duct (Equation 5.148)
Δt_k	Time interval $(t_k - t_0)$
$\Delta\sigma_{p,c+s+r}$	Time-dependent change of stress in the tendon due to creep, shrinkage and relaxation (Equation 5.152)
$\Delta\sigma_{p,c}$	Change in stress in the tendon due to creep
$\Delta\sigma_{p,r}$	Change in stress in the tendon due to relaxation
$\Delta\sigma_{p,s}$	Change in stress in the tendon due to shrinkage
$\Delta\sigma_{p,0}$	Change in stress in the tendon immediately after transfer
η	Ratio of uniform compressive stress intensity of the idealised rectangular stress block to the design compressive strength of concrete (f_{cd})
ε	Strain
ε_c	Compressive strain in the concrete
ε_{ca}	Autogenous shrinkage strain
ε_{cc}	Creep strain component in the concrete
ε_{cd}	Drying shrinkage strain
ε_{ce}	Instantaneous strain component in the concrete
ε_{cs}	Shrinkage strain component in the concrete
ε_k	Strain at time t_k
ε_k	Vector of strain at time t_k (Equations 5.94 and 5.101)
$\varepsilon_{p(i),init}$	Initial strain in the i-th layer of prestressing steel produced by the initial tensile prestressing force $P_{init(i)}$
$\varepsilon_{p.rel(i),k}$	Tensile creep strain in the i-th prestressing tendon at time t_k (Equation 5.73)
ε_{pe}	Strain in the prestressing steel caused by the effective prestress (Equation 6.13)
ε_{ptd}	Concrete strain at the level of the tendon (Equation 6.14)
ε_{pud}	Strain in the bonded tendon at the design resistance (Equation 6.15)

ε_r	Strain at the level of the reference axis
$\varepsilon_{r,k}$	Strain at the level of the reference axis at time t_k
$\varepsilon_{r,0}$	Strain at the level of the reference axis at time t_0
ε_{sd}	Design strain in the non-prestressed steel reinforcement
ε_{uk}	Characteristic strain of reinforcement or prestressing steel at maximum load
ε_{yk}	Characteristic yield strain of reinforcement or prestressing steel
ε_0	Strain at time t_0
$\boldsymbol{\varepsilon}_0$	Vector of strain components at time t_0 (Equations 5.43 and 5.45)
ϕ	Diameter of a reinforcing bar or of a prestressing duct
$\varphi(t,t_0)$	Creep coefficient of concrete, defining creep between times t and t_0, related to elastic deformation at 28 days
$\varphi(\infty,t_0)$	Final value of creep coefficient of concrete
$\varphi_{p(i)}$	Creep coefficient of the prestressing steel at time t_k
κ	Curvature
$\kappa_{cc}, \kappa_{cc}(t)$	Creep-induced curvature (Equation 5.183)
κ_{cr}	Curvature at first cracking
$\kappa_{cs}, \kappa_{cs}(t)$	Curvature induced by shrinkage (Equation 5.187)
κ_{ef}	Instantaneous effective curvature on a cracked section
κ_k	Long-term curvature at time t_k
κ_p	Curvature of prestressing tendon
κ_{sus}	Curvature caused by the sustained loads
$\kappa_{sus,0}$	Curvature caused by the sustained loads at time t_0
κ_{ud}	Design curvature at the ultimate limit state (Equation 6.10)
$(\kappa_{ud})_{min}$	Minimum design curvature at the ultimate limit state (Equation 6.22)
κ_{uncr}	Curvature on the uncracked cross-section
κ_0	Initial curvature at time t_0
λ	Ratio of the depth of the rectangular compressive stress block to the depth of the neutral axis at ultimate limit state
ν	Poisson's ratio
θ	Angle or sum in radians of the absolute values of successive angular deviations of the tendon over the length x or slope
θ_p	Angle of inclination of prestressing tendon
θ_s	Rotation available at a plastic hinge
θ_v	Angle between the axis of the concrete compression strut and the longitudinal axis of the member
ρ	Reinforcement ratio for the bonded steel $(A_s + A_p)/bd_o$
ρ_{cw}	Longitudinal compressive reinforcement ratio related to the web width $A_{sc}/(b_w d)$

ρ_w	Longitudinal reinforcement ratio for the tensile steel related to the web width $(A_s + A_{pt})/(b_w d)$ *or* shear reinforcement ratio
σ	Stress
σ_c	Compressive stress in the concrete
$\sigma_{c,btm}$	Stress in the concrete at the bottom of a cross-section
σ_{c0}, $\sigma_c(t_0)$	Stress in the concrete at time t_0
$\sigma_{c,k}$, $\sigma_c(t_k)$	Stress in the concrete at time t_k
σ_{cp}	Compressive stress in the concrete from axial load or prestressing
σ_{cs}	Maximum shrinkage-induced tensile stress on the uncracked section (Equation 5.179)
$\sigma_{c,top}$	Stress in the concrete at the top of a cross-section (in positive bending)
σ_{cy}, σ_{cz}	Normal compressive stresses on the control section in the orthogonal y- and z-directions, respectively
σ_p	Stress in the prestressing steel
$\sigma_{p(i),0}$	Stress in the i-th layer of prestressing steel at time t_0
$\sigma_{p(i),k}$	Stress in the i-th layer of prestressing steel at time t_k
σ_{pi}	Initial stress in the prestressing steel immediately after tensioning
σ_{pj}	Stress in the prestressing steel at the jack (before losses)
$\sigma_{p,max}$	Maximum permissible stress in the prestressing during jacking (Equation 5.1)
σ_{pud}	Design stress in the prestressing steel
σ_{p0}	Initial stress in the prestressing steel immediately after transfer
σ_s	Stress in the non-prestressed steel reinforcement
$\sigma_{s(i),k}$	Stress in the i-th layer of non-prestressed steel at time t_k
$\sigma_{s(i),0}$	Stress in the i-th layer of non-prestressed steel at time t_0
σ_{sd}	Design stress in a steel reinforcement bar
σ_1, σ_2	Principal stresses in concrete
Ω	A factor that depends on the time-dependent loss of prestress in the concrete (Equation 5.112)
ψ, ψ_0, ψ_1, ψ_2	Factors defining representative values of variable actions
ζ	A distribution coefficient that accounts for the moment level and the degree of cracking on the effective moment of inertia (Equation 5.181)

Chapter 1

Basic concepts

1.1 INTRODUCTION

For the construction of mankind's infrastructure, reinforced concrete is the most widely used structural material. It has maintained this position since the end of the nineteenth century and will continue to do so for the foreseeable future. Because the tensile strength of concrete is low, steel bars are embedded in the concrete to carry the internal tensile forces. Tensile forces may be caused by imposed loads or deformations, or by load-independent effects such as temperature changes and shrinkage.

Consider the simple reinforced concrete beam shown in Figure 1.1a, where the external loads cause tension in the bottom of the beam leading to cracking. Practical reinforced concrete beams are usually cracked under the day-to-day service loads. On a cracked section, the applied bending moment M is resisted by compression in the concrete above the crack and tension in the bonded reinforcing steel crossing the crack (Figure 1.1b).

Although the steel reinforcement provides the cracked beam with flexural strength, it prevents neither cracking nor loss of stiffness during cracking. Crack widths are approximately proportional to the strain, and hence stress, in the reinforcement. Steel stresses must therefore be limited to some appropriately low value under in-service conditions in order to avoid excessively wide cracks. In addition, large steel strain in a beam is the result of large curvature, which in turn is associated with large deflection. There is little benefit to be gained, therefore, by using higher strength steel or concrete, since in order to satisfy serviceability requirements, the increased capacity afforded by higher strength steel cannot be utilised.

Prestressed concrete is a particular form of reinforced concrete. Prestressing involves the application of an initial compressive load to the structure to reduce or eliminate the internal tensile forces and thereby control or eliminate cracking. The initial compressive load is imposed and sustained by highly tensioned steel reinforcement (tendons) reacting on the concrete. With cracking reduced or eliminated, a prestressed concrete section is considerably stiffer than the equivalent (usually cracked) reinforced concrete section. Prestressing may also impose internal forces that are of

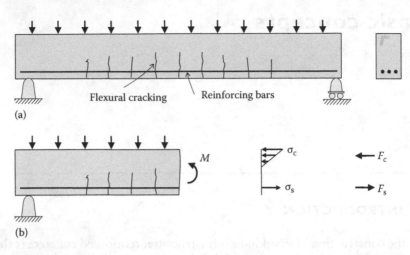

Figure 1.1 A reinforced concrete beam. (a) Elevation and section. (b) Free-body diagram, stress distribution and resultant forces F_c and F_s.

opposite sign to the external loads and may therefore significantly reduce or even eliminate deflection.

With service load behaviour improved, the use of high-strength steel reinforcement and high-strength concrete becomes both economical and structurally efficient. As we will see subsequently, only steel that can accommodate large initial elastic strains is suitable for prestressing concrete. The use of high-strength steel is therefore not only an advantage to prestressed concrete, it is a necessity. Prestressing results in lighter members, longer spans and an increase in the economical range of application of reinforced concrete.

Consider an unreinforced concrete beam of rectangular section, simply-supported over a span l, and carrying a uniform load w, as shown in Figure 1.2a. When the tensile strength of concrete (f_{ct}) is reached in the bottom fibre at mid-span, cracking and a sudden brittle failure will occur. If it is assumed that the concrete possesses zero tensile strength (i.e. $f_{ct} = 0$), then no load can be carried and failure will occur at any load greater than zero. In this case, the collapse load w_u is zero. An axial compressive force P applied to the beam, as shown in Figure 1.2b, induces a uniform compressive stress of intensity P/A on each cross-section. For failure to occur, the maximum moment caused by the external collapse load w_u must now induce an extreme fibre tensile stress equal in magnitude to P/A. In this case, the maximum moment is located at mid-span and, if linear-elastic material behaviour is assumed, simple beam theory gives (Figure 1.2b):

$$\frac{M}{Z} = \frac{w_u l^2}{8Z} = \frac{P}{A}$$

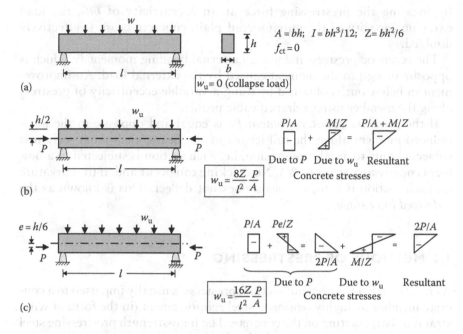

Figure 1.2 Effect of prestress on the load carrying capacity of a plain concrete beam. (a) Zero prestress. (b) Axial prestress (e = 0). (c) Eccentric prestress (e = h/6).

based on which the collapse load can be determined as:

$$w_\mathrm{u} = \frac{8Z}{l^2}\frac{P}{A}$$

If the prestressing force P is applied at an eccentricity of $h/6$, as shown in Figure 1.2c, the compressive stress caused by P in the bottom fibre at midspan is equal to:

$$\frac{P}{A} + \frac{Pe}{Z} = \frac{P}{A} + \frac{Ph/6}{bh^2/6} = \frac{2P}{A}$$

and the external load at failure w_u must now produce a tensile stress of $2P/A$ in the bottom fibre. This can be evaluated as follows (Figure 1.2c):

$$\frac{M}{Z} = \frac{w_\mathrm{u} l^2}{8Z} = \frac{2P}{A}$$

and rearranging gives:

$$w_\mathrm{u} = \frac{16Z}{l^2}\frac{P}{A}$$

By locating the prestressing force at an eccentricity of $h/6$, the load carrying capacity of the unreinforced plain concrete beam is effectively doubled.

The eccentric prestress induces an internal bending moment Pe which is opposite in sign to the moment caused by the external load. An improvement in behaviour is obtained by using a variable eccentricity of prestress along the member using a draped cable profile.

If the prestress *counter-moment Pe* is equal and opposite to the load-induced moment along the full length of the beam, each cross-section is subjected only to axial compression, i.e. each section is subjected to a uniform compressive stress of P/A. No cracking can occur and, if the curvature on each section is zero, the beam does not deflect. This is known as the *balanced load stage*.

1.2 METHODS OF PRESTRESSING

As mentioned in the previous section, prestress is usually imparted to a concrete member by highly tensioned steel reinforcement (in the form of wire, strand or bar) reacting on the concrete. The high-strength prestressing steel is most often tensioned using hydraulic jacks. The tensioning operation may occur before or after the concrete is cast and, accordingly, prestressed members are classified as either *pretensioned* or *post-tensioned*. More information on prestressing systems and prestressing hardware is provided in Chapter 3.

1.2.1 Pretensioned concrete

Figure 1.3 illustrates the procedure for pretensioning a concrete member. The prestressing tendons are initially tensioned between fixed abutments and anchored. With the formwork in place, the concrete is cast around the highly stressed steel tendons and cured. When the concrete has reached its required strength, the wires are cut or otherwise released from the abutments. As the highly stressed steel attempts to contract, it is restrained by the concrete and the concrete is compressed. Prestress is imparted to the concrete via bond between the steel and the concrete.

Pretensioned concrete members are often precast in pretensioning beds that are long enough to accommodate many identical units simultaneously. To decrease the construction cycle time, steam curing may be employed to facilitate rapid concrete strength gain, and the prestress is often transferred to the concrete within 24 hours of casting. Because the concrete is usually stressed at such an early age, elastic shortening of the concrete and

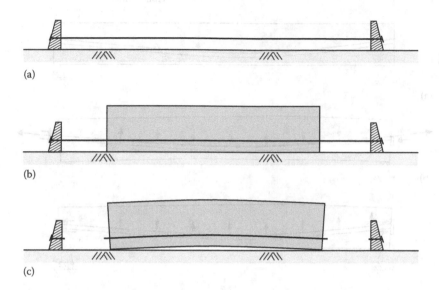

Figure 1.3 Pretensioning procedure. (a) Tendons stressed between abutments. (b) Concrete cast and cured. (c) Tendons released and prestress transferred.

subsequent creep strains tend to be high. This relatively high time-dependent shortening of the concrete causes a significant reduction in the tensile strain in the bonded prestressing steel and a relatively high loss of prestress occurs with time.

1.2.2 Post-tensioned concrete

The procedure for post-tensioning a concrete member is shown in Figure 1.4. With the formwork in position, the concrete is cast around hollow ducts which are fixed to any desired profile. The steel tendons are usually in place, unstressed in the ducts during the concrete pour, or alternatively may be threaded through the ducts at some later time. When the concrete has reached its required strength, the tendons are tensioned. Tendons may be stressed from one end with the other end anchored or may be stressed from both ends, as shown in Figure 1.4b. The tendons are then anchored at each stressing end. The concrete is compressed during the stressing operation, and the prestress is maintained after the tendons are anchored by bearing of the end anchorage plates onto the concrete. The post-tensioned tendons also impose a transverse force on the member wherever the direction of the cable changes.

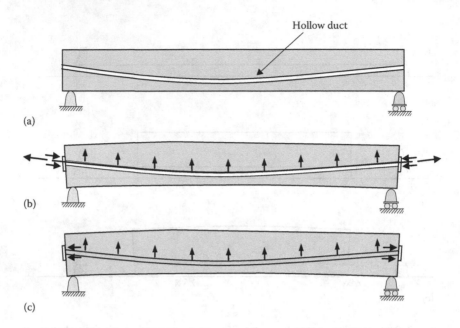

Figure 1.4 Post-tensioning procedure. (a) Concrete cast and cured. (b) Tendons stressed and prestress transferred. (c) Tendons anchored and subsequently grouted.

After the tendons have been anchored and no further stressing is required, the ducts containing the tendons are often filled with grout under pressure. In this way, the tendons are bonded to the concrete and are more efficient in controlling cracks and providing ultimate strength. Bonded tendons are also less likely to corrode or lead to safety problems if a tendon is subsequently lost or damaged. In some situations, however, tendons are not grouted for reasons of economy and remain permanently unbonded. In this form of construction, the tendons are coated with grease and encased in a plastic sleeve. Although the contribution of unbonded tendons to the ultimate strength of a beam or slab is only about 75% of that provided by bonded tendons, unbonded post-tensioned slabs are commonly used in North America and Europe.

Most in-situ prestressed concrete is post-tensioned. Relatively light and portable hydraulic jacks make on-site post-tensioning an attractive proposition. Post-tensioning is also used for segmental construction of large-span bridge girders.

1.2.3 Other methods of prestressing

Prestress may also be imposed on new or existing members using external tendons or such other devices as *flat jacks*. These systems are useful for temporary prestressing operations but may be subject to high time-dependent losses. External prestressing is discussed further in Section 3.7.

1.3 TRANSVERSE FORCES INDUCED BY DRAPED TENDONS

In addition to the longitudinal force P exerted on a prestressed member at the anchorages, transverse forces are also exerted on the member wherever curvature exists in the tendons. Consider the simply-supported beam shown in Figure 1.5a. It is prestressed by a cable with a *kink* at mid-span. The eccentricity of the cable is zero at each end of the beam and equal to e at mid-span, as shown. The slope of the two straight segments of cable is θ. Because θ is small, it can be calculated as:

$$\theta \approx \sin\theta \approx \tan\theta = \frac{e}{l/2} \tag{1.1}$$

In Figure 1.5b, the forces exerted by the tendon on the concrete are shown. At mid-span, the cable exerts an upward force F_P on the concrete equal to the sum of the vertical component of the prestressing force in the tendon on both sides of the *kink*. From statics:

$$F_P = 2P\sin\theta \approx \frac{4Pe}{l} \tag{1.2}$$

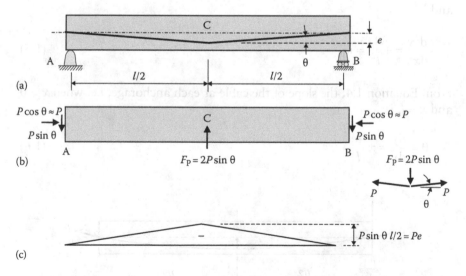

Figure 1.5 Forces and actions exerted by prestress on a beam with a centrally depressed tendon. (a) Elevation. (b) Forces imposed by prestress on concrete. (c) Bending moment diagram due to prestress.

At each anchorage, the cable has a horizontal component of $P \cos \theta$ (which is approximately equal to P for small values of θ) and a vertical component equal to $P \sin \theta$ (approximated by $2Pe/l$).

Under this condition, the beam is said to be *self-stressed*. No external reactions are induced at the supports. However, the beam exhibits a non-zero curvature along its length and deflects upward owing to the internal bending moment caused by the prestress. As illustrated in Figure 1.5c, the internal bending moment at any section can be calculated from statics and is equal to the product of the prestressing force P and the eccentricity of the tendon at that cross-section.

If the prestressing cable has a curved profile, the cable exerts trans-verse forces on the concrete throughout its length. Consider the pre-stressed beam with the parabolic cable profile shown in Figure 1.6. With the x- and y-coordinate axes in the directions shown, the shape of the parabolic cable is:

$$y = -4e \left[\frac{x}{l} - \left(\frac{x}{l} \right)^2 \right] \tag{1.3}$$

and its slope and curvature are, respectively:

$$\frac{dy}{dx} = -\frac{4e}{l} \left(1 - \frac{2x}{l} \right) \tag{1.4}$$

and

$$\frac{d^2y}{dx^2} = +\frac{8e}{l^2} = \kappa_p \tag{1.5}$$

From Equation 1.4, the slope of the cable at each anchorage, i.e. when $x = 0$ and $x = l$, is:

$$\theta = \frac{dy}{dx} = \pm \frac{4e}{l} \tag{1.6}$$

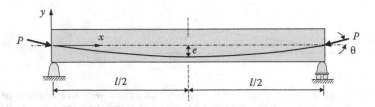

Figure 1.6 A simple beam with parabolic tendon profile.

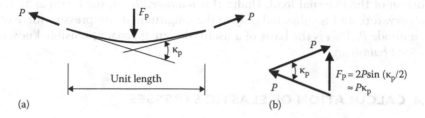

Figure 1.7 Forces on a curved cable of unit length. (a) Tendon segment of unit length. (b) Triangle of forces.

and, provided the tendon slope is small, the horizontal and vertical components of the prestressing force at each anchorage may therefore be taken as P and $4Pe/l$, respectively.

Equation 1.5 indicates that the curvature of the parabolic cable is constant along its length. The curvature κ_p is the angular change in direction of the cable per unit length, as illustrated in Figure 1.7a. From the free-body diagram in Figure 1.7b, for small tendon curvatures, the cable exerts an upward transverse force $w_p = P\kappa_p$ per unit length over the full length of the cable. This upward force is an equivalent distributed load along the member and, for a parabolic cable with the constant curvature of Equation 1.5, w_p is given by:

$$w_p = P\kappa_p = +\frac{8Pe}{l^2} \tag{1.7}$$

With the sign convention adopted in Figure 1.6, a positive value of w_p depicts an upward load. If the prestressing force is constant along the beam, which is never quite the case in practice, w_p is uniformly distributed and acts in an upward direction.

A free-body diagram of the concrete beam showing the forces exerted by the cable is illustrated in Figure 1.8. The zero reactions induced by the prestress imply that the beam is self-stressed. With the maximum eccentricity usually known, Equation 1.7 may be used to calculate the value of P required to cause an upward force w_p that exactly balances a selected

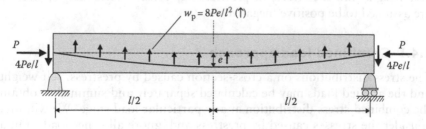

Figure 1.8 Forces exerted on a concrete beam by a tendon with a parabolic profile.

portion of the external load. Under this *balanced load*, the beam exhibits no curvature and is subjected only to the longitudinal compressive force of magnitude *P*. This is the basis of a useful design approach, sensibly known as *load balancing*.

1.4 CALCULATION OF ELASTIC STRESSES

The components of stress on a prestressed cross-section caused by the pre-stress, the self-weight and the external loads are usually calculated using simple beam theory and assuming linear-elastic material behaviour. In addition, the properties of the gross concrete section are usually used in the calculations, provided the section is not cracked. Indeed, these assump-tions have already been made in the calculations of the stresses illustrated in Figure 1.2.

Concrete, however, does not behave in a linear-elastic manner. At best, linear-elastic calculations provide only an approximation of the state of stress on a concrete section immediately after the application of the load. Creep and shrinkage strains that gradually develop in the concrete usually cause a substantial redistribution of stresses with time, particularly on a section containing significant amounts of bonded reinforcement.

Elastic calculations are useful, however, in determining, for example, if tensile stresses occur at service loads, and therefore if cracking is likely, or if compressive stresses are excessive and large time-dependent shortening may be expected. Elastic stress calculations may therefore be used to indi-cate potential serviceability problems.

If an elastic calculation indicates that cracking may occur at service loads, the cracked section analysis presented subsequently in Section 5.8.3 should be used to determine appropriate section properties for use in serviceabil-ity calculations. A more comprehensive picture of the variation of concrete stresses with time can be obtained using the time analyses described in Sections 5.7 and 5.9 to account for the time-dependent deformations caused by creep and shrinkage of the concrete.

In the following sections, several different approaches for calculating *elastic* stresses on an uncracked concrete cross-section are described to pro-vide insight into the effects of prestressing. Tensile (compressive) stresses are assumed to be positive (negative).

1.4.1 Combined load approach

The stress distributions on a cross-section caused by prestress, self-weight and the applied loads may be calculated separately and summed to obtain the combined stress distribution at any particular load stage. We will first consider the stresses caused by prestress and ignore all other loads. On a cross-section, such as that shown in Figure 1.9, equilibrium requires that

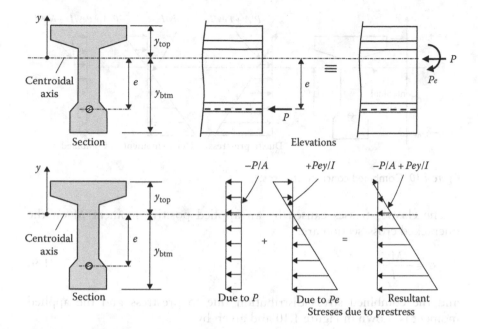

Figure 1.9 Concrete stress resultants and stresses caused by prestress.

the resultant of the concrete stresses is a compressive force that is equal and opposite to the tensile force in the steel tendon and located at the level of the steel, i.e. at an eccentricity e below the centroidal axis. This is statically equivalent to an axial compressive force P and a moment Pe located at the centroidal axis, as shown.

The stresses caused by the prestressing force of magnitude P and the hogging (−ve) moment Pe are also shown in Figure 1.9. The resultant stress induced by the prestress is given by:

$$\sigma = -\frac{P}{A} + \frac{Pey}{I} \tag{1.8}$$

where A and I are the area and second moment of area about the centroidal axis of the cross-section, respectively, and y is the distance from the centroidal axis (positive upwards).

It is common in elastic stress calculations to ignore the stiffening effect of the reinforcement and to use the properties of the gross cross-section. Although this simplification usually results in only small errors, it is not encouraged here. For cross-sections containing significant amounts of bonded steel reinforcement, the steel should be included in the determination of the properties of the transformed cross-section.

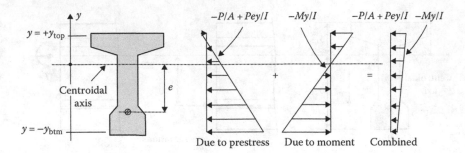

Figure 1.10 Combined concrete stresses.

The elastic stresses caused by an applied positive moment M on the uncracked cross-section are:

$$\sigma = -\frac{My}{I} \qquad (1.9)$$

and the combined stress distribution due to prestress and the applied moment is shown in Figure 1.10 and given by:

$$\sigma = -\frac{P}{A} + \frac{Pey}{I} - \frac{My}{I} \qquad (1.10)$$

1.4.2 Internal couple concept

The resultant of the combined stress distribution shown in Figure 1.10 is a compressive force of magnitude P located at a distance z_p above the level of the steel tendon, as shown in Figure 1.11. The compressive force in the concrete and the tensile force in the steel together form a couple, with magnitude equal to the applied bending moment and calculated as:

$$M = Pz_p \qquad (1.11)$$

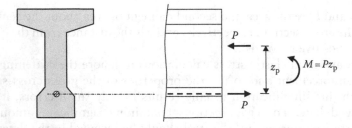

Figure 1.11 Internal couple.

When the applied moment $M = 0$, the lever arm z_p is zero and the resultant concrete compressive force is located at the steel level. As M increases, the compressive stresses in the top fibres increase and those in the bottom fibres decrease, and the location of the resultant compressive force moves upward.

It is noted that provided the section is uncracked, the magnitude of P does not change appreciably as the applied moment increases and, as a consequence, the lever arm z_p is almost directly proportional to the applied moment. If the magnitude and position of the resultant of the concrete stresses are known, the stress distribution can be readily calculated.

1.4.3 Load balancing approach

In Figure 1.8, the forces exerted on a prestressed beam by a parabolic tendon with equal end eccentricities are shown and the uniformly distributed transverse load w_p is calculated from Equation 1.7. In Figure 1.12, all the loads acting on such a beam, including the external gravity loads w, are shown.

If $w = w_p$, the bending moment and shear force on each cross-section caused by the gravity load w are balanced by the equal and opposite values caused by w_p. With the transverse loads balanced, the beam is subjected only to the longitudinal prestress P applied at the anchorage. If the anchorage is located at the centroid of the section, a uniform stress distribution of intensity P/A occurs on each section and the beam does not deflect.

If $w \neq w_p$, the bending moment M_{unbal} caused by the unbalanced load $(w - w_p)$ must be calculated and the resultant stress distribution (given by Equation 1.9) must be added to the stresses caused by the axial prestress $(-P/A)$.

1.4.4 Introductory example

The elastic stress distribution at mid-span of the simply-supported beam shown in Figure 1.13 is to be calculated. The beam spans 12 m and is post-tensioned by a single cable with zero eccentricity at each end and $e = 250$ mm at mid-span. The prestressing force in the tendon is assumed to be constant along the length of the beam and equal to $P = 1760$ kN.

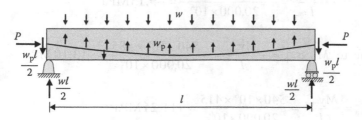

Figure 1.12 Forces on a concrete beam with a parabolic tendon profile.

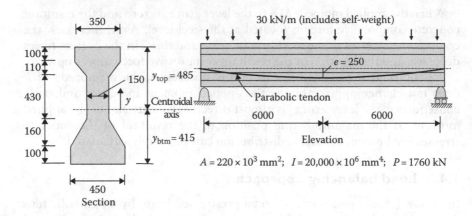

Figure 1.13 Beam details (Introductory example). (*Notes: P* is assumed constant on every section; all dimensions are in millimetres.)

Each of the procedures discussed in the preceding sections is illustrated in the following calculations.

1.4.4.1 Combined load approach

The extreme fibre stresses at mid-span (σ_{top}, σ_{btm}) due to *P*, *Pe* and *M* are calculated separately in the following and then summed.

At mid-span: $P = 1760$ kN; $Pe = 1760 \times 250 \times 10^{-3} = 440$ kNm and

$$M = \frac{wl^2}{8} = \frac{30 \times 12^2}{8} = 540 \text{ kNm}$$

Due to *P* : $\sigma_{top} = \sigma_{btm} = -\frac{P}{A} = -\frac{1760 \times 10^3}{220 \times 10^3} = -8.0 \text{ MPa}$

Due to *Pe* : $\sigma_{top} = \frac{Pey_{top}}{I} = \frac{440 \times 10^6 \times 485}{20,000 \times 10^6} = +10.67 \text{ MPa}$

$\sigma_{btm} = -\frac{Pey_{btm}}{I} = -\frac{440 \times 10^6 \times 415}{20,000 \times 10^6} = -9.13 \text{ MPa}$

Due to *M* : $\sigma_{top} = -\frac{My_{top}}{I} = -\frac{540 \times 10^6 \times 485}{20,000 \times 10^6} = -13.10 \text{ MPa}$

$\sigma_{btm} = \frac{My_{btm}}{I} = \frac{540 \times 10^6 \times 415}{20,000 \times 10^6} = +11.21 \text{ MPa}$

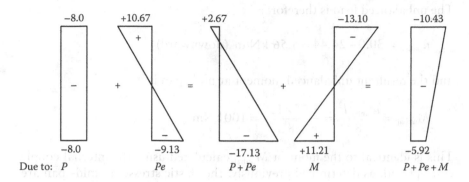

Figure 1.14 Component stress distributions in introductory example.

The corresponding concrete stress distributions and the combined elastic stress distribution on the concrete section at mid-span are shown in Figure 1.14.

1.4.4.2 Internal couple concept

From Equation 1.11:

$$z_p = \frac{M}{P} = \frac{540 \times 10^6}{1760 \times 10^3} = 306.8 \, \text{mm}$$

The resultant compressive force on the concrete section is 1760 kN, and it is located 306.8 − 250 = 56.8 mm above the centroidal axis. This is statically equivalent to an axial compressive force of 1760 kN (applied at the centroid) plus a moment $M_{unbal} = 1760 \times 56.8 \times 10^{-3} = 100$ kNm. The extreme fibre concrete stresses are therefore:

$$\sigma_{top} = -\frac{P}{A} - \frac{M_{unbal}y_{top}}{I} = -\frac{1,760 \times 10^3}{220 \times 10^3} - \frac{100 \times 10^6 \times 485}{20,000 \times 10^6} = -10.43 \, \text{MPa}$$

$$\sigma_{btm} = -\frac{P}{A} + \frac{M_{unbal}y_{btm}}{I} = -\frac{1,760 \times 10^3}{220 \times 10^3} + \frac{100 \times 10^6 \times 415}{20,000 \times 10^6} = -5.92 \, \text{MPa}$$

and, of course, these are identical with the extreme fibre stresses calculated using the combined load approach and shown in Figure 1.14.

1.4.4.3 Load balancing approach

The transverse force imposed on the concrete by the parabolic cable is obtained using Equation 1.7 as:

$$w_p = \frac{8Pe}{l^2} = \frac{8 \times 1,760 \times 10^3 \times 250}{12,000^2} = 24.44 \, \text{kN/m (upward)}$$

The unbalanced load is therefore:

$$w_{unbal} = 30.0 - 24.44 = 5.56 \text{ kN/m (downward)}$$

and the resultant unbalanced moment at mid-span is:

$$M_{unbal} = \frac{w_{unbal} l^2}{8} = \frac{5.56 \times 12^2}{8} = 100 \text{kNm}$$

This is identical to the moment M_{unbal} calculated using the internal couple concept and, as determined previously, the elastic stresses at mid-span are obtained by adding the P/A stresses to those caused by M_{unbal}:

$$\sigma_{top} = -\frac{P}{A} - \frac{M_{unbal} y_{top}}{I} = -10.43 \text{MPa}$$

$$\sigma_{btm} = -\frac{P}{A} + \frac{M_{unbal} y_{btm}}{I} = -5.92 \text{MPa}$$

1.5 INTRODUCTION TO STRUCTURAL BEHAVIOUR: INITIAL TO ULTIMATE LOADS

The choice between reinforced and prestressed concrete for the construction of a particular structure is essentially one of economics. Aesthetics may also influence the choice. For relatively short-span beams and slabs, reinforced concrete is usually the most economical alternative. As spans increase, however, reinforced concrete design is more and more controlled by the serviceability requirements. Strength and ductility can still be economically achieved but, in order to prevent excessive deflection, cross-sectional dimensions become uneconomically large. Excessive deflection is usually the governing *limit state*.

For medium- to long-span beams and slabs, the introduction of prestress improves both serviceability and economy. The optimum level of prestress depends on the span, the load history and the serviceability requirements. The level of prestress is often selected so that cracking at service loads does not occur. However, in many situations, there is no valid reason why controlled cracking should not be permitted. Insisting on enough prestress to eliminate cracking frequently results in unnecessarily high initial prestressing forces and, consequently, uneconomical designs. In addition, the high initial prestress often leads to excessively large camber and/or axial shortening. Members designed to remain uncracked at service loads are commonly termed *fully prestressed*.

In building structures, there are relatively few situations in which it is necessary to avoid cracking under the full service loads. In fact, the most economic design often results in significantly less prestress than is required for a *fully-prestressed* member. Frequently, such members are designed to remain uncracked under the sustained or permanent load, with cracks opening and closing as the variable live load is applied and removed. Prestressed concrete members generally behave satisfactorily in the post-cracking load range, provided they contain sufficient bonded reinforcement to control the cracks. A cracked prestressed concrete section under service loads is significantly stiffer than a cracked reinforced concrete section of similar size and containing similar quantities of bonded reinforcement. Members that are designed to crack at the full service load are often called *partially-prestressed*.

The elastic stress calculations presented in the previous section are applicable only if material behaviour is linear-elastic and the principle of superposition is valid. These conditions may be assumed to apply on a prestressed section prior to cracking, but only immediately after the loads are applied. As was mentioned in Section 1.4, the gradual development of creep and shrinkage strains with time in the concrete can cause a marked redistribution of stress between the bonded steel and the concrete on the cross-section. The greater the quantity of bonded reinforcement, the greater is the time-dependent redistribution of stress. This is demonstrated subsequently in Section 5.7.3 and discussed in Section 5.7.4. For the determination of the long-term stress and strain distributions, elastic stress calculations are not meaningful and may be misleading.

A typical moment versus instantaneous curvature relationship for a prestressed concrete cross-section is shown in Figure 1.15. Prior to the application of moment (i.e. when $M = 0$), if the prestressing force P acts at an eccentricity e from the centroidal axis of the uncracked cross-section, the curvature is $\kappa_0 = -Pe/(E_{cm}I_{uncr})$, corresponding to point A in Figure 1.15, where E_{cm} is the elastic modulus of the concrete and I_{uncr} is the second moment of area of the uncracked cross-section. The curvature κ_0 is negative because the internal moment caused by prestress is negative $(-Pe)$. When the applied moment M is less than the cracking moment M_{cr}, the section is uncracked and the moment–curvature relationship is linear (from point A to point B in Figure 1.15) and $\kappa = (M - Pe)/(E_{cm}I_{uncr}) \leq \kappa_{cr}$. It is only in this region (i.e. when $M < M_{cr}$) that elastic stress calculations may be used and then only for short-term calculations.

If the external loads are sufficient to cause cracking (i.e. when the extreme fibre stress calculated from elastic analysis exceeds the tensile strength of concrete), the short-term behaviour becomes non-linear and the principle of superposition is no longer applicable.

As the applied moment on a cracked prestressed section increases (i.e. as the moment increases above M_{cr} from point B to point C in Figure 1.15), the crack height gradually increases from the tension surface towards the compression zone and the size of the uncracked part of the cross-section in compression above the crack decreases. This is different to the post-cracking behaviour of a non-prestressed reinforced concrete section, where at first cracking the crack suddenly propagates deep into the beam and

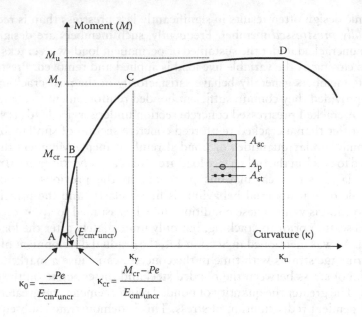

Figure 1.15 Typical moment versus instantaneous curvature relationship.

the crack height and the depth of the concrete compression zone remain approximately constant as the applied moment is subsequently varied.

As the moment on a prestressed concrete section increases further into the overload region (approaching point D in Figure 1.15), the material behaviour becomes increasingly nonlinear. Permanent deformation occurs in the bonded prestressing tendons as the stress approaches its ultimate value, the non-prestressed conventional reinforcement yields (at or near point C where there is a change in direction of the moment curvature graph) and the compressive concrete in the ever decreasing region above the crack enters the nonlinear range. The external moment is resisted by an internal couple, with tension in the reinforcement crossing the crack and compression in the concrete and in any reinforcement in the compressive zone. At the *ultimate load stage* (i.e. when the moment reaches the ultimate resistance M_u at a curvature κ_u), the prestressed section behaves in the same way as a reinforced concrete section, except that the stress in the high-strength steel tendon is very much higher than in conventional reinforcement. A significant portion of the very high steel stress and strain is due to the initial prestress. For modern prestressing steels, the initial stress in the tendon immediately after the transfer of prestress is often about 1400 MPa. If the same higher strength steel were to be used without being initially prestressed, excessive deformation and unacceptably wide cracks may result at only a small fraction of the ultimate load (well below normal service loads).

The ultimate strength of a prestressed section depends on the quantity and strength of the steel reinforcement and tendons. The level of pre-stress, however, and therefore the quantity of prestressing steel are deter-mined from serviceability considerations. In order to provide a suitable factor of safety for strength, additional conventional reinforcement may be required to supplement the prestressing steel in the tension zone. This is particularly so in the case of *partially-prestressed* members and may even apply for fully-prestressed construction. The avoidance of cracking at service loads and the satisfaction of selected elastic stress limits do not ensure adequate strength. Strength must be determined from a rational analysis which accounts for the nonlinear material behaviour of both the steel and the concrete. Flexural strength analysis is described and illus-trated in Chapter 6, and analyses for shear and torsional strength are presented in Chapter 7.

Chapter 2

Design procedures and applied actions

2.1 LIMIT STATES DESIGN PHILOSOPHY

The broad design objective for a prestressed concrete structure is that it should satisfy the needs for which it was designed and built. In doing so, the structural designer must ensure that it is safe and serviceable so that the chances of it failing during its design lifetime are sufficiently small. The structure must be strong enough and sufficiently ductile to resist, without collapsing, the overloads and environmental extremes that may be imposed on it. It must also be serviceable by performing satisfactorily under the day-to-day service loads without deforming, cracking or vibrating excessively. The two primary structural design objectives are therefore *strength* and *serviceability*.

Other structural design objectives are *stability* and *durability*. A structure must be stable and resist overturning or sliding, reinforcement must not corrode, concrete must resist abrasion and spalling and the structure must not suffer a significant reduction of strength or serviceability with time. Further, it must have adequate fire protection, and it must be robust, resist fatigue loading and satisfy any special requirements that are related to its intended use. A non-structural, but important, objective is *aesthetics* and, of course, an overarching design objective is *economy*. Ideally, the structure should be in harmony with, and enhance, the environment, and this often requires collaboration between the structural engineer, the environmental engineer, the architect and other members of the design team. The aim is to achieve, at minimum cost, an aesthetically pleasing and functional structure that satisfies the structural objectives of strength, serviceability, stability and durability.

For structural design calculations, the design objectives must be translated into quantitative terms called *design criteria*. For example, the maximum acceptable deflection for a particular beam or slab may be required, or the maximum crack width that can be tolerated in a concrete floor or wall. Also required are minimum numerical values for the strength of individual elements and connections. It is also necessary to identify and quantify appropriate design loads for the structure depending on the probability of their occurrence. Reasonable maximum values are required so that a suitable compromise is reached between the risk of overload and consequent failure and the requirement for economical construction.

Codes of practice specify design criteria that provide a suitable margin of safety (called the *safety index*) against a structure becoming unfit for service in any way. The specific form of the design criteria depends on the philosophy and method of design adopted by the code and the manner in which the inherent variability in both the load and the structural performance is considered. Modern design codes for structures have generally adopted the *limit states method* of design, whereby a structure must be designed to simultaneously satisfy a number of different *limit states* or design requirements, including adequate strength and serviceability. Minimum performance requirements are specified for each of these limit states, and any one may become critical and govern the design of a particular member.

If a structure becomes unfit for service in any way, it is said to have entered a *limit state*. Limit states are the undesirable consequences associated with each possible mode of failure. In order to satisfy the design criteria set down in codes of practice, methods of design and analysis should be used which are appropriate to the limit state being considered. For example, if the strength of a cross-section is to be calculated, *ultimate strength* analysis and design procedures are usually adopted. Collapse load methods of analysis and design (plastic methods) may be suitable for calculating the strength of ductile indeterminate structures. If the serviceability limit states of excessive deflection (or camber) or excessive cracking are considered, an analysis that accounts for the non-linear and inelastic nature of concrete is usually required. The sources of these material non-linearities include cracking, tension stiffening, creep and shrinkage. In addition, creep of the highly stressed, high-strength prestressing steel (more commonly referred to as relaxation) may affect in-service structural behaviour.

Each limit state must be considered and designed for separately. Satisfaction of the requirements for one does not ensure satisfaction of the requirements for others. All undesirable consequences must be avoided. In this chapter, the design requirements for prestressed concrete in Europe are discussed, including the loads and load combinations for each limit state specified in the loading codes EN 1991 Parts 1-1, 1-3 and 1-4 [1–3] and the relevant design criteria in EN 1990 [4] and EN 1992-1-1 [5]. If required, other design actions may need to be considered in design as specified in EN 1991 Part 1-2 [6] for structures exposed to fire, EN 1991 Part 1-5 [7] for thermal actions, EN 1991 Part 1-6 [8] for actions during construction and EN 1991 Part 1-7 [9] for accidental actions.

2.2 STRUCTURAL MODELLING AND ANALYSIS

2.2.1 Structural modelling

Structural modelling involves the development of simplified analytical models of the load distribution, the geometry of the structure and its supports and the deformation of the structure. It is an essential part of the design process. A concrete structure is extremely complex, if it is considered from a microscopic point of view. Even from a more macroscopic viewpoint, the degree of complexity is very high. The cross-sectional dimensions vary along individual members, with relatively high tolerances on the specified dimensions being accepted. Concrete material properties depend on the degree of compaction of the concrete and the location in the structure and vary significantly through the thickness of beams and slabs and over the height of columns. The location of reinforcement and the concrete cover vary from point to point in a structure. Some regions are cracked and others are not. It is not possible to account for all these variations and uncertainties in the modelling and analysis of structures and, fortunately, it is not necessary.

An idealised analytical model of a concrete structure must be simple enough to allow the structural analysis to proceed and the mathematics to be tractable. It must also be accurate enough to provide a reasonable approximation of the behaviour of the real structure, including the flow of forces through the structure. This includes a reasonable quantitative estimate of the internal actions and deformations of the real structure.

As an example of a simplified structural model, consider the design of a two-way floor slab in a framed building under gravity loads using the so-called *equivalent frame method*. The floor is divided into wide *design strips* centred on column lines in each orthogonal direction. Each one-way design strip, together with the columns immediately above and below the floor, is then modelled as a two-dimensional frame. The columns above and below the slab are usually assumed to be fixed at their far ends. The horizontal design strip consists of the slab of width l_t (equal to the transverse distance between the centrelines of adjacent panels) plus any beams located along the column line in question. The *effective length* of a typical span of the design strip l_{eff} is defined in EN 1992-1-1 [5] as shown in Figure 2.1. The frame is then analysed, usually using a linear-elastic frame analysis and appropriate estimates of member stiffness, and moments and shears in the design strip are determined. The design strip is next divided into the *column strip* (of width usually about $l_t/2$ and centred on the column line) and the *middle strips* (those portions of the design strip outside the column strip). For more details, refer to Section 12.9 (Figure 12.14). The design strip moments are then distributed to the column and middle strip, with the column strip attracting a greater share than the middle strip (see Table 12.4).

Depending on the degree of accuracy required, a range of possible analytical models of varying complexity are generally available. Different analytical

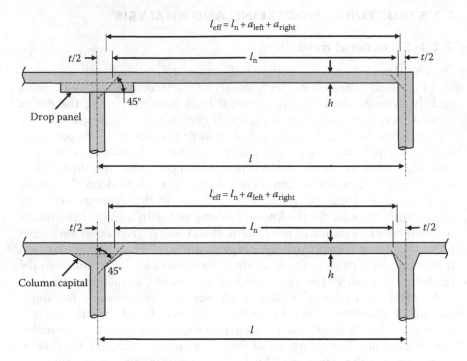

Figure 2.1 Effective span length for a flat slab. *Notes:* $a_{left} = \min(t/2, h/2)$, $a_{right} = \min(t/2, h/2)$.

models may need to be adopted depending on the limit state under consideration. For example, for the two-way floor slab, the simplified structural model consisting of one-way design strips spanning in orthogonal directions described earlier with actions determined from a linear-elastic analysis may be appropriate for an assessment of flexural strength (provided the individual cross-sections are appropriately ductile). A more complex model, such as a finite element model that accounts for cracking, creep and shrinkage of concrete, may be more appropriate for an accurate prediction of deformation at service loads.

Irrespective of the method chosen for the structural analysis or the type of structural model selected, when considering and interpreting the results of the analysis, the simplifications, idealisations and assumptions implied in the analysis and inherent in the structural model must be considered in relation to the real three-dimensional concrete structure.

2.2.2 Structural analysis

The structural analysis of an idealised structural model is the process by which the distribution of internal actions and the deformational response of the model are determined (see Reference [11]). Classical methods of structural analysis were developed for hand calculation, but today the analysis of members and frames is commonly carried out in design offices using commercial software involving the stiffness method or the finite element approach. Most often, the analysis is based on the assumption that material behaviour is linear and elastic, even though the response of a concrete structure is not linear with respect to load, even under in-service conditions. Concrete cracks at relatively low load levels causing a non-linear structural response. Under service loads, concrete undergoes creep and shrinkage deformations and, at the ultimate limit state, steel yields and concrete behaviour in compression is non-linear and inelastic. Nevertheless, *linear-elastic analysis* still forms the basis for the analysis of most concrete structures, with adjustments made in design to include both material and geometric non-linearities when predicting either the ultimate strength or the in-service deformations.

Over the last few decades, significant advances have been made in the development of methods for including both material and geometric non-linearities in the analysis of concrete structures. Numerous commercial software packages are now available for the *non-linear analysis* of concrete structures under both service loads and ultimate loads. For many applications, these packages are unnecessarily complex and time-consuming, but in some situations, non-linear analysis is appropriate and its use for the analysis of complex structures is likely to increase.

If linear-elastic analysis is used in the design for adequate strength, the internal actions caused by the design ultimate loads are determined at the critical cross-sections of each structural member. Each critical section is then designed to ensure that its design ultimate strength exceeds the design ultimate actions. Although a linear-elastic analysis of a particular structural model will give a single distribution of internal actions that is in equilibrium, it must be understood that this is not the distribution of actions in the real structure, where material behaviour under ultimate loads will be non-linear and inelastic. The use of linear-elastic analysis is only valid provided individual cross-sections and members are ductile so that redistribution of internal actions can take place as the ultimate load is approached and the linear-elastic distribution of actions assumed in design can develop in the actual structure.

Alternative methods for the design for adequate strength include plastic methods of analysis (or collapse load analysis) and strut-and-tie modelling. These are simple forms of non-linear analysis. In plastic analysis, plastic hinges with large rotational capacities are assumed to occur at the critical regions in a continuous beam or frame, and either upper or lower bound estimates of the collapse load may be made. If plastic methods are used, ductility of the critical regions is an essential requirement. Strut-and-tie modelling is a lower bound method of plastic analysis where a designer selects an internal load path and then designs the internal concrete struts and steel ties that have been selected to transfer the applied loads through the structure. A range of other methods of analysis, including methods based on stress analysis, are also available for use in strength design.

Linear-elastic analysis and various forms of non-linear analysis may also be used to predict the actions and deformations of a concrete structure at service loads, but account must be taken of the loss of stiffness caused by cracking (due to both the applied loads and the restraint to shrinkage and temperature changes), the tension stiffening effect and the time-dependent effects of creep and shrinkage. These sources of material non-linearity complicate structural analysis at service loads. A comprehensive treatment of the serviceability of concrete structures is provided in Reference [12], and Chapter 5 outlines in some detail appropriate methods for the analysis of prestressed concrete structures at service loads.

Codes of practice, including Eurocode 2 [5], provide guidance for design using linear-elastic analysis, non-linear analysis, plastic methods, strut-and-tie modelling, stress analysis and more approximate methods based on moment coefficients.

2.3 ACTIONS AND COMBINATIONS OF ACTIONS

2.3.1 General

In the design of concrete structures, the internal actions arising from applicable combinations of the applied loads should be considered, including where appropriate, permanent or dead loads, imposed or live loads, wind loads, snow loads, prestressing forces, and loads caused by earthquake, earth pressure and liquid pressure. In addition, possible accidental loading due to impact or blast should be considered where necessary. Loads arising during construction should also be considered where they may adversely affect the various limit states' requirements. Other actions that may cause either stability, strength or serviceability failures include creep of concrete; shrinkage of concrete; other imposed deformations, such as may result from temperature changes and gradients, support settlements and foundation movements; fire and dynamic effects.

In the Eurocodes [1–4], the term *action* is used to describe any effect that influences the performance of a structure, and appropriate combinations of *actions* are specified to check the structure at each limit state. Also specified are appropriate representative values for each action at each limit state. The main actions to be considered are classified as either [4]:

- *permanent actions* (given the symbol G and including the self-weight of the structure and its permanently supported elements);
- *variable actions* (given the symbol Q and including imposed or live loads on floors, wind loads on walls and roofs, snow loads and so on); or
- *accidental actions* (given the symbol A and including impact loads, blast loads and fire).

The magnitude of a particular load (*action*) for use in design is its characteristic value (G_k, Q_k or A_k). Where the statistical distribution of the loading is known, the upper (lower) characteristic load is the value with a

5% (95%) probability of being exceeded. For a particular load combination, either the upper or lower characteristic load is used, whichever produces the most adverse effect. Where the statistical distribution of a particular load is not known, the characteristic load is an appropriate estimate of either the maximum value that should not be exceeded or the minimum value that should definitely occur during the lifetime of the structure.

The *permanent actions* are the permanent dead loads imposed by both the structural and non-structural components of the structure, such as the self-weight of the structure and the forces imposed by all walls, floors, roofs, ceilings, permanent partitions, service machinery and other permanent construction. Permanent actions are usually fixed in position and can be estimated reasonably accurately from the mass of the relevant material or type of construction. For example, normal-weight concrete weighs about 24 kN/m³ and lightweight concrete can be anywhere between 15 and 20 kN/m³. Unit weights of a range of materials are provided in Table 2.1.

The *variable actions* are the imposed (live) loads that are attributed to the intended use or purpose of the structure, and all other externally imposed loads that may reasonably be assumed to act on the structure during its lifetime. The *specified* live load depends on the expected use or occupancy of the structure and usually includes allowances for impact and inertia loads (where applicable) and for possible overload. Both uniformly distributed and concentrated variable loads for a variety of activities/occupancies are specified in EN 1991-1-1 [1] and the characteristic values for some common usages are given in Table 2.2.

Table 2.1 Weights of common construction materials and building materials [1]

Material	Weight (kN/m³)
Aluminium	27.0
Bitumin	10.0–14.0
Concrete – normal weight	24.0[a,b]
Glass, in sheets	25.0
Granite	27.0–30.0
Iron, cast	71.0–72.5
Lead	112.0–114.0
Limestone (dense)	20.0–29.0
Masonry (solid brick)	19.0
Plywood (softwood)	5.0
Sand (dry)	15.0–16.0
Sandstone	22.5
Steel	77.0–78.5
Timber – softwood (hardwood)	5–7 (8–11)

[a] Increase by 1 kN/m³ for normal percentage of reinforcing and prestressing steel.
[b] Increase by 1 kN/m³ for unhardened (wet) concrete.

Table 2.2 Some characteristic imposed (live) loads specified in EN 1991-1-1 [1]

Type of activity/occupancy	Uniformly distributed load, q_k (kN/m²)	Concentrated load, Q_k (kN)
Category A: areas for domestic and residential activities		
Floors	1.5–<u>2.0</u>	<u>2.0</u>–3.0
Stairs	<u>2.0</u>–4.0	<u>2.0</u>–4.0
Balconies	<u>2.5</u>–4.0	<u>2.0</u>–3.0
Category B: office areas	2.0–<u>3.0</u>	1.5–<u>4.5</u>
Category C: areas where people may congregate (not considered in Categories A, B and D)		
C1: areas with tables	2.0–<u>3.0</u>	3.0–<u>4.0</u>
C2: areas with fixed seats	3.0–<u>4.0</u>	2.5–7.0 (<u>4.0</u>)
C3: areas without obstacles for moving people	3.0–<u>5.0</u>	<u>4.0</u>–7.0
C4: areas with possible physical activities	4.5–<u>5.0</u>	3.5–<u>7.0</u>
C5: areas susceptible to large crowds	<u>5.0</u>–7.5	3.5–<u>4.5</u>
Category D: shopping areas		
D1: areas in general retail shops	<u>4.0</u>–5.0	3.5–7.0 (<u>4.0</u>)
D2: areas in department stores	4.0–<u>5.0</u>	3.5–<u>7.0</u>
Category E: Storage areas	6.5	7.0
Category F: traffic and parking areas for light vehicles (≤30 kN gross vehicle weight and ≤8 seats not including driver)	1.0–<u>2.5</u>	10.0–<u>20.0</u>
Category G: traffic and parking areas for medium vehicles (>30, ≤160 kN gross weight on two axles)	5.0	40.0–<u>90.0</u>
Category H: roofs not accessible except for normal maintenance and repair (roof slope <30°)	0.0–1.0 (<u>0.4</u>)	0.9–1.5 (<u>1.0</u>)

Notes: When a range is given in the table, the value may be set by the National Annex. Recommended values, intended for separate applications, are underlined. Category E loads vary depending on the storage height and the nature of what is being stored, refer EN 1991-1-1 [1]. The minimum values for general storage are shown.

The distributed and concentrated characteristic live loads should be considered separately and design carried out for the most adverse effect.

At the time the structure is being designed, the magnitude and distribution of the actual imposed load are never known exactly and it is not certain that the specified imposed load will not be exceeded at some stage during the lifetime of the structure. Imposed loads may or may not be present at any particular time, they are not constant and their position can vary. Although part of the imposed load is transient (short-term), some portion may be applied for extended periods (long-term) and have effects similar to dead loads. Imposed loads also arise during construction due to stacking of building materials, equipment or the construction procedure (such as the loads induced by floor-to-floor propping in multi-storey buildings). These construction loads must be anticipated and considered by the designer. Other variable actions include the specified wind, earthquake, snow and temperature loads, and these depend on the geographical location and the relative importance of the structure (the mean return period). Wind loads on structures also depend on the surrounding terrain and the height of the structure above the ground, and characteristic values are specified in EN 1991-1-4 [3].

In design according to the Eurocodes, when different types of variable actions are combined and assumed to act simultaneously, the characteristic values are multiplied by *combination factors*, ψ_0, ψ_1 or ψ_2. The action $\psi_0 Q_k$ is the *combination value* used for checking ultimate limit states and the irreversible serviceability limit states to account for the reduced probability that two or more independent variable actions will act simultaneously. The action $\psi_1 Q_k$ is termed the *frequent value* used to check both the ultimate limit states when variable actions are combined with accidental actions and the serviceability limit states under frequent levels of variable load.

The action $\psi_2 Q_k$ is the *quasi-permanent value* representing that portion of Q_k that is acting permanently or over a considerable part of each day (more than 50%). It is used to check the ultimate limit states involving accidental loads and for long-term serviceability limit states. Values of the combination factors for some variable actions are given in Table 2.3.

2.3.2 Load combinations for the strength limit states

The *design loads* used in the design for strength are the specified values discussed earlier multiplied by specified minimum load factors. With the built-in allowance for overloads, the specified loads will not often be exceeded in

Table 2.3 Combination factors for some variable actions on buildings [4]

Variable action	ψ_0	ψ_1	ψ_2
Imposed loads in buildings			
Category A: domestic, residential areas	0.7	0.5	0.3
Category B: office areas	0.7	0.5	0.3
Category C: areas where people may congregate	0.7	0.7	0.6
Category D: shopping areas	0.7	0.7	0.6
Category E: storage areas, industrial use, access areas	1.0	0.9	0.8
Category F: traffic areas – light vehicles (≤30 kN)	0.7	0.7	0.6
Category G: traffic areas – heavy vehicles (≤160 kN)	0.7	0.5	0.3
Category H: roofs that are not accessible (except for normal maintenance and repair)	0.0	0.0	0.0
Snow loads on buildings			
For sites at altitudes >1000 m	0.7	0.5	0.2
For sites at altitudes ≤1000 m	0.5	0.2	0.0
Wind loads on buildings	0.6	0.2	0.0

the life of a structure. The load factors applied to each load type, together with factors of safety applied to the material strength (discussed subsequently), ensure that the probability of strength failure is extremely low. The load factors depend on the type of load and the load combination under consideration. For example, the load factors associated with permanent actions (dead loads) are generally less than those for the variable actions (such as live or wind loads) because the dead load is known more reliably and therefore less likely to be exceeded. Load factors for the strength limit states are generally greater than 1.0. The exception is where one load type opposes another load type. For example, when considering uplift caused by wind on a roof member, the load factor on the weight of the roof is equal to 1.0.

The most common *design load* combination used to assess the strength of a structural member in a building is the factored combination of the permanent action G_k and the imposed action (termed the *primary variable action* $Q_{k,1}$) given by:

$$\gamma_G G_k + \gamma_{Q,1} Q_{k,1} \tag{2.1}$$

where γ_G is the partial safety factor to be applied to the permanent actions and equals 1.35 when the permanent action is in the same direction as the variable action (i.e. the *unfavourable* case) and 1.0 when the permanent action opposes the variable action (i.e. the *favourable* case) and $\gamma_{Q,1}$ is the partial safety factor to be applied to the primary variable action and equals 1.5 for the *unfavourable* case and 0.0 for the *favourable* case.

More generally, EN 1990 [4] specifies that the design load at the ultimate limits state should be taken as:

$$\sum_{j \geq 1} \gamma_{G,j} G_{k,j} + \gamma_P P + \gamma_{Q,1} Q_{k,1} + \sum_{i > 1} \gamma_{Q,i} \psi_{0,i} Q_{k,i} \tag{2.2}$$

where:

+ implies 'to be combined with'

Σ implies 'the combined effect of'

$\gamma_{G,j}$ is the partial safety factor for the j-th permanent action ($G_{k,j}$)

$\quad \gamma_{G,j}$ = 1.35 (unfavourable)

$\quad \gamma_{G,j}$ = 1.0 (favourable)

γ_P is the partial safety factor for the prestress (P)

$\quad \gamma_P$ = 1.3 for local effects (unfavourable)

$\quad \gamma_P$ = 1.0 for persistent and transfer design situations (favourable)

$\gamma_{Q,1}$ is the partial safety factor for the primary variable action ($Q_{k,1}$)

$\quad \gamma_{Q,1}$ = 1.5 (unfavourable)

$\quad \gamma_{Q,1}$ = 0.0 (favourable)

$\gamma_{Q,i}$ is the partial safety factor for the i-th accompanying variable action ($Q_{k,i}$)

$\quad \gamma_{Q,i}$ = 1.5 (unfavourable)

$\quad \gamma_{Q,i}$ = 0.0 (favourable)

$\psi_{0,i}$ is the combination factor associated with i-th accompanying variable action

Where the magnitudes of the permanent actions vary from place to place in a structure, the unfavourable and favourable parts should be considered individually. EN 1990 [4] notes that the characteristic values of all permanent actions from one source are multiplied by $\gamma_G = 1.35$, if the total resulting action effect is unfavourable, and by $\gamma_G = 1.0$, if the total resulting action effect is favourable. It is further noted that all actions originating from the self-weight of the structure may be considered as coming from one source even if different construction materials are involved.

An alternative to Equation 2.2, EN 1990 [4] specifies that the design load may be taken as the lesser of the values given by the following equations:

$$\sum_{j \geq 1} \gamma_{G,j} G_{k,j} + \gamma_P P + \gamma_{Q,1} \psi_{0,1} Q_{k,1} + \sum_{i>1} \gamma_{Q,i} \psi_{0,i} Q_{k,i} \tag{2.3}$$

$$\sum_{j \geq 1} \xi_j \gamma_{G,j} G_{k,j} + \gamma_P P + \gamma_{Q,1} Q_{k,1} + \sum_{i>1} \gamma_{Q,i} \psi_{0,i} Q_{k,i} \tag{2.4}$$

where $\xi_j = 0.85$ is a reduction factor for any unfavourable permanent action so that $\xi_j \gamma_{G,j} = 1.15$.

2.3.3 Load combinations for the stability or equilibrium limit states

All structures should be designed so that the factor of safety against instability due to overturning, uplift or sliding is suitably high. EN 1990 [4] requires that the structure remains stable under the design load given in Equation 2.2, except that for the stability limit states:

$\gamma_{G,j} = 1.10$ (unfavourable), $\gamma_{G,j} = 0.90$ (favourable)
$\gamma_{Q,1} = 1.5$ (unfavourable), $\gamma_{Q,1} = 0.0$ (favourable)
$\gamma_{Q,i} = 1.5$ (unfavourable), $\gamma_{Q,i} = 0.0$ (favourable)
$\gamma_P = 1.3$ for global analysis (unfavourable), $\gamma_P = 1.0$ (favourable)

The loads causing instability should be separated from those tending to resist it. The *design action effect* is then calculated from the loads and forces tending to cause a destabilising effect (with unfavourable factors). The *design resistance effect* is calculated from the permanent actions or loads tending to cause a stabilising effect, i.e. resisting instability (with favourable factors). The structure should be so proportioned that its design resistance effect is not less than the design action effect.

Consider, for example, the case of a standard cantilever retaining wall. The overturning moment caused by both the lateral earth pressure and the lateral thrust of any permanent and imposed action surcharge is calculated using load factors 1.10 for all permanent actions and 1.50 for all imposed loads. To provide a suitable margin of safety against stability failure, the factored overturning moment should not be greater than 0.9 times the sum of the restoring moments caused by the self-weight of the wall, the weight of the backfill and any other permanent surcharge above the wall.

2.3.4 Load combinations for the serviceability limit states

The design actions to be used in serviceability calculations are the day-to-day *service loads,* and these may be less than the *specified actions.* For example, the specified imposed actions Q have a built-in allowance for overload and impact. There is a low probability that they will be exceeded. It is usually not necessary, therefore, to ensure acceptable deflections and crack widths under the full specified loads. The use of the actual load combinations under *normal* conditions of service (i.e. the *expected* loads) is more appropriate.

The combinations of actions for the serviceability limit states also utilise the combination factors ψ_0, ψ_1 and ψ_2 as specified in Table 2.3 and are as follows [4]:

1. Characteristic combination:

$$\sum_{j \geq 1} G_{k,j} + P + Q_{k,1} + \sum_{i > 1} \psi_{0,i} Q_{k,i} \tag{2.5}$$

The characteristic combination is used for the irreversible limit states, i.e. the serviceability limit states where some consequences of actions exceeding the specified service requirements will remain when the actions are removed.

2. Frequent combination:

$$\sum_{j \geq 1} G_{k,j} + P + \psi_{1,1} Q_{k,1} + \sum_{i > 1} \psi_{2,i} Q_{k,i} \tag{2.6}$$

The frequent combination is used for the reversible limit states, i.e. the serviceability limit states where no consequence of actions exceeding the specified service requirements will remain when the actions are removed.

3. Quasi-permanent combination:

$$\sum_{j \geq 1} G_{k,j} + P + \sum_{i \geq 1} \psi_{2,i} Q_{k,i} \tag{2.7}$$

The quasi-permanent combination is used for long-term effects and the appearance of the structure.

2.4 DESIGN FOR THE STRENGTH LIMIT STATES

2.4.1 General

The *design strength* of a member or connection must always be greater than the *design action effect* produced by the most severe factored load combination (as outlined in Section 2.3.2). On a particular cross-section, the design action effect may be the design axial load N_{Ed}, the design shear force V_{Ed}, the design bending moment M_{Ed}, a design twisting moment T_{Ed} or a combination of these. The design strength of a cross-section is a conservative estimate of the actual strength of the cross-section obtained using partial safety factors applied to the strengths of the concrete, tendons and conventional reinforcement.

2.4.2 Partial factors for materials

The design value X_d of a material property is expressed in general terms as:

$$X_d = \eta \frac{X_k}{\gamma_m} \tag{2.8}$$

where X_k is the characteristic material property (see Chapter 4); η is the mean value of the conversion factor taking into account volume and scale effects, the effects of moisture and temperature and any other relevant parameter and γ_m is the partial factor for the material property taking account of the possibility of an unfavourable deviation of a material property from its characteristic value and the random part of the conversion factor η.

Alternatively, in appropriate cases, the conversion factor η may be implicitly taken into account within the characteristic value itself or by using γ_M instead of γ_m:

$$X_d = \frac{X_k}{\gamma_M} \tag{2.9}$$

The design value X_d can be determined by using values of X_k obtained from measured physical properties, empirical estimates from known chemical composition, previous experience, or from values given in European Standards or other appropriate documents.

The partial factors for materials account for the inherent uncertainties in the estimation of material strengths, variations in member sizes and steel positions, uncertainties in the accuracy of the method used to predict member or cross-sectional behaviour, uncertainties in construction and workmanship and the ductility of the material. The recommended partial factors for materials specified in EN 1992-1-1 [5] for use in Equation 2.9

Table 2.4 Partial factors for materials for ultimate limit states [5]

Design situation	γ_C for concrete	γ_S for reinforcing steel	γ_S for prestressing steel
Persistent and transient	1.5	1.15	1.15
Accidental	1.2	1.0	1.0

for 'persistent and transient' design situations and for 'accidental' design situations are given in Table 2.4.

In certain situations, where measures are taken to reduce the uncertainty in the calculated design strength, reduced values for γ_C and γ_S may be used. For example, for persistent and transient design situations, γ_C may be reduced to 1.4 and γ_S reduced to 1.1, if the construction is to be subject to quality control sufficient to satisfy the tolerances given in Table A.1 in EN 1992-1-1 [5].

2.5 DESIGN FOR THE SERVICEABILITY LIMIT STATES

2.5.1 General

When designing for serviceability, the designer must ensure that the structure behaves satisfactorily and can perform its intended function at service loads. Deflection (or camber) must not be excessive, cracks must be adequately controlled and no portion of the structure should suffer excessive vibration. Design for serviceability usually first involves a linear-elastic analysis of the structure to determine the internal actions under the most severe combination of actions for the serviceability limit states (see Section 2.3.4) and then the determination of the deformation of the structure accounting for the non-linear and inelastic behaviour of concrete under in-service conditions.

The design for serviceability is possibly the most difficult and least well-understood aspect of the design of concrete structures. Service load behaviour depends primarily on the properties of the concrete, which are often not known reliably. Moreover, concrete behaves in a non-linear manner at service loads. The non-linear behaviour of concrete that complicates serviceability calculations is caused by cracking, tension stiffening, creep and shrinkage.

In modern concrete structures, serviceability failures are relatively common. The tendency towards higher-strength materials and the use of ultimate strength design procedures for the proportioning of structures has led to shallower, more slender elements, and consequently, an increase in deformations at service loads. As far back as 1967 [13], the most common cause of damage in concrete structures was due to excessive slab deflections. If the incidence of serviceability failure is to decrease, the design for serviceability must play a more significant part in routine structural design, and the structural designer must resort more often to analytical tools that are more accurate than those found in most building codes. The analytical models for the estimation of in-service deformations outlined in Chapter 5 provide designers with reliable and rational means for predicting both the short-term and time-dependent deformations in prestressed concrete structures. A comprehensive treatment of the in-service and time-dependent analysis of concrete structures is provided in Reference [12].

The level of prestress in beams and slabs is generally selected to satisfy the serviceability requirements. The control of cracking in a prestressed concrete structure is usually achieved by limiting the stress increment in the bonded reinforcement (caused by the application of the full service load) to some appropriately low value and ensuring that the bonded reinforcement is suitably distributed.

2.5.2 Deflection limits

The design for serviceability, particularly the control of deflections, is frequently the primary consideration when determining the cross-sectional dimensions of beams and floor slabs in concrete structures. This is particularly so in the case of slabs, as they are typically thin in relation to their spans and are therefore deflection sensitive. It is stiffness rather than strength that usually controls the design of slabs.

EN 1992-1-1 [5] specifies two basic approaches for deflection control. The first and simplest approach is deflection control by the satisfaction of a maximum span-to-depth ratio. However, the code only provides guidance on limiting span-to-depth ratios for reinforced concrete beams and slabs in buildings and does not cover prestressed concrete. The second approach is deflection control by the calculation of deflection (or camber) using appropriate models of material and structural behaviour. This calculated deflection should not exceed the *deflection limits* that are appropriate to the structure and its intended use. The deflection limits should be selected by the designer and are often a matter of engineering judgement.

There are three main types of deflection problem that may affect the serviceability of a concrete structure:

1. where excessive deflection causes either aesthetic or functional problems;
2. where excessive deflection results in unintended load paths or damage to either structural or non-structural elements attached or adjacent to the member; and
3. where dynamic effects due to insufficient stiffness cause discomfort to occupants.

Examples of deflection problems of Type 1 include visually unacceptable sagging (or hogging) of slabs and beams and ponding of water on roofs.

Type 1 problems are generally overcome by limiting the magnitude of the final long-term deflection (here called *total deflection*) to some appropriately low value. The total deflection of a beam or slab in a building is usually the sum of the short-term and time-dependent deflections caused by the permanent actions (including self-weight), the prestress (if any), the expected imposed actions and the load-independent effects of shrinkage and temperature change.

EN 1992-1-1 [5] supports the ISO 4356 [14] recommendation that the appearance of a structure or its ability to perform its intended function may be impaired, if the calculated final sag of a beam or slab under the quasi-permanent loads exceeds span/250. However, the total deflection limits that are appropriate for a particular member and its intended function will vary from member to member and structure to structure depending on its use and purpose. For example, a total deflection limit of span/250 may be appropriate for the floor of a car park but would be entirely inadequate for a gymnasium floor that is required to remain essentially plane under service conditions and where functional problems arise at very small total deflections.

Examples of Type 2 problems include, among others, deflection-induced damage to ceiling or floor finishes, cracking of masonry walls and other brittle partitions, improper functioning of sliding windows and doors, tilting of storage racking and so on. To avoid these problems, a limit must be placed on that part of the total deflection that occurs after the attachment of the non-structural elements in question, i.e. the *incremental deflection*. The incremental deflection that occurs after construction is the sum of the long-term deflection due to all the sustained loads and shrinkage, the short-term deflection due to the transitory imposed actions and the short-term deflection due to any permanent actions applied to the structure after the attachment of the non-structural elements under consideration, together with any temperature-induced deflection.

EN 1992-1-1 [5] confirms the recommendation of ISO 4356 [14] that the deflection that occurs after construction under the quasi-permanent loads should normally not exceed span/500. However, the deflection that could damage adjacent parts of the structure (including brittle partitions and finishes) should be assessed and limited depending on the sensitivity of the adjacent elements. In many cases, a limiting deflection of span/500 will be acceptable, but in some situations a more onerous requirement may be appropriate depending on the provisions made to minimise the effect of movement. Incremental deflections of span/500 can, in fact, cause cracking in supported masonry walls, particularly when doorways or corners prevent arching and when no provisions are made to minimise the effect of movement.

Type 3 deflection problems include the perceptible *springy* vertical motion of floor systems and other vibration-related problems. Very little quantitative information for limiting this type of deflection problem is available in codes of practice. For a member subjected to vehicular or pedestrian traffic, the Australian Standard AS3600-2009 [15] requires that the deflection caused by imposed actions be less than span/800. For a floor that is not supporting or attached to non-structural elements likely to be damaged by large deflection, ACI 318M-14 [16] places a limit of span/360 on the short-term deflection due to imposed actions. These limits provide minimum requirements for the stiffness of members that may, in some cases, be sufficient to avoid Type 3 problems. Such problems are potentially the most common for prestressed concrete floors, where load balancing is often employed to produce a nearly horizontal floor under the permanent actions and the bulk of the final deflection is due to the transient imposed actions. Such structures are generally uncracked at service loads, the total deflection is small and Types 1 and 2 deflection problems are easily avoided.

2.5.3 Vibration control

Where a structure supports vibrating machinery or is subjected to any other significant dynamic load (such as pedestrian traffic), or where a structure may be subjected to ground motion caused by earthquake, blast or adjacent road or rail traffic, vibration control becomes an important design requirement. This is particularly so for slender structures, such as tall buildings or long-span beams or slabs.

Vibration is best controlled by isolating the structure from the source of vibration. Where this is not possible, vibration may be controlled by limiting the frequency of the fundamental mode of vibration of the structure to a value that is significantly higher than the frequency of the source of vibration. The minimum frequency of the fundamental mode of vibration of a beam or slab depends on the function of the building and the source of the vibration. For example, when the only source of vibration is pedestrian traffic, 5 Hz is often taken as the minimum frequency of the fundamental mode of vibration of a beam or slab [17,18]. For detailed design guidance on dealing with floor vibrations, reference should be made to specialist literature such as [19–22].

2.5.4 Crack width limits

In the design of a prestressed concrete structure, the calculation of crack widths is rarely required. Crack control is deemed to be provided by appropriate detailing of the reinforcement and by limiting the stress in the reinforcement crossing the crack to some appropriately low value (see Section 5.12). The limiting steel stress depends on the maximum acceptable crack width for the structure and that in turn depends on the structural requirements and the local environment. The maximum crack widths recommended in EN 1992-1-1 [5] are given in Table 2.5 for the various exposure classes defined in the code and outlined in Table 2.6.

Table 2.5 Recommended values for maximum final design crack width, w_{max} (mm) [5]

Exposure class	Reinforced members and prestressed members with unbonded tendons Quasi-permanent load combination	Prestressed members with bonded tendons Frequent load combination
X0, XC1	0.4[a]	0.2
XC2, XC3, XC4	0.3	0.2[b]
XD1, XD2, XS1, XS2, XS3	0.3	Decompression[c]

[a] In exposure classes X0 and XC1, the crack width does not influence durability. This limit is set to ensure acceptable appearance. In the absence of appearance requirements, this limit may be relaxed.

[b] For exposure classes XC2, XC3 and XC4, in addition to this limit, decompression should be checked under the quasi-permanent load combination.

[c] The decompression limit requires that all parts of the bonded tendons or duct lie at least 25 mm within the compressive concrete.

2.5.5 Partial factors for materials

EN 1992-1-1 [5] recommends that the values of the partial factors to be applied to the material properties used in calculations to verify the serviceability limit states should be taken as 1.0 for both concrete and steel, except if specified differently elsewhere in the code [5] or in the National Annex of the relevant country.

2.6 DESIGN FOR DURABILITY

A durable structure is defined in EN 1992-1-1 [5] as one that meets 'the requirements of serviceability, strength and stability throughout its design working life, without significant loss of utility or excessive unforeseen maintenance'. Design for durability is an important part of the design process. The design life for most concrete structures is typically 50–100 years, but prestressed concrete structures are often required to have a design life in excess of 100 years, particularly for large bridges and monumental structures. It is important to ensure that the concrete and the steel do not deteriorate significantly during the design life of the structure so that maintenance and repair costs are kept to a minimum. Of course, for a structure in service for 100 years, some maintenance costs are inevitable, but excessive repair and maintenance can result in uneconomical life-cycle costs.

Accurate and reliable models for the deterioration of concrete with time and for the initiation and propagation of corrosion in the steel reinforcement and tendons are difficult to codify. According to EN 1992-1-1 [5], the design for durability begins with the selection of an *exposure class* for the various regions and surfaces of the structure (see Table 2.6). This is then

Table 2.6 Exposure classes for reinforced and prestressed concrete [5]

Class	Environment
No risk of corrosion	
X0	Very dry, e.g. inside buildings with very low air humidity
Corrosion induced by carbonation	
XC1	Dry or permanently wet, e.g. inside buildings with low air humidity; permanently submerged in water
XC2	Wet, rarely dry, e.g. surfaces subjected to long-term water contact; many foundations
XC3	Moderate humidity, e.g. inside buildings with moderate to high air humidity; external (sheltered from rain)
XC4	Cyclic wet and dry, e.g. surfaces subjected to water contact (not within exposure class XC2)
Corrosion induced by chlorides (not from sea water)	
XD1	Moderate humidity, e.g. surfaces exposed to airborne chlorides
XD2	Wet, rarely dry, e.g. swimming pools; surfaces exposed to industrial waters containing chlorides
XD3	Cyclic wet and dry, e.g. parts of bridges exposed to spray containing chlorides; pavements; car park slabs
Corrosion induced by chlorides from sea water	
XS1	Exposed to airborne salt but not in direct contact with sea water, e.g. structures near to or on the coast
XS2	Permanently submerged, e.g. parts of marine structures
XS3	Tidal, splash and spray zones, e.g. parts of marine structures
Freeze/thaw attack	
XF1	Moderate water saturation, without de-icing agents, e.g. vertical surfaces exposed to rain and freezing
XF2	Moderate water saturation, with de-icing agents, e.g. vertical surfaces of road structures exposed to freezing and airborne de-icing agents
XF3	High water saturation, without de-icing agents, e.g. horizontal surfaces exposed to rain and freezing
XF4	High water saturation, with de-icing agents or sea-water, e.g. road and bridge decks exposed to de-icing agents; surfaces exposed to direct spray containing de-icing agents and freezing; splash zones of marine structures and freezing
Chemical attack	
XA1	Slightly aggressive environment, e.g. natural soils and ground water
XA2	Moderately aggressive environment, e.g. natural soils and ground water
XA3	Highly aggressive environment, e.g. natural soils and ground water

followed by the satisfaction of a range of deemed-to-comply design requirements, including requirements relating to concrete quality (both in terms of minimum compressive strength and restrictions on the chemical content), concrete cover to the steel reinforcement (and tendons) and curing of the concrete.

Good quality, well-compacted concrete, with low permeability and adequate cover to the reinforcement and tendons are the essential prerequisites for durable structures. In Table 2.7, the minimum strength class for concrete (see Section 4.3.2) specified in EN 1992-1-1 [5] is provided for each of the exposure classes in Table 2.6 for a design life of 50 years and a design life of 100 years. Also provided in Table 2.7 is the minimum cover $c_{min,dur}$ necessary to protect the steel reinforcement and tendons against corrosion.

These durability requirements may result in higher concrete strength and higher cover to the reinforcement and tendons than are required for other aspects of the structural design. In addition, design requirements are also specified in the code to ensure structures resist unacceptable deterioration due to abrasion from traffic, cycles of freezing and thawing and contact with aggressive soils.

2.7 DESIGN FOR FIRE RESISTANCE

In addition to satisfying the other design requirements, prestressed concrete structures must be able to fulfil their required functions when exposed to fire for at least a specified period. This means that a structure should maintain its *structural adequacy* (load carrying capacity) for a specified period (called the *Standard fire resistance period*). In the case of walls and slabs, the structure must also maintain its *integrity* for a specified period so as to prevent the passage of flames or hot gases through the structure. In addition, a wall or slab must maintain its ability to prevent ignition of combustible material in the compartment beyond the surface exposed to the fire (*structural insulation*).

Fire resistance depends on the cover to the steel reinforcement and tendons and the member thickness. The requirements for the fire resistance of concrete structures are specified in EN 1992-1-2 [10], where the minimum fire resistance periods for concrete structures and structural components are specified, together with a range of deemed-to-comply provisions that ensure their satisfaction. For example, Table 2.8 provides possible combinations of minimum values of the beam width b at the level of the bottom tensile reinforcement and the average axis distance a for the bottom tensile reinforcing bars in beams exposed to fire for various fire resistance periods, together with the minimum web width b_w (for webs of varying width).

Table 2.7 Minimum concrete strength and concrete cover for durability for normal weight concrete [5]

Exposure class	Minimum concrete strength class	Minimum concrete cover for durability, $c_{min,dur}$ (mm)			
		50-year design life		100-year design life	
		Reinforcing steel	Tendons	Reinforcing steel	Tendons
X0	C12/15	10	10	20	20
XC1	C20/25	15	25	25	35
XC2	C25/30	25	35	35	45
XC3	C30/37	25	35	35	45
XC4	C30/37	30	40	40	50
XD1	C30/37	35	45	45	55
XD2	C30/37	40	50	50	60
XD3	C35/45	45	55	55	65
XS1	C30/37	35	45	45	55
XS2	C35/45	40	50	50	60
XS3	C35/45	45	55	55	65
XF1	C30/37	Special attention should be given to the concrete composition and to the selection of concrete cover. In most situations, concrete covers specified for class XD1 will be satisfactory for classes XF1, XF2 and XA1, concrete covers specified for class XD2 will be satisfactory for classes XF3, XF4, XA2 and concrete covers specified for class XD3 will be satisfactory for classes XA3.			
XF2	C30/37				
XF3	C30/37				
XF4	C30/37				
XA1	C30/37				
XA2	C30/37				
XA3	C35/45				

Similarly, Table 2.9 provides possible combinations of minimum values for the smallest dimension b of a column cross-section and average axis distance a for the reinforcing bars in columns exposed to fire. The axis distance is the minimum distance from the axis of a reinforcing bar to the nearest concrete surface. In preparing Table 2.8, it has been assumed that the flexural resistance in a fire is not less than 70% of the flexural resistance under normal conditions of service. Table 2.10 shows the minimum dimension h_s of a slab exposed to fire, where h_s is the sum of the thickness of the slab and the thickness of any non-combustible flooring on top of the slab. Also shown in Table 2.10 is the minimum axis distance to the bottom tensile reinforcement in the slab. In the absence of more detailed calculations, the minimum axis distances specified in Tables 2.8 and 2.10 should be increased by 10 mm for prestressing bars and 15 mm for prestressing wires and strand.

Table 2.8 Minimum dimensions b, b_w and a (in mm) for beams exposed to fire [5]

Standard fire resistance (min)	Simply-supported beams			Continuous beams		
	b	a	b_w	b	a	b_w
30	80	25	80	80	15	80
	160	15		160	12	
60	120	40	100	120	25	100
	200	30		200	12	
90	150	55	110	150	35	110
	300	40		250	25	
120	200	65	130	200	45	130
	300	55		300	35	
	500	50		500	30	
180	240	80	150	240	60	150
	300	70		400	50	
	600	60		600	40	
240	280	90	170	280	75	170
	350	80		500	60	
	700	70		700	50	

Table 2.9 Minimum dimensions b and a (in mm) for columns in a fire [5]

Standard fire resistance (min)	Columns exposed on more than one side						Columns exposed on one side	
	$\mu_{fi}{}^a = 0.2$		$\mu_{fi}{}^a = 0.51$		$\mu_{fi}{}^a = 0.71$		$\mu_{fi}{}^a = 0.7$	
	b	a	b	a	b	a	b	a
30	200	25	200	25	200	32	155	25
					300	27		
60	200	25	200	36	250	46	155	25
			300	31	350	40		
90	200	31	300	45	350	53	155	25
	300	25	400	38	450	40		
120	250	40	350[b]	45[b]	350[b]	57[b]	175	35
	350	35	450[b]	40[b]	450[b]	51[b]		
180	350[b]	45[b]	350[b]	63[b]	450[b]	70[b]	230	55
240	350[b]	61[b]	450[b]	75[b]	–	–	295	70

[a] μ_{fi} is the ratio of axial load in a fire to design resistance of the column at normal temperatures.
[b] Column must contain a minimum of eight bars.

Table 2.10 Minimum dimensions h_s and a (in mm) for a slab exposed to fire [5]

Standard fire resistance (min)	Floor thickness, h_s	One-way slab	Two-way slab ($l_y \geq l_x$)	
			$l_y/l_x \leq 1.5$	$1.5 < l_y/l_x \leq 2.0$
30	60	10	10	10
60	80	20	10	15
90	100	30	15	20
120	120	40	20	25
180	150	55	30	40
240	175	65	40	50

In design, the cover to the reinforcing bars and tendons must satisfy both the requirements for fire and the requirements for durability.

2.8 DESIGN FOR ROBUSTNESS

Robustness is the requirement that a structure should be able to withstand damage to an element without total collapse of the structure or a significant part of the structure in the vicinity of the damaged element. A structure should be designed so that should a local accident occur, then the damage should be contained within an area local to the accident. Should one member be removed, for example, the remainder of the structure should hang together and not precipitate a progressive collapse. For this requirement to be satisfied, the members and materials of construction must have adequate ductility. In particular, the reinforcement and tendons assumed in design to constitute the ties in the structure, must obviously be highly ductile. Robustness reduces the consequences of gross errors or of local structural failures.

EN 1992-1-1 [5] requires that structures that are not specifically designed to withstand accidental actions should have a suitable tying system to prevent progressive collapse and to provide an alternative load path after accidental damage. Structures should be designed and detailed such that adjacent parts of the structure are tied together in both the horizontal and vertical planes so that the structure can withstand an event without being damaged to an extent disproportionate to that event.

REFERENCES

1. EN 1991-1-1. 2002. Eurocode 1: Actions on structures – Part-1-1: General actions – Densities, self-weight, imposed loads on buildings. British Standards Institution, London, UK.
2. EN 1991-1-3. 2003. Eurocode 1: Actions on structures – Part 1-3: General actions – Snow loads. British Standards Institution, London, UK.
3. EN 1991-1-4. 2005 (+A1: 2010). Eurocode 1: Actions on structures – Part 1-4: General actions – Wind actions. British Standards Institution, London, UK.
4. EN 1990. 2002 (+A1: 2005). Eurocode: Basis of structural design. British Standards Institution, London, UK.
5. EN 1992-1-1. 2004. Eurocode 2: Design of concrete structures – Part 1-1: General rules and rules for buildings. British Standards Institution, London, UK.
6. EN 1991-1-2. 2002. Eurocode 1: Actions on structures – Part-1-2: General actions – Actions on structures exposed to fire. British Standards Institution, London, UK.
7. EN 1991-1-5. 2003. Eurocode 1: Actions on structures – Part 1-5: General actions – Thermal actions. British Standards Institution, London, UK.
8. EN 1991-1-6. 2005. Eurocode 1: Actions on structures – Part 1-6: General actions – Actions during execution. British Standards Institution, London, UK.
9. EN 1991-1-7. 2006. Eurocode 1: Actions on structures – Part 1-7: General actions – Accidental actions. British Standards Institution, London, UK.
10. EN 1992-1-2. 2004. Eurocode 2: Design of concrete structures – Part 1-2: General rules – Structural fire design. British Standards Institution, London, UK.
11. Ranzi, G. and Gilbert, R.I. 2014. *Structural Analysis – Principles, Methods and Modelling*. Boca Raton, FL: CRC Press.
12. Gilbert, R.I. and Ranzi, G. 2011. *Time-Dependent Behaviour of Concrete Structures*. London, UK: Spon Press.
13. Mayer, H. and Rüsch, H. 1967. Building damage caused by deflection of reinforced concrete building components. Technical Translation 1412, National Research Council, Ottawa, Ontario, Canada (Berlin, West Germany, 1967).
14. ISO 4356. 1977. Bases for the design of structures – Deformations of buildings at the serviceability limit states. International Standards Organisation, Geneva, Switzerland.
15. AS3600-2009. 2009. Australian standard for concrete structures. Standards Australia, Sydney, New South Wales, Australia.
16. ACI318M-14. 2014. Building code requirements for structural concrete and commentary. American Concrete Institute, Detroit, MI.
17. Irwin, A.W. 1978. Human response to dynamic motion of structures. *The Structural Engineer*, 56A(9), 237–244.
18. Mickleborough, N.C. and Gilbert, R.I. 1986. Control of concrete floor slab vibration by L/D limits. *Proceedings of the 10th Australasian Conference on the Mechanics of Structures and Materials*, University of Adelaide, Adelaide, South Australia, Australia.
19. Smith, A.L., Hicks, S.J. and Devine, P.J. 2009. Design of floors for vibration: A new approach – Revised edition (SCI P354). Steel Construction Institute, Berkshire, UK.

20. Willford, M.R. and Young, P. 2006. A design guide for footfall induced vibration of structures (CCIP-016). The Concrete Centre, Surrey, UK.
21. Murray, T.M., Allen, D.E. and Ungar, E.E. 1997. Steel Design Guide Series 11 – Floor vibrations due to human activity. American Institute of Steel Construction, Chicago, IL.
22. ISO 10137. 2007. Bases for design of structures – Serviceability of buildings and walkways against vibrations. International Standards Organisation, Geneva, Switzerland.

Chapter 3

Prestressing systems

3.1 INTRODUCTION

Prestressing systems used in the manufacture of prestressed concrete have developed over the years, mainly through research and development by specialist companies associated with the design and execution of prestressed concrete structures. These companies are often involved in other associated works, including soil and rock anchors, lifting of heavy structures, cable-stayed bridges and suspension bridges that require specialist expertise and patented materials, equipment and design. Information on the products of each company is generally available directly from the respective company or from its website.

This chapter describes and illustrates the basic forms of prestressing and the components used for prestressing. These include the steel tendons used to prestress the concrete, namely the wire, strand and bar, together with the required anchorages, ducts and couplers. The specialised equipment required for stressing and grouting of the ducts in post-tensioning applications is illustrated, and the basic principles and concepts of the various systems are provided, including illustrations of the various prestressing operations. The material properties of both the concrete and steel used in the design of prestressed concrete are detailed and discussed in Chapter 4.

3.2 TYPES OF PRESTRESSING STEEL

There are three basic types of high-strength steel commonly used as tendons in prestressed concrete construction.

1. Cold-drawn stress-relieved round wire;
2. Stress-relieved strand; and
3. High-strength alloy steel bars.

The term *tendon* is generally defined as a wire, strand or bar (or any discrete group of wires, strands or bars) that is intended to be either pretensioned or post-tensioned.

Wires are cold-drawn solid steel elements, circular in cross-section, with diameter usually in the range of 2.5–12.5 mm. Cold-drawn wires are produced by drawing hot-rolled medium to high carbon steel rods through dies to produce wires of the required diameter. The drawing process cold works the steel, thereby altering its mechanical properties and increasing its strength. The wires are then stress-relieved by a process of continuous heat treatment and straightening to improve ductility and produce the required material properties (such as low-relaxation). The typical characteristic tensile strength f_{pk} for wires is in the range of 1570–1860 MPa. Wires are sometimes indented or crimped to improve their bond characteristics. Wires' diameters vary from country to country, but most commonly, wire diameters are in the range of 4–8 mm. In recent years, the use of wires in prestressed concrete construction has declined, with 7-wire strand being preferred in most applications.

Stress-relieved strand is the most commonly used prestressing steel. Both 7-wire and 19-wire strands are available. Seven-wire strand consists of six wires tightly wound around a seventh, slightly larger diameter, central core wire, as shown in Figure 3.1a. The pitch of the six spirally wound wires is between 12 and 18 times the nominal strand diameter. The nominal

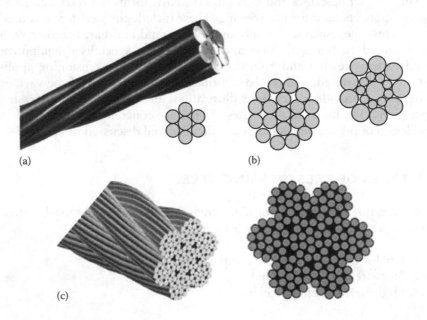

(a)   (b)

(c)

Figure 3.1 Types of strand. (a) 7-wire strand. (b) 19-wire strand – alternative cross-sections. (c) Cable consisting of seven 19-wire strands.

diameters of the 7-wire strands in general use are in the range of 7–15.2 mm, with typical characteristic tensile strength in the range of 1760–2060 MPa. Seven-wire strand is widely used in both pretensioned and post-tensioned applications. Nineteen-wire strand consists of two layers of 9 wires or alternatively two layers of 6 and 12 wires spirally wound around a central wire. The pitch of the spirally wound wires is 12–22 times the nominal strand diameter. The nominal diameters of 19-wire strands in general use are in the range of 17–22 mm, and typical cross-sections are shown in Figure 3.1b. Nineteen-wire strand is used in post-tensioned applications, but because of its relatively low surface area to volume ratio, it is not recommended for pretensioned applications, where the transfer of prestress relies on the surface area of the strand available for bond to the concrete.

Strand may be compacted by drawing the strand through a compacting die, thereby reducing the diameter, while maintaining the same cross-sectional area of steel. Compacting strand also facilitates the gripping of the strand at its anchorage.

The mechanical properties of the strand are slightly different from those of the wire from which it is made. This is because the stranded wires tend to straighten slightly when subjected to tension, thus reducing the apparent elastic modulus. For design purposes, the yield stress of stress-relieved strand is about $0.86f_{pk}$ and the elastic modulus is $E_p = 195 \times 10^3$ MPa.

Cables consist of a group of tendons often formed by multiwire strands woven together as shown in Figure 3.1c. Stay cables used extensively in cable-stayed and suspension bridges are generally made directly from strands.

High-strength alloy steel bars are hot rolled with alloying elements introduced into the steel making process. Some bars are ribbed to improve bond. Bars are single straight lengths of solid steel of greater diameter than wire, with diameters typically in the range of 20–50 mm and with typical characteristic minimum breaking stresses in the range of 1030–1230 MPa.

3.3 PRETENSIONING

As the name implies, *pretensioning* involves the tensioning of steel strands prior to casting of the concrete and was introduced in Section 1.2.1. The prestressing operation requires an appropriate tensioning bed for the precast elements, bulk heads at both ends to anchor the individual strands and formwork for the precast concrete elements. A typical pretensioning bed is shown in Figure 3.2a. Pretensioning is often carried out in a factory environment where the advantages of quality control and mass production can be achieved. Pretensioning the strands can be achieved by stressing multiple strands or wires simultaneously or by stressing each strand or wire individually. Set-ups for multi-strand stressing and single-strand stressing are shown in Figure 3.2a and b, respectively.

(a)

(b)

Figure 3.2 Pretensioning beds. (Courtesy of VSL International Limited, Hong Kong, China.) (a) Multi-strand pretensioning. (b) Single-strand pretensioning.

The application of prestress to the structure or structural element, for most practical cases, involves the use of a hydraulic jack to stress either single-strand tendons or groups of strands in a tendon (multi-strand tendons). Prestressing jacks are hydraulically operated by pumping oil under pressure into a piston device, thereby elongating the tendon and increasing the tension in the tendon. When the required tendon elongation is achieved, each end of the tendon is anchored to the bulkhead using wedges that grip the strand at the anchorage. The following sequence of operations is typical for multi-strand pretensioning:

1. The strands are laid out in a pretensioning bed, as illustrated in Figure 3.2, where ram jacks are positioned next to the bulkhead and are extended to ensure that sufficient distance is available for the deformations to take place at transfer. The stroke of the ram jacks must be longer than the desired elongation of the strands.
2. The formwork moulds are closed and the strands are then stressed.
3. Concrete is poured.
4. When the concrete reaches its transfer strength, the load from the strands is transferred to the concrete by allowing the ram jacks to gradually retract, and the load is transferred to the concrete members. The strands are cut after the ram jacks are fully retracted. The prestressing force is transferred by a combination of friction and bond between the concrete and the steel.

Whilst not now in common use, deflection or harping of the strands, if required in design, can be achieved by either anchoring the strand or wire in the bed of the unit, or by the use of a hydraulic ram or harping device to hold the pretensioned strand in the desired position whilst the concrete is cast and cured, prior to the transfer of the prestressing force to the precast elements. The harping device deflects the strands and provides a varying eccentricity of the prestressing force within the concrete member.

3.4 POST-TENSIONING

Post-tensioning of concrete was introduced in Section 1.2.2 and is used in a wide range of structures to apply prestress, usually on-site. Post-tensioning offers significant flexibility in the way the prestress is applied to a structure, with the tendon profiles readily adjusted to suit the applied loading and the support conditions. Post-tensioning lends itself to *stage stressing*, whereby increments of prestress are applied as required at different stages of construction as the external loads progressively increase.

Post-tensioned systems consist of corrugated galvanised steel or plastic ducts (with grout vents for bonded tendons), prestressing strands, anchorages and grout for bonded tendons. The post-tensioned tendon profile is achieved by fixing the ducts to temporary supports (often attached to the non-prestressed reinforcement in a beam) at appropriate intervals within the formwork. For slabs on ground, the strands are generally supported on bar chairs, as can be seen in Figure 3.3. The ducts that house the prestressing tendons may be fabricated from corrugated steel sheathing or, in more recent developments, plastic ducting, as shown in Figure 3.4a and b, respectively.

Figure 3.3 Examples of post-tensioned slabs before concrete casting. (Courtesy of VSL Australia Limited, Sydney, New South Wales, Australia.)

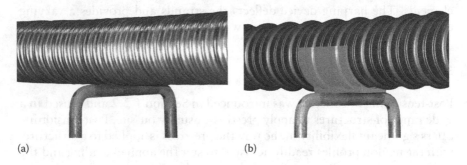

(a)   (b)

Figure 3.4 Post-tensioning ducts. (Courtesy of VSL International Limited, Hong Kong, China.) (a) Corrugated metal duct. (b) Plastic ducting.

Figure 3.5 illustrates a schematic of the layout of a post-tensioning strand in a typical continuous floor slab. The details would also apply for a continuous beam. The prestressing tendon follows a profile determined from the design loading, and the location and type of supports. After casting, the concrete is allowed to cure until it reaches the required transfer strength. Depending on the system being used, or the requirement of the structural design, an initial prestressing force is sometimes applied when the concrete compressive strength reaches about 10 MPa (to facilitate the removal of forms), with the

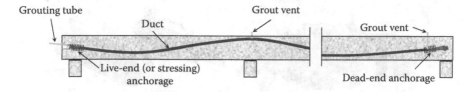

Figure 3.5 Tendon layout and details in a continuous post-tensioned slab.

strands re-stressed up to the initial jacking force at a later stage when the concrete has gained its required strength at transfer.

In many parts of the world, it is a usual practice for the ducts to be grouted after the post-tensioning operation has been completed. Grout is pumped into the duct at one end under pressure. Grout vents are located at various locations along the duct (as shown in Figure 3.5) to ensure that the wet grout completely fills the duct during the grouting operation.

After the grout has set, the post-tensioned tendon is effectively bonded to the surrounding concrete. The grout serves several purposes including higher utilisation of the prestressing steel in bending under ultimate limit state conditions, better corrosion protection of the tendon and, importantly, the prevention of failure of the entire tendon due to localised damage at the anchorage or an accidental cutting of the strand. Further discussion of the advantages and disadvantages of both bonded and unbonded prestressed concrete is presented in Section 3.5.

The prestress is applied by a hydraulic jack (Figures 3.6a and b) reacting against the concrete at the stressing anchorage located at one end of the member, usually referred to as the live end. A small handheld jack stressing the individual strands in a slab duct is shown in Figure 3.6c. The live end of a post-tensioning anchorage system has several basic components comprising an anchor head, associated wedges required to anchor the strands and an anchorage casting or bearing plate. While these anchorages come in many shapes and sizes, the load transfer mechanism of these anchorages remains essentially the same. The stressing operation involves the hydraulic jack pulling the strands protruding behind the anchorage until the required jacking force is reached. Typical live-end anchorages for a flat ducted tendon are shown in Figures 3.6d and e.

Prestressing strands at the live end of a slab tendon before post-tensioning are shown in Figure 3.7a, and the wedges used to clamp the strand are shown in Figure 3.7b. It is common practice to paint the strands before post-tensioning (as shown in Figure 3.7c) to enable the elongation of each strand to be readily measured after the stressing operation (Figure 3.7d). After jacking, the post-tensioned strands are anchored by the wedges in the anchor head, and the load is transferred from the jack to the structure via the anchor casting or bearing plate.

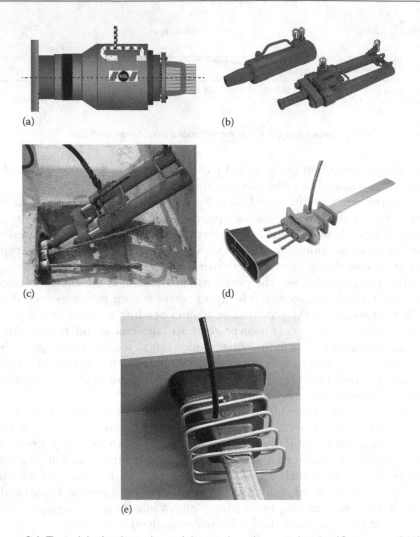

(a)

(b)

(c)

(d)

(e)

Figure 3.6 Typical hydraulic jacks and live-end anchorage details. (Courtesy of VSL International Limited, Hong Kong, China; VSL Australia Limited, Sydney, New South Wales, Australia.) (a) Hydraulic jacks for multi-strand stressing. (b) Hydraulic jacks for individual strand stressing. (c) Post-tensioning individual strands in a slab duct. (d) Live-end anchorage components. (e) Live-end anchorage with confining steel and grout vent.

Although the live anchorage can also be used at an external non-stressing end, when stressing is only required from one end of the member, the non-stressing end often takes the form of an internal dead-end anchorage where the ends of the strands are cast in the concrete. Whilst many forms of this anchorage exist, the principle is to create a passive anchorage block by either spreading out the exposed strand bundle to form local anchor

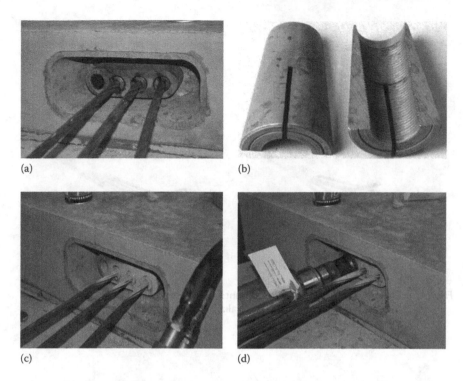

(a)

(b)

(c)

(d)

Figure 3.7 Post-tensioning of strands. (Courtesy of VSL International Limited, Hong Kong, China; VSL Australia Limited, Sydney, New South Wales, Australia.) (a) Strands before post-tensioning. (b) Anchorage wedge components. (c) Painting of strands before post-tensioning. (d) Stressing the first of the painted strands.

nodules/bulbs at the extremities beyond the duct (Figure 3.8a) or by means of swaged barrels clamped on the strands and bearing against a steel plate (Figure 3.8b). The duct is sealed to prevent the ingress of concrete during construction. The tendon is stressed only after the surrounding concrete has reached its required transfer strength.

Typical anchorage systems for use with multi-strand arrangements are shown in Figure 3.9, and a typical post-tensioning installation is shown in Figure 3.10a, with the end anchorage after completion of the prestressing operation shown in Figure 3.10b. Figure 3.10c shows the grinding of the strands after the stressing is completed.

Tendon couplers and intermediate anchorages can be used to connect tendons within a member. Typical examples of coupling and intermediate anchorages are shown in Figure 3.11.

A well-designed grout mix and properly grouted tendons are important to the durability of the structure. The success of a grouting operation depends on many factors, including the correct placement of the grout vents for the

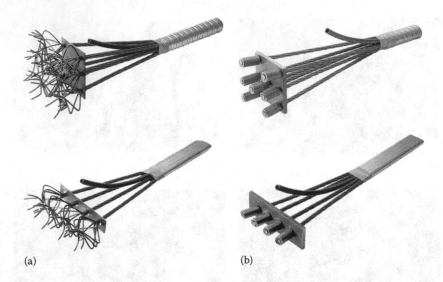

Figure 3.8 Dead-end anchorage arrangements. (Courtesy of VSL Australia Limited, Sydney, New South Wales, Australia.) (a) Strands with crimped wires (onion end). (b) Swaged barrel end plate.

Figure 3.9 Typical multi-strand tendon anchorages. (Courtesy of VSL Australia Limited, Sydney, New South Wales, Australia.)

injection of grout and for expelling the air in the duct at the grout outlets. Vents are required at the high points of the tendon profile to expel the air in the ducts as the grout is injected into the duct at the tendon end or anchorage point of the tendon. Vents at or near the high points allow air and water to be removed from the crest of the duct profile. Grout emitted from the vent at the far end of the duct signals that the duct is completely filled with grout. The use of temporary or permanent grout caps ensures complete filling of the anchorages and permits the verification of the grouting at a

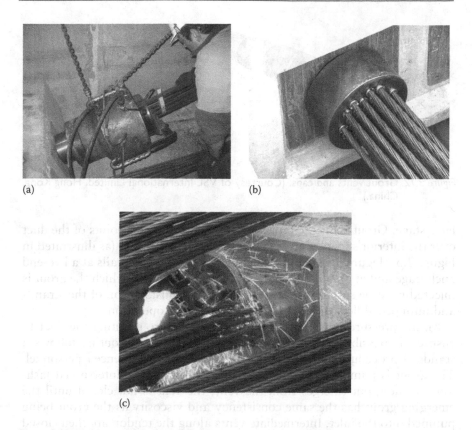

(a) (b)

(c)

Figure 3.10 Multi-strand jack and anchorage operations. (Courtesy of VSL Australia Limited, Sydney, New South Wales, Australia.) (a) Jacking the strands. (b) Anchorage before prestressing. (c) Cutting the strands after completion of post-tensioning.

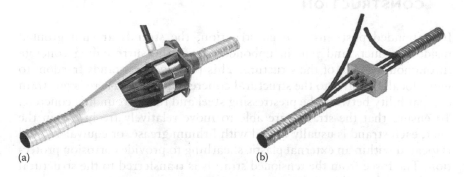

(a) (b)

Figure 3.11 Coupling and intermediate anchorages for multi-strand systems. (Courtesy of VSL Australia Limited, Sydney, New South Wales, Australia.) (a) Coupling anchorage. (b) Intermediate anchorage.

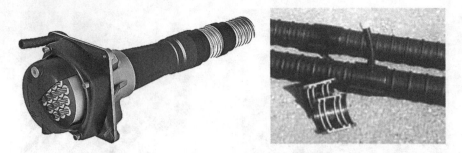

Figure 3.12 Grout vents and caps. (Courtesy of VSL International Limited, Hong Kong, China.)

later stage. Grout vents are typically located at the high points of the duct over the interior supports as well as at the end anchorages (as illustrated in Figure 3.5). Figure 3.12 shows typical grout vents and details at a live-end anchorage and at a point along the duct. The ducts into which the grout is injected must be sufficiently large to allow easy installation of the strands and unimpeded flow of grout during the grouting operation.

An air pressure test is usually undertaken before grouting the duct to ensure the possibility of grout leakage is minimised. Grouting follows a standard procedure and, to be effective, requires experienced personnel. The grout is pumped into duct inlet in a continuous uninterrupted fashion. As the grout emerges from the vent, the vent is not closed until the emerging grout has the same consistency and viscosity as the grout being pumped into the inlet. Intermediate vents along the tendon are then closed in sequence after ensuring that the grout has the required consistency and viscosity.

3.5 BONDED AND UNBONDED POST-TENSIONED CONSTRUCTION

In unbonded post-tensioned construction, the strands are not grouted inside the ducts and remain unbonded from the surrounding concrete throughout the life of the structure. This permits the strands freedom to move locally relative to the structural concrete member. There is no strain compatibility between the prestressing steel and the surrounding concrete. To ensure that the strands are able to move relatively freely within the duct, each strand is usually coated with lithium grease, or equivalent, and is located within an external plastic sheathing to provide corrosion protection. The force from the tensioned strands is transferred to the structural member at the end anchorages.

There are advantages and disadvantages of bonded or unbonded construction, and the use of either is dependent on the design and construction requirements. In some countries, the disadvantages of unbonded construction are considered to outweigh the advantages, and the use of unbonded post-tensioning is not permitted, except for slabs on ground.

Durability is an important consideration for all forms of construction. Therefore, the provision of active corrosion protection is of significant importance. By grouting the tendons, an alkaline environment is provided around the steel, thus providing active corrosion protection (passivation).

Bonded prestressing steel ensures that any change in strain at the tendon level is the same in both the tendon and the surrounding concrete. At overloads, as the concrete member deforms and the strain at the tendon level increases, the full capacity of the bonded tendon can be realised, and the ultimate capacity of the cross-section is increased substantially by grouting. The steel is capable of developing additional force, due to bond, in a relatively short distance. The effect of tendon or anchorage failure is localised after grouting, and the remainder of the tendon is largely unaffected and remains functional. Bonded tendons are also better than unbonded tendons for controlling cracking and for resisting progressive collapse if local failure occurs.

With appropriate design consideration, the prestressing forces in the unbonded tendons can theoretically be adjusted throughout the life of the structure. Tendons may be able to be inspected, re-stressed or even replaced. For example, some tendons for nuclear works are unbonded, as they need to be monitored and as necessary, re-stressed. In unbonded construction, since the prestressing force in the tendon is transmitted to the beam only at the end anchorages, there is an almost uniform distribution of strain in the tendon under load. Changes in the force in the tendon are only possible due to friction and due to deformation of the member, thereby increasing the overall length of the tendon between anchorages. At overloads, the full strength of an unbonded tendon may not be achieved and the ultimate strength of unbonded construction is therefore generally less than that of bonded construction. The anchorage of the unbonded tendons is therefore a critical component, since the entire prestressing force is transmitted at this point throughout the life of the structure.

3.6 CIRCULAR PRESTRESSING

The term *circular prestressing* is applied to structures with a circular form such as cylindrical water tanks, liquid and natural gas tanks, storage silos, tunnels, digesters and nuclear containment vessels. In general, the term is applied when the direction of the prestress at any point is circumferential, i.e. in the direction of the tangent to the circumference of the circular prestressed concrete surface structure. The circular prestressing compresses the structure to counteract the tensile bursting forces or loads from within the structure. Circular prestressing is also appropriate in cylindrical shells. The ring

Figure 3.13 Prestressing anchorages at a buttress of a prestressed tank. (Courtesy of VSL International Limited, Hong Kong, China.)

beams around the edges of long-span shell structures develop significant tension forces and these can be balanced by prestressing.

Circular prestressing can take several forms, depending on the process of prestressing and the type of structure. Circumferential prestressing may be applied using individual tendons and multiple anchorages, or by using continuous wrapping, whereby a single tendon is wrapped around the circular structure. For tanks and silos, it is common practice to have buttresses in the walls, permitting easier detailing, installation and stressing. Figure 3.13 shows the prestressing buttress of a circularly prestressed tank.

3.7 EXTERNAL PRESTRESSING

Whilst the standard internal prestressing discussed in previous sections remains the basic procedure for the majority of structures, external prestressing of concrete structures has become much more popular with certain forms of structural members. As the name suggests, external prestressing is the application of prestress from a prestressing tendon or cable placed externally to the concrete elements. The cable or tendon may be placed on the outer side of the structure or, in the case of box-type girders, inside the structure. Since the prestressing steel is located external to the concrete, there is no bond between the structural concrete and the prestressing components (unlike pretensioned concrete and bonded post-tensioned concrete). Early prestressed bridges were often steel structures with external steel bars used to impart the prestress (to stiffen and strengthen the structure), but these structures are not considered here.

Since the tendons are external and unbonded to the concrete structure, they can be removed and, if required, replaced at any time during the life of the structure.

A significant use of external prestressing is in the construction of concrete box girder bridge decks. The external tendons are typically anchored in the concrete diaphragms within the box and are deviated at carefully designed saddles located at the bottom of the structure at the mid-spans and at the top of the structure at the supports. These deviators can be made of steel pipes or void formers that are integrated with the concrete box section. Figure 3.14 shows the external tendons inside a box girder bridge, including a saddle to locate the tendons near the bottom of the section at mid-span.

As the tendons are placed outside the concrete section, pouring of concrete in the web is made easier and, as the web compression area is not reduced by the voids created by internal tendon ducts, the web thickness can be kept to a minimum.

Figure 3.14 External tendons in bridge box girder section. (Courtesy of VSL International Limited, Hong Kong, China.)

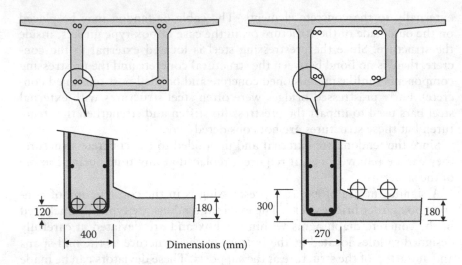

Dimensions (mm)

Figure 3.15 Illustration of web thickness reduction possible with external prestressing compared to internal prestressing.

Figure 3.15 illustrates a typical reduction in web thickness of a concrete box girder due to external prestressing. External prestressing cables also generally have lower prestress losses. Another major advantage of external prestressing is that it can be used in new structures as well as strengthening or retrofitting of existing structures.

The disadvantages of external prestressing include the reduction in tendon eccentricity when the tendons are to be kept inside the concrete box structure, i.e. above the bottom flange slab of the box section, and the slight additional costs of providing replaceable anchorages, ducts and deviation saddles for external tendons.

Chapter 4

Material properties

4.1 INTRODUCTION

The deformation of a prestressed concrete member throughout the full range of loading depends on the properties and behaviour of the constituent materials. In order to satisfy the design objective of adequate structural strength, the ultimate strengths of both concrete and steel need to be known. In addition, factors affecting material strength and the non-linear behaviour of each material in the overload range must be considered. In order to check for serviceability, the instantaneous and time-dependent properties of concrete and steel at typical in-service stress levels are required.

As was mentioned in Chapter 1, the prestressing force in a prestressed concrete member gradually decreases with time. This *loss* of prestress, which is usually 10%–25% of the initial value, is mainly caused by creep and shrinkage strains that develop with time in the concrete at the level of the bonded steel, as well as relaxation of the tendons. Reasonable estimates of the creep and shrinkage characteristics of concrete and procedures for the *time analysis* of prestressed structures are essential for an accurate prediction and a clear understanding of in-service behaviour. The loss of prestress caused by relaxation of the prestressing steel is caused by creep in the tendon. With the relatively low relaxation of modern prestressing steels, however, this component of prestress loss is usually relatively small (less than 5%).

The intention in this chapter is to present a broad outline of material behaviour and to provide sufficient quantitative information on material properties to complete most design tasks, with specific reference to European guidelines [1–13].

4.2 CONCRETE

More comprehensive treatments of the properties of concrete and the factors affecting them are given by others, including Neville [14] and Metha and Monteiro [15].

4.2.1 Composition of concrete

Concrete is a mixture of cement, water and aggregates. It may also contain one or more chemical admixtures. Within hours of mixing and placing, concrete sets and begins to develop strength and stiffness as a result of chemical reactions between the cement and the water. These reactions are known as hydration. Calcium silicates in the cement react with water to produce calcium silicate hydrate and calcium hydroxide. The resultant alkalinity of the concrete helps to provide corrosion protection for the reinforcement.

The relative proportions of cement, water and aggregates may vary considerably depending on the chemical properties of each component and the desired properties of the concrete. A typical mix used for prestressed concrete by weight might be coarse aggregate 45%, fine aggregate 30%, cement 18% and water 7%.

Cement is made from silica, alumina, lime and iron oxide, crushed and blended, and then burnt in a rotary kiln. The resulting *clinker* is cooled, mixed with gypsum and some other cementitious materials and ground to a fine powder. In most countries, several different types of cement are available, including general-purpose cements, high early strength cements, low heat of hydration cements and cements that provide enhanced sulphate resistance. Various cement replacement materials are often used, including silica fume, siliceous fly ash, calcareous fly ash, blast furnace slag, limestone, burnt shale and natural pozzolans. EN 197-1 [3] specifies five groups of cements depending on their composition.

1. CEM I Portland cement (contains mainly ground clinker and up to 5% of minor additional materials);
2. CEM II Portland composite cement (seven types are specified containing ground clinker and up to 35% of another single material);
3. CEM III blast furnace cement (contains ground clinker and 36%–95% of blast furnace slag);
4. CEM IV pozzolanic cement (comprising ground clinker and a mixture of silica fume, pozzolans and fly ash); and
5. CEM V composite cement (containing clinker and a high percentage of blast furnace slab and pozzolans or fly ash).

The standard strength of a cement is taken as the compressive strength of mortar specimens cast, cured and tested in accordance with EN 196-1 [2]. Six different strength classes of cement are specified in EN 197-1 [3] conforming to the requirements in Table 4.1. Two classes are specified for each of the three strength grades (32.5, 42.5 and 52.5 MPa), one associated with ordinary early strength (denoted by N) and one for high early strength (denoted by R) as indicated in Table 4.1.

The ratio of water to cement by weight required to hydrate the cement completely is about 0.25, although larger quantities of water are often

Table 4.1 Requirements for the different strength classes of cement [3]

Strength class (CEM)	Compressive strength (MPa)			Initial setting time (minutes)	Soundness (expansion) (mm)
	Early strength		28-day strength		
	2 days	7 days			
32.5N	–	≥16.0	≥32.5 ≤52.5	≥75	
32.5R	≥10.0	–			
42.5N	≥10.0	–	≥42.5 ≤62.5	≥60	≤10
42.5R	≥20.0	–			
52.5N	≥20.0	–	≥52.5 –	≥45	
52.5R	≥30.0	–			

required in practice in order to produce a workable mix. For the concrete typically used in prestressed structures, the water-to-cement ratio is about 0.4. It is desirable to use as little water as possible, since water not used in the hydration reaction causes voids in the cement paste that reduce the strength and increase the permeability of the concrete.

Chemical admixtures are widely used to improve one or more properties of concrete and code requirements are specified in EN 943-2 [4]. High-strength concretes with low water-to-cement ratios are made more workable by the inclusion of superplasticisers in the mix. These polymers improve the flow of the wet concrete and allow very high-strength and low-permeability concrete to be used with conventional construction techniques.

The rock and sand aggregates used in concrete should be inert and properly graded. Expansive and porous aggregates should not be used, and aggregates containing organic matter or other deleterious substances, such as salts or sulphates, should also be avoided.

4.2.2 Strength of concrete

In structural design, the quality of concrete is usually controlled by the specification of a minimum *characteristic compressive strength* at 28 days. The characteristic strength is the strength that is exceeded by 95% of the uniaxial compressive strength measurements taken from standard compression tests on concrete cylinders and is denoted f_{ck}. Cylinders are generally either 150 mm diameter by 300 mm long or 100 mm diameter by 200 mm long. In Europe, 150 mm concrete cubes are also used in standard compression tests, and the characteristic compressive strength arising from cube tests is denoted $f_{ck,cube}$. Because the restraining effect at the loading surfaces is greater for the cube than for the longer cylinder, strength measurements taken from cubes are higher than those taken from cylinders. The ratio between cylinder and cube strengths is about 0.8 for low-strength concrete (i.e. cylinder strengths of 20–30 MPa) and increases slightly as the strength increases, as indicated subsequently in Table 4.2.

In practice, the concrete used in prestressed construction is usually of better quality and higher strength than that required for ordinary reinforced concrete structures. Values of f_{ck} in the range 40–65 MPa are most often used, but higher strengths are not uncommon. Indeed, prestressed members fabricated from reactive powder concretes with compressive strengths in excess of 150 MPa have been used in a wide variety of structures.

The forces imposed on a prestressed concrete section are relatively large, and the use of high-strength concrete keeps section dimensions to a minimum. High-strength concrete also has obvious advantages in the anchorage zone of post-tensioned members where bearing stresses are large and in pretensioned members, where a higher bond strength better facilitates the transfer of prestress.

As the compressive strength of concrete increases, so too does its tensile strength. The use of higher-strength concrete may therefore delay (or even prevent) the onset of cracking in a member. High-strength concrete is considerably stiffer than low-strength concrete. The elastic modulus is higher, and elastic deformations due to both the prestress and the external loads are smaller. In addition, high-strength concrete generally creeps less than low-strength concrete. This results in smaller losses of prestress and smaller long-term deformations.

The effect of concrete strength on the shape of the stress–strain curve for concrete in uniaxial compression is shown in Figure 4.1. The modulus of elasticity (the initial slope of the tangent to the ascending portion of

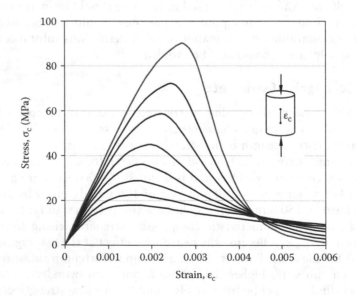

Figure 4.1 Effect of strength on the shape of the uniaxial compressive stress–strain curve for concrete.

each curve) increases with increasing strength, and each curve reaches its maximum stress at a strain in the range 0.0015–0.00285.

The shape of the unloading portion of each curve (after the peak stress has been reached) depends, among other things, on the characteristics of the testing machine. By applying deformation to a specimen, rather than load, in a testing machine that is stiff enough to absorb all the energy of a failing specimen, an extensive unloading branch of the stress–strain curve can be obtained. Concrete can undergo very large compressive strains and still carry load. This deformability of concrete tends to decrease with increasing strength.

The strength of properly placed and well-compacted concrete depends primarily on the water-to-cement ratio, the size of the specimen, the size, strength and stiffness of the aggregate, the cement type, the curing conditions and the age of the concrete. The strength of concrete increases as the water-to-cement ratio decreases. The compressive strength of concrete increases with time. A rapid initial strength gain (in the first day or so after casting) is followed by a rapidly decreasing rate of strength gain thereafter. The rate of development of strength with time depends on the type of curing, the type of cement and the temperature. In prestressed concrete construction, a rapid initial gain in strength is usually desirable so that the prestress may be applied to the structure as early as possible. This is particularly so for precast pretensioned production. Steam curing and high early strength cement are often used for this purpose.

The strength of concrete in tension is an order of magnitude less than the compressive strength and is far less reliably known. A reasonable estimate is required, however, in order to anticipate the onset of cracking and predict service-load behaviour in the post-cracking range. The tensile strength of concrete f_{ct} usually refers to the highest tensile stress reached in a concrete specimen subjected to concentric tensile loading, i.e. where the tensile stress is uniform over the cross-section. The flexural tensile strength of concrete $f_{ct,fl}$ (or modulus of rupture) is determined from the maximum extreme fibre tensile stresses calculated from the results of standard flexural strength tests on plain concrete prisms. The flexural tensile strength may be significantly higher than the direct tensile strength, particularly for members with shallow cross-sections.

In practice, concrete is often subjected to multiaxial states of stress. For example, a state of biaxial stress exists in the web of a beam, or in a shear wall, or a deep beam. Triaxial stress states exist within connections, in confined columns, in two-way slabs and other parts of a structure. A number of pioneering studies of the behaviour of concrete under multiaxial states of stress, including [16,17,18], led to the formulation of material modelling laws now used routinely in finite element software for the non-linear stress analysis of complex concrete structures. A typical biaxial strength envelope is shown in Figure 4.2, where σ_1 and σ_2 are the orthogonal stresses and σ_{cu} is the uniaxial compressive strength of concrete.

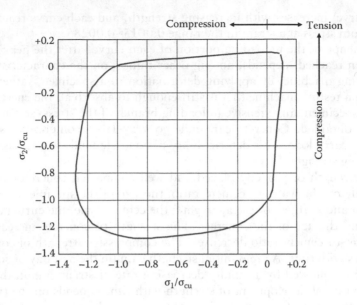

Figure 4.2 Typical biaxial strength envelope for concrete.

The strength of concrete under biaxial compression is greater than that under uniaxial compression. Transverse compression improves the longitudinal compressive strength by confining the concrete, thereby delaying (or preventing) the propagation of internal microcracks. Figure 4.2 also shows that transverse compression reduces the tensile strength of concrete, due mainly to Poisson's ratio effect. Similarly, transverse tension reduces the compressive strength. In triaxial compression, both the strength of concrete and the strain at which the peak stress is reached are greatly increased and even small confining pressures can increase strength significantly. Correctly detailed transverse reinforcement provides confinement to produce a triaxial stress state in the compressive zone of columns and beams, thereby improving both strength and ductility.

4.2.3 Strength specifications in Eurocode 2

4.2.3.1 Compressive strength

The strength of concrete is specified in EN 1992-1-1 [1] in terms of concrete *strength classes* which relate to the lower characteristic compressive strength at 28 days measured on cylinders f_{ck} or on cubes $f_{ck,cube}$. These are the values of compressive strength exceeded by 95% of all standard cylinders or cubes tested at age 28 days after curing under standard laboratory conditions in accordance with EN 12390:2 [7] and 12390:3 [8]. The *standard strength classes* are usually expressed as C $f_{ck}/f_{ck,cube}$, with the minimum strength class $C_{min} = C12/15$ and the maximum strength class C_{max} recommended as

C90/105. In EN1992-1-1 [1], the concrete strength is based on the characteristic cylinder strength f_{ck} and the strength classes specified in EN1992-1-1 [1] are shown in Table 4.2, together with the corresponding mechanical properties of concrete required for design.

The mean compressive strength of concrete f_{cm} at 28 days specified in EN 1992-1-1 [1] is also shown in Table 4.2. When curing is in accordance with EN 12390:2 [7] and when the mean temperature is 20°C, the mean compressive strength of concrete $f_{cm}(t)$ at age t (in days) may be obtained from the mean strength f_{cm} at age 28 days as follows [1]:

$$f_{cm}(t) = \beta_{cc}(t)f_{cm} \tag{4.1}$$

where:

$$\beta_{cc}(t) = \exp\left\{ s\left[1 - \left(\frac{28}{t}\right)^{0.5}\right]\right\} \tag{4.2}$$

and $s = 0.2$ for cement strength classes 42.5R, 52.5N and 52.5R (Class R), $s = 0.25$ for cement strength classes 32.5R and 42.5N (Class N) and $s = 0.38$ for cement strength classes 32.5N (Class S).

In some situations, the concrete compressive strength $f_{ck}(t)$ at some time t other than 28 days may need to be specified, such as at the time of transfer or at the time of stripping the forms. EN 1992-1-1 [1] suggests that:

$$f_{ck}(t) = f_{cm}(t) - 8\ (\text{MPa}) \quad \text{for } 3 < t < 28\ \text{days} \tag{4.3}$$

$$f_{ck}(t) = f_{ck} \quad \quad \text{for } t \geq 28\ \text{days} \tag{4.4}$$

More precise values of $f_{ck}(t)$ should be based on tests, particularly when $t \leq 3$ days.

4.2.3.2 Tensile strength

The uniaxial tensile strength f_{ct} is defined in EN 1992-1-1 [1] as the maximum stress that concrete can withstand when subjected to concentric uniaxial tension. Direct uniaxial tensile tests are difficult to perform, and the indirect tensile strength $f_{ct.sp}$ is usually measured in the so-called splitting tests on cylinders as specified in EN 12390:6 [10]. An approximate value of f_{ct} is specified in EN 1992-1-1 [1] as:

$$f_{ct} = 0.9f_{ct.sp} \tag{4.5}$$

The mean tensile strength f_{ctm} specified in EN 1992-1-1 [1] for each strength class is given in Table 4.2 and is related to the mean strength by:

$$f_{ctm} = 0.3 \times (f_{ck})^{2/3} \text{ (in MPa)} \qquad \text{for } f_{ck} \leq 50 \text{ MPa} \qquad (4.6)$$

$$f_{ctm} = 2.12 \times \ln[1 + 0.1 f_{cm}] \text{ (in MPa)} \quad \text{for } f_{ck} > 50 \text{ MPa} \qquad (4.7)$$

The mean tensile strength at any time t may be taken as:

$$f_{ctm}(t) = [\beta_{cc}(t)]^\alpha f_{ctm} \qquad (4.8)$$

where $\beta_{cc}(t)$ is given in Equation 4.2; $\alpha = 1$ for $t < 28$ days and $\alpha = 0.667$ for $t \geq 28$ days.

The upper and lower characteristic tensile strength shown in Table 4.2 are, respectively:

$$f_{ctk,0.95} = 1.3 f_{ctm} \quad \text{and} \quad f_{ctk,0.05} = 0.7 f_{ctm} \qquad (4.9)$$

The mean flexural tensile strength $f_{ctm,fl}$ specified in EN 1992-1-1 [1] depends on the mean axial tensile strength f_{ctm} (given in Table 4.2) and the depth h of the member cross-section in millimetres and may be taken as:

$$f_{ctm,fl} = \max\{(1.6 - h/1000) f_{ctm}; f_{ctm}\} \qquad (4.10)$$

4.2.3.3 Design compressive and tensile strengths

The design compressive and tensile strengths of concrete at the strength limit states are defined in EN 1992-1-1 [1] by the following equations, respectively:

$$f_{cd} = \frac{\alpha_{cc} f_{ck}}{\gamma_C} \qquad (4.11)$$

$$f_{ctd} = \frac{\alpha_{ct} f_{ctk,0.05}}{\gamma_C} \qquad (4.12)$$

where γ_C is the partial safety factor for concrete, with $\gamma_C = 1.5$ for persistent and transient design situations and $\gamma_C = 1.2$ for design situations involving accidental actions, as indicated in Table 2.4. The terms α_{cc} and α_{ct} are coefficients that account for unfavourable effects that may arise due to long-term effects or the manner of load application and may range between 0.8 and 1.0. In most cases, α_{cc} and α_{ct} are both equal to 1.0.

Table 4.2 Strength and deformation characteristics for concrete [1]

Strength class	C12/15	C16/20	C20/25	C25/30	C30/37	C35/45	C40/50	C45/55	C50/60	C55/67	C60/75	C70/85	C80/95	C90/105
f_{ck} (MPa)	12	16	20	25	30	35	40	45	50	55	60	70	80	90
$f_{ck,cube}$ (MPa)	15	20	25	30	37	45	50	55	60	67	75	85	95	105
f_{cm} (MPa)	20	24	28	33	38	43	48	53	58	63	68	78	88	98
f_{ctm} (MPa)	1.6	1.9	2.2	2.6	2.9	3.2	3.5	3.8	4.1	4.2	4.4	4.6	4.8	5.0
$f_{ctk,0.05}$ (MPa)	1.1	1.3	1.5	1.8	2.0	2.2	2.5	2.7	2.9	3.0	3.1	3.2	3.4	3.5
$f_{ctk,0.95}$ (MPa)	2	2.5	2.9	3.3	3.8	4.2	4.6	2.9	5.3	5.5	5.7	6.0	6.3	6.6
E_{cm} (GPa)	27	29	30	31	33	34	35	36	37	38	39	41	42	44
ε_{c1} (×10⁻³)	1.8	1.9	2.0	2.1	2.2	2.25	2.3	2.4	2.45	2.5	2.6	2.7	2.8	2.8
ε_{cu1} (×10⁻³)					3.5					3.2	3.0	2.8	2.8	2.8
ε_{c2} (×10⁻³)					2.0					2.2	2.3	2.4	2.5	2.6
ε_{cu2} (×10⁻³)					3.5					3.1	2.9	2.7	2.6	2.6
n					2.0					1.75	1.6	1.45	1.4	1.4
ε_{c3} (×10⁻³)					1.75					1.8	1.9	2.0	2.2	2.3
ε_{cu3} (×10⁻³)					3.5					3.1	2.9	2.7	2.6	2.6

4.2.3.4 Compressive stress–strain curves for concrete for non-linear structural analysis

Figure 4.3 shows the idealised relation between concrete compressive stress σ_c and strain ε_c specified in EN 1992-1-1 [1] for short-term uniaxial loading and, with σ_c and ε_c expressed as absolute values, the stress–strain relationship is:

$$\frac{\sigma_c}{f_{cm}} = \frac{k\eta - \eta^2}{1 + (k - 2)\eta} \tag{4.13}$$

where $\eta = \varepsilon_c/\varepsilon_{c1}$; ε_c is the strain corresponding to the stress σ_c; ε_{c1} is the strain corresponding to the peak stress f_{cm} (given in Table 4.2); $k = 1.05E_{cm}/E_{cp}$; E_{cm} is the secant modulus of elasticity of the concrete (given in Table 4.2 and defined in Figure 4.3) and E_{cp} is the secant modulus corresponding to peak stress (i.e. $E_{cp} = f_{cm}/\varepsilon_{c1}$).

Equation 4.13 is valid throughout the range $0 < \varepsilon_c < \varepsilon_{cu1}$, where ε_{cu1} is the nominal ultimate strain given in Table 4.2 and defined in Figure 4.3.

Numerous equations describing the curvilinear stress–strain relationship for concrete in compression are available in the literature, and EN 1992-1-1 [1] suggests any of them may be used in design 'if they adequately represent the behaviour of the concrete'.

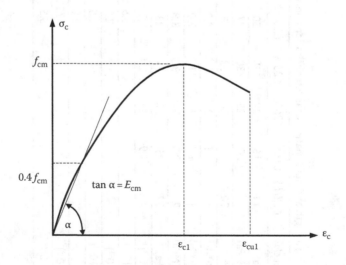

Figure 4.3 Idealised stress–strain relationship for concrete in uniaxial compression [I].

4.2.4 Deformation of concrete

4.2.4.1 Discussion

The deformation of a loaded concrete specimen is both instantaneous and time-dependent. If the load is sustained, the deformation of the specimen increases with time and may eventually be several times larger than the instantaneous value.

The gradual development of strain with time is caused by creep and shrinkage. Creep strain is produced by sustained stress. Shrinkage is independent of stress and results primarily from the loss of water as the concrete dries and from chemical reactions in the hardened concrete. Creep and shrinkage cause increases in axial deformation and curvature on reinforced and prestressed concrete cross-sections, losses of prestress, local redistribution of stress between the concrete and the steel reinforcement and redistribution of internal actions in statically indeterminate members. Creep and shrinkage are often the cause of excessive deflection (or camber) and excessive shortening of prestressed members. In addition, shrinkage may cause unsightly cracking that could lead to serviceability or durability problems. On a more positive note, creep relieves concrete of stress concentrations and imparts a measure of deformability to concrete.

Researchers have been investigating the time-dependent deformation of concrete ever since it was first observed and reported over a century ago, and a great deal of literature has been written on the topic. Detailed summaries of the time-dependent properties of concrete and the factors that affect them are contained in texts by Neville [14,19], Neville et al. [20], Gilbert [21], Gilbert and Ranzi [22], Ghali et al. [23,24] and Rüsch et al. [25] and in technical documents such as those from ACI Committee 209 [26–28].

The time-varying deformation of concrete may be illustrated by considering a concrete specimen subjected to a constant sustained stress. At any time t, the total concrete strain $\varepsilon_c(t)$ in an uncracked uniaxially loaded specimen consists of a number of components that include the instantaneous strain $\varepsilon_{ce}(t)$, the creep strain $\varepsilon_{cc}(t)$, the shrinkage strain $\varepsilon_{cs}(t)$ and the temperature strain $\varepsilon_T(t)$. Although not strictly correct, it is usually acceptable to assume that all four components are independent and may be calculated separately and combined to obtain the total strain.

When calculating the in-service behaviour of a concrete structure at constant temperature, it is usual to express the concrete strain at a point as the sum of the instantaneous, creep and shrinkage components:

$$\varepsilon_c(t) = \varepsilon_{ce}(t) + \varepsilon_{cc}(t) + \varepsilon_{cs}(t) \tag{4.14}$$

The strain components in a drying specimen held at constant temperature and subjected to a constant sustained compressive stress σ_{c0} first applied at time t_0 are illustrated in Figure 4.4. Immediately after the concrete sets or

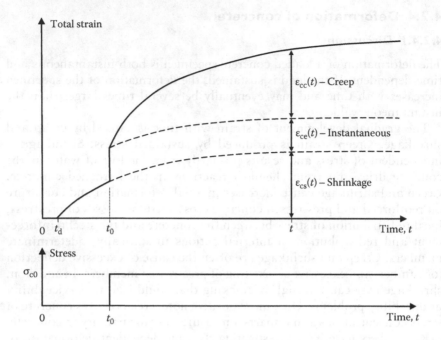

Figure 4.4 Concrete strain versus time for a specimen subjected to constant sustained stress [22].

at the end of moist curing ($t = t_d$ in Figure 4.4), shrinkage strain begins to develop and continues to increase at a decreasing rate. On application of the stress, a sudden jump in the strain diagram (instantaneous strain) is followed by an additional gradual increase in strain due to creep.

The prediction of the time-dependent behaviour of a concrete member requires an accurate estimate of each of these strain components at critical locations. This requires knowledge of the stress history, in addition to accurate data for the material properties. The stress history depends both on the applied load and on the boundary conditions of the member. Calculations are complicated by the restraint to creep and shrinkage provided by both the bonded reinforcement and the external supports and the continuously varying concrete stress history that inevitably results.

The material properties that influence each of the strain components depicted in Figure 4.4 are described in the following sections. Methods for predicting the time-dependent behaviour of prestressed concrete cross-sections and members are discussed in Section 5.7.

4.2.4.2 Instantaneous strain

The magnitude of the instantaneous strain $\varepsilon_{ce}(t)$ caused by either compressive or tensile stress depends on the magnitude of the applied stress, the rate at which the stress is applied, the age of the concrete when the stress was applied and the stress–instantaneous strain relationship for the concrete. Consider the uniaxial instantaneous strain versus compressive stress curve shown in Figure 4.3. When the applied stress is less than about half the compressive strength, the curve is essentially linear and the instantaneous strain is usually considered to be elastic (fully recoverable). In this low-stress range, the secant modulus E_{cm} does not vary significantly with stress and is only slightly smaller than the initial tangent modulus. At higher stress levels, the stress–strain curve becomes significantly non-linear and a significant proportion of the instantaneous strain is irrecoverable upon unloading.

In concrete structures, compressive concrete stresses caused by the day-to-day service loads rarely exceed half of the compressive strength. It is therefore reasonable to assume that the instantaneous behaviour of concrete at service loads is linear-elastic and that instantaneous strain is given by:

$$\varepsilon_{ce}(t) = \frac{\sigma_c(t)}{E_{cm}} \tag{4.15}$$

The secant modulus between $\sigma_c = 0$ and $\sigma_c = 0.4f_{cm}$ for concrete loaded at 28 days is the modulus of elasticity of concrete (see Figure 4.3), with the symbol E_{cm} in EN 1992-1-1 [1]. Values of E_{cm} are given in Table 4.2.

The value of the modulus of elasticity E_{cm} increases with time as the concrete gains strength and stiffness. It also depends on the rate of application of the stress and increases as the loading rate increases. For most practical purposes, these variations are often ignored, and it is common practice to assume that E_{cm} is constant with time and equal to its initial value calculated at the time of first loading t_0.

The in-service performance of a concrete structure is very much affected by the concrete's inability to carry significant tension. It is therefore necessary to consider the instantaneous behaviour of concrete in tension, as well as in compression. Prior to cracking, the instantaneous strain of concrete in tension consists of both elastic and inelastic components. In design, however, concrete is usually taken to be elastic-brittle in tension, and at stress levels less than the tensile strength of concrete the instantaneous strain versus stress relationship is assumed to be linear. Although the magnitude of the elastic modulus in tension is likely to differ from that in compression, it is usual to assume that both values are equal. Prior to cracking, the instantaneous strain in tension may be calculated using Equation 4.15. When the tensile strength is reached, cracking occurs and the concrete stress perpendicular to the crack is usually assumed to be zero. In reality, if the rate of tensile deformation is controlled and crack widths are small, concrete can carry some tension across a crack due to friction that exists on the rough mating surfaces of the crack.

Poisson's ratio for uncracked concrete ν generally lies within the range 0.15–0.22 and for most practical purposes may be taken as 0.2.

4.2.4.3 Creep strain

For concrete subjected to a constant sustained stress, the gradual development of creep strain is illustrated in Figure 4.4. In the period immediately after first loading, creep develops rapidly, but the rate of increase slows appreciably with time. Creep has traditionally been thought to approach a limiting value as the time after first loading approaches infinity, but more recent research suggests that creep continues to increase indefinitely, albeit at a slower rate [38]. After several years under load, the rate of change of creep with time is small. Creep of concrete has its origins in the hardened cement paste and is caused by a number of different mechanisms. A comprehensive treatment of creep in plain concrete is given by Neville et al. [20].

Many factors influence the magnitude and rate of development of creep. Some are properties of the concrete mix, while others depend on the environmental and loading conditions. In general, the capacity of concrete to creep decreases as the concrete quality increases. At a particular stress level, creep in higher-strength concrete is less than that in lower-strength concrete. Creep decreases as the water-to-cement ratio is reduced. An increase in either the aggregate content or the maximum aggregate size reduces creep, as does the use of a stiffer aggregate type.

Creep also depends on the environment. Creep increases as the relative humidity decreases. Creep is therefore greater when accompanied by drying. Creep is also greater in thin members with large surface area-to-volume ratios, such as slabs and walls. However, the dependence of creep on both the relative humidity and the size and shape of the specimen decreases as the concrete strength increases. Near the surface of a member, creep takes place in a drying environment and is therefore greater than in regions remote from a drying surface. In addition to the relative humidity, creep is dependent on the ambient temperature. A temperature rise increases the deformability of the cement paste and accelerates drying and thus increases creep. The dependence of creep on temperature is more pronounced at elevated temperatures and is far less significant for temperature variations between 0°C and 20°C. However, creep in concrete at a mean temperature of 40°C is perhaps 25% higher than that at 20°C [25].

In addition to the environment and the characteristics of the concrete mix, creep depends on the loading history, in particular the magnitude and duration of the stress and the age of the concrete when the stress is first applied. The age at first loading t_0 has a marked influence on the final magnitude of creep. Concrete loaded at an early age creeps more than concrete loaded at a later age. Concrete is therefore a time-hardening material, although even in very old concrete the tendency to creep never entirely disappears [29].

When the sustained concrete stress is less than about $0.5f_{cm}$, creep is approximately proportional to stress and is known as *linear creep*. At higher stress levels creep increases at a faster rate and becomes non-linear with respect to stress. This non-linear behaviour of creep at high stress levels is thought to be related to an increase in micro-cracking. Compressive stresses rarely exceed $0.5f_{cm}$ in concrete structures at service loads, and creep may be taken as proportional to stress in most situations in the design for serviceability.

Creep strain is made up of a recoverable component, called the delayed elastic strain $\varepsilon_{cc.d}(t)$ and an irrecoverable component called flow $\varepsilon_{cc.f}(t)$. These components are illustrated by the creep strain versus time curve in Figure 4.5a caused by the stress history shown in Figure 4.5b. The recoverable creep is thought to be caused by the elastic aggregate acting on the viscous cement paste after the applied stress is removed. If a concrete specimen is unloaded after a long period under load, the magnitude of the recoverable creep is of the order of 40%–50% of the elastic strain (between 10% and 20% of the total creep strain). Although the delayed elastic strain is observed only as recovery when the load is removed, it is generally believed to be of the same magnitude under load and to develop rapidly in the period immediately after loading. Rüsch et al. [25] suggested that the shape of the delayed elastic strain curve is independent of the age or dimensions of the specimen and is unaffected by the composition of the concrete.

The capacity of concrete to creep is usually measured in terms of the creep coefficient $\varphi(t, t_0)$. In a concrete specimen subjected to a constant sustained compressive stress σ_{c0} first applied at age t_0, the creep coefficient at time t is the ratio of creep strain to instantaneous strain and is given by:

$$\varphi(t, t_0) = \frac{\varepsilon_{cc}(t, t_0)}{\varepsilon_{ce}(t_0)} \tag{4.16}$$

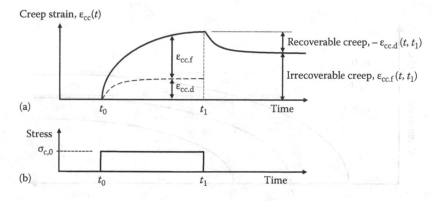

Figure 4.5 Recoverable and irrecoverable creep components. (a) Creep strain history. (b) Stress history.

and the creep strain at time t caused by a constant sustained stress σ_{c0} first applied at age t_0 is:

$$\varepsilon_{cc}(t,t_0) = \varphi(t,t_0)\varepsilon_{ce}(t_0) = \varphi(t,t_0)\frac{\sigma_{c0}}{E_c(t_0)} \qquad (4.17)$$

where $E_c(t_0)$ is the tangent modulus at time t_0 and, according to EN 1992-1-1 [1], may be taken as $1.05E_{cm}(t_0)$. For practical purposes and considering the variable nature of creep, it is usual to assume $E_c(t_0) = E_{cm}(t_0)$ in Equation 4.17. For concrete subjected to a constant sustained stress, knowledge of the creep coefficient allows the rapid determination of the creep strain at any time using Equation 4.17.

Since both the creep and the instantaneous strain components are proportional to stress for compressive stress levels less than about $0.5f_{cm}$, the creep coefficient $\varphi(t, t_0)$ is a pure time function and is independent of the applied stress. The creep coefficient increases with time at a decreasing rate. Although there is some evidence that the creep coefficient increases indefinitely, the final creep coefficient $\varphi(\infty,t_0) = \varepsilon_{cc}(\infty,t_0)/\varepsilon_{ce}(t_0)$ at $t = \infty$ is often taken as the 30-year value and its magnitude usually falls within the range of 1.5–4.0. A number of the well-known methods for predicting the creep coefficient were described and compared in Gilbert [21,37] and Gilbert and Ranzi [22]. The approach specified in EN 1992-1-1 [1] for making numerical estimates of $\varphi(t, t_0)$ is presented in Section 4.2.5.3.

The final creep coefficient is a useful measure of the capacity of concrete to creep. Since creep strain depends on the age of the concrete at the time of first loading, so too does the creep coefficient. The effect of ageing is illustrated in Figure 4.6. The magnitude of the final creep coefficient $\varphi(\infty,t_0)$ decreases as the age at first loading t_0 increases:

$$\varphi(\infty,t_i) > \varphi(\infty,t_j) \quad \text{for } t_i < t_j$$

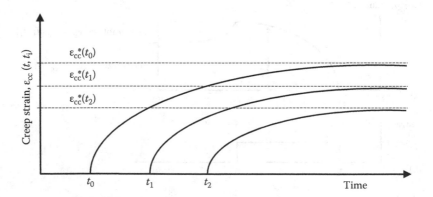

Figure 4.6 Effect of age at first loading on the creep coefficient.

This time hardening or ageing of concrete complicates the calculation of creep strain caused by a time-varying stress history.

Another frequently used time function is known as *specific creep* $C(t, t_0)$, defined as the proportionality factor relating stress to linear creep:

$$\varepsilon_{cc}(t,t_0) = C(t,t_0)\,\sigma_{c0} \tag{4.18}$$

or

$$C(t,t_0) = \frac{\varepsilon_{cc}(t,t_0)}{\sigma_{c0}} \tag{4.19}$$

$C(t, t_0)$ is the creep strain at time t produced by a sustained *unit* stress first applied at age t_0.

From Equations 4.16 through 4.18 and taking $E_c(t_0) = E_{cm}(t_0)$, the relationship between the creep coefficient and the specific creep is:

$$\varphi(t,t_0) = C(t,t_0)E_{cm}(t_0) \tag{4.20}$$

The sum of the instantaneous and creep strains at time t produced by a sustained unit stress applied at t_0 is defined as the *creep function $J(t,t_0)$* and is given by:

$$J(t,t_0) = \frac{1}{E_{cm}(t_0)} + C(t,t_0) \tag{4.21}$$

The stress-produced strains (i.e. the instantaneous plus creep strains) caused by a constant sustained stress σ_{c0} first applied at age t_0 (also called the stress-dependent strains) can therefore be determined from:

$$\varepsilon_{ce}(t) + \varepsilon_{cc}(t,t_0) = J(t,t_0)\sigma_{c0} = \frac{\sigma_{c0}}{E_{cm}(t_0)}\big[1 + \varphi(t,t_0)\big] = \frac{\sigma_{c0}}{E_{c,eff}(t,t_0)} \tag{4.22}$$

where $E_{c,eff}(t, t_0)$ is known as the *effective modulus* of concrete and is given by:

$$E_{c,eff}(t,t_0) = \frac{E_{cm}(t_0)}{1 + \varphi(t,t_0)} \tag{4.23}$$

If the stress is gradually applied to the concrete, rather than abruptly applied, the subsequent creep strain is reduced, because the concrete ages during the period of application of the stress. This can be accommodated analytically by the use of a reduced or adjusted creep coefficient. For an increment of stress $\Delta\sigma_c(t)$ applied to the concrete gradually, beginning at time t_0, the load-dependent strain may be obtained by modifying Equation 4.22 as follows:

$$\varepsilon_{ce}(t) + \varepsilon_{cc}(t,t_0) = \frac{\Delta\sigma_c(t)}{E_{cm}(t_0)}\left[1 + \chi(t,t_0)\varphi(t,t_0)\right] = \frac{\Delta\sigma_c(t)}{\overline{E}_{c,eff}(t,t_0)} \tag{4.24}$$

where:

$$\overline{E}_{c,eff}(t,t_0) = \frac{E_{cm}(t_0)}{1 + \chi(t,t_0)\varphi(t,t_0)} \tag{4.25}$$

$\overline{E}_{c,eff}(t,t_0)$ is called the *age-adjusted effective modulus* of concrete and $\chi(t, t_0)$ is an ageing coefficient first introduced by Trost [30] and later developed by Dilger and Neville [31] and Bazant [32].

Like the creep coefficient, the ageing coefficient depends on the rate of application of the gradually applied stress and the age at first loading and varies between about 0.4 and 1.0. Methods for the determination of the ageing coefficient are available, for example, in *fib* Model Code 2010 [33]. Gilbert and Ranzi [22] showed that for concrete first loaded at early ages ($t_0 < 20$ days) and where the applied load is sustained, the final long-term ageing coefficient may be taken as $\chi(\infty, t_0) = 0.65$. In situations where the deformation is held constant and the concrete stress relaxes, the final long-term ageing coefficient may be taken as $\chi(\infty, t_0) = 0.8$.

The previous discussions have been concerned with the creep of concrete in compression. However, the creep of concrete in tension is also of interest in a number of practical situations, e.g. when studying the effects of restrained or differential shrinkage. Tensile creep also plays a significant role in the analysis of suspended reinforced concrete slabs at service loads where stress levels are generally low and, typically, much of the slab is initially uncracked.

Comparatively, little attention has been devoted to the study of tensile creep [28], and only limited experimental results are available in the literature [34]. Some researchers have multiplied the creep coefficients measured for compressive stresses by factors in the range of 1–3 to produce equivalent coefficients describing tensile creep, e.g. Chu and Carreira [35] and Bazant and Oh [36].

It appears that the mechanisms of creep in tension are different to those in compression. The magnitudes of both tensile and compressive creep increase when loaded at earlier ages. However, the rate of change of tensile creep with time does not decrease in the same manner as for compressive creep, with the development of tensile creep being more linear [34]. Drying tends to increase tensile creep in a similar manner to compressive creep, and tensile creep is in part recoverable upon removal of the load. Further research is needed to provide clear design guidance. In this book, it is assumed that the magnitude and rate of development of tensile creep are similar to that of compressive creep at the same low stress levels. Although not strictly correct, this assumption simplifies calculations and does not usually introduce serious inaccuracies.

4.2.4.4 Shrinkage strain

Shrinkage of concrete is the time-dependent strain in an unloaded and unrestrained specimen at constant temperature. Shrinkage is often divided into several components, including *plastic* shrinkage, *chemical* shrinkage, *thermal* shrinkage and *drying* shrinkage. Plastic shrinkage occurs in the wet concrete before setting, whereas chemical, thermal and drying shrinkage all occur in the hardened concrete after setting. Some high-strength concretes are prone to *plastic shrinkage* that may result in significant cracking before and during the setting process. This cracking occurs due to capillary tension in the pore water and is best prevented by taking measures during construction to avoid the rapid evaporation of bleed water. Before the concrete has set, the bond between the plastic concrete and the reinforcement has not yet developed, and the steel is ineffective in controlling plastic shrinkage cracking.

Drying shrinkage is the reduction in volume caused principally by the loss of water during the drying process. It increases with time at a gradually decreasing rate and takes place in the months and years after setting. The magnitude and rate of development of drying shrinkage depend on all the factors that affect the drying of concrete, including the relative humidity, the size and shape of the member and the mix characteristics, in particular, the type and quantity of the binder, the water content and water-to-cement ratio, the ratio of fine-to-coarse aggregate and the type of aggregate.

Chemical shrinkage results from various chemical reactions within the cement paste and includes hydration shrinkage, which is related to the degree of hydration of the binder in a sealed specimen with no moisture exchange. Chemical shrinkage (often called *autogenous shrinkage*) occurs rapidly in the days and weeks after casting and is less dependent on the environment and the size of the specimen than drying shrinkage. *Thermal shrinkage* is the contraction that results in the first few hours (or days) after setting as the heat of hydration gradually dissipates. The term *endogenous shrinkage* is sometimes used to refer to that part of the shrinkage of the hardened concrete that is not associated with drying (i.e. the sum of autogenous and thermal shrinkage).

The shrinkage strain ε_{cs} is usually considered to be the sum of the drying shrinkage component (which is the reduction in volume caused principally by the loss of water during the drying process) and the endogenous shrinkage component. Drying shrinkage in high-strength concrete is smaller than in normal strength concrete due to the smaller quantities of free water after hydration. However, thermal and chemical shrinkage may be significantly higher. Although drying and endogenous shrinkage are quite different in nature, there is often no need to distinguish between them from a structural engineering point of view.

Shrinkage increases with time at a decreasing rate, as illustrated in Figure 4.4. Shrinkage is assumed to approach a final value $\varepsilon_{cs}(\infty)$ as time approaches infinity.

Drying shrinkage is affected by all the factors that affect the drying of concrete, in particular the water content and the water–cement ratio of the mix, the size and shape of the member and the ambient relative humidity. All else being equal, drying shrinkage increases when the water–cement ratio increases, the relative humidity decreases and the ratio of the exposed surface area to volume increases. Temperature rises accelerate drying and therefore increase shrinkage. By contrast, autogenous shrinkage increases as the cement content increases and the water–cement ratio decreases. In addition, autogenous shrinkage is not significantly affected by the ambient relative humidity.

The effect of a member's size on drying shrinkage should be emphasised. For a thin member, such as a slab, the drying process may be essentially complete after several years, but for the interior of a larger member, the drying process may continue throughout its lifetime. For uncracked mass concrete structures, there is no significant drying (shrinkage) except for about 300 mm from each exposed surface. By contrast, the autogenous shrinkage is less affected by the size and shape of the specimen.

Shrinkage is also affected by the volume and type of aggregate. Aggregate provides restraint to shrinkage of the cement paste so that an increase in the aggregate content reduces shrinkage. Shrinkage is also smaller when stiffer aggregates are used, i.e. aggregates with higher elastic moduli. Thus, shrinkage is considerably higher in lightweight concrete than in normal weight concrete (by up to 50%).

The approach specified in EN 1992-1-1 [1] for making numerical estimates of shrinkage strain is presented in Section 4.2.5.4.

4.2.5 Deformational characteristics specified in Eurocode 2

4.2.5.1 Introduction

Great accuracy in the prediction of the creep coefficient and the shrinkage strain is not possible. The variability of these material characteristics is high. Design predictions are most often made using one of many numerical models that are available for predicting the creep coefficient and shrinkage strain. These models vary in complexity, ranging from relatively complicated methods, involving the determination of numerous coefficients that account for the many factors affecting creep and shrinkage, to much simpler procedures. A description of and comparison between some of the more well-known methods is provided in ACI 209.2R-08 [28]. Although the properties of concrete vary from country to country as the mix characteristics and environmental conditions vary, the agreement between the procedures for estimating both creep and shrinkage is still remarkably poor, particularly for shrinkage. In addition, the comparisons between predictive models show that the accuracy of a particular model is not directly proportional to its complexity, and predictions made using several of the best-known methods differ widely.

In the following sections, the models contained in EN 1992-1-1 [1] are presented for predicting the elastic modulus, the creep coefficient and the shrinkage strain for concrete.

4.2.5.2 Modulus of elasticity

The secant modulus between $\sigma_c = 0$ and $\sigma_c = 0.4f_{cm}$ for concrete loaded at 28 days is shown in Figure 4.3 as E_{cm}. The numerical values specified in EN 1992-1-1 [1] for concrete containing quartzite aggregate are shown in Table 4.2 for each strength class. For concrete with limestone and sandstone aggregates, these values should be reduced by 10% and 30%, respectively, and for concrete with basalt aggregate, the value should be increased by 20%. For structures likely to be sensitive to variations in the design value of E_{cm}, the effect of variations in the specified values by up to $\pm20\%$ should be assessed. These variations are due to factors other than aggregate type and include the aggregate quantity, the aggregate to binder ratio, the curing regime and the rate of application of the load.

When the stress is applied slowly, say over a period of several hours, significant additional deformation occurs owing to the rapid early development of creep. For the estimation of short-term deformation in such a case, it is recommended that the specified value of E_{cm} should be reduced by about 20% [21].

According to EN 1992-1-1 [1], variations in the elastic modulus with time can be determined from:

$$E_{cm}(t) = \left[f_{cm}(t)/f_{cm} \right]^{0.3} E_{cm} \tag{4.26}$$

where $E_{cm}(t)$ is the elastic modulus at an age of t days, $f_{cm}(t)$ is the compressive strength at an age of t days (and obtained from Equation 4.1) and E_{cm} and f_{cm} are the values at age 28 days.

EN 1992-1-1 [1] specifies that Poisson's ratio should be taken as 0.2 for uncracked concrete and zero for cracked concrete.

4.2.5.3 Creep coefficient

In Section 4.2.4.3, the creep coefficient $\varphi(t, t_0)$ at time t associated with a constant stress first applied at age t_0 was defined as the ratio of the creep strain at time t to the (initial) elastic strain. The most accurate way of determining the final creep coefficient is by testing or by using results obtained from measurements on similar local concretes. However, testing is often not a practical option for the structural designer, and a relatively simple approach is provided in EN 1992-1-1 [1] for routine use in structural design.

The creep coefficient specified in EN 1992-1-1 [1] is related to the tangent modulus E_c which may be taken as $1.05E_{cm}$. For situations where great accuracy is not required and provided the concrete is not subjected to a compressive stress exceeding $0.45f_{ck}(t_0)$ at the age at first loading t_0, the final creep coefficient $\varphi(\infty, t_0)$ can be obtained from Figure 4.7. These values are valid for temperatures between $-40°C$ and $+40°C$, and for a mean relative humidity between RH = 40% and RH = 100%. The term h_0 is the notional size or the hypothetical thickness equal to $2A_c/u$, where A_c is the concrete cross-sectional area and u is that part of the cross-sectional perimeter exposed to drying. The terms S, N and R are the cement strength classes defined under Equation 4.2.

The final creep deformation at $t = \infty$ caused by a constant stress σ_{c0} first applied at age t_0 is obtained from Equation 4.17 as:

$$\varepsilon_{cc}(\infty, t_0) = \varphi(\infty, t_0)(\sigma_{c0}/E_{cm}) \tag{4.27}$$

If the compressive stress exceeds $0.45f_{ck}(t_0)$, such as that may occur in a precast, pretensioned member stressed at early age, EN 1992-1-1 [1] specifies that non-linear creep should be considered using the non-linear notional creep coefficient $\varphi_k(\infty, t_0)$ given by:

$$\varphi_k(\infty, t_0) = \varphi(\infty, t_0) \exp[1.5(k_\sigma - 0.45)] \tag{4.28}$$

where $\varphi(\infty, t_0)$ is the final linear creep coefficient; k_σ is the stress-strength ratio $\sigma_{c0}/f_{cm}(t_0)$; and $f_{cm}(t_0)$ is the mean concrete compressive strength at the time of loading.

It must be emphasised that creep of concrete is highly variable with significant differences in the measured creep strains in seemingly identical specimens tested under identical conditions (both in terms of load and environment). The creep coefficient obtained from Figure 4.7 should be regarded as a ball park approximation with a range of $\pm 20\%$.

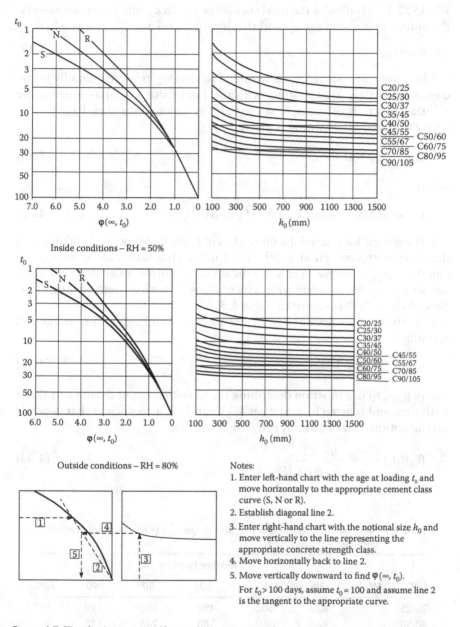

Inside conditions – RH = 50%

Outside conditions – RH = 80%

Notes:

1. Enter left-hand chart with the age at loading t_0 and move horizontally to the appropriate cement class curve (S, N or R).

2. Establish diagonal line 2.

3. Enter right-hand chart with the notional size h_0 and move vertically to the line representing the appropriate concrete strength class.

4. Move horizontally back to line 2.

5. Move vertically downward to find $\varphi(\infty, t_0)$.

For $t_0 > 100$ days, assume $t_0 = 100$ and assume line 2 is the tangent to the appropriate curve.

Figure 4.7 The final creep coefficient for concrete $\varphi(\infty, t_0)$ under normal environmental conditions.

4.2.5.4 Shrinkage strain

The model for estimating the magnitude of shrinkage strain specified in EN 1992-1-1 [1] divides the total shrinkage strain ε_{cs} into two components, the autogenous shrinkage ε_{ca} and the drying shrinkage ε_{cd}, as given by:

$$\varepsilon_{cs} = \varepsilon_{ca} + \varepsilon_{cd} \tag{4.29}$$

The autogenous shrinkage is assumed to develop relatively rapidly in the days immediately after casting, with the final value $\varepsilon_{ca}(\infty)$ assumed to be a linear function of concrete strength. The autogenous shrinkage at any time t (in days) after casting is given by:

$$\varepsilon_{ca}(t) = \varepsilon_{ca}(\infty)(1.0 - e^{-0.2t^{0.5}}) \tag{4.30}$$

where:

$$\varepsilon_{ca}(\infty) = 2.5(f_{ck} - 10) \times 10^{-6} \quad (f_{ck} \text{ in MPa}) \tag{4.31}$$

Drying shrinkage develops more slowly than autogenous shrinkage and decreases with concrete strength. The final drying shrinkage strain $\varepsilon_{cd,\infty}$ is equal to $k_h\varepsilon_{cd,0}$. The mean value of the nominal unrestrained drying shrinkage strain $\varepsilon_{cd,0}$ for concrete with cement class N (with a coefficient of variation of about 30%) is given in Table 4.3.

If drying commences at age t_s, the drying shrinkage strain at age t is given by:

$$\varepsilon_{cd}(t) = \beta_{ds}(t, t_s)k_h\varepsilon_{cd,0} \tag{4.32}$$

where $\beta_{ds}(t, t_s)$ is a function describing the development of drying shrinkage with time and is given by Equation 4.33 and k_h is a coefficient that depends on the notional size h_0 and is given in Table 4.4:

$$\beta_{ds}(t, t_s) = \frac{(t - t_s)}{(t - t_s) + 0.04\sqrt{h_0^3}} \tag{4.33}$$

Table 4.3 Nominal unrestrained drying shrinkage $\varepsilon_{cd,0}$ ($\times 10^{-6}$) for concrete with cement class N [1]

$f_{ck}/f_{ck,cube}$ (MPa)	Relative humidity (%)					
	20	40	60	80	90	100
20/25	620	580	490	300	170	0
40/50	480	460	380	240	130	0
60/75	380	360	300	190	100	0
80/95	300	280	240	150	80	0
90/105	270	250	210	130	70	0

Table 4.4 Values for k_h [1]

h_0 (mm)	100	200	300	≥500
k_h	1.0	0.85	0.75	0.7

4.2.5.5 Thermal expansion

The coefficient of thermal expansion for concrete depends on the coefficient of thermal expansion of the coarse aggregate and on the mix proportions. For most types of coarse aggregate, the coefficient lies within the range of $5 \times 10^{-6}/°C$ to $13 \times 10^{-6}/°C$ [14]. For design purposes and in the absence of more detailed information (test data), a coefficient of thermal expansion for concrete of $10 \times 10^{-6}/°C \pm 20\%$ is usually satisfactory in design.

4.3 STEEL REINFORCEMENT

The strength of a reinforced or prestressed concrete element in bending, shear, torsion or direct tension depends on the properties of the steel reinforcement and tendons, and it is necessary to adequately model the various types of steel reinforcement and their material properties.

Steel reinforcement is used in concrete structures to provide strength, ductility and serviceability. Steel reinforcement can also be strategically placed to reduce both immediate and time-dependent deformations. Adequate quantities of bonded reinforcement will also provide crack control, wherever cracks occur in the concrete.

4.3.1 General

Conventional, non-prestressed reinforcement in the form of bars, de-coiled rods, cold-drawn wires or welded wire mesh is used in prestressed concrete structures for the same reasons as it is used in conventional reinforced concrete construction and these include:

- To provide additional tensile strength and ductility in regions of the structure where sufficient tensile strength and ductility are not provided by the prestressing steel. Non-prestressed longitudinal bars, for example, are often included in the tensile zone of beams to supplement the prestressing steel and increase the flexural strength. Non-prestressed reinforcement in the form of stirrups is most frequently used to carry the diagonal tension caused by shear and torsion in the webs of prestressed concrete beams.
- To control flexural cracks at service loads in prestressed concrete beams and slabs where some degree of cracking under full service loads is expected.

- To control cracking induced by restraint to shrinkage and temperature changes in regions and directions of low (or no) prestress.
- To carry compressive forces in regions where the concrete alone may not be adequate, such as in columns or in the compressive zone of heavily reinforced beams.
- Lateral ties or helices are used to provide restraint to bars in compression (i.e. to prevent lateral buckling of compressive reinforcement prior to the attainment of full strength) and to provide confinement for the compressive concrete in columns, beams and connections, thereby increasing both the strength and deformability of the confined concrete.
- To reduce long-term deflection and shortening due to creep and shrinkage by the inclusion of longitudinal bars in the compression region of the member.
- To provide resistance to the transverse tension that develops in the anchorage zone of post-tensioned members and to assist the concrete to carry the high bearing stresses immediately behind the anchorage plates.
- To reinforce the overhanging flanges in T-, I- or L-shaped cross-sections in both the longitudinal and transverse directions.

Types and sizes of non-prestressed reinforcement vary from country to country. Most reinforcement used in prestressed structures is made up of deformed bars or wires, although some plain round bars and wires are used as fitments. Regularly spaced rib-shaped deformations on the surface of a deformed bar improve the bond between the concrete and the steel and greatly improve the anchorage potential of the bar. It is for this reason that deformed bars are used as longitudinal reinforcement in most reinforced and prestressed concrete members.

In design calculations, non-prestressed steel is usually assumed to be elastic–plastic, as shown in Figure 4.8. Before yielding, the reinforcement is elastic, with steel stress σ_s proportional to the steel strain ε_s, i.e. $\sigma_s = E_s \varepsilon_s$, where E_s is the elastic modulus of the steel. After yielding, the stress–strain curve is often assumed to be horizontal (perfectly plastic) and the steel stress $\sigma_s = f_{yd}$ $(=f_{yk}/\gamma_S)$ at all values of strain exceeding the strain at first yield $\varepsilon_{yd} = f_{yd}/E_s$. The design yield stress f_{yd} is taken to be the *design strength* of the material and strain hardening is often ignored. The stress–strain curve in compression is assumed to be similar to that in tension.

4.3.2 Specification in Eurocode 2

4.3.2.1 Strength and ductility

EN 10080 [12] specifies general requirements and definitions for the performance characteristics of weldable reinforcing steel used in concrete

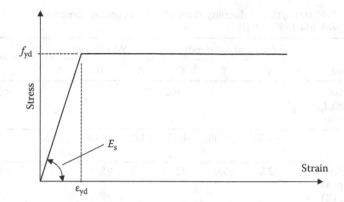

Figure 4.8 Idealised stress–strain relationship for non-prestressed steel.

structures designed to EN 1992-1-1 [1]. The following properties are specified to establish the required strength and ductility of steel reinforcement:

For strength: the characteristic yield strength, f_{yk} or $f_{0.2k}$; the characteristic tensile strength, f_{tk}

For ductility: the tensile strength to yield stress ratio, (f_{tk}/f_{yk}); the strain corresponding to the peak stress (i.e. the uniform elongation), ε_{uk}

The values of f_{yk} and f_{tk} are, respectively, the characteristic values of the yield load and the maximum direct axial tensile load, each divided by the nominal cross-sectional area of the bar.

Three ductility classes are recognised by EN 1992-1-1 [1]: Class A (low ductility); Class B (medium ductility) and Class C (high ductility). The requirements for strength and ductility of each class are shown in Table 4.5.

Class A reinforcement should not be used in situations where the reinforcement is required to undergo large plastic deformation under strength limit state conditions (i.e. strains in excess of 0.025). It should not be used if plastic methods of design are adopted, and it should not be used if the analysis has relied on some measure of moment redistribution.

4.3.2.2 Elastic modulus

The modulus of elasticity of reinforcing steel E_s is the slope of the initial elastic part of the stress–strain curve, when the stress is less than f_{yk}, and, in the absence of test data, the design value may be taken as equal to 200×10^3 MPa, irrespective of the type and ductility class of the steel [1]. Alternatively, E_s may be determined from standard tests. The elastic modulus in compression is taken to be identical to that in tension.

Table 4.5 Yield strength and ductility class of reinforcement compliant with EN 1992-1-1 [1]

Product	Bars and coiled rods			Wire fabrics			Non-compliant (%)[a]
Ductility class	A	B	C	A	B	C	
Characteristic yield stress, f_{yk} or $f_{0.2k}$ (MPa)			400–600				≤5.0
Minimum value of $k = (f_t/f_y)_k$	≥1.05	≥1.08	≥1.15 <1.35	≥1.05	≥1.08	≥1.15 <1.35	≤10.0
Characteristic strain at peak stress, ε_{uk} (%)	≥2.5	≥5.0	≥7.5	≥2.5	≥5.0	≥7.5	≤10.0

[a] The maximum percentage of test results falling below the specified minimum characteristic values of f_{yk}, $(f_t/f_y)_k$ and ε_{uk}.

4.3.2.3 Stress–strain curves: Design assumptions

The shape of the stress–strain curve for a typical hot rolled bar is shown in Figure 4.9a. For non-linear and other refined methods of analysis, actual stress–strain curves determined from testing may be used using mean rather than characteristic values.

In design, EN 1992-1-1 [1] allows the idealised bilinear relationship shown in Figure 4.9b to be used with a recommended strain limit of ε_{ud} = 0.9 ε_{uk} and a maximum stress of kf_{yk}/γ_s. Alternatively, an elastic–plastic relationship with a horizontal top branch at $f_{yd} = f_{yk}/\gamma_s$ may be used without the need to check the strain limit.

The partial safety factor for steel reinforcement is γ_S = 1.15 for persistent and transient design situations and γ_S = 1.0 for design situations involving accidental actions, as indicated in Table 2.4. The properties of two

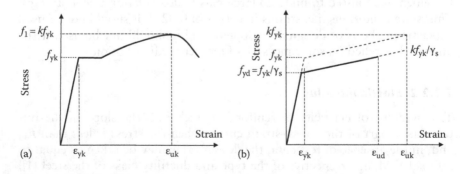

Figure 4.9 Actual and idealised stress–strain curves for reinforcing steel. (a) Actual stress–strain curve of a hot-rolled bar. (b) Idealised stress–strain curve of a hot-rolled bar.

Table 4.6 Material properties for reinforcing steels

Steel type	Ductility class	f_{tk} (MPa)	f_{td} (MPa)	f_{yk} (MPa)	f_{yd} (MPa)	ε_{uk} (%)	E_s (GPa)
B500A	A	525	457	500	435	2.5[a]	200
B500B	B	540	470	500	435	5.0	200

[a] ε_{uk} = 2.0% for 5.0 and 5.5 mm diameter bar.

Table 4.7 Preferred bar sizes in Europe [12]

Metric bar size	Linear mass density (kg/m)	Nominal diameter (mm)	Cross-sectional area (mm²)
6.0	0.222	6	28.3
8.0	0.395	8	50.3
10.0	0.617	10	78.5
12.0	0.888	12	113
14.0	1.21	14	154
16.0	1.58	16	201
20.0	2.47	20	314
25.0	3.85	25	491
28.0	4.83	28	616
32.0	6.31	32	804
40.0	9.86	40	1257
50.0	15.4	50	1963

commonly used steel types are given in Table 4.6, with design values determined for persistent and transient design situations. Preferred bar sizes in Europe complying with EN 10080 [12] are given in Table 4.7.

4.3.2.4 Coefficient of thermal expansion and density

In the absence of test data, the coefficient of thermal expansion of reinforcement may be taken as $10 \times 10^{-6}/°C$, and the density of the steel is 7850 kg/m³.

4.4 STEEL USED FOR PRESTRESSING

4.4.1 General

The shortening of the concrete caused by creep and shrinkage in a prestressed member causes a corresponding shortening of the prestressing steel that is physically attached to the concrete either by bond or by anchorages at the ends of the tendon. This shortening can be significant and usually results in a loss of stress in the steel of between 150 and 300 MPa.

In addition, creep in the highly stressed prestressing steel also causes a loss of stress through relaxation. Significant additional losses of prestress can result from other sources, such as friction along a post-tensioned tendon or draw-in at an anchorage at the time of prestressing.

For an efficient and practical design, the total loss of prestress should be a relatively small portion of the initial prestressing force. The steel used to prestress concrete must therefore be capable of carrying a very high initial stress. A tensile strength of between 1000 and 1900 MPa is typical for modern prestressing steels. The early attempts to prestress concrete with low-strength steels failed because almost the entire prestressing force was rapidly lost due to the time-dependent deformations of the poor-quality concrete in use at that time.

There are three basic forms of high-strength prestressing steels (as detailed in Chapter 3): cold-drawn stress-relieved round wire; stress-relieved strand and high-strength alloy steel bars. The stress–strain curves for the various types of prestressing steel exhibit similar characteristics, as illustrated in Figure 4.10. There is no well-defined yield point (as exists for some lower-strength steels). Each curve is initially linear-elastic (with an elastic modulus E_p similar to that for lower-strength steels) and with a relatively high proportional limit. When the curves become non-linear as deformation increases, the stress gradually increases monotonically until the steel fractures. The elongation at fracture is usually between 3.5% and 7%. High-strength steel is therefore considerably less ductile than conventional, hot-rolled non-prestressed reinforcing steel. For design purposes, the *yield stress* $f_{p0.1k}$ is usually taken as the stress corresponding to the 0.1% offset strain and is generally taken to be between 80% and 88% of the minimum tensile strength (i.e. $0.80f_{pk} - 0.88f_{pk}$). Some available sizes and properties of prestressing steel are given in Table 4.8.

The initial stress level in the prestressing steel after the prestress is transferred to the concrete is usually high, often in the range 70%–80% of the tensile strength of the material. At such high stress levels, high-strength steel creeps. At lower stress levels, such as is typical for non-prestressed steel, the creep of steel is negligible. If a tendon is stretched and held at a constant length (constant strain), the development of creep strain in the steel is exhibited as a loss of elastic strain and hence a loss of stress. This loss of stress in a specimen subjected to constant strain is known as *relaxation*. Creep, and hence relaxation, in steel is highly dependent on the stress level and increases at an increasing rate as the stress level increases. Relaxation in steel also increases rapidly as temperature increases. In recent years, low-relaxation steel has normally been used in order to minimise the losses of prestress resulting from relaxation.

Table 4.8 Types, sizes and properties of prestressing steels [13]

Type	Steel name	f_{pk} (MPa)	$f_{p0.1k}$ (MPa)	$f_{pd} = f_{p0.1k}/\gamma_S$ (MPa)	E_p (GPa)	ε_{uk} (%)
Wires	Y1860C	1860	1600	1391	205	3.5
	Y1770C	1770	1520	1322	205	3.5
	Y1670C	1670	1440	1252	205	3.5
	Y1570C	1570	1300	1130	205	3.5
Strands	Y2060S	2060	1770	1540	195	3.5
	Y1960S	1960	1680	1461	195	3.5
	Y1860S	1860	1600	1391	195	3.5
	Y1760S	1760	1520	1322	195	3.5
Bars	Y1030	1030	830	722	205	4.0
	Y1100	1100	900	783	205	4.0
	Y1230	1230	1080	939	205	4.0

Nominal diameter (mm)	Steel type	Mass (g/m)	Cross-sectional area (mm²)	Characteristic breaking strength (kN)	Maximum breaking strength (kN)	Yield strength at 0.1% proof strain (kN)
Wires						
4	Y1860C	98.4	12.6	23.4	26.9	20.8
5	Y1860C	153.1	19.6	36.5	42.0	32.5
6	Y1770C	221.0	28.3	50.1	57.6	44.5
7	Y1770C	300.7	38.5	68.1	78.3	59.9
8	Y1670C	392.8	50.3	83.9	96.5	73.9
Seven-wire strands						
7	Y2060S	234.3	30.0	61.8	71.1	54.4
8	Y1860S	296.8	38.0	70.7	81.3	60.8
9.6	Y1960S	429.6	55.0	107.8	124.0	94.0
12.5	Y1860S	726.3	93.0	173.0	199.0	149.0
12.9	Y1860S	781.0	100.0	186.0	213.9	160.0
15.2	Y1860S	1086.0	139.0	258.5	297.3	223.0
15.2	Y1760S	1086.0	139.0	244.6	281.3	212
Bars						
20	Y1030	2.57	314	323	371	261
25	Y1030	4.17	491	506	582	408
26.5	Y1030	4.49	552	569	654	458
32	Y1030	6.65	804	828	952	667
36	Y1030	8.44	1018	1049	1206	845
40	Y1030	10.36	1257	1295	1489	1043
50	Y1030	15.66	1963	2022	2325	1929

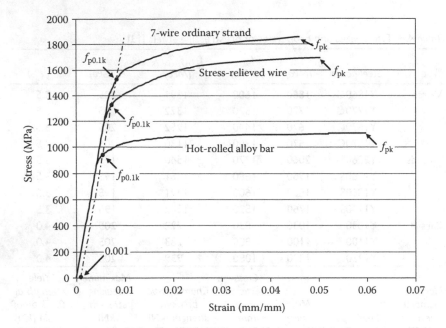

Figure 4.10 Typical stress–strain curves for tendons.

4.4.2 Specification in Eurocode 2

4.4.2.1 Strength and ductility

Prestressing steels must comply with the requirements of EN 10138 Parts 1-4 [13]. The terms used to define the strength and ductility of prestressing steel in EN 1992-1-1 [1] are illustrated on the typical stress–strain curve shown in Figure 4.11. In design calculations, the characteristic breaking strength f_{pk} is taken as the strength of the tendon. In practice, the breaking stress of 95% of all test samples will exceed f_{pk}. The strain corresponding to f_{pk} is the uniform elongation ε_{uk}. The tensile strength and ductility of some commonly used types of prestressing steel are given in Table 4.8.

The yield stress $f_{p0.1k}$ is taken as the 0.1% proof stress and may be determined by testing. In the absence of test data, the prescribed values given in Table 4.8 may be used. The design values of the steel strength f_{pd} are taken as $f_{p0.1k}/\gamma_S$, as shown in Figure 4.12 and given in Table 4.8.

According to EN 1992-1-1 [1], a prestressing tendon may be assumed to have adequate ductility if $f_{pk}/f_{p0.1k} \geq 1.1$.

4.4.2.2 Elastic modulus

The elastic modulus can be obtained by measuring the elongation of sample pieces of tendon in direct tension tests. The prescribed values of

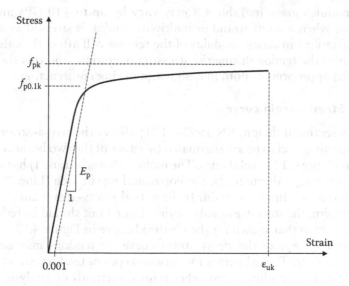

Figure 4.11 Stress–strain curve for prestressing steel [1].

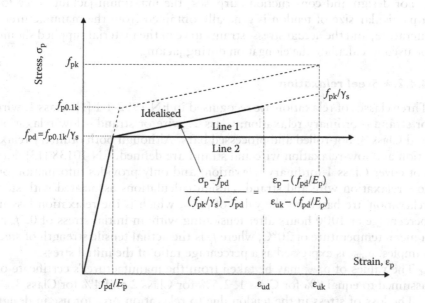

Figure 4.12 Idealised and design stress–strain diagrams for prestressing steel.

elastic modulus given in Table 4.8 may vary by up to ±10 GPa and possibly more when a multi-strand or multiwire tendon is stressed as a single cable. Variations in elastic modulus of the tendon will affect the calculated extension of the tendon during the stressing operation, and this should be considered appropriately both in design and during construction.

4.4.2.3 Stress–strain curve

For cross-sectional design, EN 1992-1-1 [1] allows the stress–strain curve for prestressing steel to be approximated by either of the two bilinear curves shown in Figure 4.12 as solid lines. The inclined branch (Line 1) has a strain limit $\varepsilon_{ud} = 0.9\ \varepsilon_{uk}$. Alternatively, the horizontal top branch (Line 2) can be used without any limit on strain. If the actual stress–strain curve for the steel is known, the steel stresses above the elastic limit should be reduced in a similar way to that shown for the idealised curve in Figure 4.12.

The actual shape of the stress–strain curve for tendons may be determined from tests. Typical curves for various types of tendons are shown in Figure 4.10. For non-linear and other refined methods of analysis, actual stress–strain curves for the steel, using mean rather than characteristic values, should be used. Alternatively, a simplified equation, such as that described by Loov [39], suitably calibrated to approximate the shape of the actual curve may be used in design.

For design and construction purposes, the maximum jacking force for a particular size of tendon is generally obtained from the manufacturer's literature, and the actual stress–strain curve of the material supplied should be used to calculate the elongation during jacking.

4.4.2.4 Steel relaxation

Three classes of relaxation are recognised in EN 1992-1-1 [1]: Class 1: wire or strand – ordinary relaxation; Class 2: wire or strand – low relaxation and Class 3: hot-rolled and processed bars. Although both ordinary relaxation and low-relaxation wire and strand are defined, EN 10138 [13] does not cover Class 1 (ordinary relaxation) and only provides information on low-relaxation wire and strand. Design calculations associated with steel relaxation are based on the value of ρ_{1000}, which is the relaxation loss in percentage at 1000 hours after tensioning with an initial stress of $0.7f_p$ at a mean temperature of 20°C, where f_p is the actual tensile strength of steel samples. ρ_{1000} is expressed as a percentage ratio of the initial stress.

The values of ρ_{1000} may be taken from the manufacturer's certificate or assumed to equal 8% for Class 1, 2.5% for Class 2 and 4% for Class 3.

The loss of stress in the tendon due to relaxation $\Delta\sigma_{pr}$ for use in design may be obtained from one of the following expressions:

$$\text{Class 1:}\quad \frac{\Delta\sigma_{pr}}{\sigma_{pi}} = -5.39\ \rho_{1000}\ e^{6.7\mu}\left(\frac{t}{1000}\right)^{0.75(1-\mu)} \times 10^{-5} \tag{4.34}$$

$$\text{Class 2:} \quad \frac{\Delta\sigma_{pr}}{\sigma_{pi}} = -0.66 \, \rho_{1000} \, e^{9.1\mu} \left(\frac{t}{1000}\right)^{0.75(1-\mu)} \times 10^{-5} \qquad (4.35)$$

$$\text{Class 3:} \quad \frac{\Delta\sigma_{pr}}{\sigma_{pi}} = -1.98 \, \rho_{1000} \, e^{8\mu} \left(\frac{t}{1000}\right)^{0.75(1-\mu)} \times 10^{-5} \qquad (4.36)$$

where σ_{pi} is the stress in a pretensioned or post-tensioned tendon immediately after anchoring the tendon (i.e. before transfer for pretensioned tendons and after transfer for post-tensioned tendons); t is the time after tensioning (in hours) and $\mu = \sigma_{pi}/f_{pk}$.

The final (long-term) values of the relaxation losses are taken at $t = 500{,}000$ hours, and values for Classes 1, 2 and 3 tendons are given in Table 4.9.

When elevated temperatures exist during curing (i.e. steam curing), relaxation is increased and occurs rapidly during the curing cycle. In this case, the equivalent time t_{eq} (in hours) given in Equation 4.37 should be added to the time after tensioning t in the relaxation time functions in Equations 4.34 through 4.36:

$$t_{eq} = \frac{1.14^{(T_{max}-20)}}{T_{max}-20} \sum_{t_i=1}^{n} (T_{(\Delta t_i)} - 20)\Delta t_i \qquad (4.37)$$

where $T_{(\varnothing_i)}$ is the temperature in °C during the time interval Δt_i and T_{max} is the maximum temperature in °C during the heat treatment.

Creep in the prestressing steel may also be defined in terms of a creep coefficient rather than as a relaxation loss. If the creep coefficient for the prestressing steel $\varphi_p(t, \sigma_{pi})$ is the ratio of creep strain in the steel to the initial elastic strain immediately after tensioning, then the final creep coefficients for wire, strand and bar are also given in Table 4.9 and have been

Table 4.9 Long-term relaxation losses (at $t = 500{,}000$ hours) and corresponding final creep coefficients for wire, strand and bar ($T = 20$°C)

Type of tendon		Tendon stress σ_{pi} as a proportion of f_{pk}		
		$\mu = 0.6$	$\mu = 0.7$	$\mu = 0.8$
Class 1	Relaxation loss (%)	15.5	19.0	23.3
	Creep coefficient, $\varphi_p(t, \sigma_{pi})$	0.155	0.190	0.233
Class 2	Relaxation loss (%)	2.5	3.9	6.1
	Creep coefficient, $\varphi_p(t, \sigma_{pi})$	0.025	0.039	0.061
Class 3	Relaxation loss (%)	6.2	8.7	12.1
	Creep coefficient, $\varphi_p(t, \sigma_{pi})$	0.062	0.087	0.121

approximated using Equation 4.38 (remembering that $\Delta\sigma_{pr}$ is the loss of stress and is therefore negative):

$$\varphi_p(t, \sigma_{pi}) = \frac{-\Delta\sigma_{pr}}{\sigma_{pi}} \tag{4.38}$$

As has already been emphasised, creep (relaxation) of the prestressing steel depends on the stress level. In a prestressed concrete member, the stress in a tendon is gradually reduced with time due to creep and shrinkage in the concrete. This gradual decrease in stress results in a reduction of creep in the steel and hence smaller relaxation losses. To determine relaxation losses in a concrete structure therefore, the final relaxation loss obtained from Equations 4.34 through 4.36 (or Table 4.9) should be multiplied by a reduction factor λ_r that accounts for the time-dependent shortening of the concrete due to creep and shrinkage. The factor λ_r depends on the creep and shrinkage characteristics of the concrete, the initial prestressing force and the stress in the concrete at the level of the steel and can be determined by iteration [24]. However, because relaxation losses in modern prestressed concrete structures (employing low-relaxation steels) are relatively small, it is usually sufficient to take $\lambda_r \approx 0.8$.

REFERENCES

1. EN 1992-1-1. 2004. Eurocode 2: Design of concrete structures – Part 1-1: General rules and rules for buildings. British Standards Institution, London, UK.
2. EN 196-1. 1995. Method of testing cement – Part 1: Determination of strength. British Standards Institution, London, UK.
3. EN 197-1. 2011. Cement – Part 1: Composition, specifications and conformity criteria for common cements. British Standards Institution, London, UK.
4. EN 943-2. 2009. Admixtures for concrete, mortar and grout. British Standards Institution, London, UK.
5. EN 206-1. 2000. Concrete. Specification, performance, production and conformity. British Standards Institution, London, UK.
6. BS 8500-1. 2006. Part 1: Method of specifying and guidance for the specifier. British Standards Institution, London, UK.
7. EN 12390:2. 2009. Testing hardened concrete: Making and curing specimens for strength tests. British Standards Institution, London, UK.
8. EN 12390:3. 2009. Testing hardened concrete: Compressive strength of test specimens. British Standards Institution, London, UK.
9. EN 12390:5. 2009. Testing hardened concrete: Flexural strength of test specimens. British Standards Institution, London, UK.
10. EN 12390:6. 2009. Testing hardened concrete: Tensile splitting strength of test specimens. British Standards Institution, London, UK.

11. BS 1881-121. 1983. Testing concrete: Methods for determination of static modulus of elasticity in compression. British Standards Institution, London, UK.
12. EN 10080. 2005. Steel for the reinforcement of concrete, weldable, ribbed reinforcing steel. British Standards Institution, London, UK.
13. EN 10138-1. 2005. Prestressing steel. Part 1: General requirements 1, EN 10138-2: Prestressing steels – Wire, EN 10138-3: Prestressing steels – Strand, EN 10138-4: Prestressing steels – Bars. British Standards Institution, London, UK.
14. Neville, A.M. 1996. *Properties of Concrete*, 4th edn. London, UK: Wiley.
15. Metha, P.K. and Monteiro, P.J. 2014. *Concrete: Microstructure, Properties and Materials*, 4th edn. New York: McGraw-Hill Education.
16. Kupfer, H.B., Hilsdorf, H.K. and Rüsch, H. 1975. Behaviour of concrete under biaxial stresses. *ACI Journal*, 66, 656–666.
17. Tasuji, M.E., Slate, F.O. and Nilson, A.H. 1978. Stress-strain response and fracture of concrete in biaxial loading. *ACI Journal*, 75, 306–312.
18. Darwin, D. and Pecknold, D.A. 1977. Nonlinear biaxial stress-strain law for concrete. *Journal of the Engineering Mechanics Division*, 103, 229–241.
19. Neville, A.M. 1970. *Creep of Concrete: Plain, Reinforced and Prestressed*. Amsterdam, the Netherlands: North-Holland.
20. Neville, A.M., Dilger, W.H. and Brooks, J.J. 1983. *Creep of Plain and Structural Concrete*. London, UK: Construction Press.
21. Gilbert, R.I. 1988. *Time Effects in Concrete Structures*. Amsterdam, the Netherlands: Elsevier.
22. Gilbert, R.I. and Ranzi, G. 2011. *Time-Dependent Behaviour of Concrete Structures*. London, UK: Spon Press, 426pp.
23. Ghali, A. and Favre, R. 1986. *Concrete Structures: Stresses and Deformations*. London, UK: Chapman and Hall.
24. Ghali, A., Favre, R. and Eldbadry, M. 2002. *Concrete Structures: Stresses and Deformations*, 3rd edn. London, UK: Spon Press, 584pp.
25. Rüsch, H., Jungwirth, D. and Hilsdorf, H.K. 1983. *Creep and Shrinkage – Their Effect on the Behaviour of Concrete Structures*. New York: Springer-Verlag, 284pp.
26. ACI 209R-92. 1992. Prediction of creep, shrinkage and temperature effects in concrete structures. ACI Committee 209, American Concrete Institute, Farmington Hills, MI, reapproved 2008.
27. ACI 209.1R-05. 2005. Report on factors affecting shrinkage and creep of hardened concrete. ACI Committee 209, American Concrete Institute, Farmington Hills, MI.
28. ACI Committee 209. 2008. Guide for modeling and calculating shrinkage and creep in hardened concrete (ACI 209.2R-08). American Concrete Institute, Farmington Hills, MI, 44pp.
29. Trost, H. 1978. Creep and creep recovery of very old concrete. RILEM Colloquium on Creep of Concrete, Leeds, UK.
30. Trost, H. 1967. Auswirkungen des Superpositionsprinzips auf Kriech- und Relaxations Probleme bei Beton und Spannbeton. *Beton- und Stahlbetonbau*, 62, 230–238, 261–269.

31. Dilger, W. and Neville, A.M. 1971. Method of creep analysis of structural members. *Australasian Conference on Information Security and Privacy*, 27–17, 349–379.
32. Bazant, Z.P. April 1972. Prediction of concrete creep effects using age-adjusted effective modulus method. *ACI Journal*, 69, 212–217.
33. fib. 2013. *Fib Model Code for Concrete Structures 2010*. Fib – International Federation for Structural Concrete, Ernst & Sohn, Lausanne, Switzerland, 434pp.
34. Ostergaard, L., Lange, D.A., Altouabat, S.A. and Stang, H. 2001. Tensile basic creep of early-age concrete under constant load. *Cement and Concrete Research*, 31, 1895–1899.
35. Chu, K.H. and Carreira, D.J. 1986. Time-dependent cyclic deflections in R/C beams. *Journal of Structural Engineering, ASCE*, 112(5), 943–959.
36. Bazant, Z.P. and Oh, B.H. 1984. Deformation of progressively cracking reinforced concrete beams. *ACI Journal*, 81(3), 268–278.
37. Gilbert, R.I. 2002. Creep and Shrinkage Models for High Strength Concrete – Proposals for inclusion in AS3600. *Australian Journal of Structural Engineering*, 4(2), 95–106.
38. Brooks, J.J. 2005. 30-year creep and shrinkage of concrete. *Magazine of Concrete Research*, 57(9), 545–556.
39. Loov, R.E. 1988. A general equation for the steel stress for bonded prestressed concrete members. *Journal of the Prestressed Concrete Institute*, 33, 108–137.

Chapter 5

Design for serviceability

5.1 INTRODUCTION

The level of prestress and the layout of tendons in a member are usually determined from the serviceability requirements for that member. For example, if a water-tight and crack-free slab is required, tension in the slab must be eliminated or limited to some appropriately low value. If, on the other hand, the deflection under a particular service load is to be minimised, a load-balancing approach may be used to determine the prestressing force and cable drape (see Section 1.4.3).

For the serviceability requirements to be satisfied in each region of a member at all times after first loading, a reasonably accurate estimate of the magnitude of prestress is needed in design. This requires reliable procedures for the determination of both the immediate and time-dependent losses of prestress. Immediate losses of prestress occur during the stressing (and anchoring) operation and include elastic shortening of concrete, the short-term relaxation of the tendon, friction along a post-tensioned cable and slip at the anchorages. As mentioned in previous chapters, the time-dependent losses of prestress are caused by creep and shrinkage of the concrete and relaxation of steel. Procedures for calculating both the immediate and time-dependent losses of prestress are presented in Section 5.10.

There are two critical stages in the design of prestressed concrete for serviceability. The first stage is immediately after the prestress is transferred to the concrete, i.e. when the member is subjected to the maximum prestress and the external load is usually at a minimum. Immediate losses have taken place, but no time-dependent losses have yet occurred. The prestressing force immediately after transfer is designated in EN 1992-1-1 [1] as P_{m0}. At this stage, the concrete is usually young and the concrete strength may be relatively low. The second critical stage is after time-dependent losses have taken place and the full service load is applied, i.e. at time t when the prestressing force is at a minimum and the external service load is at a maximum. The prestressing force at this stage is designated in EN 1992-1-1 [1] as $P_{m,t}$ and is often referred to as the *effective* prestress.

At each of these stages (and at all intermediate stages), it is necessary to ensure that both the strength and the serviceability requirements of the member are satisfied. Strength depends on the cross-sectional area and position of both the steel tendons and the non-prestressed reinforcement. However, it is not strength that determines the level of prestress, but serviceability. When the prestressing force and the amount and distribution of the prestressing steel have been determined, the flexural strength may be readily increased, if necessary, by the addition of non-prestressed conventional reinforcement. This is discussed in more detail in Chapter 6. Shear strength may be improved by the addition of transverse stirrups (as discussed in Chapter 7). As will be seen throughout this chapter, the presence of bonded conventional reinforcement also greatly influences both the short- and long-term behaviour at service loads, both for cracked and uncracked prestressed members. The design for strength and serviceability therefore cannot be performed independently, as the implications of one affect the other.

General design requirements for the serviceability limit states, including combinations of actions, were discussed in Chapter 2. It is necessary to ensure that the instantaneous and time-dependent deflection and the axial shortening under service loads are acceptably small and that cracking is well controlled by suitably detailed bonded reinforcement. To determine the in-service behaviour of a member, it is therefore necessary to establish the extent of cracking, if any, by checking the magnitude of elastic tensile stresses. If a member remains uncracked (i.e. the maximum tensile stress at all stages is less than the tensile strength of concrete), the properties of the uncracked section may be used in all deflection and camber calculations (see Sections 5.6 and 5.7). If cracking occurs, a cracked section analysis may be performed to determine the properties of the cracked section and the post-cracking behaviour of the member (see Sections 5.8 and 5.9).

5.2 CONCRETE STRESSES AT TRANSFER AND UNDER FULL SERVICE LOADS

In the past, codes of practice have set mandatory maximum limits on the magnitude of the concrete stresses, both tensile and compressive. In reality, concrete stresses calculated by a linear-elastic analysis are often not even close to those that exist after a short period of creep and shrinkage, particularly in members containing significant quantities of bonded reinforcement. It makes little sense to limit concrete stresses in compression and tension, unless they are determined based on non-linear analysis, in which the time-varying constitutive relationship for concrete is accurately modelled. Even if non-linear analysis is undertaken, limiting the concrete stresses in compression to a maximum prescribed value, or making sure the concrete tensile stresses are less than the tensile strength of concrete, does not ensure either adequate strength of a structural member or satisfactory behaviour at service loads.

Notwithstanding the above, some codes still classify prestressed members in terms of the calculated maximum tensile stress in the concrete σ_{ct} in the precompressed tensile zone of a member. For example, using the notation adopted in this book, ACI318M-14 [2] classifies prestressed members as:

  a. *Uncracked* (Class U) if $\sigma_{ct} \le 0.62\sqrt{f_{ck}}$;
  b. *Transitional* (Class T) if $0.62\sqrt{f_{ck}} < \sigma_{ct} \le 1.0\sqrt{f_{ck}}$; and
  c. *Cracked* (Class C) if $\sigma_{ct} > 1.0\sqrt{f_{ck}}$.

where σ_{ct} and f_{ck} are expressed in MPa. As we shall see subsequently in Section 5.7.4, limiting the maximum concrete tensile stress calculated in an elastic analysis to $0.62\sqrt{f_{ck}}$ certainly does not mean that the member will remain uncracked.

ACI318M-14 [2] imposes a limit of $0.6f_{ck}(t_0)$ on the calculated extreme fibre-compressive stress at transfer, except that this limit can be increased to $0.7f_{ck}(t_0)$ at the ends of simply-supported members (where $f_{ck}(t_0)$ is the specified characteristic strength of concrete at the time of transfer). ACI318M-14 [2] also requires that where the concrete tensile stress exceeds $0.5\sqrt{f_{ck}(t_0)}$ at the ends of a simply-supported member, or $0.25\sqrt{f_{ck}(t_0)}$ elsewhere, additional bonded reinforcement should be provided in the tensile zone to resist the total tensile force computed with the assumption of an uncracked cross-section.

In addition, for Class U and Class T flexural members, ACI318M-14 [2] specifies that the extreme fibre-compressive stress, calculated assuming uncracked cross-sectional properties and after all losses of prestress, should not exceed the following:

Due to prestress plus sustained load: $0.45f_{ck}$
Due to prestress plus total load: $0.6f_{ck}$

These limits are imposed to decrease the probability of fatigue failure in beams subjected to repeated loads and to avoid the development of non-linear creep that develops under high compressive stresses.

EN 1992-1-1 [1] limits the compressive stress in the concrete immediately after transfer to $0.6f_{ck}(t_0)$, but if the compressive stress in the concrete that is sustained permanently exceeds $0.45f_{ck}(t_0)$, EN 1992-1-1 [1] suggests that the effects of non-linear creep should be considered. Immediately after the transfer of prestress, the prestressing force is at its maximum value and time-dependent losses have not yet occurred. Satisfaction of the compressive stress limit will usually, although not necessarily, lead to an adequate factor of safety against compressive failure at transfer.

It is also important to ensure that cracking does not occur immediately after transfer in locations where there is no (or insufficient) bonded reinforcement. The regions of a member that are subjected to tension at transfer are often those that are later subjected to compression when the full service load is applied. If these regions are unreinforced and uncontrolled cracking is permitted at transfer, an immediate serviceability problem exists. When the region is later compressed, cracks may not close completely, local spalling may occur and even a loss of shear strength could result. If cracking is permitted at transfer, bonded reinforcement should be provided to carry all the tension and to ensure that the cracks are fine and well controlled. For the calculation of concrete stresses immediately after transfer, an elastic analysis using gross cross-sectional properties is usually satisfactory.

In some cases, concrete stresses may need to be checked under full service loads when all prestress losses have taken place. If cracking is to be avoided, concrete tensile stresses must not exceed the tensile strength of concrete. Care should be taken when calculating the maximum tensile stress to accurately account for the load-independent tension induced by restraint to shrinkage or temperature effects and the relaxation of stress caused by creep of the concrete. However, even if the tensile stress does reach the tensile strength of concrete and some minor cracking occurs, the cracks will be well controlled and the resulting loss of stiffness will not be significant, provided sufficient bonded reinforcement or tendons are provided near the tensile face.

For many prestressed concrete situations, there are no valid reasons why cracking should be avoided at service loads and, therefore, no reason why a limit should be placed on the maximum tensile stress in the concrete. Indeed, in modern prestressed concrete building structures, many members crack under normal service loads. If cracking does occur, the resulting loss of stiffness must be accounted for in deflection calculations and a nonlinear cracked section analysis is required to determine behaviour in the post-cracking range. Crack widths must also be controlled. Crack control may be achieved by limiting both the spacing of the bonded reinforcement and the change of stress in the reinforcement after cracking (see Section 5.12).

Under full service loads, which occur infrequently, there is often no practical reason why compressive stress limits should be imposed. Separate checks for flexural strength, ductility and shear strength are obviously necessary. Some members, such as trough girders or inverted T-beams, are prone to high concrete compressive stresses under full service loads and, in the design of these members, care should be taken to limit the extreme fibre-compressive stress at service loads. If a large portion of the total service load is permanent, compressive stress levels in excess of about $0.45f_{ck}$ should be avoided. With this upper limit on compressive stresses, the probability of fatigue failure in uncracked or lightly cracked members subjected to repeated loads will be reduced and excessive non-linear creep deformations will not occur.

The primary objective in calculating and perhaps setting limits on the concrete stresses, both at transfer and under full loads, is to obtain a serviceable structure. As was discussed in Section 1.4, elastic stress calculations are not strictly applicable to prestressed concrete. Creep and shrinkage cause a gradual transfer of compression from the concrete to the bonded steel. Nevertheless, elastic stress calculations may indicate potential serviceability problems and the satisfaction of concrete tensile stress limits is a useful procedure to control the extent of cracking. It should be understood, however, that the satisfaction of a set of elastic concrete stress limits does not, in itself, ensure serviceability and certainly does not ensure adequate strength. The designer must check both strength and serviceability separately, irrespective of the stress limits selected.

5.3 MAXIMUM JACKING FORCE

The mean prestressing force at any time t at a distance x from the active end of a tendon is designated in EN 1992-1-1 [1] as $P_{m,t}(x)$ and is equal to the maximum force P_{max} applied at the active end of the tendon during tensioning minus the immediate and time-dependent losses. The maximum force P_{max} (also referred to as the jacking force P_j) must satisfy:

$$P_{max} \leq A_p \, \sigma_{p.max} \tag{5.1}$$

where A_p is the cross-sectional area of the tendon and $\sigma_{p.max}$ is the smaller of 80% of the characteristic tensile strength (i.e. $0.8 f_{pk}$) and 90% of the characteristic 0.1% proof stress (i.e. $0.9 f_{p0.1k}$). Overstressing up to $\sigma_{p.max} = 0.95 f_{p0.1k}$ is permitted if the force in the jack can be measured to an accuracy of $\pm 5\%$ of the final value of the prestressing force. Any such overstressing is intended to deal only with unforeseen problems during construction and should not be assumed during the design stage.

The value of the initial prestressing force at x immediately after transfer in a pretensioned member or immediately after tensioning and anchoring in a post-tensioned member (i.e. at $t = t_0$) is designated $P_{m0}(x)$ (or simply P_{m0}) and is equal to P_{max} minus the immediate losses of prestress. According to EN 1992-1-1 [1], $P_{m0}(x)$ should satisfy:

$$P_{m0}(x) \leq A_p \, \sigma_{pm0.max} \tag{5.2}$$

where $\sigma_{pm0.max}$ is the smaller of $0.75 f_{pk}$ or $0.85 f_{p0.1k}$.

When tensioning a tendon, the stressing procedure should ensure that the force in the tendon increases at a uniform rate. The prestressing force should be measured at the jack, and the tendon extension should be measured during tensioning. A check should also be made to ensure that the measured extension of each tendon agrees with the calculated extension based on the measured prestressing force and knowledge of the cable profile and the load–extension curve for the tendon. Any difference between the two figures greater than 10% should be investigated. Differences could be due to problems arising for a variety of reasons, including blockage of the duct due to the ingress of cement paste, the wrong size tendon being used, slip at the dead-end anchorage, variations of tendon profile and hence different friction losses in the duct or anchorage from the assumed values and variations in strand properties, including differences in strands due to worn dies used in drawing the strand wires during manufacture.

5.4 DETERMINATION OF PRESTRESS AND ECCENTRICITY IN FLEXURAL MEMBERS

There are a number of possible starting points for the determination of the prestressing force P and eccentricity e required at a particular cross-section. The starting point depends on the particular serviceability requirements for the member. The prestressing force and the cable layout for a member may be selected to minimise deflection under some portion of the applied load, i.e. a load-balancing approach to design. With such an approach, cracking may occur when the applied load is substantially different from the selected *balanced load*, such as at transfer or under the full service loads after all losses, and this possibility needs to be checked and accounted for in serviceability calculations.

The quantities P and e are often determined to satisfy preselected stress limits. Cracking may or may not be permitted under service loads. As was mentioned in the previous section, satisfaction of concrete stress limits does not necessarily ensure that deflection, camber or axial shortening are within acceptable limits. Separate checks are required for each of these serviceability limit states.

5.4.1 Satisfaction of stress limits

Numerous design approaches have been proposed for the satisfaction of concrete stress limits, including analytical and graphical techniques, e.g. Magnel [3], Lin [4] and Warner and Faulkes [5]. A simple and convenient approach is described here.

If the member is required to remain uncracked throughout, suitable stress limits should be selected for the tensile stress at transfer $f_{ct,0}$ and the tensile stress under full load $f_{ct,t}$. In addition, limits should also be placed on the concrete compressive stress at transfer $f_{cc,0}$ and under full loads $f_{cc,t}$. If cracking under the full loads is permitted, the stress limit $f_{ct,t}$ is relaxed and the remaining three limits are enforced.

In Figure 5.1, the uncracked cross-section of a beam at the critical moment location is shown, together with the concrete stresses at transfer caused by the initial prestress of magnitude P_{m0} (located at an eccentricity e below the centroidal axis of the concrete section) and by the external moment M_0 resulting from the loads acting at transfer. Often self-weight is the only load

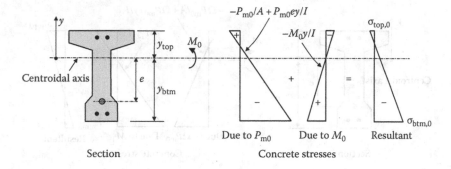

Figure 5.1 Concrete stresses at transfer.

(other than prestress) acting at transfer. In Figure 5.1, we have assumed that the cross-section is uncracked and that concrete stresses are calculated assuming linear-elastic material behaviour.

At transfer, the concrete stress in the top fibre must not exceed the tensile stress limit $f_{ct,0}$. If tensile (compressive) stress is assumed to be positive (negative), we have:

$$\sigma_{top,0} = -\frac{P_{m0}}{A} + \frac{P_{m0}e\, y_{top}}{I} - \frac{M_0\, y_{top}}{I} \leq f_{ct,0}$$

Rearranging and introducing the term $\alpha_{top} = Ay_{top}/I = A/Z_{top}$, we get:

$$P_{m0} \leq \frac{Af_{ct,0} + \alpha_{top}M_0}{\alpha_{top}e - 1} \tag{5.3}$$

where A is the area of the transformed cross-section, I is the second moment of area of the transformed section about the centroidal axis and Z_{top} is the elastic modulus of the cross-section with respect to the top fibre (equal to I/y_{top}).

Similarly, the concrete stress in the bottom fibre must be greater than the negative compressive stress limit at transfer:

$$\sigma_{btm,0} = -\frac{P_{m0}}{A} - \frac{P_{m0}e\, y_{btm}}{I} + \frac{M_0 y_{btm}}{I} \geq f_{cc,0}$$

Rearranging and introducing the term $\alpha_{btm} = Ay_{btm}/I = A/Z_{btm}$, we get:

$$P_{m0} \leq \frac{-Af_{cc,0} + \alpha_{btm}M_0}{\alpha_{btm}e + 1} \tag{5.4}$$

where Z_{btm} is the elastic modulus of the cross-section with respect to the bottom fibre $(=I/y_{btm})$ and the compressive stress limit $f_{cc,0}$ is a negative quantity.

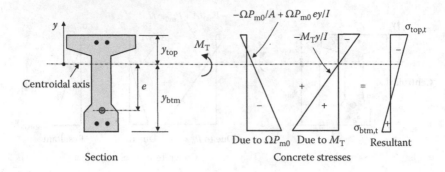

Figure 5.2 Concrete stresses under full loads (after all prestress losses).

Figure 5.2 shows the concrete stresses on an uncracked cross-section caused by both the effective prestressing force $P_{m,t}$ acting on the concrete after all losses have taken place and the applied moment M_T resulting from the full service load. The effective prestressing force acting on the concrete part of the cross-section is taken as ΩP_{m0}, where Ω depends on the time-dependent loss of prestress in the tendon and the amount of force transferred from the concrete into the bonded non-prestressed reinforcement as it restrains the development of creep and shrinkage in the concrete with time. For cross-sections containing no conventional reinforcement, Ω is typically about 0.80 but may be significantly smaller for sections containing conventional reinforcement (see Section 5.7.4, where in Tables 5.1 and 5.2, Ω varies between 0.438 and 0.851 depending on the amount and position of bonded reinforcement on the cross-section under consideration).

For an uncracked member, the concrete stress in the bottom fibre must be less than the selected tensile stress limit $f_{ct,t}$:

$$\sigma_{btm,t} = -\frac{\Omega P_{m0}}{A} - \frac{\Omega P_{m0} e\, y_{btm}}{I} + \frac{M_T y_{btm}}{I} \le f_{ct,t}$$

and rearranging gives:

$$P_{m0} \ge \frac{-A f_{ct,t} + \alpha_{btm} M_T}{\Omega(\alpha_{btm} e + 1)} \tag{5.5}$$

The compressive stress in the top fibre must also satisfy the appropriate stress limit $f_{cc,t}$:

$$\sigma_{top,t} = -\frac{\Omega P_{m0}}{A} + \frac{\Omega P_{m0} e\, y_{top}}{I} - \frac{M_T y_{top}}{I} \ge f_{cc,t}$$

and rearranging gives:

$$P_{m0} \geq \frac{Af_{cc,t} + \alpha_{top}M_T}{\Omega(\alpha_{top}e - 1)} \tag{5.6}$$

If the eccentricity e at the cross-section is known, satisfaction of Equations 5.3 and 5.4 will ensure that the desired stress limits at transfer are not exceeded. Equations 5.3 and 5.4 provide an upper limit on the magnitude of P_{m0}. Equations 5.5 and 5.6 provide a lower limit on the magnitude of P_{m0}. Satisfaction of all four equations will ensure that the selected stress limits at transfer and under full loads are all satisfied.

If a particular cross-section is too small, it may not be possible to satisfy all four stress limits and either a larger cross-section can be selected or the offending stress limit(s) can be relaxed and the effect of this variation assessed in the subsequent design. Even though separate checks are required to ensure satisfaction of the strength and serviceability requirements, Equations 5.3 through 5.6 provide a useful starting point in design for sizing both the cross-sectional dimensions and the prestressing details.

If the maximum value of P_{m0} that satisfies Equation 5.4 is the same as the minimum value required to satisfy Equation 5.5, information is obtained about the properties of the smallest cross-section that can be selected to ensure satisfaction of both the stress limits $f_{cc,0}$ at transfer and $f_{ct,t}$ under full loads (i.e. the smallest sized cross-section that will ensure that cracking does not occur under full service loads). Equating the right-hand sides of Equations 5.4 and 5.5, we get the following expression for the section modulus (Z_{btm}) of the minimum sized cross-section:

$$(Z_{btm})_{min} = \frac{M_T - \Omega M_0}{f_{ct,t} - \Omega f_{cc,0}} \tag{5.7}$$

Similarly, if we equate the right-hand sides of Equations 5.3 and 5.6, we get the following expression for the section modulus (Z_{top}) of the smallest sized cross-section required to satisfy both the stress limits $f_{ct,0}$ at transfer and $f_{cc,t}$ under full loads (i.e. the smallest sized cross-section to ensure that cracking does not occur at transfer):

$$(Z_{top})_{min} = \frac{M_T - \Omega M_0}{\Omega f_{ct,0} - f_{cc,t}} \tag{5.8}$$

It must be remembered that $f_{cc,0}$ and $f_{cc,t}$ represent compressive stress limits and are negative quantities. Equations 5.7 and 5.8 are useful starting points in the selection of an initial cross-section.

In order to make use of Equations 5.7 and 5.8 in preliminary design, an estimate of the time-dependent loss of prestress in the concrete must be made. Usually, a first estimate of Ω of about 0.8 is reasonable if low-relaxation presetressing steel is used and the cross-section does not contain significant quantities of non-prestressed steel. As already mentioned, if the cross-section contains significant quantities of bonded reinforcement, Ω may be significantly smaller. However, if this is the case, it may not be necessary to enforce a no cracking requirement and Equations 5.7 and 5.8 would no longer be relevant. Any initial estimate of Ω must be checked after the prestress, the eccentricity and the quantity of bonded reinforcement have been determined. A suitable procedure for determining the time-dependent loss of stress in the concrete is described in Section 5.7.3, and simplified methods are described in Section 5.10.3.

EN 1992-1-1 [1] assumes that flexural cracking is controlled if the maximum tensile concrete stress in a member caused by the short-term service loads and prestress does not exceed the effective tensile strength $f_{ct,eff}(t)$ taken as either the mean value of the tensile strength of the concrete $f_{ctm}(t)$ or the mean value of the flexural tensile strength of the concrete $f_{ctm,fl}(t)$ at the time when the cracks may first be expected to occur. The calculation for the minimum tension reinforcement (see Equation 5.192 discussed subsequently) must be based on the same value of $f_{ct,eff}(t)$. If cracking is permitted under full service loads, a tensile stress limit $f_{ct,t}$ is not specified and Equations 5.5 and 5.7 do not apply. Tensile and compressive stress limits at transfer are usually enforced and, therefore, Equations 5.3 and 5.4 are still applicable and continue to provide an upper limit on the level of prestress. The only minimum limit on the level of prestress is that imposed by Equation 5.6 and, for most practical cases, this does not influence the design.

When there is no need to satisfy a tensile stress limit under full loads, any level of prestress that satisfies Equations 5.3, 5.4 and 5.6 may be used, including $P_{m0} = 0$ (which corresponds to a reinforced concrete member). Often members that are designed to crack under the full service loads are proportioned so that no tension exists in the concrete under the sustained load. Cracks open and close as the variable live load is applied and removed. The selection of prestress in such a case can still be made conveniently using Equation 5.5, if the maximum total service moment M_T is replaced in Equation 5.5 by the sustained or permanent moment M_{sus}.

If cracking occurs, the cross-section required for the cracked prestressed member may need to be larger than that required for a fully-prestressed, uncracked member for the same deflection limit. In addition, the quantity of non-prestressed reinforcement is usually significantly greater. Often, however, the reduction in prestressing costs more than compensates for the additional concrete and non-prestressed reinforcement costs and cracked *partially-prestressed* members are the most economical structural solution in a wide range of applications.

EXAMPLE 5.1

A one-way slab is simply-supported over a span of l = 12 m and is to be designed to carry a maximum superimposed design service load of w_s = 5 kPa (kN/m^2) in addition to its own self-weight. The slab is post-tensioned by regularly spaced tendons with parabolic profiles (with zero eccentricity at each support and a maximum eccentricity at mid-span). Each tendon contains four 12.9 mm diameter strands in a flat duct. The material properties are:

$$f_{ck}(t_0) = 25\,\text{MPa}; \quad f_{ck} = 40\,\text{MPa}; \quad E_{cm}(t_0) = 31,000\,\text{MPa};$$

$$E_{cm} = 35,000\,\text{MPa}; \quad f_{pk} = 1860\,\text{MPa}$$

Assume that the loss of prestress at mid-span immediately after transfer is 8% and the total time-dependent losses due to creep, shrinkage and relaxation are 15%. Determine the prestressing force and eccentricity required to satisfy the following concrete stress limits:

At transfer: $f_{ct,0} = 0.5f_{ctm}(t_0) = 1.3$ MPa   and $f_{cc,0} = -0.5f_{ck}(t_0) = -12.5$ MPa

After all losses: $f_{ct,t} = 0.5f_{ctm} = 1.75$ MPa   and $f_{cc,t} = -0.5f_{ck} = -20.0$ MPa

Also determine the required number and spacing of tendons, and the initial deflection of the slab at mid-span immediately after transfer.

In order to obtain an estimate of the slab self-weight (which is the only load other than the prestress at transfer), a trial slab thickness of 300 mm (span/40) is assumed initially. Assuming the concrete weighs 24 kN/m³, the self-weight is:

$$w_{sw} = 24 \times 0.3 = 7.2\ \text{kN/m}^2$$

and the moments at mid-span of the slab both at transfer and under the full service load (evaluated for a 1 m wide strip of slab) are:

$$M_0 = \frac{w_{sw}l^2}{8} = \frac{7.2 \times 12^2}{8} = 129.6\ \text{kNm/m}$$

$$M_T = \frac{(w_{sw} + w_s)l^2}{8} = \frac{(7.2 + 5.0) \times 12^2}{8} = 219.6\ \text{kNm/m}$$

From Equation 5.7:

$$(Z_{btm})_{min} = \frac{M_T - \Omega M_0}{f_{ct,t} - \Omega f_{cc,0}} = \frac{(219.6 - 0.85 \times 129.6) \times 10^6}{1.75 - [0.85 \times (-12.5)]} = 8.84 \times 10^6\ \text{mm}^3/\text{m}$$

and from Equation 5.8:

$$(Z_{top})_{min} = \frac{M_T - \Omega M_0}{\Omega f_{ct,0} - f_{cc,t}} = \frac{(219.6 - 0.85 \times 129.6) \times 10^6}{(0.85 \times 1.3) - (-20.0)} = 5.19 \times 10^6 \text{ mm}^3/\text{m}$$

For the rectangular slab cross-section, the minimum section modulus must exceed $(Z_{btm})_{min}$ = 9.39 × 10^6 mm³/m and the corresponding minimum slab depth is therefore:

$$h_{min} = \sqrt{6(Z_b)_{min}/1000} = 230.3 \text{ mm}$$

If we select a slab thickness h = 230 mm, the revised self-weight is w_{sw} = 5.52 kN/m², the revised moments are M_0 = 99.4 kNm/m and M_T = 189.4 kNm/m, and the revised section properties are $(Z_{btm})_{min}$ = 8.48 mm³/m and h_{min} = 225.5 mm.

Taking h = 230 mm, the relevant section properties are:

A = 230 × 10^3 mm²/m, I = 1014 × 10^6 mm⁴/m,

$Z_{btm} = Z_{top}$ = 8.817 × 10^6 mm³/m, and $\alpha_{btm} = \alpha_{top}$ = 26.1 × 10^{-3} mm⁻¹.

If we take the minimum concrete cover to the strand as 30 mm, the eccentricity at mid-span of the 12.9 mm diameter strands is:

$$e = e_{max} = h/2 - 30 - 0.5 \times 12.9 = 78.6 \text{ mm}$$

From Equation 5.3:

$$P_{m0} \leq \frac{Af_{ct,0} + \alpha_{top}M_0}{\alpha_{top}e - 1}$$

$$= \frac{230 \times 10^3 \times 1.3 + 26.1 \times 10^{-3} \times 99.4 \times 10^6}{(26.1 \times 10^{-3} \times 78.6) - 1} = 2752 \text{ kN/m}$$

From Equation 5.4:

$$P_{m0} \leq \frac{-Af_{cc,0} + \alpha_{btm}M_0}{\alpha_{btm}e + 1}$$

$$= \frac{-230 \times 10^3 \times (-12.5) + 26.1 \times 10^{-3} \times 99.4 \times 10^6}{(26.1 \times 10^{-3} \times 78.6) + 1} = 1792 \text{ kN/m}$$

The prestressing force immediately after transfer P_{m0} must not exceed 1792 kN/m.

From Equation 5.5:

$$P_{m0} \geq \frac{-Af_{ct,t} + \alpha_{btm}M_T}{\Omega(\alpha_{btm}e + 1)}$$

$$= \frac{-230 \times 10^3 \times 1.75 + 26.1 \times 10^{-3} \times 189.4 \times 10^6}{0.85 \times (26.1 \times 10^{-3} \times 78.6 + 1)} = 1751 \, kN/m$$

From Equation 5.6:

$$P_{m0} \geq \frac{Af_{cc,t} + \alpha_{top}M_T}{\Omega(\alpha_{top}e - 1)} = \frac{230 \times 10^3 \times (-20.0) + 26.1 \times 10^{-3} \times 189.4 \times 10^6}{0.85 \times (26.1 \times 10^{-3} \times 78.6 - 1)}$$

$$= 384 \, kN/m$$

The minimum prestressing force P_{m0} is therefore 1751 kN/m, and this value is used in the following calculations. With 8% immediate losses between mid-span and the jacking point at one end of the span, the required jacking force is:

$$P_i = \frac{P_{m0}}{(1 - 0.08)} = \frac{1751}{0.92} = 1903 \, kN/m$$

From Table 4.8, a 12.9 mm diameter 7-wire low-relaxation strand has a cross-sectional area of $A_p = 100.0 \, mm^2$, a characteristic breaking load of $f_{pk}A_p = 186.0$ kN and a 0.1% proof load of $f_{p0.1k}A_p = 160.0$ kN. According to EN 1992-1-1 [1], the maximum jacking force in a strand is the smaller of 80% of the characteristic tensile strength (i.e. $0.8\, f_{pk}A_p = 148.8$ kN) and 90% of the 0.1% proof stress (i.e. $0.90\, f_{p0.1k}A_p = 144$ kN) (see Section 5.3). A flat duct containing four 12.9 mm strands can therefore be stressed with a maximum jacking force of $4 \times 144 = 576$ kN.

The minimum number of ducts required in each metre width of slab is therefore:

$$\frac{P_i}{576} = \frac{1903}{576} = 3.30$$

and the maximum spacing between cables is therefore 1000/3.30 = 303 mm.

A 4-strand tendon every 300 mm is specified with a jacking force per tendon of 1903 × 0.30 = 571 kN.

For this slab, provided the initial estimates of losses are correct, the properties of the uncracked cross-section can be used in all deflection calculations, since stress limits have been selected to ensure that cracking does not occur either at transfer or under the full service loads. At transfer,

$E_{cm}(t_0) = 31,000$ MPa, $I = 1014 \times 10^6$ mm^4/m and the uniformly distributed upward load caused by the parabolic tendons with drape equal to 0.0786 m is obtained from Equation 1.7 as follows:

$$w_p = \frac{8P_{m0}e}{l^2} = \frac{8 \times 1751 \times 0.0786}{12^2} = 7.65 \text{ kN/m}$$

The resultant upward load is $w_p - w_{sw} = 7.65 - 5.52 = 2.13$ kN/m, and the initial deflection (camber) at mid-span at transfer (before any creep and shrinkage deformations have taken place) is:

$$v_i = \frac{5}{384} \frac{(w_p - w_{sw})l^4}{E_{cm}(t_0)I} = \frac{5 \times 2.13 \times 12,000^4}{384 \times 31,000 \times 1014 \times 10^6} = 18.3 \text{ mm (upwards)}$$

This may or may not be acceptable depending on the serviceability (deflection) requirements for the slab.

To complete this design, the effects of creep and shrinkage will need to be considered under permanent loads. The final long-term deflections will need to be calculated and checked against the deflection criteria, and the actual long-term losses must be checked. A reliable procedure for undertaking the long-term deflection calculations is outlined in Section 5.11.4. Of course, the ultimate strength in bending and in shear will also need to be checked and the anchorage zones must be designed.

5.4.2 Load balancing

Using the load-balancing approach, the effective prestress after losses $P_{m,t}$ and the eccentricity e are selected such that the transverse load imposed by the prestress w_p balances a selected portion of the external load. The effective prestress $P_{m,t}$ in a parabolic cable of drape e required to balance a uniformly distributed external load w_b is obtained using Equation 1.7 as follows:

$$P_{m,t} = \frac{w_b l^2}{8 e} \tag{5.9}$$

Concrete stresses are checked under the remaining unbalanced service loads to identify regions of possible cracking and regions of high compression. Deflection under the unbalanced loads may need to be calculated and controlled. Losses are calculated and stresses immediately after transfer are also checked. Having determined the amount and layout of the prestressing steel (and the prestressing force) to satisfy serviceability requirements, the design for adequate strength can then proceed.

Load balancing is widely used for the design of indeterminate members and also for simple determinate beams and slabs. It is only strictly applicable, however, prior to cracking when the member behaves linearly and the principle of superposition, on which load balancing relies, is valid.

EXAMPLE 5.2

Reconsider the 12 m span, 230 mm thick one-way slab of Example 5.1 and determine the prestress required to balance the slab self-weight (5.52 kN/m²). The parabolic tendons have zero eccentricity at each support and e = 78.6 mm at mid-span. As in Example 5.1, time-dependent losses at mid-span are assumed to be 15% and the instantaneous losses between mid-span and the jacking end are taken to be 8%.

With w_b = 5.52 kN/m² and e = 0.0786 m, Equation 5.9 gives:

$$P_{m,t} = \frac{w_b l^2}{8e} = \frac{5.52 \times 12^2}{8 \times 0.0786} = 1264 \text{ kN/m}$$

The prestressing force at mid-span immediately after transfer and the jacking force are:

$$P_{m0} = \frac{P_{m,t}}{(1-0.15)} = \frac{1264}{0.85} = 1487 \text{ kN/m} \quad \text{and} \quad P_i = \frac{P_{m0}}{(1-0.08)} = \frac{1487}{0.92} = 1616 \text{ kN/m}$$

As in Example 5.1, the maximum jacking force in a duct containing four 12.9 mm diameter strands is 4 × 0.9 $f_{p0.1k}A_p$ = 576 kN and so the required maximum duct spacing is 356 mm.

Using a four-strand tendon every 350 mm, the jacking force per tendon is:

1616 × 0.35 = 566 kN

In Example 5.1, the tensile stress limit under full loads was $f_{ct,t}$ = 1.75 MPa. In this example, the prestress is significantly lower and, therefore, $f_{ct,t}$ will be exceeded. Assuming gross cross-sectional properties, the bottom fibre stress at mid-span after all losses and under the full service loads, i.e. when M_T = 189.4 kNm/m (see Example 5.1), is:

$$\sigma_{btm} = -\frac{P_{m,t}}{A} - \frac{P_{m,t}e}{Z_{btm}} + \frac{M_T}{Z_{btm}}$$

$$= -\frac{1264 \times 10^3}{230 \times 10^3} - \frac{1264 \times 10^3 \times 78.6}{8.817 \times 10^6} + \frac{189.4 \times 10^6}{8.817 \times 10^6} = 4.72 \text{ MPa}$$

and this will almost certainly cause cracking. The resulting loss of stiffness must be included in subsequent deflection calculations using the procedures outlined in Sections 5.11.3 and 5.11.4. In addition, the smaller quantity of prestressing steel required in this example, in comparison with the slab in Example 5.1, will result in reduced flexural strength. A layer of non-prestressed bottom reinforcement may be required to satisfy strength requirements.

5.5 CABLE PROFILES

When the prestressing force and eccentricity are determined at the critical sections, the location of the cable at every section along the member must be specified. For a member that has been designed using concrete stress limits, the tendons may be located so that the stress limits are satisfied on every cross-section. Equations 5.3 through 5.6 may be used to establish a range of values for eccentricity at any particular cross-section that satisfies the selected stress limits.

At any cross-section, if M_o and M_T are the moments caused by the external loads at transfer and under full service loads, respectively, and P_{m0} and $P_{m,t}$ are the prestressing forces before and after time-dependent losses at the same section, the extreme fibre stresses must satisfy the following:

$$-\frac{P_{m0}}{A} + \frac{(P_{m0}e - M_0)}{Z_{top}} \leq f_{ct,0} \tag{5.10}$$

$$-\frac{P_{m0}}{A} - \frac{(P_{m0}e - M_0)}{Z_{btm}} \geq f_{cc,0} \tag{5.11}$$

$$-\frac{P_{m,t}}{A} - \frac{(P_{m,t}e - M_T)}{Z_{btm}} \leq f_{ct,t} \tag{5.12}$$

$$-\frac{P_{m,t}}{A} + \frac{(P_{m,t}e - M_T)}{Z_{top}} \geq f_{cc,t} \tag{5.13}$$

Equations 5.10 through 5.13 are equivalent to Equations 5.3 through 5.6 and can be rearranged to provide limits on the tendon eccentricity, as follows:

$$e \leq \frac{M_0}{P_{m0}} + \frac{Z_{top}f_{ct,0}}{P_{m0}} + \frac{1}{\alpha_{top}} \tag{5.14}$$

$$e \le \frac{M_0}{P_{m0}} - \frac{Z_{btm}f_{cc,0}}{P_{m0}} - \frac{1}{\alpha_{btm}} \tag{5.15}$$

$$e \ge \frac{M_T}{P_{m,t}} - \frac{Z_{btm}f_{ct,t}}{P_{m,t}} - \frac{1}{\alpha_{btm}} \tag{5.16}$$

$$e \ge \frac{M_T}{P_{m,t}} + \frac{Z_{top}f_{cc,t}}{P_{m,t}} + \frac{1}{\alpha_{top}} \tag{5.17}$$

It should be remembered that $f_{cc,0}$ and $f_{cc,t}$ are negative numbers and that $\alpha_{top} = A/Z_{top}$ and $\alpha_{btm} = A/Z_{btm}$.

After P_{m0} and $P_{m,t}$ have been determined at the critical sections, the friction, draw-in and time-dependent losses along the member are estimated (see Sections 5.10.2 and 5.10.3), and the corresponding prestressing forces at intermediate sections are calculated. At each intermediate section, the maximum eccentricity that will satisfy both stress limits at transfer is obtained from either Equation 5.14 or 5.15. The minimum eccentricity required to satisfy the tensile and compressive stress limits under full loads is obtained from either Equation 5.16 or 5.17. A region of the member is thus established in which the line of action of the resulting prestressing force should be located. Such a permissible region is shown in Figure 5.3. Relatively few intermediate sections need to be considered to determine an acceptable cable profile.

When the prestress and eccentricity at the critical sections are selected using the load-balancing approach, the cable profile should match, as closely as practicable, the bending moment diagram caused by the balanced load. For cracked, partially prestressed members, Equations 5.14 and 5.15 are usually applicable and allow the maximum eccentricity to be defined. The cable profile should then be selected according to the loading type and the bending moment diagram.

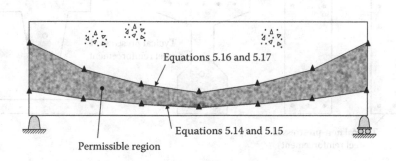

Figure 5.3 Typical permissible region for determination of cable profile.

5.6 SHORT-TERM ANALYSIS OF UNCRACKED CROSS-SECTIONS

5.6.1 General

The short-term behaviour of an uncracked prestressed concrete cross-section can be determined by transforming the bonded reinforcement into equivalent areas of concrete and performing an elastic analysis on the equivalent concrete section. The concrete is assumed to be linear-elastic in both tension and compression, and so too is the non-prestressed reinforcement and the prestressing tendons. The following mathematical formulation of the short-term analysis of an uncracked cross-section forms the basis of the time-dependent analysis described in Section 5.7 and was described by Gilbert and Ranzi [6]. The procedure can be applied to cross-sections with a vertical axis of symmetry, such as those shown in Figure 5.4.

The contribution of each reinforcing bar or bonded prestressing tendon is included in the calculations according to its location within the cross-section, as shown in Figure 5.5. The numbers of *layers* of non-prestressed and prestressed reinforcement are m_s and m_p, respectively. In Figure 5.5b, $m_s = 3$ and $m_p = 2$. The properties of each layer of non-prestressed reinforcement are defined by its area, elastic modulus and location with respect to the arbitrarily chosen x-axis and are labelled as $A_{s(i)}$, $E_{s(i)}$ and $y_{s(i)}$, respectively, where $i = 1, ..., m_s$. Similarly, $A_{p(i)}$, $E_{p(i)}$ and $y_{p(i)}$ represent, respectively, the area, elastic modulus and location of the prestressing steel with respect to the x-axis and $i = 1, ..., m_p$.

The geometric properties of the concrete part of the cross-section are A_c, B_c and I_c, where A_c is the concrete area, B_c is the first moment of area of the concrete about the x-axis and I_c is the second moment of area of the concrete about the x-axis. In this analysis, the orientation of the x- and y-axes is as shown in Figure 5.5.

If the cross-section is subjected to an axial force N_{ext} applied at the origin of the x- and y-axes and a bending moment M_{ext} applied about the x-axis,

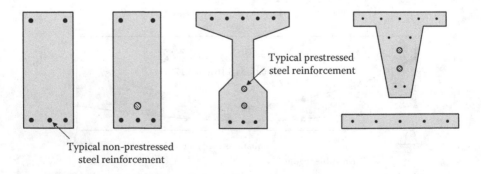

Figure 5.4 Typical reinforced and prestressed concrete sections.

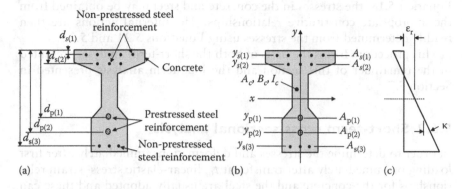

Figure 5.5 Generic cross-section, arrangement of reinforcement and strain. (a) Cross-section. (b) Reinforcement bar and tendon areas ($A_{s(i)}$ and $A_{p(i)}$) located $y_{s(i)}$ and $y_{p(i)}$ from the reference axis (x-axis). (c) Strain.

the strain diagram is as shown in Figure 5.5c and the strain at any depth y above the reference axis is given by:

$$\varepsilon = \varepsilon_r - y\kappa \tag{5.18}$$

The two unknowns in the problem, i.e. ε_r and κ, are then determined by enforcing horizontal and rotational equilibrium at the cross-section:

$$N_{int} = N_{ext} \tag{5.19}$$

and

$$M_{int} = M_{ext} \tag{5.20}$$

where N_{int} and M_{int} are the internal axial force and moment, respectively, given by:

$$N_{int} = \int_A \sigma \, dA \tag{5.21}$$

and

$$M_{int} = \int_A -y\sigma \, dA \tag{5.22}$$

When the two unknowns (ε_r and κ) are calculated from the two equilibrium equations (Equations 5.19 and 5.20) and the strain is determined using

Equation 5.18, the stresses in the concrete and steel may be obtained from the appropriate constitutive relationships. The internal actions are then readily determined from the stresses using Equations 5.21 and 5.22.

This procedure forms the basis of both the short-term analysis presented in the remainder of this section and the long-term analysis presented in Section 5.7.

5.6.2 Short-term cross-sectional analysis

In order to determine the stresses and deformations immediately after first loading or immediately after transfer at t_0, linear-elastic stress–strain relationships for the concrete and the steel are usually adopted and these can be expressed by:

$$\sigma_{c,0} = E_{cm,0}\varepsilon_0 \tag{5.23}$$

$$\sigma_{s(i),0} = E_{s(i)}\varepsilon_{s(i),0} \tag{5.24}$$

$$\sigma_{p(i),0} = E_{p(i)}(\varepsilon_{cp(i),0} + \varepsilon_{p(i),\text{init}}) \quad \text{if } A_{p(i)} \text{ is bonded} \tag{5.25}$$

$$\sigma_{p(i),0} = E_{p(i)}\varepsilon_{p(i),\text{init}} \quad \text{if } A_{p(i)} \text{ is unbonded} \tag{5.26}$$

in which $\sigma_{c,0}$, $\sigma_{s(i),0}$ and $\sigma_{p(i),0}$ represent the stresses in the concrete, in the i-th layer of non-prestressed reinforcement (with $i = 1, ..., m_s$) and in the i-th layer of prestressing steel (with $i = 1, ..., m_p$), respectively, immediately after first loading at time t_0, while $\varepsilon_{cp(i),0}$ is the strain in the concrete at the level of the i-th layer of prestressing steel at time t_0 and $\varepsilon_{p(i),\text{init}}$ is the initial strain in the i-th layer of prestressing steel produced by the initial tensile prestressing force $P_{\text{init}(i)}$ and is given by:

$$\varepsilon_{p(i),\text{init}} = \frac{P_{\text{init}(i)}}{A_{p(i)}E_{p(i)}} \tag{5.27}$$

For a post-tensioned cross-section, $P_{\text{init}(i)}$ is the prestressing force immediately after stressing the tendon in the i-th layer of prestressing steel and, for a pretensioned member, $P_{\text{init}(i)}$ is the prestressing force immediately before transfer. With this method, the prestressing force is included in the analysis by means of an induced strain $\varepsilon_{p(i),\text{init}}$, rather than an external action [7,8].

For unbonded tendons, Equation 5.26 is strictly only applicable immediately before transfer. After transfer, as the beam deforms under load, the strain in the tendon will increase. However, at service loads, the change in geometry of the member is relatively small and the resulting change in the tendon strain is usually less than 0.5% of $\varepsilon_{p(i),\text{init}}$ and can be ignored.

At time t_0, the internal axial force resisted by the cross-section and the internal moment about the reference axis are denoted $N_{\text{int},0}$ and $M_{\text{int},0}$, respectively. The internal axial force $N_{\text{int},0}$ is the sum of the axial forces resisted by the component materials forming the cross-section and is given by:

$$N_{\text{int},0} = N_{c,0} + N_{s,0} + N_{p,0} \tag{5.28}$$

where $N_{c,0}$, $N_{s,0}$ and $N_{p,0}$ represent the axial forces resisted by the concrete, the non-prestressed reinforcement and the prestressing steel, respectively. The axial force resisted by the concrete is calculated from:

$$N_{c,0} = \int_{A_c} \sigma_{c,0} \, dA = \int_{A_c} E_{cm,0}\varepsilon_0 \, dA$$

$$= \int_{A_c} E_{cm,0}(\varepsilon_{r,0} - y\kappa_0)\,dA = A_c E_{cm,0}\ \varepsilon_{r,0} - B_c E_{cm,0}\ \kappa_0 \tag{5.29}$$

The axial force resisted by the non-prestressed steel is:

$$N_{s,0} = \sum_{i=1}^{m_s}(A_{s(i)}E_{s(i)})(\varepsilon_{r,0} - y_{s(i)}\kappa_0) = \sum_{i=1}^{m_s}(A_{s(i)}E_{s(i)})\varepsilon_{r,0} - \sum_{i=1}^{m_s}(y_{s(i)}A_{s(i)}E_{s(i)})\kappa_0$$

$$\tag{5.30}$$

and the axial force resisted by the prestressing steel, if bonded, is given in Equation 5.31 and, if unbonded, may be approximated by Equation 5.32:

$$N_{p,0} = \sum_{i=1}^{m_p}(A_{p(i)}E_{p(i)})\varepsilon_{r,0} - \sum_{i=1}^{m_p}(y_{p(i)}A_{p(i)}E_{p(i)})\kappa_0 + \sum_{i=1}^{m_p}(A_{p(i)}E_{p(i)}\varepsilon_{p(i),\text{init}}) \tag{5.31}$$

$$N_{p,0} = \sum_{i=1}^{m_p}(A_{p(i)}E_{p(i)}\varepsilon_{p(i),\text{init}}) \tag{5.32}$$

The additional subscripts '0' used for the strain at the level of the reference axis ($\varepsilon_{r,0}$) and the curvature (κ_0) highlight that these are calculated at time t_0 immediately after the application of $N_{\text{ext},0}$ and $M_{\text{ext},0}$ and after the transfer of prestress.

By substituting Equations 5.29 through 5.32 into Equation 5.28, the equation for $N_{int,0}$ is expressed in terms of the actual geometry and elastic moduli of the materials forming the cross-section.

When the prestressing steel is bonded to the concrete:

$$N_{int,0} = \left(A_c E_{cm,0} + \sum_{i=1}^{m_s} A_{s(i)} E_{s(i)} + \sum_{i=1}^{m_p} A_{p(i)} E_{p(i)} \right) \varepsilon_{r,0}$$

$$- \left(B_c E_{cm,0} + \sum_{i=1}^{m_s} y_{s(i)} A_{s(i)} E_{s(i)} + \sum_{i=1}^{m_p} y_{p(i)} A_{p(i)} E_{p(i)} \right) \kappa_0 + \sum_{i=1}^{m_p} (A_{p(i)} E_{p(i)} \varepsilon_{p(i),init})$$

$$= R_{A,0} \varepsilon_{r,0} - R_{B,0} \kappa_0 + \sum_{i=1}^{m_p} (A_{p(i)} E_{p(i)} \varepsilon_{p(i),init}) \tag{5.33}$$

When the prestressing steel is unbonded:

$$N_{int,0} = \left(A_c E_{cm,0} + \sum_{i=1}^{m_s} A_{s(i)} E_{s(i)} \right) \varepsilon_{r,0} - \left(B_c E_{cm,0} + \sum_{i=1}^{m_s} y_{s(i)} A_{s(i)} E_{s(i)} \right) \kappa_0$$

$$+ \sum_{i=1}^{m_p} \left(A_{p(i)} E_{p(i)} \varepsilon_{p(i),init} \right)$$

$$= R_{A,0} \varepsilon_{r,0} - R_{B,0} \kappa_0 + \sum_{i=1}^{m_p} (A_{p(i)} E_{p(i)} \varepsilon_{p(i),init}) \tag{5.34}$$

In Equations 5.33 and 5.34, $R_{A,0}$ and $R_{B,0}$ represent, respectively, the axial rigidity and the stiffness related to the first moment of area about the reference axis calculated at time t_0 and, for a cross-section containing bonded tendons, are given by:

$$R_{A,0} = A_c E_{cm,0} + \sum_{i=1}^{m_s} A_{s(i)} E_{s(i)} + \sum_{i=1}^{m_p} A_{p(i)} E_{p(i)} \tag{5.35}$$

$$R_{B,0} = B_c E_{cm,0} + \sum_{i=1}^{m_s} y_{s(i)} A_{s(i)} E_{s(i)} + \sum_{i=1}^{m_p} y_{p(i)} A_{p(i)} E_{p(i)} \tag{5.36}$$

For unbonded construction, the contribution of the prestressing steel $A_{p(i)}$ to the rigidities $R_{A,0}$ and $R_{B,0}$ is ignored.

Similarly, the equation for $M_{int,0}$ may be expressed by Equations 5.37 and 5.38. When the prestressing steel is bonded to the concrete:

$$M_{int,0} = -\left(B_c E_{cm,0} + \sum_{i=1}^{m_s} y_{s(i)} A_{s(i)} E_{s(i)} + \sum_{i=1}^{m_p} y_{p(i)} A_{p(i)} E_{p(i)} \right) \varepsilon_{r,0}$$

$$+ \left(I_c E_{cm,0} + \sum_{i=1}^{m_s} y_{s(i)}^2 A_{s(i)} E_{s(i)} + \sum_{i=1}^{m_p} y_{p(i)}^2 A_{p(i)} E_{p(i)} \right) \kappa_0 - \sum_{i=1}^{m_p} \left(y_{p(i)} A_{p(i)} E_{p(i)} \varepsilon_{p(i),init} \right)$$

$$= -R_{B,0} \varepsilon_{r,0} + R_{I,0} \kappa_0 - \sum_{i=1}^{m_p} \left(y_{p(i)} A_{p(i)} E_{p(i)} \varepsilon_{p(i),init} \right) \tag{5.37}$$

When the prestressing steel is not bonded to the concrete:

$$M_{int,0} = -\left(B_c E_{cm,0} + \sum_{i=1}^{m_s} y_{s(i)} A_{s(i)} E_{s(i)} \right) \varepsilon_{r,0}$$

$$+ \left(I_{cm} E_{cm,0} + \sum_{i=1}^{m_s} y_{s(i)}^2 A_{s(i)} E_{s(i)} \right) \kappa_0 - \sum_{i=1}^{m_p} \left(y_{p(i)} A_{p(i)} E_{p(i)} \varepsilon_{p(i),init} \right)$$

$$= -R_{B,0} \varepsilon_{r,0} + R_{I,0} \kappa_0 - \sum_{i=1}^{m_p} \left(y_{p(i)} A_{p(i)} E_{p(i)} \varepsilon_{p(i),init} \right) \tag{5.38}$$

in which $R_{I,0}$ is the flexural rigidity at time t_0 and, for a cross-section containing bonded tendons, is given by:

$$R_{I,0} = I_c E_{cm,0} + \sum_{i=1}^{m_s} y_{s(i)}^2 A_{s(i)} E_{s(i)} + \sum_{i=1}^{m_p} y_{p(i)}^2 A_{p(i)} E_{p(i)} \tag{5.39}$$

For unbonded construction, the contribution of the prestressing steel $A_{p(i)}$ to the flexural rigidity $R_{I,0}$ is ignored.

Substituting the expressions for $N_{int,0}$ and $M_{int,0}$ (Equation 5.33 or 5.34 and Equation 5.37 or 5.38) into Equations 5.19 and 5.20 produces a system of equilibrium equations that may be written in compact form as follows:

$$r_{ext,0} = D_0 \varepsilon_0 + f_{p,init} \tag{5.40}$$

where:

$$r_{ext,0} = \begin{bmatrix} N_{ext,0} \\ M_{ext,0} \end{bmatrix} \tag{5.41}$$

$$\mathbf{D}_0 = \begin{bmatrix} R_{A,0} & -R_{B,0} \\ -R_{B,0} & R_{I,0} \end{bmatrix} \tag{5.42}$$

$$\boldsymbol{\varepsilon}_0 = \begin{bmatrix} \varepsilon_{r,0} \\ \kappa_0 \end{bmatrix} \tag{5.43}$$

$$\mathbf{f}_{p,init} = \sum_{i=1}^{m_p} \begin{bmatrix} A_{p(i)}E_{p(i)}\varepsilon_{p(i),init} \\ -y_{p(i)}A_{p(i)}E_{p(i)}\varepsilon_{p(i),init} \end{bmatrix} = \sum_{i=1}^{m_p} \begin{bmatrix} P_{init(i)} \\ -y_{p(i)}P_{init(i)} \end{bmatrix} \tag{5.44}$$

The vector $\mathbf{r}_{ext,0}$ is the vector of the external actions at first loading (at time t_0), i.e. axial force $N_{ext,0}$ and moment $M_{ext,0}$; the matrix $\mathbf{D}_0$ contains the cross-sectional material and geometric properties calculated at t_0; the strain vector $\boldsymbol{\varepsilon}_0$ contains the unknown independent variables describing the strain diagram at time t_0 ($\varepsilon_{r,0}$ and κ_0); the vector $\mathbf{f}_{p,init}$ contains the actions caused by the initial prestressing.

The vector $\boldsymbol{\varepsilon}_0$ is readily obtained by solving the equilibrium equations (Equation 5.40):

$$\boldsymbol{\varepsilon}_0 = \mathbf{D}_0^{-1}(\mathbf{r}_{ext,0} - \mathbf{f}_{p,init}) = \mathbf{F}_0(\mathbf{r}_{ext,0} - \mathbf{f}_{p,init}) \tag{5.45}$$

where:

$$\mathbf{F}_0 = \frac{1}{R_{A,0}R_{I,0} - R_{B,0}^2} \begin{bmatrix} R_{I,0} & R_{B,0} \\ R_{B,0} & R_{A,0} \end{bmatrix} \tag{5.46}$$

The stress distribution related to the concrete and reinforcement can then be calculated from the constitutive equations (Equations 5.23 through 5.26) re-expressed here as:

$$\sigma_{c,0} = E_{cm,0}\varepsilon_0 = E_{cm,0}[1-y]\boldsymbol{\varepsilon}_0 \tag{5.47}$$

$$\sigma_{s(i),0} = E_{s(i)}\varepsilon_{s(i),0} = E_{s(i)}[1-y_{s(i)}]\boldsymbol{\varepsilon}_0 \tag{5.48}$$

If $A_{p(i)}$ is bonded:

$$\sigma_{p(i),0} = E_{p(i)}(\varepsilon_{cp(i),0} + \varepsilon_{p(i),init})$$

$$= E_{p(i)}[1-y_{p(i)}]\boldsymbol{\varepsilon}_0 + E_{p(i)}\varepsilon_{p(i),init} \tag{5.49}$$

If $A_{p(i)}$ is unbonded:

$$\sigma_{p(i),0} = E_{p(i)}\varepsilon_{p(i),init} \tag{5.50}$$

where $\varepsilon_0 = \varepsilon_{r,0} - y\kappa_0 = [1-y]\boldsymbol{\varepsilon}_0$.

Although this procedure is presented here assuming linear-elastic material properties, it is quite general and also applicable to non-linear material representations, in which case the integrals of Equations 5.21 and 5.22 might have to be evaluated numerically. However, when calculating the short-term response of uncracked reinforced and prestressed concrete cross-sections under typical in-service loads, material behaviour is essentially linear-elastic.

In reinforced and prestressed concrete design, it is common to calculate the cross-sectional properties by transforming the section into equivalent areas of one of the constituent materials. For example, for the cross-section of Figure 5.5a, the transformed concrete cross-section for the short-term analysis is shown in Figure 5.6, with the area of each layer of bonded steel reinforcement and tendons ($A_{s(i)}$ and $A_{p(i)}$, respectively) transformed into equivalent areas of concrete ($\alpha_{s(i),0} A_{s(i)}$ and $\alpha_{p(i),0} A_{p(i)}$, respectively), where $\alpha_{s(i),0} = E_{s(i)}/E_{cm,0}$ is the modular ratio of the ith layer of non-prestressed steel and $\alpha_{p(i),0} = E_{p(i)}/E_{cm,0}$ is the modular ratio of the ith layer of prestressing steel.

For the transformed section of Figure 5.6, the cross-sectional rigidities defined in Equations 5.35, 5.36 and 5.39 can be recalculated as:

$$R_{A,0} = A_0 E_{cm,0} \tag{5.51}$$

$$R_{B,0} = B_0 E_{cm,0} \tag{5.52}$$

$$R_{I,0} = I_0 E_{cm,0} \tag{5.53}$$

where A_0 is the area of the transformed concrete section, and B_0 and I_0 are the first and second moments of the transformed area about the reference x-axis at first loading.

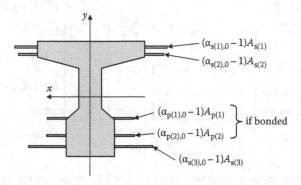

Figure 5.6 Transformed section with bonded reinforcement transformed into equivalent areas of concrete.

Substituting Equations 5.51 through 5.53 into Equation 5.46 enables $\mathbf{F_0}$ to be expressed in terms of the properties of the transformed concrete section as

$$\mathbf{F_0} = \frac{1}{E_{cm,0}\left(A_0 I_0 - B_0^2\right)}\begin{bmatrix} I_0 & B_0 \\ B_0 & A_0 \end{bmatrix} \tag{5.54}$$

The two approaches proposed for the calculation of the cross-sectional rigidities, i.e. the one based on Equations 5.35, 5.36 and 5.39 and the other using the properties of the transformed section (Equations 5.51 through 5.53), are equivalent. The procedure based on the transformed section (Equations 5.51 through 5.53) is often preferred for the analysis of reinforced and prestressed concrete sections. The use of both approaches is illustrated in the following example.

EXAMPLE 5.3

The short-term behaviour of the post-tensioned beam cross-section shown in Figure 5.7a is to be determined immediately after transfer (at time t_0). The section contains a single unbonded cable, containing ten 12.9 mm diameter strands (from Table 4.8, $A_p = 1000$ mm^2 and $f_{pb} = 1870$ MPa) located within a 60 mm diameter duct, and two layers of non-prestressed reinforcement. The force in the prestressing steel is $P_{m0} = P_{init} = 1350$ kN, and the external moment acting on the cross-section at transfer is $M_o = 100$ kNm ($=M_{ext}$). The elastic moduli for concrete at transfer and the steel are $E_{cm,0} = 30,000$ MPa, $E_s = 200,000$ MPa and $E_p = 195,000$ MPa, from which $\alpha_s = E_s/E_{cm,0} = 6.67$ and $\alpha_p = E_p/E_{cm,0} = 6.5$.

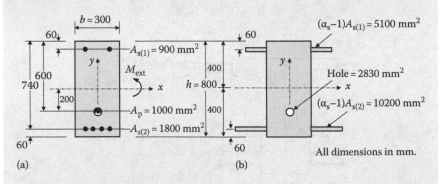

Figure 5.7 Post-tensioned cross-section (Example 5.3). (a) Section. (b) Transformed section.

In this example, the reference x-axis is taken at mid-depth. The transformed section is shown in Figure 5.7b. Because the prestressing steel is not bonded to the concrete, it does not form part of the transformed section. In addition, the hole created in the concrete section by the hollow duct must also be taken into account. The properties of the transformed section with respect to the reference x-axis are:

$$A_0 = bh + (\alpha_s - 1)A_{s(1)} + (\alpha_s - 1)A_{s(2)} - A_{hole}$$

$$= 300 \times 800 + (6.67 - 1) \times 900 + (6.67 - 1) \times 1800 - 2830$$

$$= 252,500 \text{ mm}^2$$

$$B_0 = bhy_c + (\alpha_s - 1)A_{s(1)}y_{s(1)} + (\alpha_s - 1)A_{s(2)}y_{s(2)} - A_{hole}y_{hole}$$

$$= 300 \times 800 \times 0 + (6.67 - 1) \times 900 \times (+340) + (6.67 - 1)$$

$$\times 1800 \times (-340) - 2830 \times (-200)$$

$$= -1.168 \times 10^6 \text{ mm}^3$$

$$I_0 = \frac{bh^3}{12} + bhy_c^2 + (\alpha_s - 1)A_{s(1)}y_{s(1)}^2 + (\alpha_s - 1)A_{s(2)}y_{s(2)}^2 - A_{hole}y_{hole}^2$$

$$= \frac{300 \times 800^3}{12} + 300 \times 800 \times 0 + (6.67 - 1) \times 900 \times 340^2 + (6.67 - 1)$$

$$\times 1800 \times (-340)^2 - 2830 \times (-200)^2$$

$$= 14,455 \times 10^6 \text{ mm}^4$$

From Equation 5.54:

$$\mathbf{F}_0 = \frac{1}{30,000 \times (252,500 \times 14,455 \times 10^6 - (-1.168 \times 10^6)^2)}$$

$$\times \begin{bmatrix} 14,455 \times 10^6 & -1.168 \times 10^6 \\ -1.168 \times 10^6 & 252,470 \end{bmatrix}$$

$$= \begin{bmatrix} 132.1 \times 10^{-12} & -10.67 \times 10^{-15} \\ -10.67 \times 10^{-15} & 2.307 \times 10^{-15} \end{bmatrix}$$

From Equation 5.41, the vector of internal actions is:

$$\mathbf{r}_{ext,0} = \begin{bmatrix} N_{ext,0} \\ M_{ext,0} \end{bmatrix} = \begin{bmatrix} 0 \\ 100 \times 10^6 \end{bmatrix}$$

and from Equation 5.27, the initial strain in the prestressing steel due to the initial prestressing force is:

$$\varepsilon_{p,init} = \frac{P_{init}}{A_p E_p} = \frac{1,350 \times 10^3}{1,000 \times 195,000} = 0.00692$$

The vector of internal actions caused by the initial prestress $\mathbf{f}_{p,init}$, containing the initial prestressing force ($P_{init} = A_p E_p \varepsilon_{p,init}$) and its moment about the reference x-axis ($-y_p P_{init} = -y_p A_p E_p \varepsilon_{p,init}$), is given by Equation 5.44:

$$\mathbf{f}_{p,init} = \begin{bmatrix} P_{init} \\ -y_p P_{init} \end{bmatrix} = \begin{bmatrix} 1350 \times 10^3 \\ -(-200) \times 1350 \times 10^3 \end{bmatrix} = \begin{bmatrix} 1350 \times 10^3 \\ 270 \times 10^6 \end{bmatrix}$$

where the dimension y_p is the distance from the x-axis to the centroid of the prestressing steel, i.e. $y_p = 200$ mm in this example. The strain vector ε_0 containing the unknown strain variables is determined from Equation 5.45:

$$\varepsilon_0 = \mathbf{F}_0 \ (\mathbf{r}_{ext,0} - \mathbf{f}_{p,init})$$

$$= \begin{bmatrix} 132.1 \times 10^{-12} & -10.67 \times 10^{-15} \\ -10.67 \times 10^{-15} & 2.307 \times 10^{-15} \end{bmatrix} \left(\begin{bmatrix} 0 \\ 100 \times 10^6 \end{bmatrix} - \begin{bmatrix} 1350 \times 10^3 \\ 270 \times 10^6 \end{bmatrix} \right)$$

$$= \begin{bmatrix} -176.5 \times 10^{-6} \\ -0.3778 \times 10^{-6} \end{bmatrix}$$

The strain at the reference axis and the curvature are therefore:

$$\varepsilon_{r,0} = -176.5 \times 10^{-6} \quad \text{and} \quad \kappa_0 = -0.3778 \times 10^{-6} \text{ mm}^{-1}$$

and, from Equation 5.18, the top fibre strain (at $y = +400$ mm) and the bottom fibre strain (at $y = -400$ mm) are:

$$\varepsilon_{0(top)} = \varepsilon_{r,0} - 400 \times \kappa_0 = [-176.5 - 400 \times (-0.3778)] \times 10^{-6}$$

$$= -25.4 \times 10^{-6}$$

$$\varepsilon_{0(btm)} = \varepsilon_{r,0} - (-400) \times \kappa_0 = [-176.5 + 400 \times (-0.3778)] \times 10^{-6}$$

$$= -327.6 \times 10^{-6}$$

The top and bottom fibre stresses in the concrete and the stresses in the two layers of reinforcement and in the prestressing steel are obtained from Equations 5.47 through 5.50:

$$\sigma_{c,0(top)} = E_{cm,0} [1 - y_{top}] \varepsilon_0 = E_{cm,0} \varepsilon_{0(top)} = 30,000 \times (-25.4 \times 10^{-6}) = -0.76 \text{ MPa}$$

$$\sigma_{c,0(btm)} = E_{cm,0} [1 - y_{btm}] \varepsilon_0 = E_{cm,0} \varepsilon_{0(btm)} = 30,000 \times (-327.6 \times 10^{-6}) = -9.83 \text{ MPa}$$

$$\sigma_{s(1),0} = E_s\,[1 - y_{s(1)}]\,\varepsilon_0 = 200{,}000 \times [1 - 340]\begin{bmatrix} -176.5 \times 10^{-6} \\ -0.3778 \times 10^{-6} \end{bmatrix} = -9.61\,\text{MPa}$$

$$\sigma_{s(2),0} = E_s\,[1 - y_{s(2)}]\,\varepsilon_0 = 200{,}000 \times [1 + 340]\begin{bmatrix} -176.5 \times 10^{-6} \\ -0.3778 \times 10^{-6} \end{bmatrix} = -61.0\,\text{MPa}$$

$$\sigma_{p,0} = E_p\varepsilon_{p,\text{init}} = 195{,}000 \times 0.00692 = +1350\,\text{MPa}$$

The distributions of strain and stress on the cross-section immediately after transfer are shown in Figure 5.8.

Alternatively, instead of analysing the transformed cross-section, the cross-sectional rigidities could have been calculated using Equations 5.35, 5.36 and 5.39. The properties of the concrete part of the cross-section (with respect to the x-axis) are:

$$A_c = bh - A_{s(1)} - A_{s(2)} - A_{\text{hole}}$$
$$= 300 \times 800 - 900 - 1800 - 2830 = 234{,}470\,\text{mm}^2$$

$$B_c = bhy_c - A_{s(1)}y_{s(1)} - A_{s(2)}y_{s(2)} - A_{\text{hole}}y_{\text{hole}}$$
$$= 300 \times 800 \times 0 - 900 \times 340 - 1800 \times (-340) - 2830 \times (-200)$$
$$= 872{,}000\,\text{mm}^3$$

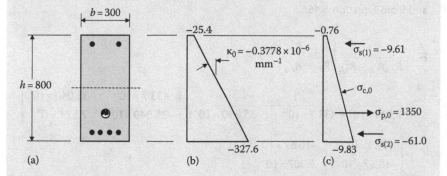

Figure 5.8 Strains and stresses immediately after transfer (Example 5.3). (a) Section. (b) Strain (×10⁻⁶). (c) Stress (MPa).

$$I_c = bh^3/12 + bhy_c^2 - A_{s(1)}y_{s(1)}^2 - A_{s(2)}y_{s(2)}^2 - A_{hole}y_{hole}^2$$

$$= 300 \times 800^3/12 + 300 \times 800 \times 0 - 900 \times 340^2 - 1800 \times (-340)^2$$
$$- 2830 \times (-200)^2$$

$$= 12,375 \times 10^6 \text{ mm}^4$$

For this member with unbonded tendons, the cross-sectional rigidities are given by:

$$R_{A,0} = A_c E_{cm,0} + \sum_{i=1}^{2} A_{s(i)} E_{s(i)}$$

$$= 234,470 \times 30,000 + (900 + 1800) \times 200,000 = 7574 \times 10^6 \text{ N}$$

$$R_{B,0} = B_c E_{cm,0} + \sum_{i=1}^{2} y_{s(i)} A_{s(i)} E_{s(i)}$$

$$= 872,000 \times 30,000 + [340 \times 900 + (-340) \times 1800] \times 200,000$$

$$= -35,040 \times 10^6 \text{ Nmm}$$

$$R_{I,0} = I_c E_{cm,0} + \sum_{i=1}^{2} y_{s(i)}^2 A_{s(i)} E_{s(i)}$$

$$= 12,375 \times 10^6 \times 30,000 + [340^2 \times 900 + (-340)^2 \times 1800] \times 200,000)$$

$$= 433.7 \times 10^{12} \text{ Nmm}^2$$

and from Equation 5.46:

$$\mathbf{F}_0 = \frac{1}{R_{A,0}R_{I,0} - R_{B,0}^2} \begin{bmatrix} R_{I,0} & R_{B,0} \\ R_{B,0} & R_{A,0} \end{bmatrix}$$

$$= \frac{1}{7,574 \times 10^6 \times 433.7 \times 10^{12} - (-35,040 \times 10^6)^2} \begin{bmatrix} 433.7 \times 10^{12} & -35,040 \times 10^6 \\ -35,040 \times 10^6 & 7,574 \times 0^6 \end{bmatrix}$$

$$= \begin{bmatrix} 132.1 \times 10^{-12} & -10.67 \times 10^{-15} \\ -10.67 \times 10^{-15} & 2.307 \times 10^{-15} \end{bmatrix}$$

This is identical to the matrix $\mathbf{F}_0$ obtained earlier from Equation 5.54.

EXAMPLE 5.4

Determine the instantaneous stress and strain distributions on the precast pretensioned concrete section shown in Figure 5.9. The area and the second moment of area of the gross cross-section about the centroidal axis are $A_{gross} = 317 \times 10^3$ mm^2 and $I_{gross} = 49,900 \times 10^6$ mm^4. The centroid of the gross cross-sectional area is located 602 mm below its top fibre, i.e. $d_c = 602$ mm.

The section is subjected to a compressive axial force $N_{ext,0} = -100$ kN and a sagging moment of $M_{ext,0} = +1000$ kNm applied with respect to the reference x-axis, that is taken in this example to be 300 mm below the top fibre of the cross-section.

Assume that all materials are linear-elastic with $E_{cm,0} = 32,000$ MPa, $E_s = 200,000$ MPa and $E_p = 195,000$ MPa. The modular ratios of the reinforcing steel and the prestressing steel are therefore $\alpha_{s(i),0} = 6.25$ and $\alpha_{p(i),0} = 6.09$. The prestressing forces in each of the three layers of tendons ($A_{p(1)} = 300$ mm^2, $A_{p(2)} = 500$ mm^2 and $A_{p(3)} = 800$ mm^2, respectively) just before the transfer of prestress are $P_{init(1)} = 375$ kN, $P_{init(2)} = 625$ kN and $P_{init(3)} = 1000$ kN.

The distances of the steel layers from the reference axis are $y_{s(1)} = +240$ mm, $y_{s(2)} = -790$ mm, $y_{p(1)} = -580$ mm, $y_{p(2)} = -645$ mm and $y_{p(3)} = -710$ mm. From Equation 5.27, the initial strains in the prestressing steel prior to the transfer of prestress to the concrete are:

$$\varepsilon_{p(1),init} = \frac{P_{init(1)}}{A_{p(1)}E_p} = \frac{375 \times 10^3}{300 \times 195,000} = 0.00641$$

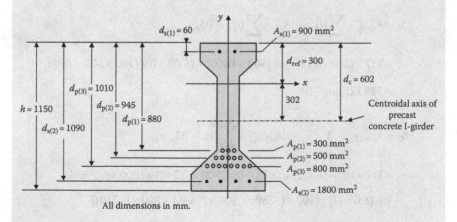

Figure 5.9 Precast prestressed concrete section (Example 5.4).

$$\varepsilon_{p(2),init} = \frac{P_{init(2)}}{A_{p(2)}E_p} = \frac{625 \times 10^3}{500 \times 195,000} = 0.00641$$

$$\varepsilon_{p(3),init} = \frac{P_{init(3)}}{A_{p(3)}E_p} = \frac{1,000 \times 10^3}{800 \times 195,000} = 0.00641$$

From Equation 5.41, the vector of internal actions at first loading is:

$$\mathbf{r}_{ext,0} = \begin{bmatrix} N_{ext,0} \\ M_{ext,0} \end{bmatrix} = \begin{bmatrix} -100 \times 10^3 \text{ N} \\ 1000 \times 10^6 \text{ Nmm} \end{bmatrix}$$

and, from Equation 5.44, the vector of initial prestressing forces is:

$$\mathbf{f}_{p,init} = \sum_{i=1}^{m_p} \begin{bmatrix} P_{init(i)} \\ -y_{p(i)}P_{init(i)} \end{bmatrix} = \begin{bmatrix} 375 \times 10^3 \\ -(-580) \times 375 \times 10^3 \end{bmatrix} + \begin{bmatrix} 625 \times 10^3 \\ -(-645) \times 625 \times 10^3 \end{bmatrix}$$

$$+ \begin{bmatrix} 1000 \times 10^3 \\ -(-710) \times 1000 \times 10^3 \end{bmatrix} = \begin{bmatrix} 2000 \times 10^3 \text{ N} \\ 1330.6 \times 10^6 \text{ Nmm} \end{bmatrix}$$

With the centroid of the concrete cross-section 302 mm below the reference axis (i.e. $y_c = -302$ mm), the properties of the transformed section (with the steel transformed into equivalent areas of concrete) with respect to the reference x-axis are:

$$A_0 = A_{gross} + \sum_{i=1}^{2}(\alpha_s - 1)A_{s(i)} + \sum_{i=1}^{3}(\alpha_p - 1)A_{p(i)}$$

$$= 317 \times 10^3 + (6.25 - 1) \times (900 + 1800) + (6.09 - 1) \times (300 + 500 + 800)$$

$$= 339,325 \text{ mm}^2$$

$$B_0 = A_{gross}y_c + \sum_{i=1}^{2}(\alpha_s - 1)A_{s(i)}y_{s(i)} + \sum_{i=1}^{3}(\alpha_s - 1)A_{p(i)}y_{p(i)}$$

$$= 317 \times 10^3 \times (-302) + (6.25 - 1) \times [900 \times (+240) + 1800 \times (-790)]$$

$$+ (6.09 - 1) \times [300 \times (-580) + 500 \times (-645) + 800 \times (-710)]$$

$$= -107.49 \times 10^6 \text{ mm}^3$$

$$I_0 = I_{gross} + A_{gross}y_c^2 + \sum_{i=1}^{2}(\alpha_s - 1)A_{s(i)}y_{s(i)}^2 + \sum_{i=1}^{3}(\alpha_p - 1)A_{p(i)}y_{p(i)}^2$$

$$= 49,900 \times 10^6 + 317 \times 10^3 \times (-302)^2 + (6.25 - 1)$$

$$\times [900 \times (+240)^2 + 1800 \times (-790)^2] + (6.09 - 1)$$

$$\times [300 \times (-580)^2 + 500 \times (-645)^2 + 800 \times (-710)^2]$$

$$= 88,609 \times 10^6 \text{ mm}^4$$

From Equation 5.54:

$$\mathbf{F}_0 = \frac{1}{32,000 \times (339,325 \times 88,609 \times 10^6 - (-107.49 \times 10^6)^2)}$$

$$\times \begin{bmatrix} 88,609 \times 10^6 & -107.49 \times 10^6 \\ -107.49 \times 10^6 & 339,325 \end{bmatrix}$$

$$= \begin{bmatrix} 149.6 \times 10^{-12} & -181.4 \times 10^{-15} \\ -181.4 \times 10^{-15} & 572.8 \times 10^{-18} \end{bmatrix}$$

and the strain vector ε_0 containing the unknown strain variables is determined from Equation 5.45:

$$\varepsilon_0 = \mathbf{F}_0(\mathbf{r}_{ext,0} - \mathbf{f}_{p,init}) = \begin{bmatrix} 149.6 \times 10^{-12} & -181.4 \times 10^{-15} \\ -181.4 \times 10^{-15} & 572.8 \times 10^{-18} \end{bmatrix}$$

$$\times \left(\begin{bmatrix} -100 \times 10^3 \\ 1000 \times 10^6 \end{bmatrix} - \begin{bmatrix} 2000 \times 10^3 \\ 1330.6 \times 10^6 \end{bmatrix} \right)$$

$$= \begin{bmatrix} -254.1 \times 10^{-6} \\ +0.1916 \times 10^{-6} \end{bmatrix}$$

The strain at the reference axis and the curvature are therefore:

$$\varepsilon_{r,0} = -254.1 \times 10^{-6} \quad \text{and} \quad \kappa_0 = +0.1916 \times 10^{-6} \text{ mm}^{-1}$$

and, from Equation 5.18, the top fibre strain (at $y = +300$ mm) and the bottom fibre strain (at $y = -850$ mm) are:

$$\varepsilon_{0(top)} = \varepsilon_{r,0} - 300 \times \kappa_0 = (-254.1 - 300 \times 0.1916) \times 10^{-6} = -311.6 \times 10^{-6}$$

$$\varepsilon_{0(btm)} = \varepsilon_{r,0} - (-850) \times \kappa_0 = (-254.1 + 850 \times 0.1916) \times 10^{-6} = -91.2 \times 10^{-6}$$

The top and bottom fibre stresses in the concrete and the stresses in the two layers of reinforcement and in the prestressing steel are obtained from Equations 5.47 through 5.49:

$$\sigma_{c,0(top)} = E_{cm,0}\varepsilon_{0(top)} = 32,000 \times (-311.6 \times 10^{-6}) = -9.97 \text{ MPa}$$

$$\sigma_{c,0(btm)} = E_{cm,0}\varepsilon_{0(btm)} = 32,000 \times (-91.2 \times 10^{-6}) = -2.92 \text{ MPa}$$

$$\sigma_{s(1),0} = E_s[1 - y_{s(1)}]\varepsilon_0 = 200,000 \times [1 - 240]\begin{bmatrix} -254.1 \times 10^{-6} \\ +0.1916 \times 10^{-6} \end{bmatrix} = -60.0 \text{ MPa}$$

$$\sigma_{s(2),0} = E_s[1 - y_{s(2)}]\varepsilon_0 = 200,000 \times [1 + 790]\begin{bmatrix} -254.1 \times 10^{-6} \\ +0.1916 \times 10^{-6} \end{bmatrix} = -20.5 \text{ MPa}$$

$$\sigma_{p(1),0} = E_p[1 - y_{p(1)}] \ \varepsilon_0 + E_p\varepsilon_{p(1),init}$$

$$= 195,000 \times [1 + 580]\begin{bmatrix} -254.1 \times 10^{-6} \\ +0.1916 \times 10^{-6} \end{bmatrix} + 195,000 \times 0.00641$$

$$= +1222.1 \text{ MPa}$$

$$\sigma_{p(2),0} = 195,000 \times [1 + 645]\begin{bmatrix} -254.1 \times 10^{-6} \\ +0.1916 \times 10^{-6} \end{bmatrix} + 195,000 \times 0.00641 = +1224.6 \text{ MPa}$$

$$\sigma_{p(3),0} = 195,000 \times [1 + 710]\begin{bmatrix} -254.1 \times 10^{-6} \\ +0.1916 \times 10^{-6} \end{bmatrix} + 195,000 \times 0.00641 = +1227.0 \text{ MPa}$$

The distributions of strain and stress on the cross-section immediately after transfer are shown in Figure 5.10.

As already outlined in Example 5.3, the cross-sectional rigidities included in $\mathbf{F}_0$ can also be calculated using Equations 5.35, 5.36 and 5.39. The properties of the concrete part of the cross-section (with respect to the x-axis) are:

$$A_c = A_{gross} - A_{s(1)} - A_{s(2)} - A_{p(1)} - A_{p(2)} - A_{p(3)} = 312,700 \text{ mm}^2$$

$$B_c = A_{gross} y_c - A_{s(1)}y_{s(1)} - A_{s(2)}y_{s(2)} - A_{p(1)}y_{p(1)} - A_{p(2)}y_{p(2)} - A_{p(3)}y_{p(3)}$$

$$= -93.46 \times 10^6 \text{ mm}^3$$

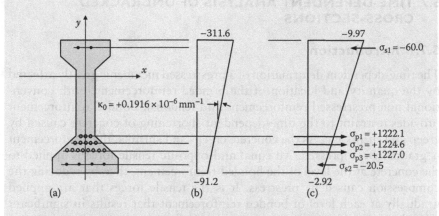

Figure 5.10 Strain and stress diagrams (Example 5.4). (a) Cross-section. (b) Strain ($\times 10^{-6}$). (c) Stress (MPa).

$$I_c = I_{gross} + A_{gross}y_c^2 - A_{s(1)}y_{s(1)}^2 - A_{s(2)}y_{s(2)}^2 - A_{p(1)}y_{p(1)}^2 - A_{p(2)}y_{p(2)}^2 - A_{p(3)}y_{p(3)}^2$$

$$= 76,924 \times 10^6 \text{ mm}^4$$

For this member with bonded tendons, Equations 5.35, 5.36 and 5.39 give:

$$R_{A,0} = A_c E_{cm,0} + \sum_{i=1}^{2} A_{s(i)}E_{s(i)} + \sum_{i=1}^{3} A_{p(i)}E_{p(i)}$$

$$= 312,700 \times 32,000 + (900 + 1,800) \times 200,000$$

$$+ (300 + 500 + 800) \times 195,000$$

$$= 10,858 \times 10^6 \text{ N}$$

$$R_{B,0} = B_c E_{cm,0} + \sum_{i=1}^{2} y_{s(i)}A_{s(i)}E_{s(i)} + \sum_{i=1}^{3} y_{p(i)}A_{p(i)}E_{p(i)} = -3440 \times 10^9 \text{ Nmm}$$

$$R_{I,0} = I_c E_{cm,0} + \sum_{i=1}^{2} y_{s(i)}^2 A_{s(i)}E_{s(i)} + \sum_{i=1}^{3} y_{p(i)}^2 A_{p(i)}E_{p(i)} = 28,359 \times 10^{12} \text{ Nmm}^2$$

and from Equation 5.46:

$$\mathbf{F}_0 = \frac{1}{R_{A,0}R_{I,0} - R_{B,0}^2}\begin{bmatrix} R_{I,0} & R_{B,0} \\ R_{B,0} & R_{A,0} \end{bmatrix} = \begin{bmatrix} 149.6 \times 10^{-12} & -181.4 \times 10^{-15} \\ -181.4 \times 10^{-15} & 572.8 \times 10^{-18} \end{bmatrix}$$

This is identical to the matrix $\mathbf{F}_0$ obtained earlier from Equation 5.54.

5.7 TIME-DEPENDENT ANALYSIS OF UNCRACKED CROSS-SECTIONS

5.7.1 Introduction

The time-dependent deformation of a prestressed member is greatly affected by the quantity and location of the bonded reinforcement (both conventional non-prestressed reinforcement and tendons). Bonded reinforcement provides restraint to the time-dependent shortening of concrete caused by creep and shrinkage. As the concrete creeps and shrinks, the reinforcement is gradually compressed. An equal and opposite tensile force is applied to the concrete at the level of the bonded reinforcement, thereby reducing the compression caused by prestress. It is the tensile forces that are applied gradually at each level of bonded reinforcement that results in significant time-dependent changes in curvature and deflection. A reliable estimate of these forces is essential if meaningful predictions of long-term behaviour are required.

Procedures specified in codes of practice for predicting losses of prestress due to creep and shrinkage are usually too simplified to be reliable and often lead to significant error, particularly for members containing non-prestressed reinforcement. In the following section, a simple analytical technique is presented for estimating the time-dependent behaviour of a general prestressed cross-section of any shape and containing any number of levels of prestressed and non-prestressed reinforcement. The procedure has been described in more detail in Gilbert and Ranzi [6] and makes use of the age-adjusted effective modulus method (AEMM) to model the effects of creep in concrete.

5.7.2 The age-adjusted effective modulus method

The age-adjusted effective modulus for concrete is often used to account for the creep strain that develops in concrete due to a gradually applied stress and was introduced in Section 4.2.4.3 (see Equations 4.22 through 4.25). The stress history shown in Figure 5.11 is typical of the change in stress

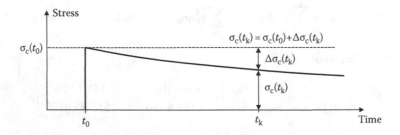

Figure 5.11 A gradually reducing stress history.

that occurs with time at many points in a reinforced or prestressed concrete member containing bonded reinforcement (or other forms of restraint) when the member is subjected to a constant sustained load.

For concrete subjected to the stress history of Figure 5.11, the total strain at time t_k may be expressed as the sum of the instantaneous and creep strains produced by $\sigma_c(t_0)$ (see Equation 4.22), the instantaneous and creep strains produced by the gradually applied stress increment $\Delta\sigma_c(t_k)$ (see Equation 4.24) and the shrinkage strain as follows:

$$\varepsilon(t_k) = \frac{\sigma_c(t_0)}{E_{cm,0}}[1 + \varphi(t_k,t_0)] + \frac{\Delta\sigma_c(t_k)}{E_{cm,0}}[1 + \chi(t_k,t_0)\,\varphi(t_k,t_0)] + \varepsilon_{cs}(t_k)$$

$$= \frac{\sigma_c(t_0)}{E_{c,eff}(t_k,t_0)} + \frac{\Delta\sigma_c(t_k)}{\bar{E}_{c,eff}(t,t_0)} + \varepsilon_{cs}(t_k) \tag{5.55}$$

where $E_{c,eff}(t_k,t_0)$ is the *effective modulus* (Equation 4.23) and $\bar{E}_{c,eff}(t_k,t_0)$ is the *age-adjusted effective modulus* (Equation 4.25), both reproduced here for convenience:

$$E_{c,eff}(t_k,t_0) = \frac{E_{cm,0}}{1 + \varphi(t_k,t_0)} \tag{5.56}$$

$$\bar{E}_{c,eff}(t_k,t_0) = \frac{E_{cm,0}}{1 + \chi(t_k,t_0)\,\varphi(t_k,t_0)} \tag{5.57}$$

In the remainder of this chapter, the simplified notations $E_{c,eff}$ and $\bar{E}_{c,eff}$ will be used instead of $E_{c,eff}(t_k,t_0)$ and $\bar{E}_{c,eff}(t_k,t_0)$, respectively.

For members subjected to constant sustained loads, where the change in concrete stress is caused by the restraint to creep and shrinkage provided by bonded reinforcement, the ageing coefficient may be approximated by $\chi(t_k,t_0) = 0.65$, when t_k exceeds about 100 days [6].

Equation 5.55 can be written in terms of the concrete stress at first loading, i.e. $\sigma_c(t_0)\ (=\sigma_{c,0})$, and the concrete stress at time t_k, i.e. $\sigma_c(t_k)\ (=\sigma_{c,k})$, as follows:

$$\varepsilon(t_k) = \varepsilon_k = \frac{\sigma_{c,0}}{E_{c,eff}} + \frac{\sigma_{c,k} - \sigma_{c,0}}{\bar{E}_{c,eff}} + \varepsilon_{cs,k}$$

$$= \frac{\sigma_{c,0}\varphi(t_k,t_0)[1 - \chi(t_k,t_0)]}{E_{cm,0}} + \frac{\sigma_{c,k}[1 + \chi(t_k,t_0)\varphi(t_k,t_0)]}{E_{cm,0}} + \varepsilon_{cs,k} \tag{5.58}$$

and rearranging Equation 5.58 gives:

$$\sigma_{c,k} = \bar{E}_{c,eff}(\varepsilon_k - \varepsilon_{cs,k}) + \sigma_{c,0}\bar{F}_{e,0} \tag{5.59}$$

where:

$$\overline{F}_{e,0} = \varphi(t_k, t_0) \frac{[\chi(t_k, t_0) - 1]}{[1 + \chi(t_k, t_0)\varphi(t_k, t_0)]} \tag{5.60}$$

Equation 5.59 is a stress–strain time relationship for concrete and can be conveniently used to determine the long-term deformations of a wide variety of concrete structures. The method of analysis is known as the *age-adjusted effective modulus method* [6].

5.7.3 Long-term analysis of an uncracked cross-section subjected to combined axial force and bending using AEMM

Cross-sectional analysis using Equation 5.59 as the constitutive relationship for concrete provides an effective tool for determining how stresses and strains vary with time due to creep and shrinkage of the concrete and relaxation of the prestressing steel. For this purpose, two instants in time are identified, as shown in Figure 5.12. One time instant is the time at first loading, i.e. $t = t_0$, and the other represents the instant in time at which stresses and strains need to be evaluated, i.e. $t = t_k$. It is usually convenient to measure time in days starting from the time when concrete is poured.

During the time interval $\Delta t_k\ (= t_k - t_0)$, creep and shrinkage strains develop in the concrete and relaxation occurs in the tendons. The gradual change of concrete strain with time causes change of stress in the bonded reinforcement. In general, as the concrete shortens due to compressive creep and shrinkage, the reinforcement is compressed and there is a gradual increase in the compressive stress in the non-prestressed reinforcement and a gradual loss of prestress in any bonded tendons. To maintain equilibrium, the gradual change of force in the steel at each bonded reinforcement level is opposed by an equal and opposite restraining force on the concrete, as shown in Figure 5.13.

These gradually applied restraining forces ($\Delta F_{ct.s(i)}$ and $\Delta F_{ct.p(i)}$) are usually tensile and, for a prestressed or partially prestressed cross-section, tend

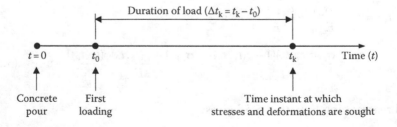

Figure 5.12 Relevant instants in time (AEMM).

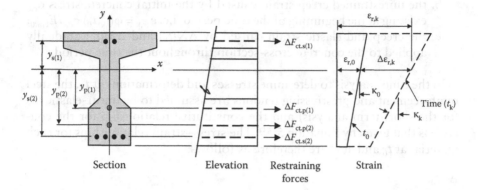

Figure 5.13 Time-dependent actions and deformations.

to relieve the concrete of its initial compression. The loss of prestress in the concrete is therefore often significantly more than the loss of prestress in the tendons.

The resultants of the creep- and shrinkage-induced internal restraining forces on the concrete are an increment of axial force $\Delta N(t_k)$ and an increment of moment about the reference axis $\Delta M(t_k)$ given by:

$$\Delta N(t_k) = \sum_{i=1}^{m_s} \Delta F_{ct.s(i)} + \sum_{i=1}^{m_p} \Delta F_{ct.p(i)} \qquad (5.61)$$

and

$$\Delta M(t_k) = -\sum_{i=1}^{m_s} \Delta F_{ct.s(i)} y_{s(i)} - \sum_{i=1}^{m_p} \Delta F_{ct.p(i)} y_{p(i)} \qquad (5.62)$$

Equal and opposite actions, i.e. $-\Delta N(t_k)$ and $-\Delta M(t_k)$, are applied to the bonded steel parts of the cross-section.

The strain at time t_k at any distance y from the reference axis (i.e. the x-axis in Figure 5.13) may be expressed in terms of the strain at the reference axis $\varepsilon_{r,k}$ and the curvature κ_k:

$$\varepsilon_k = \varepsilon_{r,k} - y\,\kappa_k \qquad (5.63)$$

The magnitude of the change of strain $\Delta\varepsilon_k$ ($=\varepsilon_k - \varepsilon_0$) that occurs with time at any point on the cross-section is the sum of each of the following components:

a. the free shrinkage strain $\varepsilon_{cs}(t_k) = \varepsilon_{cs,k}$ (which is usually considered to be uniform over the section);

b. the unrestrained creep strain caused by the initial concrete stress $\sigma_{c,0}$ existing at the beginning of the time period, i.e. $\varepsilon_{cr,k} = \varphi(t_k,t_0)\,\sigma_{c,0}/E_{cm,0}$;

c. the creep and elastic strain caused by $\Delta N(t_k)$ and $\Delta M(t_k)$ gradually applied to the concrete cross-section throughout the time period.

In the time analysis to determine stresses and deformations at t_k, the steel reinforcement and prestressing tendons are assumed to be linear-elastic (as for the short-term analysis) and the constitutive relationship for the concrete is that given by Equation 5.59. The stress–strain relationships for each material at t_0 and at t_k are therefore as follows:

At t_0:

Concrete: $\sigma_{c,0} = E_{cm,0}\varepsilon_0$ (5.64)

Steel reinforcement: $\sigma_{s(i),0} = E_{s(i)}\varepsilon_{s(i),0}$ (5.65)

Tendons:

    Bonded: $\sigma_{p(i),0} = E_{p(i)}(\varepsilon_{cp(i),0} + \varepsilon_{p(i),init})$ (5.66)

    Unbonded: $\sigma_{p(i),0} = E_{p(i)}\varepsilon_{p(i),init}$ (5.67)

At t_k:

Concrete: $\sigma_{c,k} = \bar{E}_{c,eff}(\varepsilon_k - \varepsilon_{cs,k}) + \bar{F}_{e,0}\sigma_{c,0}$ (5.68)

Steel reinforcement: $\sigma_{s(i),k} = E_{s(i)}\varepsilon_{s(i),k}$ (5.69)

Pretensioned or post-tensioned tendons bonded at t_0:

    $\sigma_{p(i),k} = E_{p(i)}(\varepsilon_{cp(i),k} + \varepsilon_{p(i),init} - \varepsilon_{p.rel(i),k})$ (5.70)

Post-tensioned tendons unbonded at t_0:

    Bonded: $\sigma_{p(i),k} = E_{p(i)}(\varepsilon_{cp(i),k} - \varepsilon_{cp(i),0} + \varepsilon_{p(i),init} - \varepsilon_{p.rel(i),k})$ (5.71)

    Unbonded: $\sigma_{p(i),k} = E_{p(i)}(\varepsilon_{p(i),init} - \varepsilon_{p.rel(i),k})$ (5.72)

where $\bar{F}_{e,0}$ is given by Equation 5.60 and $\bar{E}_{c,eff}$ is the age-adjusted effective modulus at $t = t_k$ (Equation 5.57). For a pretensioned tendon, the strain at time t_k is the concrete strain at the tendon level $(\varepsilon_{cp(i),k})$ plus the tensile strain in the tendon prior to transfer $(\varepsilon_{p(i),init})$ minus the tensile relaxation strain $(\varepsilon_{p.rel(i),k})$.

For a bonded post-tensioned tendon (initially unbonded at t_0), the strain at time t_k is the change in concrete strain at the level of the tendon during the time interval t_0 to t_k (i.e. $\varepsilon_{cp(i),k} - \varepsilon_{cp(i),0}$) plus the tensile strain in the tendon prior to transfer $(\varepsilon_{p(i),init})$ minus the tensile relaxation strain $(\varepsilon_{p.rel(i),k})$. For an unbonded post-tensioned tendon, slip occurs between the tendon and the concrete as the concrete deforms with time and the tendon strain at time t_k is the tensile strain in the tendon prior to transfer $(\varepsilon_{p(i),init})$ minus the tensile relaxation strain $(\varepsilon_{p.rel(i),k})$.

The relaxation strain $\varepsilon_{p.rel(i),k}$ is the tensile creep strain that has developed in the i-th prestressing tendon at time t_k and may be calculated from:

$$\varepsilon_{p.rel(i),k} = \frac{\sigma_{p(i),init}}{E_{p(i)}}\varphi_{p(i)} = \varepsilon_{p(i),init}\,\varphi_{p(i)} \tag{5.73}$$

where $\varphi_{p(i)}$ is the creep coefficient for the prestressing steel at time t_k due to an initial stress $\sigma_{p(i),init}$ in the ith prestressing tendon just after anchoring (given as $\varphi_p(t,\sigma_{pi})$ in Table 4.9).

The governing equations describing the long-term behaviour of a cross-section are obtained by enforcing equilibrium at the cross-section at time t_k following the approach already presented in the previous section for the instantaneous analysis at time t_0 (Equations 5.23 through 5.54). Restating the equilibrium equations (Equations 5.19 and 5.20) at time t_k gives:

$$\mathbf{r}_{ext,k} = \mathbf{r}_{int,k} \tag{5.74}$$

where:

$$\mathbf{r}_{ext,k} = \begin{bmatrix} N_{ext,k} \\ M_{ext,k} \end{bmatrix} \tag{5.75}$$

and

$$\mathbf{r}_{int,k} = \begin{bmatrix} N_{int,k} \\ M_{int,k} \end{bmatrix} \tag{5.76}$$

and $N_{int,k}$ and $M_{int,k}$ are the internal axial force and moment resisted by the cross-section at time t_k, while $N_{ext,k}$ and $M_{ext,k}$ are the external applied actions at this time. As in Equation 5.28, the axial force $N_{int,k}$ is the sum of the axial forces carried by the concrete, reinforcement and tendons:

$$N_{int,k} = N_{c,k} + N_{s,k} + N_{p,k} \tag{5.77}$$

Considering the time-dependent constitutive relationship for the concrete (Equation 5.68), the axial force resisted by the concrete at time t_k can be expressed as:

$$N_{c,k} = \int_{A_c} \sigma_{c,k}\, dA = \int_{A_c} \left[\bar{E}_{c,eff}(\varepsilon_{r,k} - y\kappa_k - \varepsilon_{cs,k}) + \bar{F}_{e,0}\sigma_{c,0} \right] dA$$

$$= A_c\bar{E}_{c,eff}\varepsilon_{r,k} - B_c\bar{E}_{c,eff}\kappa_k - A_c\bar{E}_{c,eff}\varepsilon_{cs,k} + \bar{F}_{e,0}N_{c,0} \tag{5.78}$$

where $\varepsilon_{r,k}$ and κ_k are the strain at the level of the reference axis and the curvature at time t_k, while $N_{c,0}$ is the axial force resisted by the concrete at time t_0. For the time analysis, $N_{c,0}$ is assumed to be known having been determined from the instantaneous analysis and may be calculated from Equation 5.29.

Using the constitutive equation for the reinforcing steel (Equation 5.69), the force carried by the reinforcing bars at time t_k is:

$$N_{s,k} = \sum_{i=1}^{m_s}(A_{s(i)}E_{s(i)})\varepsilon_{r,k} - \sum_{i=1}^{m_s}(y_{s(i)}A_{s(i)}E_{s(i)})\kappa_k \tag{5.79}$$

The force carried by the tendons at time t_k is determined using Equations 5.70, 5.71 and 5.72, as appropriate. For pretensioned tendons and post-tensioned tendons bonded at time t_0:

$$N_{p,k} = \sum_{i=1}^{m_p}(A_{p(i)}E_{p(i)})\varepsilon_{r,k} - \sum_{i=1}^{m_p}(y_{p(i)}A_{p(i)}E_{p(i)})\kappa_k + \sum_{i=1}^{m_p}[A_{p(i)}E_{p(i)}(\varepsilon_{p(i),\text{init}} - \varepsilon_{p.\text{rel}(i),k})]$$

(5.80)

For bonded post-tensioned tendons unbonded at time t_0:

$$N_{p,k} = \sum_{i=1}^{m_p}(A_{p(i)}E_{p(i)})\varepsilon_{r,k} - \sum_{i=1}^{m_p}(y_{p(i)}A_{p(i)}E_{p(i)})\kappa_k$$

$$+ \sum_{i=1}^{m_p}[A_{p(i)}E_{p(i)}(\varepsilon_{p(i),\text{init}} - \varepsilon_{p.\text{rel}(i),k} - \varepsilon_{cp(i),0})]$$

(5.81)

where $\varepsilon_{cp(i),0}$ is the strain in the concrete at the i-th level of post-tensioned tendon at time t_0. For unbonded post-tensioned tendons:

$$N_{p,k} = \sum_{i=1}^{m_p}[A_{p(i)}E_{p(i)}(\varepsilon_{p(i),\text{init}} - \varepsilon_{p.\text{rel}(i),k})]$$

(5.82)

By substituting Equations 5.78 through 5.80 into Equation 5.77, the internal axial force in a pretensioned member (or in a member with post-tensioned tendons that were bonded at time t_0) is given by:

$$N_{\text{int},k} = \left(A_c\bar{E}_{c,\text{eff}} + \sum_{i=1}^{m_s}A_{s(i)}E_{s(i)} + \sum_{i=1}^{m_p}A_{p(i)}E_{p(i)} \right)\varepsilon_{r,k}$$

$$- \left(B_c\bar{E}_{c,\text{eff}} + \sum_{i=1}^{m_s}y_{s(i)}A_{s(i)}E_{s(i)} + \sum_{i=1}^{m_p}y_{p(i)}A_{p(i)}E_{p(i)} \right)\kappa_k - A_c\bar{E}_{c,\text{eff}}\varepsilon_{cs,k} + \bar{F}_{e,0}N_{c,0}$$

$$+ \sum_{i=1}^{m_p}[A_{p(i)}E_{p(i)}(\varepsilon_{p(i),\text{init}} - \varepsilon_{p.\text{rel}(i),k})]$$

$$= R_{A,k}\varepsilon_{r,k} - R_{B,k}\kappa_k - A_c\bar{E}_{c,\text{eff}}\varepsilon_{cs,k} + \bar{F}_{e,0}N_{c,0}$$

$$+ \sum_{i=1}^{m_p}[A_{p(i)}E_{p(i)}(\varepsilon_{p(i),\text{init}} - \varepsilon_{p.\text{rel}(i),k})]$$

(5.83)

where the axial rigidity and the stiffness related to the first moment of area calculated at time t_k have been referred to as $R_{A,k}$ and $R_{B,k}$, respectively, and are given by:

$$R_{A,k} = A_c \bar{E}_{c,eff} + \sum_{i=1}^{m_s} A_{s(i)} E_{s(i)} + \sum_{i=1}^{m_p} A_{p(i)} E_{p(i)} \qquad (5.84)$$

$$R_{B,k} = B_c \bar{E}_{c,eff} + \sum_{i=1}^{m_s} y_{s(i)} A_{s(i)} E_{s(i)} + \sum_{i=1}^{m_p} y_{p(i)} A_{p(i)} E_{p(i)} \qquad (5.85)$$

Similarly, for post-tensioned members with bonded tendons (unbonded at t_0):

$$N_{int.k} = R_{A,k} \varepsilon_{r,k} - R_{B,k} \kappa_k - A_c \bar{E}_{c,eff} \varepsilon_{cs,k}$$
$$+ \bar{F}_{e,0} N_{c,0} + \sum_{i=1}^{m_p} [A_{p(i)} E_{p(i)} (\varepsilon_{p(i),init} - \varepsilon_{p.rel(i),k} - \varepsilon_{cp(i),0})] \qquad (5.86)$$

and for post-tensioned members with unbonded tendons:

$$N_{int.k} = R_{A,k} \varepsilon_{r,k} - R_{B,k} \kappa_k - A_c \bar{E}_{c,eff} \varepsilon_{cs,k} + \bar{F}_{e,0} N_{c,0} + \sum_{i=1}^{m_p} [A_{p(i)} E_{p(i)} (\varepsilon_{p(i),init} - \varepsilon_{p.rel(i),k})]$$
$$(5.87)$$

For unbonded tendons, the contribution of the prestressing steel $A_{p(i)}$ to the rigidities $R_{A,k}$ and $R_{B,k}$ is ignored, i.e. $A_{p(i)}$ is set to zero in Equations 5.84 and 5.85.

In a similar manner, the internal moment $M_{int,k}$ resisted by the cross-section at time t_k in a pretensioned member or in a post-tensioned member with bonded tendons at t_0 can be expressed as:

$$M_{int,k} = -\left(B_c \bar{E}_{c,eff} + \sum_{i=1}^{m_s} y_{s(i)} A_{s(i)} E_{s(i)} + \sum_{i=1}^{m_p} y_{p(i)} A_{p(i)} E_{p(i)} \right) \varepsilon_{r,k}$$

$$+ \left(I_c \bar{E}_{c,eff} + \sum_{i=1}^{m_s} y_{s(i)}^2 A_{s(i)} E_{s(i)} + \sum_{i=1}^{m_p} y_{p(i)}^2 A_{p(i)} E_{p(i)} \right) \kappa_k$$

$$+ B_c \bar{E}_{c,eff} \varepsilon_{cs,k} + \bar{F}_{e,0} M_{c,0} - \sum_{i=1}^{m_p} [y_{p(i)} A_{p(i)} E_{p(i)} (\varepsilon_{p(i),init} - \varepsilon_{p.rel(i),k})]$$

$$= -R_{B,k} \varepsilon_{r,k} + R_{I,k} \kappa_k + B_c \bar{E}_{c,eff} \varepsilon_{cs,k} + \bar{F}_{e,0} M_{c,0}$$

$$- \sum_{i=1}^{m_p} [y_{p(i)} A_{p(i)} E_{p(i)} (\varepsilon_{p(i),init} - \varepsilon_{p.rel(i),k})] \qquad (5.88)$$

where the flexural rigidity $R_{I,k}$ of the cross-section calculated at time t_k is given by:

$$R_{I,k} = I_c \bar{E}_{c,eff} + \sum_{i=1}^{m_s} y_{s(i)}^2 A_{s(i)} E_{s(i)} + \sum_{i=1}^{m_p} y_{p(i)}^2 A_{p(i)} E_{p(i)} \tag{5.89}$$

and $M_{c,0}$ is the moment resisted by the concrete component at time t_0. From the instantaneous analysis:

$$M_{c,0} = -\int_{A_c} y\sigma_{c,0}\, dA = -\int_{A_c} yE_{cm,0}(\varepsilon_{r,0} - y\kappa_0)dA = -B_c E_{cm,0}\varepsilon_{r,0} + I_c E_{cm,0}\kappa_0 \tag{5.90}$$

Similarly, for bonded post-tensioned members (unbonded at t_0):

$$M_{int.k} = -R_{B,k}\varepsilon_{r,k} + R_{I,k}\kappa_k + B_c \bar{E}_{c,eff}\varepsilon_{cs,k} + \bar{F}_{e,0}M_{c,0}$$

$$- \sum_{i=1}^{m_p} [y_{p(i)}A_{p(i)}E_{p(i)}(\varepsilon_{p(i),init} - \varepsilon_{p.rel(i),k} - \varepsilon_{cp(i),0})] \tag{5.91}$$

and for unbonded post-tensioned members:

$$M_{int.k} = -R_{B,k}\varepsilon_{r,k} + R_{I,k}\kappa_k + B_c \bar{E}_{c,eff}\varepsilon_{cs,k} + \bar{F}_{e,0}M_{c,0}$$

$$- \sum_{i=1}^{m_p} \left[y_{p(i)}A_{p(i)}E_{p(i)}(\varepsilon_{p(i),init} - \varepsilon_{p.rel(i),k}) \right] \tag{5.92}$$

For unbonded tendons, the contribution of the prestressing steel $A_{p(i)}$ to the rigidity $R_{I,k}$ is ignored, i.e. $A_{p(i)}$ is set to zero in Equation 5.89.

After substituting Equations 5.86 and 5.91 into Equation 5.74, the equilibrium equations at time t_k for a post-tensioned member with bonded tendons (unbonded at t_0) may be written in compact form as:

$$r_{ext,k} = D_k\varepsilon_k + f_{cr,k} - f_{cs,k} + f_{p,init} - f_{p.rel,k} - f_{cp,0} \tag{5.93}$$

where:

$$\varepsilon_k = \begin{bmatrix} \varepsilon_{r,k} \\ \kappa_k \end{bmatrix} \tag{5.94}$$

$$D_k = \begin{bmatrix} R_{A,k} & -R_{B,k} \\ -R_{B,k} & R_{I,k} \end{bmatrix} \tag{5.95}$$

The vector $f_{cr,k}$ represents a portion of the effects of creep produced by the initial stress $\sigma_{c,0}$ resisted by the concrete at time t_0 and is given by:

$$f_{cr,k} = \overline{F}_{e,0} \begin{bmatrix} N_{c,0} \\ M_{c,0} \end{bmatrix} = \overline{F}_{e,0} E_{cm,0} \begin{bmatrix} A_c \varepsilon_{r,0} - B_c \kappa_0 \\ -B_c \varepsilon_{r,0} + I_c \kappa_0 \end{bmatrix} \tag{5.96}$$

and $\overline{F}_{e,0}$ is given in Equation 5.60. The vector $f_{cs,k}$ accounts for the uniform (unrestrained) shrinkage strain that develops in the concrete over the time period and is given by:

$$f_{cs,k} = \begin{bmatrix} A_c \\ -B_c \end{bmatrix} \overline{E}_{c,eff} \varepsilon_{cs,k} \tag{5.97}$$

The vector $f_{p,init}$ in Equation 5.93 accounts for the initial prestress and the vector $f_{p,rel,k}$ accounts for the resultant actions caused by the loss of prestress in the tendon due to relaxation. These are given by:

$$f_{p,init} = \sum_{i=1}^{m_p} \begin{bmatrix} P_{init(i)} \\ -y_{p(i)} P_{init(i)} \end{bmatrix} \tag{5.98}$$

$$f_{p,rel,k} = \sum_{i=1}^{m_p} \begin{bmatrix} P_{init(i)} \varphi_{p(i)} \\ -y_{p(i)} P_{init(i)} \varphi_{p(i)} \end{bmatrix} \tag{5.99}$$

where $P_{init(i)}$ is the prestressing force at the time from which $\varphi_{p(i)}$ is measured.

For a member with post-tensioned tendons that were unbonded at t_0, the vector $f_{cp,0}$ is given by:

$$f_{cp,0} = \sum_{i=1}^{m_p} \begin{bmatrix} A_{p(i)} E_{p(i)} \varepsilon_{cp(i),0} \\ -y_{p(i)} A_{p(i)} E_{p(i)} \varepsilon_{cp(i),0} \end{bmatrix} \tag{5.100}$$

For a pretensioned member or a post-tensioned member with bonded tendons at the time of the short-term analysis (t_0), Equation 5.93 applies except that the vector $f_{cp,0}$ is set to zero. For a post-tensioned member with all tendons unbonded throughout the time period t_0 to t_k, the vector $f_{cp,0}$ in Equation 5.93 is also set to zero.

Equation 5.93 can be solved for ε_k as:

$$\varepsilon_k = D_k^{-1}(r_{ext,k} - f_{cr,k} + f_{cs,k} - f_{p,init} + f_{p,rel,k} + f_{cp,0})$$

$$= F_k(r_{ext,k} - f_{cr,k} + f_{cs,k} - f_{p,init} + f_{p,rel,k} + f_{cp,0}) \tag{5.101}$$

where:

$$\mathbf{F_k} = \frac{1}{R_{A,k}\,R_{I,k} - R_{B,k}^2}\begin{bmatrix} R_{I,k} & R_{B,k} \\ R_{B,k} & R_{A,k} \end{bmatrix} \tag{5.102}$$

The stress distribution in the concrete at time t_k can then be calculated as:

$$\sigma_{c,k} = \bar{E}_{c,eff}(\varepsilon_k - \varepsilon_{cs,k}) + \bar{F}_{e,0}\sigma_{c,0} = \bar{E}_{c,eff}\{[1-y]\varepsilon_k - \varepsilon_{cs,k}\} + \bar{F}_{e,0}\sigma_{c,0} \tag{5.103}$$

where at any point y from the reference axis $\varepsilon_k = \varepsilon_{r,k} - y\kappa_k = [1-y]\varepsilon_k$.

The stress in the non-prestressed reinforcement at time t_k is:

$$\sigma_{s(i),k} = E_{s(i)}\varepsilon_{s(i),k} = E_{s(i)}[1 - y_{s(i)}]\varepsilon_k \tag{5.104}$$

and the stress in any pretensioned tendons, or any post-tensioned tendons that were bonded to the concrete at the time of the short-term analysis (t_0), is:

$$\sigma_{p(i),k} = E_{p(i)}(\varepsilon_{p(i),k} + \varepsilon_{p(i),init} - \varepsilon_{p.rel(i),k})$$

$$= E_{p(i)}[1 - y_{p(i)}]\varepsilon_k + E_{p(i)}\varepsilon_{p(i),init} - E_{p(i)}\varepsilon_{p.rel(i),k} \tag{5.105}$$

For bonded post-tensioned tendons (initially unbonded at t_0), the stress at time t_k is:

$$\sigma_{p(i),k} = E_{p(i)}(\varepsilon_{cp(i),k} - \varepsilon_{cp(i),0} + \varepsilon_{p(i),init} - \varepsilon_{p.rel(i),k})$$

$$= E_{p(i)}[1 - y_{p(i)}]\varepsilon_k - E_{p(i)}[1 - y_{p(i)}]\varepsilon_0 + E_{p(i)}\varepsilon_{p(i),init} - E_{p(i)}\varepsilon_{p.rel(i),k} \tag{5.106}$$

while for unbonded tendons:

$$\sigma_{p(i),k} = E_{p(i)}(\varepsilon_{p(i),init} - \varepsilon_{p.rel(i),k}) \tag{5.107}$$

The cross-sectional rigidities, i.e. $R_{A,k}$, $R_{B,k}$ and $R_{I,k}$, required for the solution at time t_k can also be calculated from the properties of the age-adjusted transformed section, obtained by transforming the bonded steel areas (reinforcement and tendons) into equivalent areas of the aged concrete at time t_k, as follows:

$$R_{A,k} = \bar{A}_k\bar{E}_{c,eff} \tag{5.108}$$

$$R_{B,k} = \bar{B}_k\bar{E}_{c,eff} \tag{5.109}$$

$$R_{I,k} = \bar{I}_k\bar{E}_{c,eff} \tag{5.110}$$

where $\bar{E}_{c,eff}$ is the age-adjusted effective modulus, $\bar{A}_k$ is the area of the age-adjusted transformed section and $\bar{B}_k$ and $\bar{I}_k$ are the first and second moments of the area of the age-adjusted transformed section about the reference axis. For the determination of $\bar{A}_k$, $\bar{B}_k$ and $\bar{I}_k$, the areas of the bonded steel are transformed into equivalent areas of concrete by multiplying by the age-adjusted modular ratio $\bar{\alpha}_{es(i),k} = E_{s(i)}/\bar{E}_{c,eff}$ or $\bar{\alpha}_{ep(i),k} = E_{p(i)}/\bar{E}_{c,eff}$, as appropriate.

Based on Equations 5.108 through 5.110, the expression for F_k (in Equation 5.102) can be rewritten as:

$$F_k = \frac{1}{\bar{E}_{c,eff}\left(\bar{A}_k\bar{I}_k - \bar{B}_k^2\right)}\begin{bmatrix} \bar{I}_k & \bar{B}_k \\ \bar{B}_k & \bar{A}_k \end{bmatrix} \tag{5.111}$$

The calculation of the time-dependent stresses and deformations using the earlier-mentioned procedure is illustrated in Examples 5.5 and 5.6.

EXAMPLE 5.5

For the post-tensioned concrete cross-section shown in Figure 5.7, the strain and stress distributions at t_0 were calculated in Example 5.3, immediately after the transfer of prestress and the application of an external bending moment of $M_{ext,0} = 100$ kNm (see Figure 5.8). Soon after transfer, the post-tensioned duct was filled with grout, thereby bonding the tendon to the concrete and ensuring compatibility of concrete and steel strains at all times after t_0. If the applied moment remains constant during the time interval t_0 to t_k (i.e. $M_{ext,k} = M_{ext,0}$), calculate the strain and stress distributions at time t_k using the AEMM.

As in Example 5.3, $E_{cm,0} = 30,000$ MPa; $E_s = 200,000$ MPa; $E_p = 195,000$ MPa; $\alpha_{s,0} = E_s/E_{cm,0} = 6.67$; $\alpha_{p,0} = E_p/E_{cm,0} = 6.5$; $f_{pk} = 1,860$ MPa and, with $P_{m0} = 1,350$ kN, the initial strain in the tendon is $\varepsilon_{p,init} = P_{m0}/(A_p E_p) = 0.00692$.

Take $\varphi(t_k,t_0) = 2.5$, $\chi(t_k,t_0) = 0.65$, $\varepsilon_{cs}(t_k) = -600 \times 10^{-6}$ and (from Table 4.9, with a Class 2 low-relaxation strand stressed to $\sigma_{pi} = P_{m0}/A_p = 1350$ MPa $= 0.726 f_{pk}$) $\varphi_p = 0.0459$.

From Example 5.3, the strain at the reference axis and the curvature at t_0 are:

$$\varepsilon_{r,0} = -176.5 \times 10^{-6} \quad \text{and} \quad \kappa_0 = -0.3778 \times 10^{-6} \text{ mm}^{-1}$$

and the strain in the concrete at the tendon level at t_0 before the tendon is grouted is:

$$\varepsilon_{cp,0} = \varepsilon_{r,0} - y_p\kappa_0 = -176.5 \times 10^{-6} - (-200) \times (-0.3778 \times 10^{-6})$$

$$= -252.1 \times 10^{-6}$$

The vector of external actions on the cross-section at t_k (expressed in N and Nmm) is:

$$\mathbf{r}_{ext,k} = \begin{bmatrix} N_{ext,k} \\ M_{ext,k} \end{bmatrix} = \begin{bmatrix} 0 \\ 100 \times 10^6 \end{bmatrix}$$

From Equation 5.57:

$$\bar{E}_{c,eff} = \frac{E_{cm,0}}{1 + \chi(t_k,t_0)\varphi(t_k,t_0)} = \frac{30,000}{1 + 0.65 \times 2.5} = 11,430 \text{ MPa}$$

and therefore $\bar{\alpha}_{es,k} = 17.5$ and $\bar{\alpha}_{ep,k} = 17.1$.

From Equation 5.60:

$$\bar{F}_{e,0} = \frac{\varphi(t_k,t_0)[\chi(t_k,t_0)-1]}{1 + \chi(t_k,t_0)\varphi(t_k,t_0)} = \frac{2.5 \times (0.65 - 1.0)}{1.0 + 0.65 \times 2.5} = -0.333$$

From Example 5.3, the properties of the concrete part of the section before the duct is grouted are:

$$A_c = bh - A_{s(1)} - A_{s(2)} - A_{hole} = 234,470 \text{ mm}^2$$

$$B_c = bhy_c - A_{s(1)}y_{s(1)} - A_{s(2)}y_{s(2)} - A_{hole}y_{hole} = 872,000 \text{ mm}^3$$

$$I_c = bh^3/12 - A_{s(1)}y_{s(1)}^2 - A_{s(2)}y_{s(2)}^2 - A_{hole}y_{hole}^2 = 12,375 \times 10^6 \text{ mm}^4$$

and the properties of the age-adjusted transformed section after the duct is grouted are:

$$\bar{A}_k = bh + (\bar{\alpha}_{es,k} - 1)A_{s(1)} + (\bar{\alpha}_{es,k} - 1)A_{s(2)} + (\bar{\alpha}_{ep,k} - 1)A_p$$

$$= 300 \times 800 + (17.5-1) \times 900 + (17.5-1) \times 1800 + (17.1-1) \times 1000$$

$$= 300,613 \text{ mm}^2$$

$$\bar{B}_k = bhy_c + (\bar{\alpha}_{es,k} - 1)A_{s(1)}y_{s(1)} + (\bar{\alpha}_{es,k} - 1)A_{s(2)}y_{s(2)} + (\bar{\alpha}_{ep,k} - 1)A_py_p$$

$$= 300 \times 800 \times 0 + (17.5-1) \times [900 \times (+340) + 1800 \times (-340)]$$

$$\quad + (17.1-1) \times 1000 \times (-200)$$

$$= -8.262 \times 10^6 \text{ mm}^3$$

$$\bar{I}_k = \frac{bh^3}{12} + bhy_c^2 + (\bar{\alpha}_{es,k} - 1)\left[A_{s(1)}y_{s(1)}^2 + A_{s(2)}y_{s(2)}^2\right] + (\bar{\alpha}_{ep,k} - 1)A_py_p^2$$

$$= \frac{300 \times 800^3}{12} + 300 \times 800 \times 0 + (17.5 - 1)[900 \times 340^2 + 1800 \times (-340)^2]$$

$$\quad + (17.1 - 1) \times 1000 \times (-200)^2$$

$$= 18,592 \times 10^6 \text{ mm}^4$$

From Equation 5.96:

$$f_{cr,k} = \bar{F}_{e,0} E_{cm,0} \begin{bmatrix} A_c \varepsilon_{r,0} - B_c \kappa_0 \\ -B_c \varepsilon_{r,0} + I_c \kappa_0 \end{bmatrix}$$

$$= -0.333 \times 30,000$$

$$\times \begin{bmatrix} 234,470 \times (-176.5 \times 10^{-6}) - 872,000 \times (-0.3778 \times 10^{-6}) \\ -872,000 \times (-176.5 \times 10^{-6}) + 12,375 \times 10^6 \times (-0.3778 \times 10^{-6}) \end{bmatrix}$$

$$= \begin{bmatrix} +410.5 \times 10^3 \text{ N} \\ +45.21 \times 10^6 \text{ Nmm} \end{bmatrix}$$

and from Equation 5.97:

$$f_{cs,k} = \begin{bmatrix} A_c \\ -B_c \end{bmatrix} \bar{E}_{c,eff} \varepsilon_{cs,k} = \begin{bmatrix} 234,470 \times 11,430 \times (-600 \times 10^{-6}) \\ -872,000 \times 11,430 \times (-600 \times 10^{-6}) \end{bmatrix}$$

$$= \begin{bmatrix} -1,608 \times 10^3 \text{ N} \\ 5.98 \times 10^6 \text{ Nmm} \end{bmatrix}$$

The relaxation strain in the prestressing tendon is:

$$\varepsilon_{p.rel,k} = \varepsilon_{p,init} \, \varphi_P = 0.00692 \times 0.0459 = 0.0003178$$

and the vectors of internal actions caused by the initial prestress and by relaxation are given by Equations 5.98 and 5.99, respectively:

$$f_{p,init} = \begin{bmatrix} P_{init} \\ -P_{init} y_p \end{bmatrix} = \begin{bmatrix} 1350 \times 10^3 \text{ N} \\ 270 \times 10^6 \text{ Nmm} \end{bmatrix}$$

$$f_{p.rel,k} = \begin{bmatrix} P_{init} \, \varphi_p \\ -y_p P_{init} \, \varphi_p \end{bmatrix} = \begin{bmatrix} 1350 \times 10^3 \times 0.0459 \\ 270 \times 10^6 \times 0.0459 \end{bmatrix} = \begin{bmatrix} 62.0 \times 10^3 \text{ N} \\ 12.4 \times 10^6 \text{ Nmm} \end{bmatrix}$$

For this post-tensioned member with a single bonded tendon, Equation 5.100 gives:

$$f_{cp,0} = \begin{bmatrix} A_p E_p \varepsilon_{cp,0} \\ -y_p A_p E_p \varepsilon_{cp,0} \end{bmatrix} = \begin{bmatrix} 1,000 \times 195,000 \times (-252.1 \times 10^{-6}) \\ -(-200) \times 1,000 \times 195,000 \times (-252.1 \times 10^{-6}) \end{bmatrix}$$

$$= \begin{bmatrix} -49.1 \times 10^3 \text{ N} \\ -9.83 \times 10^6 \text{ Nmm} \end{bmatrix}$$

Equation 5.111 gives:

$$\mathbf{F}_k = \frac{1}{\bar{E}_{c,\text{eff}}\left(\bar{A}_k \bar{I}_k - \bar{B}_k^2\right)}\begin{bmatrix} \bar{I}_k & \bar{B}_k \\ \bar{B}_k & \bar{A}_k \end{bmatrix}$$

$$= \frac{1}{11{,}430 \times (300{,}613 \times 18{,}592 \times 10^6 - (-8.262 \times 10^6)^2)}\begin{bmatrix} 18{,}592 \times 10^6 & -8.262 \times 10^6 \\ -8.262 \times 10^6 & 300{,}613 \end{bmatrix}$$

$$= \begin{bmatrix} 294.7 \times 10^{-12}\ \text{N}^{-1} & -130.9 \times 10^{-15}\ \text{N}^{-1}\text{mm}^{-1} \\ -130.9 \times 10^{-15}\ \text{N}^{-1}\text{mm}^{-1} & 4.764 \times 10^{-15}\ \text{N}^{-1}\text{mm}^{-2} \end{bmatrix}$$

and the strain ε_k at time t_k is determined using Equation 5.101:

$$\varepsilon_k = \mathbf{F}_k(\mathbf{r}_{\text{ext},k} - \mathbf{f}_{\text{cr},k} + \mathbf{f}_{\text{cs},k} - \mathbf{f}_{p,\text{init}} + \mathbf{f}_{p.\text{rel},k} + \mathbf{f}_{\text{cp.0}})$$

$$= \begin{bmatrix} 294.7 \times 10^{-12} & -130.9 \times 10^{-15} \\ -130.9 \times 10^{-15} & 4.764 \times 10^{-15} \end{bmatrix}\begin{bmatrix} (0 - 410.5 - 1608 - 1350 + 62.0 - 49.1) \times 10^3 \\ (100 - 45.2 + 5.98 - 270 + 12.4 - 9.83) \times 10^6 \end{bmatrix}$$

$$= \begin{bmatrix} -961.7 \times 10^{-6} \\ -0.5453 \times 10^{-6}\ \text{mm}^{-1} \end{bmatrix}$$

The strain at the reference axis and the curvature at time t_k are, respectively:

$$\varepsilon_{r,k} = -961.7 \times 10^{-6} \quad \text{and} \quad \kappa_k = -0.5453 \times 10^{-6}\ \text{mm}^{-1}$$

From Equation 5.63, the top ($y = +400$ mm) and bottom ($y = -400$ mm) fibre strains are:

$$\varepsilon_{k(\text{top})} = \varepsilon_{r,k} - 400 \times \kappa_k = [-961.7 - 400 \times (-0.5453)] \times 10^{-6} = -743.6 \times 10^{-6}$$

$$\varepsilon_{k(\text{btm})} = \varepsilon_{r,k} - (-400) \times \kappa_k = [-961.7 + 400 \times (-0.5453)] \times 10^{-6} = -1180 \times 10^{-6}$$

The concrete stress distribution at time t_k is calculated using Equation 5.103:

$$\sigma_{c,k(\text{top})} = \bar{E}_{c,\text{eff}}(\varepsilon_{k(\text{top})} - \varepsilon_{\text{cs},k}) + \bar{F}_{e,0}\sigma_{c,0(\text{top})}$$

$$= 11{,}430 \times [-743.6 - (-600)] \times 10^{-6} + (-0.333) \times (-0.762) = -1.39\ \text{MPa}$$

$$\sigma_{c,k(\text{btm})} = \bar{E}_{c,\text{eff}}(\varepsilon_{k(\text{btm})} - \varepsilon_{\text{cs},k}) + \bar{F}_{e,0}\sigma_{c,0(\text{btm})}$$

$$= 11{,}430 \times [-1180 - (-600)] \times 10^{-6} + (-0.333) \times (-9.828) = -3.35\ \text{MPa}$$

and, from Equation 5.104, the stresses in the non-prestressed reinforcement are:

$$\sigma_{s(1),k} = E_{s(1)}[1 - y_{s(1)}]\varepsilon_k = 200 \times 10^3 [1 - 340]\begin{bmatrix} -961.7 \times 10^{-6} \\ -0.5453 \times 10^{-6} \end{bmatrix} = -155\ \text{MPa}$$

$$\sigma_{s(2),k} = E_s[1 - y_{s(2)}]\varepsilon_k = 200 \times 10^3 [1 - (-340)]\begin{bmatrix} -961.7 \times 10^{-6} \\ -0.5453 \times 10^{-6} \end{bmatrix} = -229 \text{ MPa}$$

The final stress in the prestressing steel at time t_k is given by Equation 5.106:

$$\sigma_{p,k} = E_p[1 - y_p]\varepsilon_k - E_p[1 - y_p]\varepsilon_0 + E_p\varepsilon_{p,init} - E_p\varepsilon_{p.rel,k}$$

$$= 195 \times 10^3 [1 - (-200)]\begin{bmatrix} -961.7 \times 10^{-6} \\ -0.5453 \times 10^{-6} \end{bmatrix}$$

$$- 195 \times 10^3 [1 - (-200)]\begin{bmatrix} -176.5 \times 10^{-6} \\ -0.3778 \times 10^{-6} \end{bmatrix}$$

$$+ 195 \times 10^3 \times (0.006923 - 0.0003178)$$

$$= 1128 \text{ MPa}$$

The stress and strain distributions at t_0 (from Example 5.3) and t_k are shown in Figure 5.14.

Note that the time-dependent loss of prestress in the tendon is 16.4%, but the time-dependent loss in the compressive stress in the concrete at the bottom fibre is 65.9%.

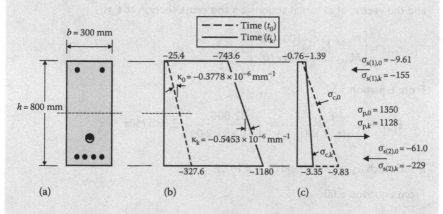

Figure 5.14 Strains and stresses at times at t_0 and t_k (Example 5.5). (a) Section. (b) Strain (×10⁻⁶). (c) Stress (MPa).

With time, much of the compressive force exerted by the tendon on the cross-section in Example 5.5 has been transferred from the concrete to the bonded steel reinforcement as a direct result of creep and shrinkage in the concrete. The very significant change in stress in the non-prestressed reinforcement with time is typical of the behaviour of uncracked regions of reinforced and prestressed concrete members in compression.

EXAMPLE 5.6

A time-dependent analysis of the pretensioned concrete cross-section of Example 5.4 (see Figure 5.9) is to be undertaken using the AEMM. The strain and stress distributions at t_0 immediately after the application of an axial force $N_{ext,0} = -100$ kN and a bending moment of $M_{ext,0} = 1000$ kNm were calculated in Example 5.4 and shown in Figure 5.10. If the applied actions remain constant during the time interval t_0 to t_k, determine the strain and stress distributions at time t_k.

As in Example 5.4: $E_{cm,0} = 32,000$ MPa, $E_s = 200,000$ MPa, $E_p = 195,000$ MPa and, therefore, $\alpha_{s(i),0} = 6.25$ and $\alpha_{p(i),0} = 6.09$. For the time interval t_0 to t_k, the creep and shrinkage input for the concrete is $\varphi(t_k,t_0) = 2.0$, $\chi(t_k,t_0) = 0.65$, $\varepsilon_{cs}(t_k) = -400 \times 10^{-6}$ and, for the prestressing steel, the creep coefficients associated with the time after transfer and time t_k are $\varphi_{p(1)} = \varphi_{p(2)} = \varphi_{p(3)} = 0.03$.

From Example 5.4, the strain at the reference axis and the curvature at t_0 are:

$$\varepsilon_{r,0} = -254.1 \times 10^{-6} \quad \text{and} \quad \kappa_0 = +0.1916 \times 10^{-6} \text{ mm}^{-1}.$$

and the vector of external actions on the cross section at t_k is:

$$\mathbf{r}_{ext,k} = \begin{bmatrix} N_{ext,k} \\ M_{ext,k} \end{bmatrix} = \begin{bmatrix} -100 \times 10^3 \\ 1000 \times 10^6 \end{bmatrix}$$

From Equation 5.57:

$$\bar{E}_{c,eff} = \frac{E_{cm,0}}{1 + \chi(t_k,t_0)\varphi(t_k,t_0)} = \frac{32,000}{1 + 0.65 \times 2.0} = 13,910 \text{ MPa}$$

from which $\bar{\alpha}_{es,k} = 14.37$ and $\bar{\alpha}_{ep,k} = 14.02$.

From Equation 5.60:

$$\bar{F}_{e,0} = \frac{\varphi(t_k,t_0)[\chi(t_k,t_0) - 1]}{1 + \chi(t_k,t_0)\varphi(t_k,t_0)} = \frac{2.0 \times (0.65 - 1.0)}{1.0 + 0.65 \times 2.0} = -0.304$$

The properties of the concrete part of the cross-section are:

$$A_c = A_g - A_{s(1)} - A_{s(2)} - A_{p(1)} - A_{p(2)} - A_{p(3)} = 312{,}700 \text{ mm}^2$$

$$B_c = A_g (d_{ref} - d_c) - (A_{s(1)} y_{s(1)} + A_{s(2)} y_{s(2)}) - (A_{p(1)} y_{p(1)} + A_{p(2)} y_{p(2)} + A_{p(3)} y_{p(3)})$$

$$= -93.46 \times 10^6 \text{ mm}^3$$

$$I_c = I_g + A_g (d_{ref} - d_c)^2 - \left(A_{s(1)} y_{s(1)}^2 + A_{s(2)} y_{s(2)}^2\right)$$

$$- \left(A_{p(1)} y_{p(1)}^2 + A_{p(2)} y_{p(2)}^2 + A_{p(3)} y_{p(3)}^2\right) = 76{,}920 \times 10^6 \text{ mm}^4$$

and the properties of the age-adjusted transformed section in equivalent concrete areas are:

$$\bar{A}_k = A_g + \left(\bar{\alpha}_{es,k} - 1\right)(A_{s(1)} + A_{s(2)}) + \left(\bar{\alpha}_{ep,k} - 1\right)(A_{p(1)} + A_{p(2)} + A_{p(3)})$$

$$= 373.9 \times 10^3 \text{ mm}^2$$

$$\bar{B}_k = A_g (d_{ref} - d_c) + (\bar{\alpha}_{es,k} - 1)(A_{s(1)} y_{s(1)} + A_{s(2)} y_{s(2)})$$

$$+ (\bar{\alpha}_{ep,k} - 1)(A_{p(1)} y_{p(1)} + A_{p(2)} y_{p(2)} + A_{p(3)} y_{p(3)}) = -125.7 \times 10^6 \text{ mm}^3$$

$$\bar{I}_k = I_g + A_g (d_{ref} - d_c)^2 + (\bar{\alpha}_{es,k} - 1)\left(A_{s(1)} y_{s(1)}^2 + A_{s(2)} y_{s(2)}^2\right)$$

$$+ (\bar{\alpha}_{ep,k} - 1)\left(A_{p(1)} y_{p(1)}^2 + A_{p(2)} y_{p(2)}^2 + A_{p(3)} y_{p(3)}^2\right) = 103{,}800 \times 10^6 \text{ mm}^4$$

From Equation 5.96:

$$f_{cr,k} = \begin{bmatrix} +599.4 \times 10^3 \text{ N} \\ +87.73 \times 10^6 \text{ Nmm} \end{bmatrix}$$

and from Equation 5.97:

$$f_{cs,k} = \begin{bmatrix} -1740 \times 10^3 \text{ N} \\ -520.1 \times 10^6 \text{ Nmm} \end{bmatrix}$$

The relaxation strains in the prestressing tendons between transfer and time t_k are:

$$\varepsilon_{p(1).rel,k} = \varepsilon_{p(1),0}\, \varphi_{p(1)} = 0.0001880$$

$$\varepsilon_{p(2).rel,k} = \varepsilon_{p(2),0} \; \varphi_{p(2)} = 0.0001884$$

$$\varepsilon_{p(3).rel,k} = \varepsilon_{p(3),0} \; \varphi_{p(3)} = 0.0001888$$

and the vectors of internal actions caused by the initial prestress and by relaxation are given by Equations 5.98 and 5.99, respectively:

$$\mathbf{f}_{p,init} = \sum_{i=1}^{3} \begin{bmatrix} P_{init(i)} \\ -y_{p(i)} P_{init(i)} \end{bmatrix} = \begin{bmatrix} 2000 \times 10^3 \, \text{N} \\ 1330.6 \times 10^6 \, \text{Nmm} \end{bmatrix}$$

$$\mathbf{f}_{p.rel,k} = \sum_{i=1}^{3} \begin{bmatrix} P_{m0(i)} \, \varphi_{p(i)} \\ -y_{p(i)} P_{m0(i)} \, \varphi_{p(i)} \end{bmatrix} = \begin{bmatrix} 58.82 \times 10^3 \, \text{N} \\ 39.14 \times 10^6 \, \text{Nmm} \end{bmatrix}$$

Note that in this pretensioned member, $P_{m0(i)}$ is used in Equation 5.99 since $\varphi_{p(i)}$ in this example is measured from the time at transfer t_0.

The terms in the vector $\mathbf{f}_{cp,0}$ are both zero for this pretensioned cross-section since all tendons were bonded at transfer.

Equation 5.99 gives:

$$\mathbf{F}_k = \frac{1}{\bar{E}_{c,eff}(\bar{A}_k \bar{I}_k - \bar{B}_k^2)} \begin{bmatrix} \bar{I}_k & \bar{B}_k \\ \bar{B}_k & \bar{A}_k \end{bmatrix}$$

$$= \begin{bmatrix} 324.2 \times 10^{-12} \, \text{N}^{-1} & -392.7 \times 10^{-15} \, \text{N}^{-1}\text{mm}^{-1} \\ -392.7 \times 10^{-15} \, \text{N}^{-1}\text{mm}^{-1} & 1.1687 \times 10^{-15} \, \text{N}^{-1}\text{mm}^{-2} \end{bmatrix}$$

The strain ε_k at time t_k is determined using Equation 5.91:

$$\varepsilon_k = \mathbf{F}_k (\mathbf{r}_{ext,k} - \mathbf{f}_{cr,k} + \mathbf{f}_{cs,k} - \mathbf{f}_{p,init} + \mathbf{f}_{p.rel,k} + \mathbf{f}_{cp,0})$$

$$= \begin{bmatrix} 324.2 \times 10^{-12} & -392.7 \times 10^{-15} \\ -392.7 \times 10^{-15} & 1.168 \times 10^{-15} \end{bmatrix}$$

$$\times \begin{bmatrix} (-100 - 599.4 - 1740 - 2000 + 58.82 + 0) \times 10^3 \\ (1000 - 87.73 - 520.1 - 1330.6 + 39.14 + 0) \times 10^6 \end{bmatrix}$$

$$= \begin{bmatrix} -1067.3 \times 10^{-6} \\ +0.6699 \times 10^{-6} \, \text{mm}^{-1} \end{bmatrix}$$

The strain at the reference axis and the curvature at time t_k are, respectively:

$$\varepsilon_{r,k} = -1067.3 \times 10^{-6} \quad \text{and} \quad \kappa_k = +0.6699 \times 10^{-6} \, \text{mm}^{-1}$$

From Equation 5.63, the top (y = +300 mm) and bottom (y = −850 mm) fibre strains are:

$$\varepsilon_{k(top)} = \varepsilon_{r,k} - 300 \times \kappa_k = -1268 \times 10^{-6}$$

$$\varepsilon_{k(btm)} = \varepsilon_{r,k} + 850 \times \kappa_k = -497.9 \times 10^{-6}$$

The concrete stress distribution at time t_k is calculated using Equation 5.103:

$$\sigma_{c,k(top)} = \bar{E}_{c,eff}(\varepsilon_{k(top)} - \varepsilon_{cs,k}) + \bar{F}_{e,0}\sigma_{c,0(top)} = -9.05 \text{ MPa}$$

$$\sigma_{c,k(btm)} = \bar{E}_{c,eff}(\varepsilon_{k(btm)} - \varepsilon_{cs,k}) + \bar{F}_{e,0}\sigma_{c,0(top)} = -0.47 \text{ MPa}$$

and, from Equation 5.104, the stresses in the non-prestressed reinforcement are:

$$\sigma_{s(1),k} = E_s[1 - y_{s(1)}]\varepsilon_k = -245.6 \text{ MPa}$$

$$\sigma_{s(2),k} = E_s[1 - y_{s(2)}]\varepsilon_k = -107.6 \text{ MPa}$$

The final stress in the prestressing steel at time t_k is given by Equation 5.105:

$$\sigma_{p(1),k} = E_p[1 - y_{p(1)}]\varepsilon_k + E_p(\varepsilon_{p(1),init} - \varepsilon_{p(1).rel,k}) = 1081.0 \text{ MPa}$$

$$\sigma_{p(2),k} = E_p[1 - y_{p(2)}]\varepsilon_k + E_p(\varepsilon_{p(2),init} - \varepsilon_{p(2).rel,k}) = 1089.4 \text{ MPa}$$

$$\sigma_{p(3),k} = E_p[1 - y_{p(3)}]\varepsilon_k + E_p(\varepsilon_{p(3),init} - \varepsilon_{p(3).rel,k}) = 1097.8 \text{ MPa}$$

The stress and strain distributions at t_0 (obtained in Example 5.4) and t_k are shown in Figure 5.15.

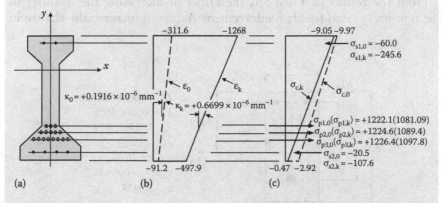

Figure 5.15 Strains and stresses at times at t_0 and t_k (Example 5.6). (a) Cross-section. (b) Strain (×10⁻⁶). (c) Stress (MPa).

5.7.4 Discussion

The results of several time analyses on the cross-section shown in Figure 5.16 are presented in Tables 5.1 and 5.2. The geometry of the cross-section and the material properties are similar to the cross-section of Examples 5.3 and 5.5. The effects of varying the quantities of the compressive and tensile non-prestressed reinforcements ($A_{s(1)}$ and $A_{s(2)}$, respectively) on the time-dependent deformation can be seen for three different values of sustained bending moment. At M_{ext} = 100 kNm, the initial concrete stress distribution is approximately triangular with higher compressive stresses in the bottom fibres (as determined in Example 5.3). At M_{ext} = 270 kNm, the initial concrete stress distribution is approximately uniform over the depth of the section and the curvature is small. At M_{ext} = 440 kNm, the initial stress distribution is again triangular with high compressive stresses in the top fibres. In each case, the prestressing force in the tendon immediately after transfer is $P_{init} = P_{m,0}$ = 1350 kN and, therefore, $\sigma_{pi} = \sigma_{p,0}$ = 1369 MPa.

In Tables 5.1 and 5.2, $\Delta F_{s(1)}$, $\Delta F_{s(2)}$ and ΔF_p are the compressive changes of force that gradually occur in the non-prestressed steel and the tendons with time. To maintain equilibrium, equal and opposite tensile forces are gradually imposed on the concrete as the bonded reinforcement restrains the time-dependent creep and shrinkage strains in the concrete. In Section 5.4.1, we defined the final compressive force acting on the concrete as ΩP_{m0} and here:

$$\Omega = 1 + \frac{\Delta F_{s(1)} + \Delta F_{s(2)} + \Delta F_p}{P_{m0}} \tag{5.112}$$

remembering that $\Delta F_{s(1)}$, $\Delta F_{s(2)}$ and ΔF_p are all negative. Values of Ω for each analysis are also given in Tables 5.1 and 5.2.

From the results in Table 5.1, the effect of increasing the quantity of the non-prestressed tensile reinforcement $A_{s(2)}$ is to increase the change in

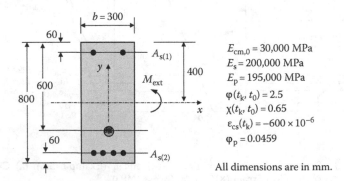

Figure 5.16 Post-tensioned cross section with tendon bonded after transfer.

Table 5.1 Effect of varying the bottom steel $A_{s(2)}$ (with $A_{s(1)} = 0$)

M_{ext} (kNm)	$A_{s(2)}$ (mm²)	$\varepsilon_{r,0}$ (×10⁻⁶)	κ_0 (×10⁻⁶) mm⁻¹	$\Delta F_{s(1)}$ (kN)	$\Delta F_{s(2)}$ (kN)	ΔF_p (kN)	Ω (Equation 5.112)	$\varepsilon_{r,k}$ (×10⁻⁶)	κ_k (×10⁻⁶) mm⁻¹
100	0	−191	−0.455	0	0	−277	0.795	−1154	−1.165
	1800	−178	−0.372	0	−288	−221	0.623	−1019	−0.252
	3600	−167	−0.306	0	−416	−190	0.551	−941	+0.271
270	0	−190	−0.008	0	0	−239	0.823	−1167	+0.336
	1800	−182	−0.038	0	−209	−200	0.697	−1075	+0.955
	3600	−177	+0.075	0	−306	−179	0.641	−1022	+1.311
440	0	−189	+0.438	0	0	−201	0.851	−1179	+1.838
	1800	−187	+0.448	0	−130	−179	0.771	−1132	+2.162
	3600	−186	+0.455	0	−197	−167	0.730	−1104	+2.350

Table 5.2 Effect of varying the top steel $A_{s(1)}$ (with $A_{s(2)} = 1800$ mm²)

M_{ext} (kNm)	$A_{s(1)}$ (mm²)	$\varepsilon_{r,0}$ (×10⁻⁶)	κ_0 (×10⁻⁶) mm⁻¹	$\Delta F_{s(1)}$ (kN)	$\Delta F_{s(2)}$ (kN)	ΔF_p (kN)	Ω (Equation 5.112)	$\varepsilon_{r,k}$ (×10⁻⁶)	κ_k (×10⁻⁶) mm⁻¹
100	0	−178	−0.372	0	−288	−221	0.623	−1019	−0.252
	900	−176	−0.378	−131	−303	−222	0.514	−962	−0.545
	1800	−176	−0.383	−222	−314	−222	0.438	−921	−0.757
270	0	−182	+0.038	0	−209	−200	0.697	−1075	+0.955
	900	−178	+0.014	−176	−231	−201	0.550	−989	+0.509
	1800	−175	−0.007	−294	−247	−201	0.450	−926	+0.188
440	0	−187	+0.448	0	−130	−179	0.771	−1132	+2.162
	900	−180	+0.406	−221	−159	−180	0.585	−1015	+1.564
	1800	−174	+0.370	−366	−179	−180	0.463	−932	+1.133

positive or sagging curvature with time. The increase is most pronounced when the initial concrete compressive stress at the level of the steel is high, i.e. when the sustained moment is relatively small and the section is initially subjected to a negative or hogging curvature. When $A_{s(2)} = 3600$ mm² and $M_{ext} = 100$ kNm (row 3 in Table 5.1), κ_k is positive despite the significant initial negative curvature κ_0. Table 5.1 also indicates that the addition of non-prestressed steel in the tensile zone will reduce the time-dependent camber which often causes problems in precast members subjected to low sustained loads. For sections on which M_{ext} is sufficient to cause an initial positive curvature, such as when $M_{ext} = 440$ kNm in Table 5.1, an increase in $A_{s(2)}$ causes an increase in time-dependent curvature and hence an increase in final deflection.

The inclusion of non-prestressed steel at the top of the section $A_{s(1)}$ increases the change in negative curvature with time, as indicated in Table 5.2. For sections where the initial curvature is positive, such as when $M_{ext} = 440$ kNm, the inclusion of $A_{s(1)}$ reduces the time-dependent change in positive curvature (and hence the deflection of the member). However, when κ_0 is negative, such as when $M_{ext} = 100$ kNm, the inclusion of $A_{s(1)}$ can cause an increase in negative curvature with time and hence an increase in the upward camber of the member with time.

The significant unloading of the concrete with time on the sections containing non-prestressed reinforcement should be noted. For example, when $M_{ext} = 270$ kNm and $A_{s(1)} = A_{s(2)} = 1800$ mm^2 (i.e. equal quantities of top and bottom non-prestressed reinforcement), the concrete is subjected to a total gradually applied tensile force of $-(\Delta F_{s(1)} + \Delta F_{s(2)} + \Delta F_p) = 742$ kN as shown in Table 5.2. This means that 55% of the initial compression in the concrete (the initial prestressing force) is transferred into the bonded reinforcement with time. The bottom fibre concrete compressive stress reduces from -5.32 to -1.10 MPa. The loss of prestress in the tendon, however, is only 201 kN (14.9%).

It is evident that an accurate picture of the time-dependent behaviour of a prestressed concrete cross-section cannot be obtained unless the restraint provided to creep and shrinkage by the non-prestressed steel is adequately accounted for. It is also evident that the presence of non-prestressed reinforcement significantly reduces the cracking moment with time and may in fact relieve the concrete of much of its initial compression.

5.8 SHORT-TERM ANALYSIS OF CRACKED CROSS-SECTIONS

5.8.1 General

In the cross-sectional analyses in Sections 5.6 and 5.7, it was assumed that concrete can carry the imposed stresses, both compressive and tensile. However, concrete is not able to carry large tensile stresses. If the tensile stress at a point reaches the tensile strength of concrete (Equations 4.6 and 4.7), cracking occurs. On a cracked cross-section, tensile stress of any magnitude cannot be carried normal to the crack surface at any time after cracking and tensile forces can only be carried across a crack by steel reinforcement. Therefore, on a cracked cross-section, internal actions can be carried only by the steel reinforcement (and tendons) and the uncracked parts of the concrete section.

In members subjected only to axial tension, caused either by external loads or by restraint to shrinkage or temperature change, *full-depth* cracks occur when the tensile stress reaches the tensile strength of the concrete at a particular location (i.e. at each crack location, the entire cross-section is cracked). When the axial tension is caused by restraint to shrinkage, cracking causes a loss of stiffness and a consequent decrease in the internal tension. The crack width and the magnitude of the restraining force, as well as the spacing between cracks, depend on the amount of bonded reinforcement. The steel carries the entire tensile force across each crack, but between the cracks in a member subjected to axial tension the concrete continues to carry tensile stress due to the bond between the steel and the concrete, and hence, the tensile concrete between the cracks continues to contribute to the member stiffness. This is known as the tension stiffening effect.

In a flexural member, cracking occurs when the tensile stress produced by the external moment at a particular section overcomes the compression caused by prestress, and the extreme fibre stress reaches the tensile strength of concrete. *Primary cracks* develop at a reasonably regular spacing on the tensile side of the member. The bending moment at which cracking first occurs is the *cracking moment* M_{cr}. If the applied moment at any time is greater than the cracking moment, cracking will occur and, at each crack, the concrete below the neutral axis on the cracked section is ineffective. In the previous section, we saw that the initial compressive stress in the concrete due to prestress is gradually relieved by creep and shrinkage and so the cracking moment decreases with time.

A loss of stiffness occurs at first cracking and the short-term moment–curvature relationship becomes non-linear. The height of primary cracks h_o depends on the quantity of tensile reinforcement and the magnitude of any axial force or prestress. For reinforced concrete members in pure bending with no axial force, the height of the primary cracks h_o immediately after cracking is usually relatively high (0.6–0.9 times the depth of the member depending on the quantity and position of tensile steel) and remains approximately constant under increasing bending moments until either the steel reinforcement yields or the concrete stress–strain relationship in the compressive region becomes non-linear. For prestressed members and members subjected to bending plus axial compression, h_o may be relatively small initially and gradually increases as the applied moment increases.

Immediately after first cracking, the intact concrete between adjacent primary cracks carries considerable tensile force, mainly in the direction of the reinforcement, due to the bond between the steel and the concrete. The average tensile stress in the concrete may be a significant percentage of the tensile strength of concrete. The steel stress is a maximum at a crack, where the steel carries the entire tensile force, and drops to a minimum between the cracks. The bending stiffness of the member is considerably greater than that based on a fully cracked section, where concrete in tension is assumed to carry zero stress. This *tension stiffening* effect is particularly significant in lightly reinforced concrete slabs under service loads.

For prestressed concrete members, or reinforced members in combined bending and compression, the effect of tension stiffening is less pronounced because the loss of stiffness caused by cracking is less significant. As the applied moment increases, the depth of the primary cracks increases gradually (in contrast to the sudden crack propagation in a reinforced member

in pure bending) and the depth of the concrete compressive zone is significantly greater than would be the case if no axial prestress was present.

5.8.2 Assumptions

The Euler–Bernoulli assumption that plane sections remain plane is not
strictly true for a cross-section in the cracked region of a beam. However, if
strains are measured over a gauge length containing several primary cracks,
the *average* strain diagram may be assumed to be linear over the depth of
a cracked cross-section.

The analysis presented here is based on the following assumptions:

1. Plane sections remain plane and, as a consequence, the strain distribution is linear over the depth of the section.
2. Perfect bond exists between the non-prestressed steel reinforcement
   and the concrete and between the bonded tendons and the concrete,
   i.e. the bonded steel and concrete strains are assumed to be compatible. This is usually a reasonable assumption at service loads in members containing deformed steel reinforcing bars and strands.
3. Strain in unbonded tendons is assumed to be unaffected by deformation of the concrete cross-section.
4. Tensile stress in the concrete is ignored, and therefore the tensile concrete does not contribute to the cross-sectional properties.
5. Material behaviour is linear-elastic. This includes concrete in compression, and both the non-prestressed and prestressed reinforcements.

5.8.3 Analysis

In the short-term analysis of fully cracked prestressed concrete cross-sections
at first loading (at time t_0), it is assumed that the axial force and bending
moment about the x-axis ($N_{ext,0}$ and $M_{ext,0}$, respectively) produce tension
of sufficient magnitude to cause cracking in the *bottom* fibres of the cross-
section and compression at the top of the section.

Consider the cracked prestressed concrete cross-section shown in
Figure 5.17. The section is symmetric about the y-axis, and the orthogonal
x-axis is selected as the reference axis. Also shown in Figure 5.17 are the
initial stress and strain distributions when the section is subjected to combined external bending and axial force ($M_{ext.0}$ and $N_{ext.0}$) sufficient to cause
cracking in the bottom fibres.

As for the analysis of an uncracked cross-section, the properties of each
layer of non-prestressed reinforcement are defined by its area, elastic modulus and location with respect to the arbitrarily chosen x-axis, i.e. $A_{s(i)}$, $E_{s(i)}$
and $y_{s(i)}$ ($= d_{ref} - d_{s(i)}$), respectively. Similarly, $A_{p(i)}$, $E_{p(i)}$ and $y_{p(i)}$ ($= d_{ref} - d_{p(i)}$)

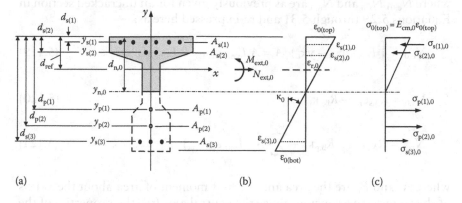

Figure 5.17 Fully cracked prestressed concrete cross-section. (a) Cross-section. (b) Strain diagram. (c) Stress.

represent the area, elastic modulus and location of the prestressing steel with respect to the x-axis, respectively.

The strain at any distance y from the reference x-axis at time t_0 is given by:

$$\varepsilon_0 = \varepsilon_{r,0} - y\kappa_0 \tag{5.113}$$

and the stresses in the concrete, the steel reinforcement and the bonded tendons are:

$$\sigma_{c,0} = E_{cm,0}\varepsilon_0 = E_{cm,0}(\varepsilon_{r,0} - y\kappa_0) \quad \text{for } y \geq y_{n,0}(= -d_{n,0} + d_{ref}) \tag{5.114}$$

$$\sigma_{c,0} = 0 \quad \text{for } y < y_{n,0} \tag{5.115}$$

$$\sigma_{s(i),0} = E_{s(i)}\varepsilon_0 = E_{s(i)}(\varepsilon_{r,0} - y_{s(i)}\kappa_0) \tag{5.116}$$

$$\sigma_{p(i),0} = E_{p(i)}(\varepsilon_0 + \varepsilon_{p(i),init}) = E_{p(i)}(\varepsilon_{r,0} - y_{p(i)}\kappa_0 + \varepsilon_{p(i),init}) \tag{5.117}$$

where $y_{n,0}$ is the y coordinate of the neutral axis, as shown in Figure 5.17a and $\varepsilon_{p(i),init}$ is the strain in the i-th layer of prestressing steel immediately before the transfer of prestress to the concrete as expressed in Equation 5.27. For unbonded tendons, the stress in the steel at t_0 is $E_{p(i)}\varepsilon_{p(i),init}$.

The internal axial force $N_{int,0}$ on the cracked cross-section is the sum of the axial forces resisted by the various materials forming the cross-section and is given by:

$$N_{int,0} = N_{c,0} + N_{s,0} + N_{p,0} \tag{5.118}$$

where $N_{c,0}$, $N_{s,0}$ and $N_{p,0}$ are as previously given for an uncracked section in Equations 5.29 through 5.31 and re-expressed here as:

$$N_{c,0} = \int_{A_c} E_{cm,0}(\varepsilon_{r,0} - y\kappa_0)dA = A_c E_{cm,0}\varepsilon_{r,0} - B_c E_{cm,0}\kappa_0 \tag{5.119}$$

$$N_{s,0} = R_{A,s}\varepsilon_{r,0} - R_{B,s}\kappa_0 \tag{5.120}$$

$$N_{p,0} = R_{A,p}\varepsilon_{r,0} - R_{B,p}\kappa_0 + \sum_{i=1}^{m_p}(A_{p(i)}E_{p(i)}\varepsilon_{p(i),\text{init}}) \tag{5.121}$$

where A_c and B_c are the area and the first moment of area about the x-axis of the compressive concrete above the neutral axis (i.e. the properties of the intact compressive concrete). The rigidities of the bonded steel are:

$$R_{A,s} = \sum_{i=1}^{m_s}(A_{s(i)}E_{s(i)}) \tag{5.122}$$

$$R_{B,s} = \sum_{i=1}^{m_s}(y_{s(i)}A_{s(i)}E_{s(i)}) \tag{5.123}$$

$$R_{I,s} = \sum_{i=1}^{m_s}\left(y_{s(i)}^2 A_{s(i)}E_{s(i)}\right) \tag{5.124}$$

$$R_{A,p} = \sum_{i=1}^{m_p}(A_{p(i)}E_{p(i)}) \tag{5.125}$$

$$R_{B,p} = \sum_{i=1}^{m_p}\left(y_{p(i)}A_{p(i)}E_{p(i)}\right) \tag{5.126}$$

$$R_{I,p} = \sum_{i=1}^{m_p}\left(y_{p(i)}^2 A_{p(i)}E_{p(i)}\right) \tag{5.127}$$

For unbonded tendons, $R_{A,p} = R_{B,p} = R_{I,p} = 0$.
   Noting that:

$$P_{\text{init}} = \sum_{i=1}^{m_p}(A_{p(i)}E_{p(i)}\varepsilon_{p(i),\text{init}}) \tag{5.128}$$

and defining the internal moment caused by P_{init} about the reference axis as:

$$M_{init} = -\sum_{i=1}^{m_p} (y_{p(i)} A_{p(i)} E_{p(i)} \varepsilon_{p(i),init})$$ (5.129)

Equation 5.118 can be rewritten as:

$$N_{int,0} = \int_{A_c} E_{cm,0}(\varepsilon_{r,0} - y\kappa_0)dA + (R_{A,s} + R_{A,p})\varepsilon_{r,0} - (R_{B,s} + R_{B,p})\kappa_0 + P_{init}$$ (5.130)

Remembering that for equilibrium $N_{ext,0} = N_{int,0}$, Equation 5.130 can be re-expressed as:

$$N_{ext,0} - P_{init} = \int_{A_c} E_{cm,0}(\varepsilon_{r,0} - y\kappa_0)dA + (R_{A,s} + R_{A,p})\varepsilon_{r,0} - (R_{B,s} + R_{B,p})\kappa_0$$ (5.131)

Similarly, the following expression based on moment equilibrium can be derived as:

$$M_{ext,0} - M_{init} = -\int_{A_c} E_{cm,0}(\varepsilon_{r,0} - y\kappa_0)y\,dA - (R_{B,s} + R_{B,p})\varepsilon_{r,0} + (R_{I,s} + R_{I,p})\kappa_0$$

(5.132)

For a reinforced concrete cross-section comprising rectangular components (e.g. rectangular flanges and webs) loaded in pure bending (i.e. $N_{ext,0} = N_{int,0} = 0$) and with no prestress, Equation 5.131 becomes a quadratic equation that can be solved to obtain the location of the neutral axis $y_{n,0}$.

If the cross-section is prestressed or the axial load $N_{ext,0}$ is not equal to zero (i.e. if $N_{ext,0} - P_{init} \neq 0$), dividing Equation 5.132 by Equation 5.131 gives:

$$\frac{M_{ext,0} - M_{init}}{N_{ext,0} - P_{init}} = \frac{-\int_{A_c} E_{cm,0}(\varepsilon_{r,0} - y\kappa_0)y\,dA - (R_{B,s} + R_{B,p})\varepsilon_{r,0} + (R_{I,s} + R_{I,p})\kappa_0}{\int_{A_c} E_{cm,0}(\varepsilon_{r,0} - y\kappa_0)dA + (R_{A,s} + R_{A,p})\varepsilon_{r,0} - (R_{B,s} + R_{B,p})\kappa_0}$$

and dividing the top and bottom of the right-hand side by κ_0 and recognising that at the axis of zero strain $y = y_{n,0} = \varepsilon_{r,0}/\kappa_0$, the previous expression becomes:

$$\frac{M_{ext,0} - M_{init}}{N_{ext,0} - P_{init}} = \frac{-\int_{A_c} E_{cm,0}(y_{n,0} - y)y\,dA - (R_{B,s} + R_{B,p})y_{n,0} + (R_{I,s} + R_{I,p})}{\int_{A_c} E_{cm,0}(y_{n,0} - y)\,dA + (R_{A,s} + R_{A,p})y_{n,0} - (R_{B,s} + R_{B,p})} \quad (5.133)$$

For a rectangular section of width b, Equation 5.133 becomes:

$$\frac{M_{ext,0} - M_{init}}{N_{ext,0} - P_{init}} = \frac{-\int_{y=y_{n,0}}^{y=d_{ref}} E_{cm,0}(y_{n,0} - y)\,b\,y\,dy - (R_{B,s} + R_{B,p})y_{n,0} + (R_{I,s} + R_{I,p})}{\int_{y=y_{n,0}}^{y=d_{ref}} E_{cm,0}(y_{n,0} - y)\,b\,dy + (R_{A,s} + R_{A,p})y_{n,0} - (R_{B,s} + R_{B,p})}$$

$$(5.134)$$

Equation 5.134 may be solved for $y_{n,0}$ relatively quickly using a simple trial-and-error search.

When $y_{n,0}$ is determined, and the depth of the intact compressive concrete above the cracked tensile zone $d_{n,0}$ is known, the properties of the compressive concrete (A_c, B_c and I_c) with respect to the reference axis may be readily calculated. Similarly, the axial rigidity and the stiffness related to the first and second moments of area of the cracked section about the reference axis (i.e. $R_{A,0}$, $R_{B,0}$ and $R_{I,0}$) are calculated at time t_0 using Equations 5.35, 5.36 and 5.39 and are re-expressed here as:

$$R_{A,0} = A_c E_{cm,0} + R_{A,s} + R_{A,p} \quad (5.135)$$

$$R_{B,0} = B_c E_{cm,0} + R_{B,s} + R_{B,p} \quad (5.136)$$

$$R_{I,0} = I_c E_{cm,0} + R_{I,s} + R_{I,p} \quad (5.137)$$

Using the same solution procedure previously adopted for an uncracked cross-section, the system of equilibrium equations governing the problem (Equation 5.40) is rewritten here as:

$$\mathbf{r}_{ext,0} = \mathbf{D}_0 \mathbf{\varepsilon}_0 + \mathbf{f}_{p,init} \quad (5.40)$$

where

$$\mathbf{r}_{ext,0} = \begin{bmatrix} N_{ext,0} \\ M_{ext,0} \end{bmatrix} \quad (5.41)$$

$$\mathbf{D}_0 = \begin{bmatrix} R_{A,0} & -R_{B,0} \\ -R_{B,0} & R_{I,0} \end{bmatrix} \tag{5.42}$$

$$\boldsymbol{\varepsilon}_0 = \begin{bmatrix} \varepsilon_{r,0} \\ \kappa_0 \end{bmatrix} \tag{5.43}$$

$$\mathbf{f}_{p,init} = \begin{bmatrix} \sum P_{init(i)} \\ \sum -y_{p(i)}P_{init(i)} \end{bmatrix} = \begin{bmatrix} P_{init} \\ M_{init} \end{bmatrix} \tag{5.44}$$

The vector $\boldsymbol{\varepsilon}_0$ is readily obtained by solving the equilibrium equations (Equation 5.40) giving:

$$\boldsymbol{\varepsilon}_0 = \mathbf{D}_0^{-1}\left(\mathbf{r}_{ext,0} - \mathbf{f}_{p,init}\right) = \mathbf{F}_0\left(\mathbf{r}_{ext,0} - \mathbf{f}_{p,init}\right) \tag{5.45}$$

where:

$$\mathbf{F}_0 = \frac{1}{R_{A,0}R_{I,0} - R_{B,0}^2}\begin{bmatrix} R_{I,0} & R_{B,0} \\ R_{B,0} & R_{A,0} \end{bmatrix} \tag{5.46}$$

The stress distribution related to the concrete and reinforcement can then be calculated from the constitutive equations specified in Equations 5.114 through 5.117.

As an alternative approach, the solution may also be conveniently obtained using the cross-sectional properties of the transformed section. For example, for the cross-section of Figure 5.17, the transformed cross-section in equivalent areas of concrete for the short-term analysis is shown in Figure 5.18.

The cross-sectional rigidities of the transformed section defined in Equations 5.135 through 5.137 can be recalculated as:

$$R_{A,0} = A_0 E_{cm,0} \tag{5.138}$$

$$R_{B,0} = B_0 E_{cm,0} \tag{5.139}$$

$$R_{I,0} = I_0 E_{cm,0} \tag{5.140}$$

where A_0 is the area of the transformed cracked concrete section and B_0 and I_0 are the first and second moments of the transformed area about the reference x-axis at first loading.

Substituting Equations 5.138 through 5.140 into Equation 5.46, the matrix $\mathbf{F}_0$ becomes:

$$\mathbf{F}_0 = \frac{1}{E_{cm,0}\left(A_0 I_0 - B_0^2\right)}\begin{bmatrix} I_0 & -B_0 \\ -B_0 & A_0 \end{bmatrix} \tag{5.141}$$

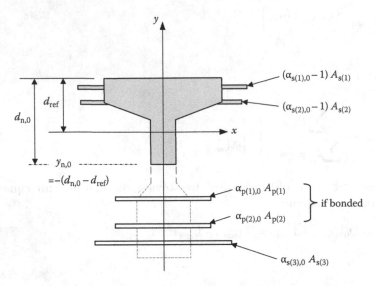

Figure 5.18 Transformed cracked section with bonded reinforcement transformed into equivalent areas of concrete.

EXAMPLE 5.7

The depth of the concrete compression zone d_n and the short-term stress and strain distributions are to be calculated on the prestressed concrete beam cross-section shown in Figure 5.19, when $M_{ext,0}$ = 400 kNm (and $N_{ext,0}$ = 0). The section contains two layers of non-prestressed reinforcement as shown (each with E_s = 200,000 MPa) and one layer of bonded prestressing steel (E_p = 195,000 MPa). The prestressing force before transfer is P_{init} = 900 kN (i.e. $\sigma_{p,init}$ = 1200 MPa). The tensile strength of the concrete is 3.5 MPa, and the elastic modulus is $E_{cm,0}$ = 30,000 MPa.

From Equations 5.122 through 5.127:

$R_{A,s} = (A_{s(1)} + A_{s(2)})\, E_s = (500 + 1,000) \times 200,000 = 300 \times 10^6$ N

$R_{B,s} = (y_{s(1)}A_{s(1)} + y_{s(2)}A_{s(2)})\, E_s = (250 \times 500 - 400 \times 1,000) \times 200,000$

$= -55.0 \times 10^9$ Nmm

$R_{I,s} = (y_{s(1)}^2 A_{s(1)} + y_{s(2)}^2 A_{s(2)})\, E_s = [250^2 \times 500 + (-400)^2 \times 1,000]$

$\times 200,000$

$= 38.25 \times 10^{12}$ Nmm^2

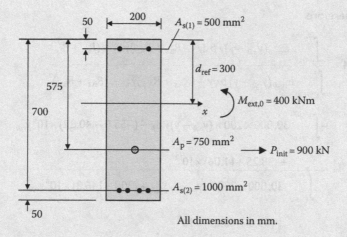

Figure 5.19 Cross-section (Example 5.7).

$R_{A,p} = A_p E_p = 750 \times 195,000 = 146.3 \times 10^6$ N

$R_{B,p} = y_p A_p E_s = -275 \times 750 \times 195,000 = -40.22 \times 10^9$ Nmm

$R_{I,p} = y_p^2 A_p E_s = (-275)^2 \times 750 \times 195,000 = 11.06 \times 10^{12}$ Nmm²

The vector of actions due to initial prestress is given by Equation 5.44:

$$\mathbf{f}_{p,init} = \begin{bmatrix} P_{init} \\ M_{init} \end{bmatrix} = \begin{bmatrix} 900 \times 10^3 \\ -(-275) \times 900 \times 10^3 \end{bmatrix} = \begin{bmatrix} 900 \times 10^3 \text{ N} \\ +247.5 \times 10^6 \text{ Nmm} \end{bmatrix}$$

If it is initially assumed that the section is uncracked, an analysis using the procedure outlined in Section 5.6.2 indicates that the tensile strength of the concrete has been exceeded in the bottom fibres of the cross-section. With the reference axis selected at $d_{ref} = 300$ mm below the top of the section, the depth of the neutral axis below the reference axis $y_{n,0}$ is determined from Equation 5.134. The left-hand side of Equation 5.134 is first calculated as:

$$\frac{M_{ext,0} - M_{init}}{N_{ext,0} - P_{init}} = \frac{400 \times 10^6 - 247.5 \times 10^6}{0 - 900 \times 10^3} = -169.4 \text{ mm}$$

and therefore

$$-169.4 = \frac{-\displaystyle\int_{y=y_{n,0}}^{y=300} E_{cm,0}(y_{n,0} - y)y\,b\,dy - (R_{B,s} + R_{B,p})y_{n,0} + (R_{I,s} + R_{I,p})}{\displaystyle\int_{y=y_{n,0}}^{y=300} E_{cm,0}(y_{n,0} - y)\,b\,dy + (R_{A,s} + R_{A,p})y_{n,0} - (R_{B,s} + R_{B,p})}$$

$$= \frac{-\displaystyle\int_{y=y_{n,0}}^{y=300} 30,000 \times 200 \times (y_{n,0} - y)y\,dy - (-55.0 - 40.22)\times 10^9 y_{n,0}}{\displaystyle\int_{y=y_{n,0}}^{y=300} 30,000 \times 200 \times (y_{n,0} - y)\,dy + (300 + 146.3)\times 10^6 y_{n,0}} \\ \frac{+ (38.25 + 11.06)\times 10^{12}}{+ (55.0 + 40.22)\times 10^9}$$

$$= \frac{-\left| 6.0\times 10^6 \times (0.5y_{n,0}y^2 - 0.3\dot{3}y^3)\right|_{y_{n,0}}^{300} + 95.22\times 10^9 y_{n,0} + 49.31\times 10^{12}}{\left| 6.0\times 10^6 \times (y_{n,0}y - 0.5y^2)\right|_{y_{n,0}}^{300} + 446.3\times 10^6 y_{n,0} + 95.22\times 10^9}$$

$$= \frac{y_{n,0}^3 - 174.8\times 10^3 y_{n,0} + 103.3\times 10^6}{-3y_{n,0}^2 + 2.246\times 10^3 y_{n,0} - 174.78\times 10^3}$$

Solving gives $y_{n,0} = -206.8$ mm and the depth of the neutral axis below the top surface is $d_{n,0} = d_{ref} - y_{n,0} = 506.8$ mm.

The properties of the compressive concrete (A_c, B_c and I_c) with respect to the reference axis are:

$A_c = 506.8 \times 200 - 500 = 100,858$ mm^2

$B_c = 506.8 \times 200 \times (300-253.4) - 500 \times (300-50) = +4.599 \times 10^6$ mm^3

$I_c = 200 \times 506.8^3/12 + 506.8 \times 200 \times (300-253.4)^2 - 500 \times (300-50)^2$

$ = 2358 \times 10^6$ mm^4

The cross-sectional rigidities $R_{A,0}$, $R_{B,0}$ and $R_{I,0}$ are obtained from Equations 5.135 through 5.137:

$R_{A,0} = A_c E_{cm,0} + R_{A,s} + R_{A,p} = 100,858 \times 30,000 + 300 \times 10^6 + 146.3 \times 10^6$

$\phantom{R_{A,0}} = 3,472 \times 10^6$ mm^2

$R_{B,0} = B_c E_{cm,0} + R_{B,s} + R_{B,p} = 4.599 \times 10^6 \times 30,000 - 55 \times 10^9 - 40.22 \times 10^9$

$\phantom{R_{B,0}} = 42.75 \times 10^9$ mm^3

$$R_{I,0} = I_c E_{cm,0} + R_{I,s} + R_{I,p} = 2{,}358 \times 10^6 \times 30{,}000 + 38.25 \times 10^{12} + 11.06 \times 10^{12}$$

$$= 120.1 \times 10^{12} \text{ mm}^4$$

From Equation 5.46:

$$\mathbf{F}_0 = \frac{1}{R_{A,0}R_{I,0} - R_{B,0}^2}\begin{bmatrix} R_{I,0} & R_{B,0} \\ R_{B,0} & R_{A,0} \end{bmatrix} = \begin{bmatrix} 289.3 \times 10^{-12} & 103.0 \times 10^{-15} \\ 103.0 \times 10^{-15} & 8.366 \times 10^{-15} \end{bmatrix}$$

and the strain vector is obtained from Equation 5.45:

$$\varepsilon_0 = \mathbf{F}_0 \ (r_{ext,0} - f_{p,init}) = \begin{bmatrix} 289.3 \times 10^{-12} & 103.0 \times 10^{-15} \\ 103.0 \times 10^{-15} & 8.366 \times 10^{-15} \end{bmatrix}\begin{bmatrix} 0 - 900 \times 10^3 \\ 400 \times 10^6 - 247.5 \times 10^6 \end{bmatrix}$$

$$= \begin{bmatrix} -244.7 \times 10^{-6} \\ 1.183 \times 10^{-6} \text{ mm}^{-1} \end{bmatrix}$$

The top ($y = 300$ mm) and bottom ($y = -450$ mm) fibre strains are:

$$\varepsilon_{0(top)} = \varepsilon_{r,0} - 300 \times \kappa_0 = (-244.7 - 300 \times 1.183) \times 10^{-6} = -600 \times 10^{-6}$$

$$\varepsilon_{0(bot)} = = \varepsilon_{r,0} + 450 \times \kappa_0 = (-244.7 + 450 \times 1.183) \times 10^{-6} = +288 \times 10^{-6}$$

The distribution of strains is shown in Figure 5.20b.

The top fibre stress in the concrete and the stress in the non-prestressed reinforcement are (Equations 5.114 and 5.116):

$$\sigma_{c,0(top)} = E_{cm,0} \ \varepsilon_{0(top)} = 30{,}000 \times (-600 \times 10^{-6}) = -17.99 \text{ MPa}$$

$$\sigma_{s(1),0} = E_s \ (\varepsilon_{r,0} - y_{s(1)} \kappa_0) = 200{,}000 \times (-244.7 - 250 \times 1.183) \times 10^{-6}$$

$$= -108.1 \text{ MPa}$$

$$\sigma_{s(2),0} = E_s \ (\varepsilon_{r,0} - y_{s(2)} \kappa_0) = 200{,}000 \times (-244.7 + 400 \times 1.183) \times 10^{-6}$$

$$= +45.7 \text{ MPa}$$

and the stress in the prestressing steel is given by Equation 5.117:

$$\sigma_{p,0} = E_p \ (\varepsilon_{r,0} - y_p \kappa_0 + \varepsilon_{p,init})$$

$$= 195{,}000 \times (-244.7 + 275 \times 1.183 + 6{,}000) \times 10^{-6}$$

$$= +1{,}216 \text{ MPa}$$

The stresses are plotted in Figure 5.20c.

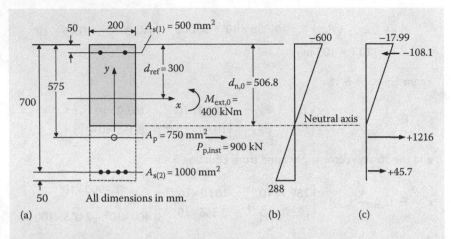

Figure 5.20 Stress and strain distributions for cracked section (Example 5.7). (a) Section. (b) Strain (×10⁻⁶). (c) Stress (MPa).

5.9 TIME-DEPENDENT ANALYSIS OF CRACKED CROSS-SECTIONS

5.9.1 Simplifying assumption

For a cracked cross-section under sustained actions, creep causes a gradual change of the position of the neutral axis and the size of the concrete compressive zone gradually increases with time. With the size of the cross-section gradually changing, and hence the sectional properties of the concrete gradually change, the principle of superposition does not apply. To accurately account for the increasing size of the compressive zone, a detailed numerical analysis is required in which time is discretised into many small steps.

However, if one assumes that the depth of the concrete compression zone d_n remains constant with time, the time analysis of a fully cracked cross-section using the AEMM is essentially the same as that outlined in Section 5.7.3. This assumption greatly simplifies the analysis and usually results in a relatively little error in the calculated deformations.

5.9.2 Long-term analysis of a cracked cross-section subjected to combined axial force and bending using the AEMM

Consider the fully cracked cross-section shown in Figure 5.21a subjected to a sustained external bending moment $M_{ext,k}$ and axial force $N_{ext,k}$. Both the short-term and time-dependent strain distributions are shown in Figure 5.21b.

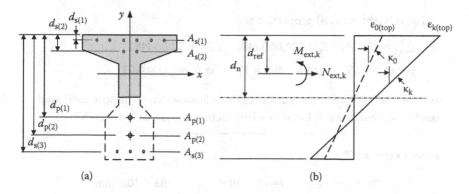

Figure 5.21 Fully cracked cross-section — time analyses (AEMM). (a) Cross-section. (b) Strain.

For the time analysis, the constitutive relationship for the concrete at t_k is that given by Equation 5.59 as follows:

$$\sigma_{c,k} = \bar{E}_{c,eff}(\varepsilon_k - \varepsilon_{cs,k}) + \bar{F}_{e,0}\sigma_{c,0} \quad \text{for } y \geq y_{n,0} \tag{5.142}$$

$$\sigma_{c,k} = 0 \quad \text{for } y < y_{n,0} \tag{5.143}$$

where $\bar{E}_{c,eff}$ and $\bar{F}_{e,0}$ are as defined in Equations 5.57 and 5.60. The stress–strain relationships for the reinforcement and tendons are as given in Equations 5.69 through 5.72.

At time t_k, the internal axial force $N_{int,k}$ and moment $M_{int,k}$ on the cross-section are given by Equations 5.83 and 5.88 and the axial rigidity and the stiffness related to the first and second moments of area ($R_{A,k}$, $R_{B,k}$, and $R_{I,k}$, respectively) are given by Equations 5.84, 5.85 and 5.89 (ignoring the cracked concrete below the neutral axis). The equilibrium equations are expressed in Equation 5.93, and solving using Equation 5.101 gives the strain vector at time t_k. The stresses in the concrete, steel reinforcement and tendons at time t_k are then calculated from Equations 5.103 through 5.107, respectively.

EXAMPLE 5.8

Calculate the change of stress and strain with time on the cracked prestressed cross-section of Example 5.7 using the AEMM. The cracked section and the initial strain distribution are shown in Figure 5.20. The actions on the section are assumed to be constant throughout the time period under consideration (i.e. t_0 to t_k) and equal to:

$$N_{ext,0} = N_{ext,k} = 0 \quad \text{and} \quad M_{ext,0} = M_{ext,k} = 400 \text{ kNm}$$

The relevant material properties are:

$$E_{cm,0} = 30 \text{ GPa} \quad E_s = 200{,}000 \text{ MPa} \quad E_p = 195{,}000 \text{ MPa} \quad \varphi(t_k, t_0) = 2.5$$

$$\chi(t_k, t_0) = 0.65 \quad \varepsilon_{cs}(t_k) = -400 \times 10^{-6} \quad \varphi_p(t_k, \sigma_{p(i),0}) = 0.02$$

and the steel reinforcement is assumed to be linear-elastic. The post-tensioned tendon was bonded at t_0 in the cracked section analysis of Example 5.7.

From Example 5.7:

$$d_{n,0} = 506.8 \text{ mm} \quad \varepsilon_{r,0} = -244.7 \times 10^{-6} \quad \kappa_0 = 1.183 \times 10^{-6} \text{ mm}^{-1}$$

$$A_c = 101{,}858 \text{ mm}^2 \quad B_c = +4.599 \times 10^6 \text{ mm}^3 \quad I_c = 2{,}358 \times 10^6 \text{ mm}^4$$

and the rigidities of the steel reinforcement and tendons are:

$$R_{A,s} = 300 \times 10^6 \text{ N} \quad R_{B,s} = -55.0 \times 10^9 \text{ Nmm} \quad R_{I,s} = 38.25 \times 10^{12} \text{ Nmm}^2$$

$$R_{A,p} = 146.3 \times 10^6 \text{ N} \quad R_{B,p} = -40.22 \times 10^9 \text{ Nmm} \quad R_{I,p} = 11.06 \times 10^{12} \text{ Nmm}^2$$

From Equations 5.57 and 5.60:

$$\bar{E}_{c,eff} = \frac{30{,}000}{1 + 0.65 \times 2.5} = 11430 \text{ MPa} \quad \text{and} \quad \bar{F}_{e,0} = \frac{2.5 \times (0.65 - 1.0)}{1.0 + 0.65 \times 2.5} = -0.33\dot{3}$$

From Equation 5.96, the axial force and moment resisted by the concrete part of the cross-section at time t_0 are:

$$N_{c,0} = A_c E_{cm,0} \varepsilon_{r,0} - B_c E_{cm,0} \kappa_0 = -903.5 \times 10^3 \text{ N}$$

$$M_{c,0} = -B_c E_{cm,0} \varepsilon_{r,0} + I_c E_{cm,0} \kappa_0 = +117.5 \times 10^6 \text{ Nmm}$$

and from Equations 5.84, 5.85 and 5.89, the cross-sectional rigidities $R_{A,k}$, $R_{B,k}$ and $R_{I,k}$ are:

$$R_{A,k} = A_c \bar{E}_{c,eff} + R_{A,s} + R_{A,p} = 1599 \times 10^6 \text{ N}$$

$$R_{B,k} = B_c \bar{E}_{c,eff} + R_{B,s} + R_{B,p} = -42.66 \times 10^9 \text{ Nmm}$$

$$R_{I,k} = I_c \bar{E}_{c,eff} + R_{I,s} + R_{I,p} = 76.26 \times 10^{12} \text{ Nmm}^2$$

Equation 5.102 gives:

$$\mathbf{F}_k = \frac{1}{R_{A,k} R_{I,k} - R_{B,k}^2} \begin{bmatrix} R_{I,k} & -R_{B,k} \\ -R_{B,k} & R_{A,k} \end{bmatrix} = \begin{bmatrix} 634.9 \times 10^{-12} & -355.2 \times 10^{-15} \\ -355.2 \times 10^{-15} & 13.31 \times 10^{-15} \end{bmatrix}$$

From Equations 5.96 through 5.100:

$$\mathbf{f}_{cr,k} = \bar{F}_{e,0}\begin{bmatrix} N_{c,0} \\ M_{c,0} \end{bmatrix} = \begin{bmatrix} +301.2 \times 10^3 \\ -39.15 \times 10^6 \end{bmatrix}$$

$$\mathbf{f}_{cs,k} = \begin{bmatrix} A_c \\ -B_c \end{bmatrix} \bar{E}_{c,eff}\varepsilon_{cs,k} = \begin{bmatrix} -461.1 \times 10^3 \\ +21.02 \times 10^6 \end{bmatrix}$$

$$\mathbf{f}_{p,init} = \sum_{i=1}^{m_p}\begin{bmatrix} A_{p(i)}E_{p(i)}\varepsilon_{p(i),init} \\ y_{p(i)}A_{p(i)}E_{p(i)}\varepsilon_{p(i),init} \end{bmatrix} = \begin{bmatrix} 900 \times 10^3 \\ 247.5 \times 10^6 \end{bmatrix} \text{ (from Example 5.7)}$$

$$\mathbf{f}_{p,rel,k} = \sum_{i=1}^{m_p}\begin{bmatrix} A_{p(i)}E_{p(i)}\varepsilon_{p(i),0}\varphi_{p(i)} \\ y_{p(i)}A_{p(i)}E_{p(i)}\varepsilon_{p(i),0}\varphi_{p(i)} \end{bmatrix} = \begin{bmatrix} 18 \times 10^3 \\ 4.95 \times 10^6 \end{bmatrix}$$

and as the tendon was bonded at transfer:

$$\mathbf{f}_{cp,0} = \begin{bmatrix} 0 \\ 0 \end{bmatrix}$$

The strain vector is obtained from Equation 5.101:

$$\varepsilon_k = \mathbf{F}_k(\mathbf{r}_{ext,k} - \mathbf{f}_{cr,k} + \mathbf{f}_{cs,k} - \mathbf{f}_{p,init} + \mathbf{f}_{p,rel,k})$$

$$\varepsilon_k = \begin{bmatrix} 634.9 \times 10^{-12} & -355.2 \times 10^{-15} \\ -355.2 \times 10^{-15} & 13.31 \times 10^{-15} \end{bmatrix}\begin{bmatrix} (0 - 301.2 - 461.1 - 900 + 18) \times 10^3 \\ (400 + 39.15 + 21.02 - 247.5 + 4.95) \times 10^6 \end{bmatrix}$$

$$= \begin{bmatrix} -1121 \times 10^{-6} \\ 3.481 \times 10^{-6} \text{ mm}^{-1} \end{bmatrix}$$

The top ($y = 300$ mm) and bottom ($y = -450$ mm) fibre strains are:

$$\varepsilon_{k(top)} = \varepsilon_{r,k} - 300 \times \kappa_k = (-1121 - 300 \times 3.481) \times 10^{-6} = -2166 \times 10^{-6}$$

$$\varepsilon_{k(btm)} = \varepsilon_{r,k} + 450 \times \kappa_k = (-1121 + 450 \times 3.481) \times 10^{-6} = +445 \times 10^{-6}$$

At the neutral axis depth at $y_{n,0} = -208.1$ mm (below the reference axis):

$$\varepsilon_{k(dn)} = \varepsilon_{r,k} - y_{n,0} \times \kappa_k = (-1121 + 206.8 \times 3.481) \times 10^{-6} = -401 \times 10^{-6}$$

The concrete stresses at time t_k at the top fibre ($y = +300$ mm) and the bottom fibre of the compressive concrete are obtained from Equation 5.103: At the top of section:

$$\sigma_{c,k} = 11,430 \times (-2,166 + 400) \times 10^{-6} - 0.33\dot{3} \times (-17.99) = -14.2 \text{ MPa}$$

At $y = -206.8$ mm:

$$\sigma_{c,k} = 11,430 \times (-1,121 + 208.1 \times 3.483 + 400) \times 10^{-6} - 0.33\dot{3} \times 0 = 0.0 \text{ MPa}$$

The stresses in the non-prestressed reinforcement are (Equation 5.104):

$$\sigma_{s(1),k} = E_s(\varepsilon_{r,k} - y_{s(1)}\kappa_k) = -398 \text{ MPa}$$

$$\sigma_{s(2),k} = E_s(\varepsilon_{r,k} - y_{s(2)}\kappa_k) = +54.2 \text{ MPa}$$

and the stress in the prestressing steel is given by (Equation 5.105)

$$\sigma_{p,k} = E_p(\varepsilon_{r,k} - y_p\kappa_k + \varepsilon_{p.init,k} - \varepsilon_{p.rel,k}) = +1144 \text{ MPa}$$

The results are plotted in Figure 5.22. It can be seen that the time-dependent loss of prestress in the tendon on this cross-section is only 72 MPa or only 6% of the initial prestress. With relaxation losses at 2%, creep and shrinkage have resulted in only 4% loss as the change in strain at the tendon level is relatively small after cracking.

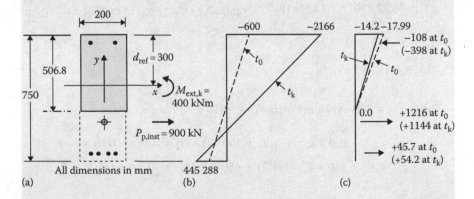

Figure 5.22 Initial and time-dependent strain and stress distributions. (a) Section. (b) Strain ($\times 10^{-6}$). (c) Stress (MPa).

5.10 LOSSES OF PRESTRESS

5.10.1 Definitions

The losses of prestress that occur in a tendon are categorised as either *immediate losses* or *time-dependent losses* and are illustrated in Figure 5.23.

Immediate losses occur when the prestress is transferred to the concrete at time t_0 and may vary along the length of the tendon. Immediate losses are the difference between the force imposed on the tendon by the hydraulic prestressing jack P_{max} (=P_j) and the force in the tendon immediately after transfer at a distance x from the active end of the tendon $P_{m0}(x)$ and can be expressed as:

$$\text{Immediate loss} = P_{max} - P_{m0}(x) \tag{5.144}$$

Time-dependent losses are the gradual losses of prestress that occur with time over the life of the structure. If $P_{m,t}(x)$ is the force in the prestressing tendon at x from the active end of the tendon after all losses, then:

$$\text{Time-dependent loss} = P_{m0}(x) - P_{m,t}(x) \tag{5.145}$$

Both immediate and time-dependent losses are made up of several components. Immediate losses depend on the method and equipment used to prestress the concrete and include losses due to elastic shortening of concrete, wedge draw-in at the prestressing anchorage, friction in the jack and along the tendon, deformation of the forms for precast members, deformation in the joints between elements of precast structures, temperature changes that may occur during this period and the relaxation of the tendon in a pretensioned member between the time of tensioning the wires before the concrete is cast and the time of transfer (particularly significant when the concrete is cured at elevated temperatures prior to transfer).

Time-dependent losses are the gradual losses of prestress that occur with time over the life of the structure. These include losses caused by the gradual shortening of concrete at the steel level due to creep and shrinkage, relaxation of the tendon after transfer and time-dependent deformation that may occur within the joints in segmental construction.

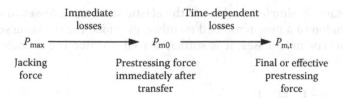

Figure 5.23 Losses of prestress in the tendons.

5.10.2 Immediate losses

The magnitude of immediate losses is taken as the sum of the losses caused by each relevant phenomenon. Where appropriate, the effects of one type of immediate loss on the magnitude of other immediate losses should be considered. For example, in a pretensioned member, the loss caused by relaxation of the tendon prior to transfer will affect the magnitude of the immediate loss caused by elastic deformation of concrete.

5.10.2.1 Elastic deformation losses

Pretensioned members: The change in strain in a tendon in a pretensioned member immediately after transfer $\Delta\varepsilon_{p,0}$ caused by elastic shortening of the concrete is equal to the instantaneous strain in the concrete at the steel level $\varepsilon_{cp,0}$:

$$\varepsilon_{cp,0} = \frac{\sigma_{cp,0}}{E_{cm,0}} = \Delta\varepsilon_{p,0} = \frac{\Delta\sigma_{p,0}}{E_p}$$

The corresponding loss of stress in a tendon at transfer is therefore the product of the modular ratio $(E_p/E_{cm,0})$, and the stress in the adjacent concrete at the tendon level $\sigma_{cp,0}$ and the loss of force in the tendon is given by:

$$\Delta P_{el} = \Delta\sigma_{p,0}A_p = \frac{E_p}{E_{cm,0}}\sigma_{cp,0}A_p \qquad (5.146)$$

Post-tensioned members: For post-tensioned members with one tendon, or with two or more tendons stressed simultaneously, the elastic deformation of the concrete occurs during the stressing operation before the tendons are anchored. In this case, elastic shortening losses are zero. In a member containing more than one tendon and where the tendons are stressed sequentially, stressing of a tendon causes an elastic shortening loss in all previously stressed and anchored tendons. Consequently, the first tendon to be stressed suffers the largest elastic shortening loss and the last tendon to be stressed suffers no elastic shortening loss at all. Elastic shortening losses in the tendons stressed early in the prestressing sequence can be reduced by re-stressing the tendons (prior to grouting of the prestressing ducts).

It is relatively simple to calculate the elastic shortening losses in an individual tendon of a post-tensioned member, provided the stressing sequence is known. For most cases, it is sufficient to determine the average loss of stress as follows:

$$\Delta\sigma_p = \frac{n-1}{2n}\frac{E_p}{E_{cm,0}}\frac{P}{A} \qquad (5.147)$$

where n is the number of tendons and P/A is the average concrete compressive stress. In post-tensioned members, the tendons are not bonded to the concrete until grouting of the duct occurs after the stressing sequence is completed. It is the shortening of the member between the anchorage plates that leads to elastic shortening, and not the strain at the steel level, as is the case for pretensioned members.

5.10.2.2 Friction in the jack and anchorage

The loss caused by friction in the jack and anchorage depends on the jack pressure and the type of jack and anchorage system used. It is usually allowed for during the stressing operation and is generally relatively small.

5.10.2.3 Friction along the tendon

In post-tensioned members, friction losses occur along the tendon during the stressing operation. Friction between the tendon and the duct causes a gradual reduction in prestress with the distance along the tendon x from the jacking end. The coefficient of friction between the tendon and the duct depends basically on the condition of the surfaces in contact, the profile of the duct, the nature of the tendon and its preparation. The magnitude of the friction loss depends on the tendon length, x, and the total angular change of the tendon over that length, as well as the size and type of the duct containing the tendon. An estimation of the loss of force in the tendon due to friction at any distance x from the jacking end may be made using [1]:

$$\Delta P_\mu(x) = P_{max} \left(1 - e^{-\mu(\theta+kx)}\right) \tag{5.148}$$

where θ is the sum in radians of the absolute values of successive angular deviations of the tendon over the length x. Care should be taken during construction to achieve the same cable profile as that assumed in the design. μ is the coefficient of friction between the tendon and its duct and depends on the surface characteristics of the tendon and the duct, the presence of rust on the surface of the tendon and the elongation of the tendon. In the absence of more specific data, the values of μ given in Table 5.3 are specified in EN 1992-1-1 [1], when all tendons in contact within the same duct are stressed simultaneously. For tendons showing a high but still acceptable amount of rusting, the value of μ may increase by 20%. If the wires or strands in contact in one duct are stressed separately, μ may be significantly greater than the values given above and should be checked by tests. For external tendons passing over machined cast-steel saddles, μ may increase markedly for large movements of tendons across the saddles. k is an estimate of the unintentional angular deviation (in radians/m) due to wobble effects in the straight or curved parts of internal

Table 5.3 Coefficient of friction μ for post-tensioned tendons [1]

	Internal tendons[a]	External unbonded tendons			
		Steel duct/ non-lubricated	HDPE duct/ non-lubricated	Steel duct/ lubricated	HDPE duct/ lubricated
Cold-drawn wire	0.17	0.25	0.14	0.18	0.12
Strand	0.19	0.24	0.12	0.16	0.10
Deformed bar	0.65	—	—	—	—
Smooth round bar	0.33	—	—	—	—

HDPE, high-density polyethylene.

[a] For tendons that fill about half the duct.

tendons and depends on the rigidity of sheaths, the spacing and fixing of their supports, the care taken in placing the prestressing tendons, the clearance of tendons in the duct, the stiffness of the tendons and the precautions taken during concreting. In segmental construction, the angular deviation per metre (k) may be greater in the event of mismatching of ducts and the designer should allow for this possibility. The most important parameter affecting the rigidity of sheaths is their diameter ϕ. In the absence of other data, EN 1992-1-1 [1] suggests that for internal tendons k will generally be in the range of $0.005 < k < 0.01$ per metre. The Australian standard AS3600-2009 [9] suggests that:

- for sheathing containing wires or strands:
  $k = 0.024{-}0.016$ rad/m　　　　when $\phi \le 50$ mm
  $k = 0.016{-}0.012$ rad/m　　　　when 50 mm $< \phi \le 90$ mm
  $k = 0.012{-}0.008$ rad/m　　　　when $\phi > 90$ mm
- for flat metal ducts containing wires or strands:
  $k = 0.024{-}0.016$ rad/m
- for sheathing containing bars:
  $k = 0.016{-}0.008$ rad/m　　　　when $\phi \le 50$ mm
- for bars with a greased-and-wrapped coating:
  $k = 0.008$ rad/m.

EXAMPLE 5.9

Calculate the friction losses in the prestressing cable in the end span of the post-tensioned girder of Figure 5.24. The jacking force $P_{max} = 2500$ kN. For this cable, $\mu = 0.19$ and $k = 0.01$.

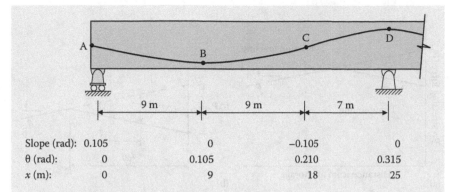

Slope (rad):	0.105		0		−0.105		0
θ (rad):	0		0.105		0.210		0.315
x (m):	0		9		18		25

Figure 5.24 Tendon profile for end span (Example 5.9).

From Equation 5.148:

At B: $\Delta P_\mu(x) = 2500(1-e^{-0.19(0.105+0.01\times9)}) = 90.9$ kN (i.e. 3.64% losses)

At C: $\Delta P_\mu(x) = 2500(1-e^{-0.19(0.210+0.01\times18)}) = 178.6$ kN (i.e. 7.14% losses)

At D: $\Delta P_\mu(x) = 2500(1-e^{-0.19(0.315+0.01\times25)}) = 254.5$ kN (i.e. 10.2% losses)

5.10.2.4 Anchorage losses

In post-tensioned members, some slip or draw-in occurs when the pre-stressing force is transferred from the jack to the anchorage. This causes an additional loss of prestress. The amount of slip depends on the type of anchorage. For wedge-type anchorages used for strand, the slip (Δ_{slip}) may be as high as 6 mm. The loss of prestress caused by Δ_{slip} decreases with distance from the anchorage owing to friction and, for longer tendons, may be negligible at the critical design section. For short tendons, this loss may be significant and should not be ignored in design.

The loss of tension in the tendon caused by slip is opposed by friction in the same way as the initial prestressing force is opposed by friction, but in the opposite direction, i.e. μ and k are the same. The variations of pre-stressing force along a member due to friction before anchoring the tendon (calculated using Equation 5.148) and after anchoring are shown in Figure 5.25, where the mirror image reduction in prestressing force in the vicinity of the anchorage is caused by slip at the anchorage. The slope of the draw-in line adjacent to the anchorage has the same magnitude as the slope of the friction loss line (β), but the opposite sign. It follows that tendons with a small drape (and therefore small β) will suffer anchorage slip losses over a longer length of tendons than tendons with a large drape (larger β).

In order to calculate the draw-in loss at the anchorage ΔP_{di}, the length of the draw-in line L_{di} must be determined. By equating the anchorage slip

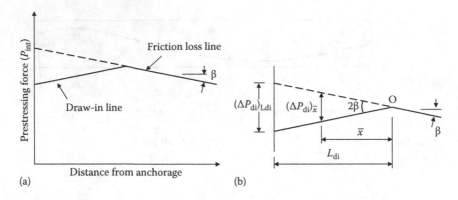

Figure 5.25 Variation of prestress adjacent to the anchorage due to draw-in. (a) Prestressing force versus distance from anchorage. (b) Loss of prestress in vicinity of anchorage.

Δ_{slip} with the integral of the change in strain in the steel tendon over the length of the draw-in line, L_{di} may be determined. If β is the slope of the friction loss line (i.e. the friction loss per unit length) as shown in Figure 5.25, the loss of prestress due to draw-in at a distance $\bar{x}$ from point O in Figure 5.25b is $(\Delta P_{di})_{\bar{x}} = 2\beta\bar{x}$ and Δ_{slip} can be estimated as follows:

$$\Delta_{slip} = \int_0^{L_{di}} \frac{2\beta\bar{x}}{E_p A_p} \, d\bar{x} = \frac{\beta L_{di}^2}{E_p A_p} \tag{5.149}$$

and rearranging gives:

$$L_{di} = \sqrt{\frac{E_p A_p \Delta_{slip}}{\beta}} \tag{5.150}$$

The immediate loss of prestress at the anchorage caused by Δ_{slip} is:

$$(\Delta P_{di})_{L_{di}} = 2\beta L_{di} \tag{5.151}$$

The immediate loss of prestress near an anchorage can be determined from geometry using Figure 5.25. At a distance of more than L_{di} from the live end anchorage, the immediate loss of prestress due to Δ_{slip} is zero.

The magnitude of the slip that should be anticipated in design is usually supplied by the anchorage manufacturer and should be checked on site. Cautious overstressing at the anchorage is often an effective means of compensating for slip.

5.10.2.5 Other causes of immediate losses

Additional immediate losses may occur due to deformation of the forms of precast members and deformation in the construction joints between the precast units in segmental construction, and these losses must be assessed in design. Any change in temperature between the time of stressing the tendon and the time of casting the concrete in a pretensioned member will also cause immediate losses, as will any difference in the temperature between the stressed tendons and the concrete during steam curing.

5.10.3 Time-dependent losses of prestress

5.10.3.1 Discussion

We have seen in Sections 5.7 and 5.9 that, in addition to causing time-dependent increases in deflection or camber, both compressive creep and shrinkage of the concrete cause gradual shortening of a concrete member and this, in turn, leads to time-dependent shortening of the prestressing tendons, and a consequent reduction in the prestressing force. These time-dependent losses of prestress are in addition to the losses caused by steel relaxation and may adversely affect the long-term serviceability of the structure and should be accounted for in design.

In Section 5.7, a time analysis was presented for determining the effects of creep and shrinkage of concrete and relaxation of the tendon on the long-term stresses and deformations of a prestressed concrete cross-section of any shape and containing any layout of prestressed and non-prestressed reinforcement.

In the following section, the approximate procedure specified in EN 1992-1-1 [1] for calculating time-dependent losses of prestress is outlined. The method may give inaccurate and sometimes misleading results because it does not adequately account for the significant loss of precompression in the concrete that occurs when non-prestressed reinforcement is present. For a realistic estimate of the time-dependent losses of prestress in the tendon, and the redistribution of stresses between the bonded reinforcement and the concrete, the method described in Section 5.7 is recommended.

For members containing only tendons, the loss in tensile force in the tendons is simply equal to the loss in compressive force in the concrete. Where the member contains a significant amount of longitudinal non-prestressed reinforcement, there is a gradual transfer of the compressive prestressing force from the concrete into the bonded reinforcement. Shortening of the concrete, due to creep and shrinkage, causes a shortening of the bonded reinforcement and therefore an increase in compressive stress in the steel. The gradual increase in compressive force in the bonded reinforcement is accompanied by an equal and opposite decrease in the compressive force in the concrete. The loss in compressive force in the concrete is therefore considerably greater than the loss in tensile force in the tendon. The redistribution of stresses with time was discussed in Section 5.7, and illustrated, for example, in Figures 5.14 and 5.15, where the immediate strain and stress distributions (at time t_0 immediately after the application of both prestress and the applied moment $M_{ext.0}$) and the long-term strain and stress distributions (after creep and shrinkage at time t_k) on prestressed concrete cross-sections are shown.

Many authorities suggest that the total time-dependent loss of prestress should be estimated by adding the calculated losses of prestress due to shrinkage, creep and relaxation. However, separate calculation of these losses is problematic as time-dependent losses interact with each other, and this interaction should be considered when the sum of all the losses is determined. For example, the loss in tendon force due to creep and shrinkage of the concrete decreases the average force in the tendon with time, and this in turn reduces the relaxation loss. Restraint to shrinkage often substantially reduces the compressive stresses in the concrete at the steel level, and this may significantly affect the creep of the concrete at this level and reduce losses due to creep.

5.10.3.2 Simplified method specified in EN 1992-1-1:2004

The time-dependent losses due to creep, shrinkage and relaxation (ΔP_{c+s+r}) at any location x under permanent loads may be approximated by:

$$\Delta P_{c+s+r} = A_p \Delta \sigma_{p,c+s+r} = A_p \frac{\varepsilon_{cs} E_p + 0.8 \Delta \sigma_{p,r} + \dfrac{E_p}{E_{cm}} \varphi(t,t_0) \sigma_{c.QP}}{1 + \dfrac{E_p A_p}{E_{cm} A_c} \left(1 + \dfrac{A_c}{I_c} z_{cp}^2\right)[1 + 0.8\varphi(t,t_0)]} \tag{5.152}$$

where:

    $\sigma_{p,c+s+r}$ is the absolute value of the variation of stress in the tendon due to creep, shrinkage and relaxation at location x at time t;

    ε_{cs} is the absolute value of the estimated shrinkage strain at the time under consideration and may be estimated using the procedures outlined in Section 4.2.5.4;

    $\Delta \sigma_{p,r}$ is the absolute value of the variation of stress in the tendon at time t due to relaxation of the tendon and should be calculated from the initial stress in the tendon caused by P_{m0} and the quasi-permanent actions $(G + \psi_2 Q)$;

    $\varphi(t,t_0)$ is the creep coefficient at time t for loads applied at t_0;

    $\sigma_{c.QP}$ is the stress in the concrete adjacent to the tendon due to self-weight, initial prestress and other quasi-permanent actions where relevant (depending on the stage of construction under consideration);

    A_p is the total area of all the tendons at location x;

    A_c is the area of the concrete cross-section;

    I_c is the second moment of the concrete cross-section about its centroidal axis; and

    z_{cp} is the distance between the centre of gravity of the concrete section and the tendons.

The denominator in Equation 5.152 accounts for the restraint to creep and shrinkage provided by the bonded tendons but largely ignores the effect of the restraint offered by any non-prestressed bonded reinforcement on the loss of prestress in the tendon.

5.10.3.3 Alternative simplified method

If a concrete member of length L contains no bonded reinforcement (and no bonded tendons) and is unrestrained at its supports and along its length, the member would shorten due to shrinkage by an amount equal to $\varepsilon_{cs}L$. If the member contained an unbonded post-tensioned tendon with an anchorage at each end of the member, the tendon would shorten by the same amount and the change of stress in the tendon due to shrinkage (ignoring the effects of friction) would be constant along its length and equal to $\Delta\sigma_{p.s} = E_p\varepsilon_{cs}$. In concrete structures, unrestrained contraction is unusual. Reinforcement and bonded tendons embedded in the concrete provide restraint to shrinkage and reduce the shortening of the member. This in turn reduces the loss of prestress in any tendon within the member.

Where the centroid of the bonded steel area (non-prestressed and prestressed) is at an eccentricity e_s from the centroidal axis of the concrete cross-section, the change of strain due to shrinkage at the centroid of the tendon can be approximated by Equation 5.153 and the corresponding change of stress in a tendon at this location is given by Equation 5.154:

$$\Delta\varepsilon_{p.s} = \frac{\varepsilon_{cs}}{1 + \bar{\alpha}_{ep.k}\rho\left[1 + \dfrac{A_c z_{cp} e_s}{I_c}\right]} \tag{5.153}$$

$$\Delta\sigma_{p.s} = \frac{\varepsilon_{cs}E_p}{1 + \bar{\alpha}_{ep.k}\rho\left[1 + \dfrac{A_c z_{cp} e_s}{I_c}\right]} \tag{5.154}$$

where $\bar{\alpha}_{ep.k}$ is the age-adjusted modular ratio ($= E_p/\bar{E}_{c,eff}$), ρ is the ratio of the area of the bonded steel ($A_s + A_p$) to the area of concrete (i.e. $\rho = (A_s + A_p)/A_c$) and z_{cp} is the distance between the centroidal axis of the concrete section and the tendons. Equation 5.154 is in fact similar to the change in stress indicated in the first term of the right-hand side of Equation 5.152, except that, in Equation 5.152, the ageing coefficient χ has been taken as 0.8 and the only bonded steel assumed to provide restraint is the tendon, A_p.

Creep strain in the concrete at the level of the bonded tendon depends on the stress history of the concrete at that level. Because the concrete stress varies with time, a reliable estimate of creep losses requires a detailed time analysis of the cross-section (such as that presented in Section 5.7). An approximate and conservative estimate of creep losses can be made by assuming that the concrete stress at the tendon level remains constant with time and equal to the short-term value $\sigma_{c.QP}$ which is calculated using the initial prestressing force (prior to any time-dependent losses) and the sustained portion of all the service loads. Under this constant stress, the creep strain that would develop in the concrete is the product of the immediate elastic strain ($\sigma_{c.QP}/E_{cm}$) and the creep coefficient $\varphi(t,t_0)$. With this assumption, the change in stress in the tendon due to creep $\Delta\sigma_{p.c}$ may be obtained from Equation 5.155:

$$\Delta\sigma_{p.c} = \frac{E_p\,\varphi(t,t_0)(\sigma_{c.QP}/E_{cm})}{1+\bar{\alpha}_{ep.k}\,\rho[1+((A_c z_{cp} e_s)/I_c)]} \tag{5.155}$$

Again, this is similar to the creep term in Equation 5.152, where χ is 0.8 and the area of bonded steel is taken as A_p, i.e. the restraint provided by any non-prestressed reinforcement has been ignored.

The loss of stress in a tendon due to relaxation depends on the sustained stress in the steel. Owing to creep and shrinkage in the concrete, the stress in the tendon decreases with time at a faster rate than would occur due to relaxation alone. Since the steel strain is reducing with time due to concrete creep and shrinkage, the relaxation losses are reduced from those that would occur in a constant strain relaxation test. With the creep coefficient for the prestressing steel given in Table 4.9 (or calculated from the design relaxation loss using Equation 4.38), the percentage loss of prestress due to relaxation may be calculated from a detailed time analysis such as described in Section 5.7. In the absence of such an analysis, the change of stress in the tendon due to relaxation is approximated by:

$$\Delta\sigma_{p.r} = -\left(1 - \frac{|\Delta\sigma_{p.s} + \Delta\sigma_{p.c}|}{\sigma_{pi}}\right)\varphi_p\,\sigma_{pi} \tag{5.156}$$

where $\Delta\sigma_{p.s}$ and $\Delta\sigma_{p.c}$ are the changes in stress in the tendon caused by shrinkage and creep as given by Equations 5.154 and 5.155, respectively, and are usually compressive; σ_{pi} is the tendon stress just after transfer under the sustained service loads ($=P_{m0}/A_p$). The absolute values of $\Delta\sigma_{p.s}$ and $\Delta\sigma_{p.c}$ are used in Equation 5.156 to convert the negative changes of stress into positive losses.

EXAMPLE 5.10

Determine the time-dependent loss of prestress in the bonded tendon in the post-tensioned concrete cross-section shown in Figure 5.26 using:

1. the approximate procedure discussed in Section 5.10.3.3 (Equations 5.154 through 5.156); and
2. the approach specified in EN 1992-1-1 [1], i.e. Equation 5.152 in Section 5.10.3.2.

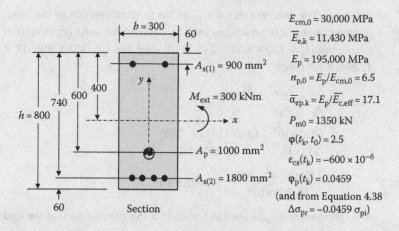

$E_{cm,0} = 30,000$ MPa

$\overline{E}_{e,k} = 11,430$ MPa

$E_p = 195,000$ MPa

$n_{p,0} = E_p/E_{cm,0} = 6.5$

$\overline{\alpha}_{ep.k} = E_p/\overline{E}_{c,eff} = 17.1$

$P_{m0} = 1350$ kN

$\varphi(t_k, t_0) = 2.5$

$\varepsilon_{cs}(t_k) = -600 \times 10^{-6}$

$\varphi_p(t_k) = 0.0459$

(and from Equation 4.38 $\Delta\sigma_{pr} = -0.0459\,\sigma_{pi}$)

Figure 5.26 Cross-section and material properties (Example 5.10).

The material properties are as shown in the figure and the prestressing force in the tendon immediately after transfer, when immediate losses had taken place and before the duct was grouted, is $P_{m0} = 1350$ kN. The sustained external bending moment is $M_{ext} = 270$ kNm.

In this example, the area of the concrete cross-section is $A_c = bh - A_{s(1)} - A_{s(2)} - A_p = 236.3 \times 10^3$ mm² and its centroid lies 397.9 mm below the top of the cross-section. The second moment of area of the concrete cross-section about its centroidal axis is $I_c = 12,450 \times 10^6$ mm⁴. The total area of bonded steel is $A_{s(1)} + A_{s(2)} + A_p = 3700$ mm² and its centroid is 536.8 mm below the top of the cross-section. The distance between the centroid of the concrete section and the centroid of the bonded steel area is $e_s = 138.9$ mm, and the distance between the centroid of the concrete section and the centroid of the tendon is $z_{cp} = 202.1$ mm. The steel ratio is $\rho = 3700/(236.3 \times 10^3) = 0.0157$.

1. *Using the simplified analysis of Section 5.10.3.3:*
   Shrinkage loss: From Equation 5.154:

$$\Delta\sigma_{p.s} = \frac{\varepsilon_{cs}E_p}{1 + \overline{\alpha}_{ep.k}\rho\left[1 + \dfrac{A_c z_{cp}e_s}{I_c}\right]} = \frac{-0.0006 \times 195,000}{1 + 17.1 \times 0.0157\left[1 + \dfrac{236.3 \times 10^3 \times 202.1 \times 138.9}{12,450 \times 10^6}\right]}$$

$$= -82.9\,\text{MPa}$$

The calculated loss of stress in the tendon determined using the time analysis of Section 5.7.3 caused by restrained shrinkage is -83.2 MPa, and this is in close agreement with the approximation of Equation 5.154.

Creep loss: The concrete stress $\sigma_{c.QP}$ at the tendon level due to the initial prestress and sustained actions may be determined using gross section properties ($A_g = 300 \times 800 = 240 \times 10^3$ mm² and $I_g = 300 \times 800^3/12 = 12,800 \times 10^6$ mm⁴):

$$\sigma_{c.QP} = -\frac{P_{mo}}{A_g} - \frac{P_{mo}y_p^2}{I_g} - \frac{M_{ext,0}y_p}{I_g}$$

$$= -\frac{1,350 \times 10^3}{240 \times 10^3} - \frac{1,350 \times 10^3 \times (-200)^2}{12,800 \times 10^6}$$

$$- \frac{270 \times 10^6 \times (-200)}{12,800 \times 10^6} = -5.625 \text{ MPa}$$

and this value of $\sigma_{c.QP}$ is used in Equation 5.155 to approximate the loss of stress in the tendon due to creep:

$$\Delta\sigma_{p.c} = \frac{E_p \, \varphi(t,t_0)(\sigma_{c.QP}/E_{cm})}{1 + \bar{\alpha}_{ep.k}\rho\left[1 + \dfrac{A_c z_{cp} e_s}{I_c}\right]}$$

$$= \frac{195,000 \times 2.5 \times (-5.625/30,000)}{1 + 17.1 \times 0.0157 \times \left[1 + \dfrac{236.3 \times 10^3 \times 202.1 \times 138.9}{12,450 \times 10^6}\right]} = -64.8 \text{ MPa}$$

Using the time analysis of Section 5.7.3, the loss of stress in the tendon due to creep is equal to −60.7 MPa and the approximation of Equation 5.155 is conservative.

Relaxation loss: With the stress in the tendon immediately after transfer $\sigma_{pi} = P_{m0}/A_p = 1350$ MPa, the loss of stress in the tendon due to relaxation is obtained from Equation 5.156:

$$\Delta\sigma_{p.r} = -\left(1 - \frac{|\Delta\sigma_{p.s} + \Delta\sigma_{p.c}|}{\sigma_{pi}}\right)\varphi_p\sigma_{pi}$$

$$= -\left(1 - \frac{|-82.9 - 64.8|}{1350}\right) \times 0.0459 \times 1350 = -55.2 \text{ MPa}$$

Using the time analysis of Section 5.7.3, the loss of stress in the tendon due to relaxation is equal to −56.8 MPa and this is in good agreement with the approximation of Equation 5.156.

Total time-dependent losses: Summing the losses caused by shrinkage, creep and relaxation, we get:

$$\Delta\sigma_{p.c+s+r} = \Delta\sigma_{p.s} + \Delta\sigma_{p.c} + \Delta\sigma_{p.r} = -202.9 \text{ MPa}$$

This is 15.0% of the initial prestress in the tendon, and this is in excellent agreement with the 14.9% losses (−200.7 MPa) determined using the more rigorous time analysis of Section 5.7.3.

2. *Simplified approach in EN 1992-1-1 [1]:*

   Equation 5.152 (with values of stresses and strains specified according to the sign convention adopted throughout this book) gives:

$$\Delta\sigma_{p.c+s+r} = \frac{\varepsilon_{cs}E_p + 0.8\Delta\sigma_{p.r} + \dfrac{E_p}{E_{cm}}\varphi(t,t_0)\,\sigma_{c.QP}}{1 + \dfrac{E_p A_p}{E_{cm} A_c}\left(1 + \dfrac{A_c}{I_c}z_{cp}^2\right)[1 + 0.8\varphi(t,t_0)]}$$

$$= \frac{-0.0006 \times 195,000 + 0.8 \times (-0.0459 \times 1,350) + \dfrac{195,000}{30,000} \times 2.5 \times (-5.625)}{1 + \dfrac{195,000 \times 1,000}{30,000 \times 236.3 \times 10^3}\left(1 + \dfrac{236.3 \times 10^3}{12,450 \times 10^6} \times 202.1^2\right)[1 + 0.8 \times 2.5]}$$

$$= -225.0 \text{ MPa}$$

This significantly overestimates time-dependent losses. Equation 5.152 fails to account for the restraining effects of the bonded non-prestressed steel on the creep and shrinkage losses and inappropriately assumes that the restraint to creep and shrinkage in the concrete provided by the tendon also applies to the development of relaxation (tensile creep) in the tendon.

5.11 DEFLECTION CALCULATIONS

5.11.1 General

If the axial strain and curvature are known at regular intervals along a member, it is a relatively simple task to determine the deformation of that member. Consider the statically determinate member AB of span l subjected to the axial and transverse loads, as shown in Figure 5.27a. The axial deformation of the member e_{AB} (either elongation or shortening) is obtained by integrating the axial strain at the centroid of the member $\varepsilon_a(z)$ over the length of the member, as shown:

$$e_{AB} = \int_0^l \varepsilon_a(z)\,dz \tag{5.157}$$

where z is measured along the member.

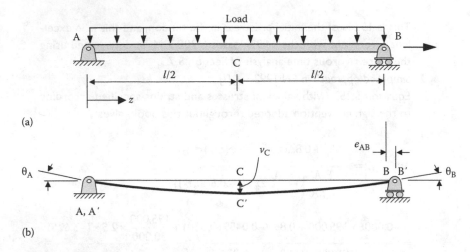

Figure 5.27 Deformation of a statically determinate member. (a) Original geometry. (b) Deformed shape.

Provided that deflections are small and that simple beam theory is applicable, the slope θ and deflection v at any point z along the member are obtained by integrating the curvature $\kappa(z)$ over the length of the member as follows:

$$\theta = \int_0^l \kappa(z)\, dz \tag{5.158}$$

$$v = \int_0^l \int_0^l \kappa(z)\, dz\, dz \tag{5.159}$$

Equations 5.158 and 5.159 are quite general and apply to both elastic and inelastic material behaviour.

If the axial strain and curvature are calculated at any time after loading at a preselected number of points along the member shown in Figure 5.27a and, if a reasonable variation of strain and curvature is assumed between adjacent points, it is a simple matter of geometry to determine the deformation of the member. For convenience, some simple equations are given below for the determination of the deformation of a single span and a cantilever. If the axial strain ε_a and the curvature κ are known at the mid-span and at each end of the member shown in Figure 5.27 (i.e. at supports A and B and at the mid-span C), the axial deformation e_{AB}, the slope at each support θ_A and θ_B and the deflection at mid-span v_C are given by Equations 5.160 through 5.167.

For a linear variation of strain and curvature:

$$e_{AB} = \frac{l}{4}(\varepsilon_{aA} + 2\varepsilon_{aC} + \varepsilon_{aB}) \tag{5.160}$$

$$\upsilon_C = \frac{l^2}{48}(\kappa_A + 4\kappa_C + \kappa_B) \tag{5.161}$$

$$\theta_A = \frac{l}{24}(5\kappa_A + 6\kappa_C + \kappa_B) \tag{5.162}$$

$$\theta_B = -\frac{l}{24}(\kappa_A + 6\kappa_C + 5\kappa_B) \tag{5.163}$$

and for a parabolic variation of strain and curvature:

$$e_{AB} = \frac{l}{6}(\varepsilon_{aA} + 4\varepsilon_{aC} + \varepsilon_{aB}) \tag{5.164}$$

$$\upsilon_C = \frac{l^2}{96}(\kappa_A + 10\kappa_C + \kappa_B) \tag{5.165}$$

$$\theta_A = \frac{l}{6}(\kappa_A + 2\kappa_C) \tag{5.166}$$

$$\theta_B = -\frac{l}{6}(2\kappa_C + \kappa_B) \tag{5.167}$$

In addition to the simple span shown in Figure 5.27, Equations 5.160 through 5.167 also apply to any member in a statically indeterminate frame, provided the strain and curvature at each end and at mid-span are known.

Consider the fixed-end cantilever shown in Figure 5.28. If the curvatures at the fixed support at A and the free end at B are known, then the slope and deflection at the free end of the member are given by Equations 5.168 through 5.171.

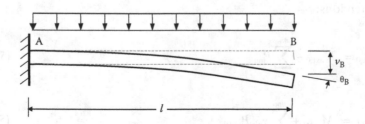

Figure 5.28 Deformation of a fixed-end cantilever.

For a linear variation of curvature:

$$\theta_B = -\frac{l}{2}(\kappa_A + \kappa_B) \tag{5.168}$$

$$v_B = \frac{l^2}{6}(2\kappa_A + \kappa_B) \tag{5.169}$$

For a parabolic variation of curvature typical of what occurs in a uniformly loaded cantilever:

$$\theta_B = -\frac{l}{3}(\kappa_A + 2\kappa_B) \tag{5.170}$$

$$v_B = -\frac{l^2}{4}(\kappa_A + \kappa_B) \tag{5.171}$$

5.11.2 Short-term moment–curvature relationship and tension stiffening

For any prestressed concrete section, the instantaneous moment–curvature relationship before cracking is linear-elastic. For an uncracked cross-section, the instantaneous curvature may be calculated using the procedure of Section 5.6.2. In particular, from Equation 5.45, the instantaneous curvature is:

$$\kappa_0 = \frac{R_{B,0}N_{R,0} + R_{A,0}M_{R,0}}{R_{A,0}R_{I,0} - R_{B,0}^2} \tag{5.172}$$

where $R_{A,0}$, $R_{B,0}$ and $R_{I,0}$ are the cross-sectional rigidities given by Equations 5.35, 5.36 and 5.39; $N_{R,0}$ is the sum of the external axial force (if any) and the resultant compressive prestressing force exerted on the cross-section by the tendons and $M_{R,0}$ is the sum of the external moment and the resultant moment about the centroidal axis caused by the compressive forces exerted by the tendons:

$$N_{R,0} = N_{ext,0} - \sum_{i=1}^{m_p} P_{init(i)} \tag{5.173}$$

$$M_{R,0} = M_{ext,0} + \sum_{i=1}^{m_p} y_{p(i)} P_{init(i)} \tag{5.174}$$

For uncracked, prestressed concrete cross-sections, if the reference axis is taken as the centroidal axis of the transformed section, the flexural rigidity $E_{cm,0} I_{uncr}$ is in fact the rigidity $R_{I,0}$ calculated using Equation 5.39, where I_{uncr} is the second moment of area of the uncracked transformed cross-section about its centroidal axis. Codes of practice generally suggest that, for short-term deflection calculations, I_{uncr} may be approximated by the second moment of area of the gross cross-section about its centroidal axis. The initial curvature caused by the applied moment and prestress acting on any uncracked cross-section may therefore be approximated by:

$$\kappa_0 = \frac{M_{R,0}}{E_{cm,0} I_{uncr}} \qquad (5.175)$$

After cracking, the instantaneous moment–curvature relationship can be determined using the analysis described in Section 5.8 (and illustrated in Example 5.7) for any level of applied moment greater than the cracking moment, provided the assumption of linear-elastic material behaviour remains valid for both the steel reinforcement/tendons and the concrete in compression. The analysis of the cross-section of Figure 5.19 after cracking in pure bending (i.e. when $M_{ext,0} = 400$ kNm and $N_{ext,0} = 0$) was illustrated in Example 5.7. If the analysis is repeated for different values of applied moment ($M_{ext,0}$ greater than the cracking moment M_{cr}), the instantaneous moment versus curvature ($M_{ext,0}$ vs κ_0) relationship for the cross-section can be determined and is shown in Figure 5.29. In addition to the post-cracking

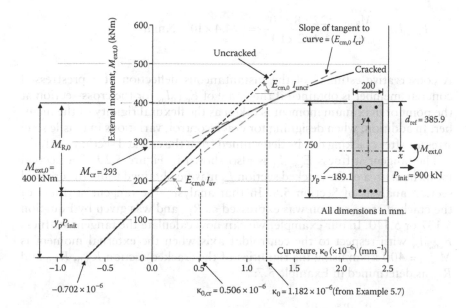

Figure 5.29 Short-term moment–curvature relationship for the prestressed concrete cross-section of Figure 5.19.

relationship, the linear relationship prior to cracking (Equation 5.175) is also shown in Figure 5.29, with $E_{cm,0}I_{uncr} = 242.4 \times 10^{12}$ Nmm^2 in this example.

It is noted that after cracking the neutral axis gradually rises as the applied moment increases. With the area of concrete above the crack becoming smaller, the second moment of area of the cracked section I_{cr} decreases as the applied moment increases. This is not the case for reinforced concrete sections where the depth to the neutral axis remains approximately constant with increasing moment and I_{cr} is constant in the post-cracking range.

For the cross-section shown in Figure 5.29, the x-axis has been taken to coincide with the centroidal axis of the transformed cross-section, and so the y coordinate of the prestressing steel (y_p) is numerically equal to the eccentricity of the prestressing force e. In this case, from Equation 5.174, $M_{R,0} = M_{ext,0} + y_p P_{init} = M_{ext,0} - eP_{init}$. In Figure 5.29, at any moment $M_{ext,0}$ greater than the cracking moment (M_{cr}), the curvature is:

$$\kappa_0 = \frac{M_{R,0}}{E_{cm,0}I_{av}} \qquad (5.176)$$

where $E_{cm,0}I_{av}$ is the secant stiffness. The secant stiffness $E_{cm,0}I_{av}$ corresponding to the external moment of $M_{ext,0} = 400$ kNm is shown in Figure 5.29, with $y_p P_{init} = -170.2$ kNm and, therefore, $M_{R,0} = 229.8$ kNm. With the value of curvature determined in Example 5.7 at this moment equal to 1.182×10^{-6}, the stiffness $E_{cm,0}I_{av}$ is obtained from Equation 5.176:

$$E_{cm,0}I_{av} = \frac{M_{R,0}}{\kappa_0} = \frac{229.8 \times 10^6}{1.182 \times 10^{-6}} = 194.4 \times 10^{12} \text{ Nmm}^2$$

A conservative estimate of the instantaneous deflection of a prestressed concrete member is obtained if the value of $E_{cm,0}I_{av}$ for the cross-section at the point of maximum moment is taken as the flexural rigidity of the member. In addition, when designing for crack control, variations in tensile steel stresses after cracking can be determined from the cracked section analysis.

The tangent stiffness $E_{cm,0}I_{cr}$ is also shown in Figure 5.29. The second moment of area of the cracked section I_{cr} may be obtained using the cracked section analysis of Section 5.8. In that analysis, the tangent stiffness of the cracked cross-section was expressed as $R_{I,0}$ and was given by Equation 5.137 or 5.140. In our example, we can now calculate the tangent stiffness $E_{cm,0}I_{cr}$ with respect to the centroidal axis when the external moment is $M_{ext,0} = 400$ kNm using the rigidities of the cracked section $R_{A,0}$, $R_{B,0}$ and $R_{I,0}$ as determined in Example 5.7:

$$E_{cm,0}I_{cr} = \frac{R_{A,0}R_{I,0} - R_{B,0}^2}{R_{A,0}} = 120.2 \times 10^{12} \text{ Nmm}^2$$

If the reference axis had corresponded to the centroidal axis of the cracked section, then $R_{B,0}$ would equal zero and $E_{cm,0}I_{cr} = R_{I,0}$.

For small variations in applied moment, curvature increments can be calculated using I_{cr}. In reinforced concrete construction, I_{cr} is constant and equal to I_{av}, but this is not so for prestressed concrete. For the moment–curvature graph of Figure 5.29, at $M_{ext,0} = 400$ kNm, we have $I_{uncr} = 8080 \times 10^6$ mm^4, $I_{av} = 6480 \times 10^6$ mm^4 and $I_{cr} = 4020 \times 10^6$ mm^4. It is noted that for this 200 mm by 750 mm rectangular cross-section, $I_g = 7030 \times 10^6$ mm^4 and this is 13% less than I_{uncr}.

For the case where $M_{ext,0} = 0$ in Figure 5.29, the internal moment caused by the resultant prestressing force about the centroidal axis of the uncracked section causes an initial negative curvature of $M_{R,0}/(E_{cm,0}I_{uncr}) = -e_p P_{init}/(E_{cm,0}I_{uncr}) = -0.702 \times 10^{-6}$ mm^{-1}. If the beam remained unloaded for a period of time after transfer and, if shrinkage occurred during this period, the restraint provided by the bonded reinforcement to shrinkage would introduce a positive change of curvature $\kappa_{cs,0}$ provided the centroid of the bonded reinforcement is below the centroidal axis of the cross-section (as is the case in Figure 5.29). Shrinkage before loading causes the curve in Figure 5.29 to shift to the right. The restraint to shrinkage also causes tensile stresses in the bottom fibres of the cross-section, and this may significantly reduce the cracking moment. The effect of a modest early shrinkage on the moment–curvature relationship of Figure 5.29 is illustrated in Figure 5.30.

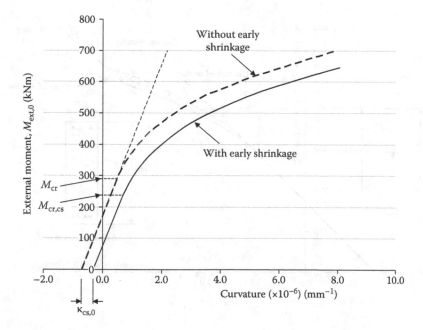

Figure 5.30 Effect of early shrinkage on the short-term moment–curvature relationship for a prestressed concrete cross section.

For cracked prestressed concrete members, the stiffness of the cracked cross-section, calculated using the procedure outlined in Section 5.8.3, may underestimate the actual stiffness of the member in the cracked region. The intact concrete between adjacent cracks carries tensile force, mainly in the direction of the reinforcement, due to the bond between the steel and the concrete. The average tensile stress in the concrete is therefore not zero and may be a significant fraction of the tensile strength of concrete. The stiffening effect of the uncracked tensile concrete is known as *tension stiffening*. The moment–curvature relationship of Figure 5.29 is reproduced in Figure 5.31. Also shown as the dashed line in Figure 5.31 is the moment versus *average curvature* relationship, with the average curvature being determined for a segment of beam containing two or more primary cracks. The hatched region between the curves at moments greater than the cracking moment M_{cr} represents the tension stiffening effect, i.e. the contribution of the tensile concrete between the primary cracks to the cross-sectional stiffness.

For conventionally reinforced members, tension stiffening contributes significantly to the member stiffness, particularly when the maximum moment is not much greater than the cracking moment. However, as the moment level increases, the tension stiffening effect decreases owing to additional secondary cracking at the level of the bonded reinforcement. For a prestressed member (or a reinforced member subjected to significant axial

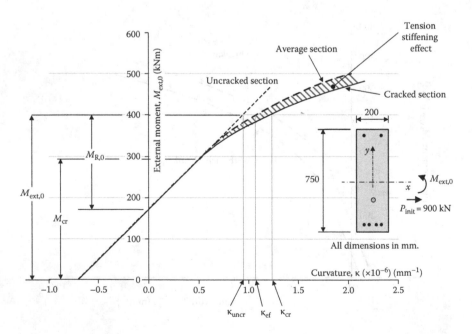

Figure 5.31 Short-term moment–average curvature relationship for the prestressed concrete cross section of Figure 5.19.

compression), the effect of tension stiffening is less pronounced because the loss of stiffness due to cracking is more gradual and significantly smaller.

Shrinkage-induced cracking and tensile creep cause a reduction of the tension stiffening effect with time. Repeated or cyclic loading also causes a gradual breakdown of tension stiffening.

Tension stiffening is usually accounted for in design by an empirical adjustment to the stiffness of the fully-cracked cross-section as discussed in the following section.

5.11.3 Short-term deflection

If the initial curvature is determined at the mid-span and at each end of the span of a beam or slab, the short-term deflection can be estimated using Equation 5.161 or 5.165, whichever is appropriate.

For uncracked cross-sections, the initial curvature is given by Equation 5.175. For cracked prestressed cross-sections, the initial curvature may be determined from Equation 5.176. This will be conservative unless an adjustment is made to include the tension stiffening effect. In codes of practice, this adjustment is often made using simplified techniques involving the determination of an effective second moment of area I_{ef} for the member. A number of empirical equations are available for estimating I_{ef}. Most have been developed specifically for reinforced concrete, where for a cracked member, I_{ef} lies between the second moments of area of the uncracked cross-section and the cracked transformed section about their centroidal axes I_{uncr} and I_{cr}, respectively. We have seen that for a prestressed concrete cross-section, I_{cr} varies with the applied moment as the depth of the crack gradually changes and its value at any load level is usually considerably less than I_{av}, as illustrated in Figure 5.29. The equations used for estimating I_{ef} for a reinforced section are not therefore directly applicable to prestressed concrete.

The empirical equation for I_{ef} proposed by Branson [10] is adopted in many codes and specifications for reinforced concrete members, including ACI318M [2]. For a prestressed concrete section, the following form of the equation can be used:

$$I_{ef} = I_{av} + (I_{uncr} - I_{av})(M_{cr}/M_{ext,0})^3 \le I_{uncr} \qquad (5.177)$$

where $M_{ext,0}$ is the maximum bending moment at the section, based on the short-term serviceability design load or the construction load and M_{cr} is the cracking moment. The cracking moment is best determined by undertaking a time analysis, as outlined in Section 5.7, to determine the effects of creep and shrinkage on the time-dependent redistribution of stresses between the concrete and the bonded reinforcement. An estimate of the cracking moment can be made from:

$$M_{cr} = Z\left(f_{ctm} - \sigma_{cs} + \frac{P}{A}\right) + \frac{P}{A} \ge 0.0 \qquad (5.178)$$

where Z is the section modulus of the uncracked section, referred to the extreme fibre at which cracking occurs; f_{ctm} is the mean tensile strength given by Equation 4.6 or 4.7; P is the effective prestressing force (after all losses); e is the eccentricity of the effective prestressing force measured to the centroidal axis of the uncracked section; A is the area of the uncracked cross-section and σ_{cs} is the maximum shrinkage-induced tensile stress on the uncracked section at the extreme fibre at which cracking occurs. In the absence of more refined calculation, σ_{cs} may be taken as [9]:

$$\sigma_{cs} = \left(\frac{2.5\rho_w - 0.8\rho_{cw}}{1 + 50\rho_w} E_s \varepsilon_{cs} \right) \tag{5.179}$$

where ρ_w is the web reinforcement ratio for the tensile steel $(A_s + A_{pt})/(b_w d)$; ρ_{cw} is the web reinforcement ratio for the compressive steel, if any, $A_{sc}/(b_w d)$; A_s is the area on non-prestressed tensile reinforcement; A_{pt} is the area of prestressing steel in the tensile zone; A_{sc} is the area of non-prestressed compressive reinforcement; E_s is the elastic modulus of the steel in MPa and ε_{cs} is the final design shrinkage strain (after 30 years).

The approach to account for the tension stiffening effect in EN 1992-1-1 [1] involves estimating the instantaneous effective curvature of a cracked prestressed section κ_{ef} as a weighted average of the values calculated on a cracked section (κ_{cr}) and on an uncracked section (κ_{uncr}) as follows:

$$\kappa_{ef} = \zeta\kappa_{cr} + (1 - \zeta)\kappa_{uncr} \tag{5.180}$$

where ζ is a distribution coefficient that accounts for the moment level and the degree of cracking. For prestressed concrete flexural members, ζ may be taken as:

$$\zeta = 1 - \beta\left(\frac{M_{cr}}{M_{ext,0}} \right)^2 \tag{5.181}$$

where β is a coefficient to account for the effects of duration of loading or repeated loading on the average deformation and equals 1.0 for a single, short-term load and 0.5 for sustained loading or many cycles of repeated loading; M_{cr} is the external moment at which cracking first occurs and $M_{ext,0}$ is the external moment at which the instantaneous curvature is to be calculated (see Figure 5.31).

The introduction of $\beta = 0.5$ in Equation 5.181 for long-term loading reduces the cracking moment by about 30% and is a crude way of accounting for shrinkage-induced tension and time-dependent cracking. If σ_{cs} is included in the calculation of M_{cr} (as in Equation 5.178), β should be taken as 1.0 for sustained loading.

If we express the curvatures in Equation 5.180 in terms of the flexural rigidities, i.e. $\kappa_{ef} = M_{R,0}/(E_{cm}I_{ef})$, $\kappa_{cr} = M_{R,0}/(E_{cm}I_{av})$ and $\kappa_{uncr} = M_{R,0}/(E_{cm}I_{uncr})$, Equation 5.180 can be rearranged to give:

$$I_{ef} = \frac{I_{av}}{1-\beta\left(1-\dfrac{I_{av}}{I_{uncr}}\right)\left(\dfrac{M_{cr}}{M_{ext,0}}\right)^2} \leq I_{uncr} \tag{5.182}$$

An expression similar to Equation 5.182 for reinforced concrete was first proposed by Bischoff [11].

Several other approaches have been developed for modelling the tension stiffening phenomenon. However, for most practical prestressed members, the maximum in-service moment is less than the cracking moment and cracking is not an issue. Even for those members that crack under service loads, the maximum moment is usually not much greater than the cracking moment and tension stiffening is not very significant. A conservative, but often quite reasonable estimate of deflection can be obtained by ignoring tension stiffening and using $E_{cm,0}I_{av}$ (from Equation 5.176) in the calculations.

EXAMPLE 5.11

Determine the *short-term or instantaneous* deflection immediately after first loading of a uniformly loaded, simply-supported post-tensioned beam of span 12 m. An elevation of the member is shown in Figure 5.32, together with details of the cross-section at mid-span (which is identical with the cross-section analysed in Example 5.3). The prestressing cable is parabolic with the depth of the tendon below the top fibre d_p at each support equal to 400 mm and at mid-span $d_p = 600$ mm, as shown. The non-prestressed reinforcement is uniform throughout the span.

Owing to friction and draw-in losses, the prestressing force at the left support is $P = 1300$ kN, at mid-span $P = 1300$ kN and at the right support $P = 1250$ kN (see Figure 5.32c). The tendon is housed inside a 60 mm diameter ungrouted duct.

The following two service load cases are to be considered:

a. a uniformly distributed load of 6 kN/m (which is the self-weight of the member); and

b. a uniformly distributed load of 40 kN/m (and this includes self-weight).

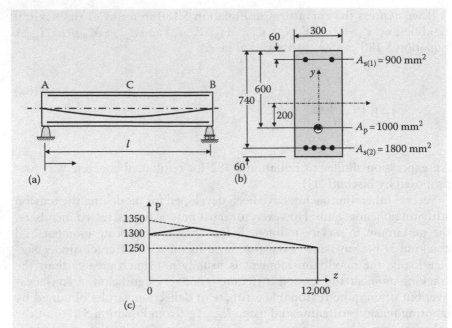

Figure 5.32 Beam details (Example 5.11). (a) Elevation. (b) Section at mid-span. (c) Prestressing force.

The material properties are $E_{cm,0}$ = 30,000 MPa, E_s = 200,000 MPa, E_p = 195,000 MPa and f_{ck} = 40 MPa, and the mean tensile strength is taken to be f_{ctm} = 3.5 MPa.

At support A: The applied moment at support A is zero for both load cases. The prestressing tendon is located at the mid-depth of the section (d_p = 400 mm) and the prestressing force P = 1300 kN. With the reference axis taken as the centroidal axis of the gross cross-section (as shown in Figure 5.32b) and using the cross-sectional analysis described in Section 5.6.2, the initial strain at the centroidal axis and the curvature are determined using Equation 5.45 as follows:

$$\varepsilon_{r,0} = -171.7 \times 10^{-6} \quad \text{and} \quad \kappa_0 = +0.0204 \times 10^{-6} \text{ mm}^{-1}$$

At support B: The prestressing force is 1250 kN and the tendon is located 400 mm below the top fibre. As at support A, $M_{ext,0}$ = 0, and solving Equation 5.45, we get:

$$\varepsilon_{r,0} = -165.1 \times 10^{-6} \quad \text{and} \quad \kappa_0 = +0.0196 \times 10^{-6} \text{ mm}^{-1}$$

At *mid-span C*: The prestressing force is 1300 kN at a depth of 600 mm below the top fibre (i.e. at an eccentricity e = 200 mm) and, assuming no shrinkage has occurred prior to loading, the cracking moment may be estimated from Equation 5.178:

$$M_{cr} = Z(f_{ctm} + P/A) + Pe = \frac{300 \times 800^2}{6} \left(3.5 + \frac{1300 \times 10^3}{300 \times 800} \right) + 1300 \times 10^3 \times 200$$

$$= 545 \times 10^6 \text{ Nmm} = 545 \text{ kNm}$$

Using the more accurate uncracked section analysis of Section 5.6.2, the second moment of area of the uncracked cross-section is I_{uncr} = 14,450 × 10^6 mm^4 and the cracking moment is determined to be M_{cr} = 570 kNm.

For load case (a): $M_{ext,0} = \dfrac{6 \times 12^2}{8} = 108$ kNm

The cross-section is uncracked and from Equation 5.45:

$$\varepsilon_{r,0} = -170.1 \times 10^{-6} \quad \text{and} \quad \kappa_0 = -0.337 \times 10^{-6} \text{ mm}^{-1}$$

For load case (b): $M_{ext,0} = \dfrac{40 \times 12^2}{8} = 720$ kNm

The cross-section has cracked and from Equation 5.134, the depth to the neutral axis is d_n = 443.6 mm. From Equation 5.45:

$$\varepsilon_{r,0} = -70.3 \times 10^{-6} \quad \text{and} \quad \kappa_0 = +1.611 \times 10^{-6} \text{ mm}^{-1}$$

The value of I_{av} is calculated from Equation 5.176 as:

$$I_{av} = \frac{M_{R,0}}{E_{cm,0}\kappa_0} = \frac{M_{ext,0} + y_p P}{E_{cm,0}\kappa_0} = \frac{720 \times 10^6 - 200 \times 1,300 \times 10^3}{30,000 \times 1.611 \times 10^{-6}} = 9,520 \times 10^6 \text{ mm}^4$$

Deflection: With the initial curvature calculated at each end of the member and at mid-span, and with a parabolic variation of curvature along the beam, the short-term deflection at mid-span for each load case is determined using Equation 5.165.

For load case (a):

$$v_C = \frac{12,000^2}{96} [0.0204 + 10 \times (-0.337) + 0.0196] \times 10^{-6} = -5.0 \text{ mm} (\uparrow)$$

For load case (b):

$$v_C = \frac{12,000^2}{96} [0.0204 + 10 \times 1.611 + 0.0196] \times 10^{-6} = +24.2 \text{ mm} (\downarrow)$$

For load case (b), tension stiffening in the cracked region of the member near mid-span has been ignored. To include the effects of tension stiffening in the calculations, the effective second moment of area given by Equation 5.182 can be used instead of I_{av} for the estimation of curvature. With $\beta = 1$ for short-term calculations, we have:

$$I_{ef} = \frac{9{,}520 \times 10^6}{1 - 1.0 \times \left(1 - \dfrac{9{,}520 \times 10^6}{14{,}450 \times 10^6}\right)\left(\dfrac{570}{720}\right)^2} = 12{,}110 \times 10^6 \ mm^4$$

The revised curvature at mid-span for load case (b) is:

$$\kappa_0 = \frac{M_{ext,0} + y_p P}{E_{cm,0} I_{ef}} = 1.266 \times 10^{-6} \ mm^{-1}$$

and the revised mid-span deflection for load case (b) is:

$$v_C = \frac{12{,}000^2}{96}(0.0204 + 10 \times 1.266 + 0.0196) \times 10^{-6} = +19.0 \ mm \ (\downarrow).$$

5.11.4 Long-term deflection

Long-term deflections due to concrete creep and shrinkage are affected by many variables, including load intensity, mix proportions, member size, age at first loading, curing conditions, total quantity of compressive and tensile reinforcing steel, level of prestress, relative humidity and temperature. To account accurately for these parameters, a time analysis similar to that described in Sections 5.7.3 and 5.9.2 is required. The change in curvature during any period of sustained load may be calculated using Equation 5.101. Typical calculations are illustrated in Examples 5.5 and 5.6 for uncracked cross-sections and in Example 5.8 for a cracked cross-section.

When the final curvature has been determined at each end of the member and at mid-span, the long-term deflection can be calculated using either Equation 5.161 or 5.165.

In prestressed concrete construction, a large proportion of the sustained external load is often balanced by the transverse force exerted by the tendons. Under this balanced load, the short-term deflection may be zero, but the long-term deflection is not zero. The restraint to creep and shrinkage offered by non-symmetrically placed bonded reinforcement on a section can cause significant time-dependent curvature and, hence, significant deflection of the member. The use of a simple deflection multiplier to calculate long-term deflection from the short-term deflection is therefore not satisfactory.

In this section, approximate procedures are presented that allow a rough estimate of long-term deflections. In some situations, this is all that is required. However, for most applications, the procedures outlined in Sections 5.7.3 and 5.9.2 are recommended.

5.11.4.1 Creep-induced curvature

The creep-induced curvature $\kappa_{cc}(t)$ of a particular cross section at any time t due to a sustained service load first applied at age t_0 may be obtained from:

$$\kappa_{cc}(t) = \kappa_{sus,0} \frac{\varphi(t,t_0)}{\alpha} \tag{5.183}$$

where $\kappa_{sus,0}$ is the instantaneous curvature due to the sustained service loads; $\varphi(t,t_0)$ is the creep coefficient at time t due to load first applied at age t_0 and α is a creep modification factor that accounts for the effects of cracking and the restraining action of the reinforcement on creep and may be estimated from Equations 5.184, 5.185 and 5.186 [6,12].

For a cracked reinforced concrete section in pure bending ($I_{ef} < I_{uncr}$), $\alpha = \alpha_1$, where:

$$\alpha_1 = (0.48\rho^{-0.5})\left(\frac{I_{cr}}{I_{ef}}\right)^{0.33}\left[1+(125\rho+0.1)\left(\frac{A_{sc}}{A_{st}}\right)^{1.2}\right] \tag{5.184}$$

For an uncracked reinforced or prestressed concrete section ($I_{ef} = I_{uncr}$), $\alpha = \alpha_2$, where:

$$\alpha_2 = 1.0+(45\rho-900\rho^2)\left(1+\frac{A_{sc}}{A_{st}}\right) \tag{5.185}$$

and A_{st} is the equivalent area of bonded reinforcement in the tensile zone (including bonded tendons); A_{sc} is the area of the bonded reinforcement in the compressive zone between the neutral axis and the extreme compressive fibre; ρ is the tensile reinforcement ratio $A_{st}/(b\,d_o)$ and d_o is the depth from the extreme compressive fibre to the centroid of the outermost layer of tensile reinforcement. The area of any bonded reinforcement in the tensile zone (including bonded tendons) not contained in the outermost layer of tensile reinforcement (i.e. located at a depth d_1 less than d_o) should be included in the calculation of A_{st} by multiplying that area by d_1/d_o. For the purpose of the calculation of A_{st}, the tensile zone is that zone that would be in tension due to the applied moment acting in isolation.

For a cracked prestressed concrete section or a cracked reinforced concrete section subjected to bending and axial compression, α may be taken as:

$$\alpha = \alpha_2 + (\alpha_1 - \alpha_2)\left(\frac{d_{n1}}{d_n}\right)^{2.4} \tag{5.186}$$

where α_1 is determined from Equation 5.184 provided I_{cr} is replaced by I_{av}, d_n is the depth of the intact compressive concrete on the cracked section and d_{n1} is the depth of the intact compressive concrete on the cracked section ignoring the axial compression and/or the prestressing force (i.e. the value of d_n for an equivalent cracked reinforced concrete section in pure bending containing the same quantity of bonded reinforcement).

5.11.4.2 Shrinkage-induced curvature

The shrinkage-induced curvature on a reinforced or prestressed concrete section is approximated by:

$$\kappa_{cs}(t) = -\left[\frac{k_r \, \varepsilon_{cs}(t)}{h}\right] \tag{5.187}$$

where h is the overall depth of the section; ε_{cs} is the shrinkage strain (note that ε_{cs} is a negative value) and k_r depends on the quantity and location of bonded reinforcement areas A_{st} and A_{sc} and may be estimated from Equations 5.188, 5.189, 5.190 and 5.191, as appropriate [6,12].

For a cracked reinforced concrete section in pure bending ($I_{ef} < I_{uncr}$), $k_r = k_{r1}$, where:

$$k_{r1} = 1.2\left(\frac{I_{cr}}{I_{ef}}\right)^{0.67}\left(1 - 0.5\frac{A_{sc}}{A_{st}}\right)\left(\frac{h}{d_o}\right) \tag{5.188}$$

For an uncracked cross-section ($I_{ef} = I_{uncr}$), $k_r = k_{r2}$, where:

$$k_{r2} = (100\rho - 2500\rho^2)\left(\frac{d_o}{0.5h} - 1\right)\left(1 - \frac{A_{sc}}{A_{st}}\right)^{1.3} \quad \text{when } \rho = A_{st}/bd_o \leq 0.01 \tag{5.189}$$

$$k_{r2} = (40\rho + 0.35)\left(\frac{d_o}{0.5h} - 1\right)\left(1 - \frac{A_{sc}}{A_{st}}\right)^{1.3} \quad \text{when } \rho = A_{st}/bd_o > 0.01 \tag{5.190}$$

In Equations 5.188, 5.189 and 5.190, A_{sc} is defined as the area of the bonded reinforcement on the compressive side of the cross-section. This is a different definition to that provided under Equation 5.185. Whilst bonded steel near the compressive face of a cracked cross-section that is located at or below the neutral axis will not restrain compressive creep, it will provide restraint to shrinkage and will be effective in reducing shrinkage-induced curvature on a cracked section.

For a cracked prestressed concrete section or for a cracked reinforced concrete section subjected to bending and axial compression, k_r may be taken as:

$$k_r = k_{r2} + (k_{r1} - k_{r2})\left(\frac{d_{n1}}{d_n}\right) \tag{5.191}$$

where k_{r1} and k_{r2} are determined from Equations 5.188 through 5.190 by replacing A_{st} with $(A_{st} + A_{pt})$ and, for a cracked prestressed section I_{cr} is replaced by I_{av} in Equation 5.188 and d_n and d_{n1} are as defined after Equation 5.186.

Equations 5.183 through 5.190 have been developed [12] as empirical fits to the results obtained from a parametric study of the creep- and shrinkage-induced changes in curvature on reinforced and prestressed concrete cross-sections under constant sustained internal actions using the AEMM of analysis presented in Sections 5.7 and 5.9.

EXAMPLE 5.12

The final time-dependent deflection of the beam described in Example 5.11 and illustrated in Figure 5.32 is to be calculated. It is assumed that the duct is grouted soon after transfer and the tendon is effectively bonded to the concrete for the time period t_0 to t_k. As in Example 5.11, the following two load cases are to be considered:

a. a uniformly distributed constant sustained load of 6 kN/m; and
b. a uniformly distributed constant sustained load of 40 kN/m.

For each load case, the time-dependent material properties are:

$$\varphi(t_k, t_0) = 2.5; \quad \chi(t_k, t_0) = 0.65; \quad \varepsilon_{cs}(t_k) = -450 \times 10^{-6}; \quad \varphi_p(t_k, \sigma_{p(i),init}) = 0.03.$$

All other material properties are as specified in Example 5.11.

i. Calculation using the refined method (AEMM analysis):

At support A: The sustained moment at support A is zero for both load cases and the prestressing force is $P = 1300$ kN at $d_p = 400$ mm. Using the procedure outlined in Section 5.7.3 and solving Equation 5.101, the strain at the reference axis and the curvature at time t_k are:

$$\varepsilon_{r,k} = -843 \times 10^{-6} \quad \text{and} \quad \kappa_k = +0.275 \times 10^{-6} \text{ mm}^{-1}$$

At support B: As at support A, the sustained moment is zero, but the prestressing force is $P = 1250$ kN at $d_p = 400$ mm. Solving Equation 5.101, the strain at the reference axis and the curvature at time t_k are:

$$\varepsilon_{r,k} = -825 \times 10^{-6} \quad \text{and} \quad \kappa_k = +0.269 \times 10^{-6} \text{ mm}^{-1}, \text{ respectively.}$$

At mid-span C: For load case (a), $M_{ext,k} = 108$ kNm and the prestressing force is 1300 kN at a depth of 600 mm below the top fibre. For this uncracked section, solving Equation 5.101 gives:

$$\varepsilon_{r,k} = -827 \times 10^{-6} \quad \text{and} \quad \kappa_k = -0.503 \times 10^{-6} \text{ mm}^{-1}$$

For load case (b), $M_{ext,0} = 720$ kNm and for this cracked cross-section with $d_n = 443.6$ mm, Equation 5.101 gives:

$$\varepsilon_{r,k} = -771 \times 10^{-6} \quad \text{and} \quad \kappa_k = +4.056 \times 10^{-6} \text{ mm}^{-1}$$

To include tension stiffening, we may use Equations 5.180 and 5.181. If the cross-section at mid-span for load case (b) was considered to be uncracked, and the uncracked cross-section reanalysed, the final curvature is $(\kappa_k)_{uncr} = +3.293 \times 10^{-6}$ mm^{-1}. With the cracking moment determined in Example 5.11 to be $M_{cr} = 570$ kNm, Equation 5.181 gives:

$$\zeta = 1 - 0.5\left(\frac{570}{720}\right)^2 = 0.687$$

and from Equation 5.180:

$$\kappa_{ef} = 0.687 \times 4.056 \times 10^{-6} + (1 - 0.687) \times 3.293 \times 10^{-6} = 3.817 \times 10^{-6} \text{ mm}^{-1}$$

Deflection: With the final curvature calculated at each end of the member and at mid-span, and with a parabolic variation of curvature along the beam, the long-term deflection at mid-span for each load case is determined using Equation 5.165.

For load case (a):

$$v_C = \frac{12,000^2}{96}[0.275 + 10 \times (-0.503) + 0.269] \times 10^{-6} = -6.6 \text{ mm } (\uparrow)$$

For load case (b), including the effects of tension stiffening:

$$v_C = \frac{12,000^2}{96}[0.275 + 10 \times 3.817 + 0.269] \times 10^{-6} = +58.1 \text{ mm } (\downarrow)$$

ii. **Calculation using the simplified method (Equations 5.183 through 5.191):**

Load case (a):

The creep- and shrinkage-induced curvatures at each support and at mid-span are estimated using Equations 5.183 and 5.187, respectively.

At supports A and B, the cross-section is uncracked, with $A_{sc} = A_{s(1)} = 900 \text{ mm}^2$, $A_{st} = A_{s(2)} = 1800 \text{ mm}^2$ (noting that the tendons are at mid-depth and therefore not in the tension zone) and $\rho = A_{st}/(bd_o) = 0.00811$. From Equation 5.185:

$$\alpha = \alpha_2 = 1.0 + (45 \times 0.00811 - 900 \times 0.00811^2)\left(1 + \frac{900}{1800}\right) = 1.46$$

and from Equation 5.189:

$$k_r = k_{r2} = (100 \times 0.00811 - 2500 \times 0.00811^2)$$

$$\times \left(\frac{740}{0.5 \times 800} - 1\right)\left(1 - \frac{900}{1800}\right)^{1.3} = 0.223$$

From Equation 5.183, the creep-induced curvatures are:
At support A:

$$\kappa_{cc}(t) = 0.0139 \times 10^{-6} \times \frac{2.5}{1.46} = 0.0238 \times 10^{-6} \text{ mm}^{-1}$$

At support B:

$$\kappa_{cc}(t) = 0.0133 \times 10^{-6} \times \frac{2.5}{1.46} = 0.0228 \times 10^{-6} \text{ mm}^{-1}$$

and, from Equation 5.187, the shrinkage-induced curvature at each support is:

$$\kappa_{cs}(t) = -\left[\frac{0.223 \times (-450 \times 10^{-6})}{800}\right] = +0.125 \times 10^{-6} \text{ mm}^{-1}$$

For load case (a), the cross-section at the mid-span C is uncracked, with $A_{sc} = 900$ mm^2, $A_{st} = A_{s(2)} + A_p(d_p/d_o) = 1800 + 1000 \times (600/740) = 2611$ mm^2 and $\rho = A_{st}/(bd_o) = 0.0117$. From Equations 5.185 and 5.190, we get, respectively:

$$\alpha = \alpha_2 = 1.0 + (45 \times 0.0117 - 900 \times 0.0117^2)\left(1 + \frac{900}{2611}\right) = 1.54$$

and

$$k_{r2} = (40 \times 0.0117 + 0.35)\left(\frac{740}{0.5 \times 800} - 1\right)\left(1 - \frac{900}{2611}\right)^{1.3} = 0.400$$

From Equations 5.183 and 5.187, the creep- and shrinkage-induced curvatures at mid-span are, respectively:

$$\kappa_{cc}(t) = -0.337 \times 10^{-6} \times \frac{2.5}{1.54} = -0.547 \times 10^{-6} \text{ mm}^{-1}$$

and

$$\kappa_{cs}(t) = -\left[\frac{0.400 \times (-450 \times 10^{-6})}{800}\right] = +0.225 \times 10^{-6} \text{ mm}^{-1}$$

The final curvature at each cross-section is the sum of the instantaneous, creep and shrinkage-induced curvatures:

At support A: $\kappa(t) = (0.0139 + 0.0238 + 0.125) \times 10^{-6}$

$= +0.163 \times 10^{-6} \text{ mm}^{-1}$

At support B: $\kappa(t) = (0.0133 + 0.0228 + 0.125) \times 10^{-6}$

$= +0.161 \times 10^{-6} \text{ mm}^{-1}$

At mid-span C: $\kappa(t) = (-0.337 - 0.547 + 0.225) \times 10^{-6}$

$= -0.659 \times 10^{-6} \text{ mm}^{-1}$

From Equation 5.165, the long-term deflection at mid-span for load case (a) is:

$$v_C = \frac{12,000^2}{96}(0.163 + 10 \times (-0.659) + 0.161) \times 10^{-6} = -9.4 \text{ mm} (\uparrow)$$

For this load case, the simplified equations (Equations 5.183 through 5.191) overestimate the upward camber of the uncracked beam calculated using the more refined AEMM method (−6.6 mm).

Load case (b):
As for load case (a), at supports A and B, the cross-sections are uncracked, with $A_{sc} = A_{s(1)} = 900$ mm^2, $A_{st} = A_{s(2)} = 1800$ mm^2 and $\rho = A_{st}/(bd_o) = 0.00811$. The final curvatures at the two supports are identical to those calculated for load case (a):

At support A: $\kappa(t) = +0.163 \times 10^{-6}$ mm^{-1}
At support B: $\kappa(t) = +0.161 \times 10^{-6}$ mm^{-1}

At mid-span C, the cross-section is cracked, with $A_{sc} = 900$ mm^2, $A_{st} = A_{s(2)} + A_p(d_p/d_o) = 1800 + 1000 \times (600/740) = 2611$ mm^2 and $\rho = A_{st}/bd_o = 0.0117$. From Example 5.11, $d_n = 443.6$ mm. If the prestress is ignored and a cracked section analysis is performed on the equivalent reinforced concrete cross-section, we determine that $d_{n1} = 227.3$ mm (where d_{n1} is defined in the text under Equation 5.186) and $I_{cr} = 5360 \times 10^6$ mm^4. In Example 5.11, the value of I_{ef} determined using Equation 5.182 was calculated as $I_{ef} = 12,110 \times 10^6$ mm^4.

From Equations 5.184 and 5.185, we get:

$$\alpha_1 = (0.48 \times 0.0117^{-0.5})\left(\frac{5360 \times 10^6}{12110 \times 10^6}\right)^{0.33}\left[1 + (125 \times 0.0117 + 0.1)\left(\frac{900}{2599}\right)^{1.2}\right] = 4.87$$

and

$$\alpha_2 = 1.0 + (45 \times 0.0117 - 900 \times 0.0117^2)\left(1 + \frac{900}{2599}\right) = 1.54$$

and from Equation 5.186, we have:

$$\alpha = 1.54 + (4.87 - 1.54)\left(\frac{227.3}{443.7}\right)^{2.4} = 2.21$$

The creep-induced curvature at mid-span is given by Equation 5.183:

$$\kappa_{cc}(t) = 1.611 \times 10^{-6} \times \frac{2.5}{2.21} = 1.822 \times 10^{-6} \text{ mm}^{-1}$$

From Equations 5.188 and 5.190, we get:

$$k_{r1} = 1.2\left(\frac{5360 \times 10^6}{12110 \times 10^6}\right)^{0.67}\left(1 - 0.5 \times \frac{900}{2599}\right)\left(\frac{800}{740}\right) = 0.621$$

$$k_{r2} = (40 \times 0.0117 + 0.35)\left(\frac{740}{0.5 \times 800} - 1\right)\left(1 - \frac{900}{2599}\right)^{1.3} = 0.400$$

and Equation 5.191 gives:

$$k_r = 0.400 + (0.621 - 0.400)\left(\frac{227.3}{443.7}\right) = 0.513$$

From Equation 5.187, the shrinkage-induced curvature at mid-span is:

$$\kappa_{cs}(t) = -\left[\frac{0.513 \times (-450 \times 10^{-6})}{800}\right] = +0.289 \times 10^{-6} \text{ mm}^{-1}$$

The final curvature at mid-span is therefore:

$$\kappa(t) = (1.611 + 1.822 + 0.289) \times 10^{-6} = +3.722 \times 10^{-6} \text{ mm}^{-1}$$

From Equation 5.165, the long-term deflection at mid-span for load case (b) (including the effects of tension stiffening) is:

$$v_C = \frac{12,000^2}{96}(0.163 + 10 \times 3.722 + 0.161) \times 10^{-6} = +56.3 \text{ mm } (\downarrow)$$

For this load case, the deflection determined using the simplified equations (Equations 5.183 through 5.191) is in good agreement with the final long-term deflection of the cracked beam calculated using the more refined AEMM method (+58.1 mm).

5.12 CRACK CONTROL

5.12.1 Minimum reinforcement

When flexural cracking occurs in a prestressed concrete beam or slab, the axial prestressing force on the concrete controls the propagation of the crack and, unlike flexural cracking in a reinforced concrete member, the crack does not suddenly propagate to its full height (usually a large percentage of the depth of the cross-section). The height of a flexural crack gradually increases as the load increases and the loss of stiffness due to cracking is far more gradual than for a reinforced concrete member. The change in strain at the tensile steel level at first cracking is much less than that in a conventionally reinforced section with similar quantities of bonded reinforcement. After cracking, therefore, a prestressed beam generally suffers less deformation than the equivalent reinforced concrete beam, with finer, less extensive cracks. EN 1992-1-1 [1] cautions that the durability of prestressed members may be more adversely affected by cracking than that of reinforced concrete members. Nevertheless, flexural crack control in prestressed concrete beams and slabs is not usually a critical design consideration, provided an appropriate distribution and quantity of bonded reinforcement is provided in the tensile zone.

According to EN 1992-1-1 [1], if the maximum tensile stress in the concrete due to the frequent combinations of service loads (after the effects of creep, shrinkage and relaxation have been appropriately considered) is less than the effective tensile strength of the concrete $f_{ct,eff}(t)$, the section may be considered to be uncracked and no further consideration needs to be given to crack control. In this context, $f_{ct,eff}(t)$ may be taken as either the mean value of the tensile strength of the concrete $f_{ctm}(t)$ or the mean value of the flexural tensile strength of the concrete at the time $f_{ctm,fl}(t)$, provided that the calculation for the minimum tension reinforcement in Equation 5.192 is based on the same value. In the calculation of the maximum tensile stress, care should be taken to consider the loss of compressive stress in the concrete due to the restraint provided by the bonded reinforcement to creep and shrinkage deformations and any restraint provided externally to shrinkage by the supports or adjacent parts of the structure.

Where cracking does occur, the maximum crack width must not impair the proper functioning of the structure or adversely affect its appearance. EN 1992-1-1 [1] specifies a minimum quantity of bonded reinforcement for members where crack control is required. This minimum quantity of reinforcement is estimated by equating the force in the concrete just before cracking with the tensile force in the reinforcement after cracking assuming a stress in the reinforcement equal to the yield stress or an appropriately lower stress required to limit the maximum crack width. In T-beams, L-beams or box girders, the minimum reinforcement in each part of the cross-section is determined by applying Equation 5.192 to each web or flange that is in tension.

$$A_{s,min}\, \sigma_s = k_c\, k\, f_{ct,eff}\, A_{ct} \tag{5.192}$$

where $A_{s,min}$ is the minimum area of bonded reinforcing steel in the tensile zone; A_{ct} is the area of the concrete in the tensile zone just before cracking; σ_s is the absolute value of the maximum stress permitted in the steel to satisfy the maximum crack width limit (see Section 5.12.2) (but should not exceed f_{yk}); $f_{ct,eff}$ is the mean tensile strength of the concrete at the time when cracking is expected to occur (either $f_{ctm}(t)$ or $f_{ctm,fl}(t)$, as appropriate); k accounts for non-uniform eigenstresses that develop due to differential shrinkage in each web or flange of the cross-section, with $k = 1.0$ when the width of a flange or the depth of a web is less than 300 mm and $k = 0.65$ when the width of a flange or the depth of a web is greater than 800 mm (intermediate values may be interpolated) and k_c depends on the shape of the stress distribution before cracking. For pure tension, $k_c = 1.0$.

For rectangular sections and the webs of box section and T-section subjected to bending or bending and axial force, k_c is given by:

$$k_c = 0.4 \left[1 - \frac{\sigma_c}{k_1(h/h^*)f_{ct,eff}} \right] \leq 1.0 \tag{5.193}$$

and for flanges of box tension and T-section:

$$k_c = 0.9 \frac{F_{cr}}{A_{ct}f_{ct,eff}} \geq 0.5 \tag{5.194}$$

where:

σ_c is the mean concrete stress acting on the part of the section under consideration $(=N_{Ed}/bh)$;

N_{Ed} is the axial force at the serviceability limit state acting on the part of the section under consideration (compression is taken as positive here), including the effect of prestress and axial force;

h^* is equal to h for $h < 1000$ mm and equal to 1000 mm for $h \geq 1000$ mm;

k_1 accounts for axial force, with $k_1 = 1.5$ if N_{Ed} is compressive and $k_1 = (2h^*)/(3h)$ if N_{Ed} is tensile; and

F_{cr} is the absolute value of the tensile force within the flange immediately prior to cracking (using f_{ctm} to calculate the cracking moment on the section).

For cross-sections containing bonded tendons in the tension zone, the bonded tendons may be assumed to contribute to crack control up to a distance of 150 mm from the centre of the tendon and the minimum reinforcement requirements within this area are obtained from the following modification to Equation 5.192:

$$A_{s,min}\sigma_s + \xi_1 A_p' \Delta\sigma_p = k_c k f_{ct,eff} A_{ct} \tag{5.195}$$

where A_p' is the area of the bonded tendons within the concrete area $A_{c,eff}$; $A_{c,eff}$ is the effective area of concrete in tension surrounding the tendon with depth $h_{c,ef}$ equal to the lesser of $2.5(h - d)$, $(h - x)/3$ or $h/2$; $\Delta\sigma_p$ is the variation in stress in the tendons from the state of zero strain in the concrete at the same level; ξ_1 is an adjusted ratio of bond strength to account for the different diameters of the tendons and the reinforcing steel given by:

$$\xi_1 = \sqrt{\xi \frac{\phi_s}{\phi_p}} \tag{5.196}$$

Table 5.4 Ratio of bond strength (ξ) between tendons and reinforcing steel [I]

| | | ξ | |
| | | Bonded, post-tensioned | |
Prestressing steel	Pretensioned tendon	≤C50/60	≥C70/85
Smooth bars and wires	Not applicable	0.3	0.15
Strands	0.6	0.5	0.25
Indented wires	0.7	0.6	0.3
Ribbed bars	0.8	0.7	0.35

Note: For intermediate values between C50/60 and C70/85, interpolation may be used.

where:

ξ is the ratio of bond strength of prestressing and reinforcing steel (as given in Table 5.4);

ϕ_s is the largest bar diameter of the reinforcing steel; and

ϕ_p is the equivalent diameter of the tendon and is given by:

$\phi_p = 1.6\sqrt{A_p}$ for bundles of tendons

$\phi_p = 1.75\phi_{wire}$ for single 7-wire strands (ϕ_{wire} = wire diameter)

$\phi_p = 1.20\phi_{wire}$ for single 3-wire strands (ϕ_{wire} = wire diameter)

If no conventional reinforcement is included and only the bonded prestressing steel is used to control cracking, $\xi_1 = \sqrt{\xi}$.

5.12.2 Control of cracking without direct calculation

According to EN 1992-1-1 [1], when the areas of bonded reinforcement and tendons exceed the minimum values obtained from Equations 5.192 and 5.195, crack widths will not be excessive, provided:

1. For cracking caused dominantly by restraint: the bar sizes given in Table 5.5 are not exceeded, where the steel stress is the value determined on the cracked section immediately after cracking; and
2. For cracking caused dominantly by loading: the bar sizes given in Table 5.5 are not exceeded or the bar spacings given in Table 5.6 are complied with, where the steel stress is the value determined on the cracked section under the relevant combination of actions.

It is recommended here that the bar spacing does not exceed 300 mm and that the cover to the bars does not exceed about 100 mm.

For other values of the variables presented in the note at the bottom of Tables 5.5 and 5.6, the maximum bar diameter may be modified as follows: In bending:

$$\phi_s = \phi_s^* \frac{f_{ct,eff}}{2.9} \frac{k_c h_{cr}}{2(h-d)} \tag{5.197}$$

Table 5.5 Maximum bar diameters φ_s^* for crack control[a] [1]

Steel stress (MPa)	Maximum bars size (mm)		
	$w_k = 0.4$ mm	$w_k = 0.3$ mm	$w_k = 0.2$ mm
160	40	32	25
200	32	25	16
240	20	16	12
280	16	12	8
320	12	10	6
360	10	8	5
400	8	6	4
450	6	5	—

[a] The values in the table were determined for $c = 25$ mm, $f_{ct,eff} = 2.9$ MPa, $h_{cr} = 0.5$, $(h - d) = 0.1\,h, k_1 = 0.8, k_2 = 0.5, k_c = 0.4, k = 1.0, k_t = 0.4$ and $k' = 1.0$.

Table 5.6 Maximum bar spacing for crack control[a] [1]

Steel stress (MPa)	Maximum bars spacing (mm)		
	$w_k = 0.4$ mm	$w_k = 0.3$ mm	$w_k = 0.2$ mm
160	300	300	200
200	300	250	150
240	250	200	100
280	200	150	50
320	250	100	—
360	100	50	—

[a] The values in the table were determined for $c = 25$ mm, $f_{ct,eff} = 2.9$ MPa, $h_{cr} = 0.5$, $(h - d) = 0.1\,h, k_1 = 0.8, k_2 = 0.5, k_c = 0.4, k = 1.0, k_t = 0.4$ and $k' = 1.0$.

In direct tension:

$$\phi_s = \phi_s^* \frac{f_{ct,eff}}{2.9} \frac{h_{cr}}{8(h-d)} \qquad (5.198)$$

where ϕ_s is the modified maximum bar diameter; ϕ_s^* is the maximum bar diameter given in Table 5.5; h is the overall depth of the cross-section; h_{cr} is the depth of the tensile zone immediately before cracking considering the characteristic values of prestress and internal actions under the quasi-permanent combination of actions and d is the effective depth to the centroid of the outer layer of reinforcement.

For pretensioned concrete, where crack control is provided by the bonded tendons, Tables 5.5 and 5.6 may be used except that the steel stress to consider is the total stress after cracking minus the prestress. For post-tensioned concrete, where crack control is provided mainly by ordinary reinforcement, Tables 5.5 and 5.6 may be used with the calculated reinforcement stress accounting for prestressing.

For beams where $h \geq 1000$ mm and the main reinforcement is concentrated at either the top or bottom of the cross-section, skin reinforcement is required to control cracking in the side faces of the beam. This side-face reinforcement should be uniformly distributed between the main tensile steel and the neutral axis and should be located inside the stirrups and tied to them. It is preferable to use small-diameter bars at spacing not exceeding 300 mm. The area of the side-face reinforcement may be determined using Equation 5.192 taking $k = 0.5$ and $\sigma_s = f_{yk}$.

5.12.3 Calculation of crack widths

For crack control by direct calculation, EN 1992-1-1 [1] permits the calculation of the crack width in a reinforced concrete member using:

$$w = s_{r,max}(\varepsilon_{sm} - \varepsilon_{cm}) \tag{5.199}$$

where $s_{r,max}$ is the maximum crack spacing; ε_{sm} is the mean strain in the reinforcement at design loads, including the effects of tension stiffening and any imposed deformations; and ε_{cm} is the mean strain in the concrete between the cracks.

The difference between the mean strain in the reinforcement and the mean strain in the concrete may be taken as:

$$\varepsilon_{sm} - \varepsilon_{cm} = \frac{\sigma_s}{E_s} - k_t \frac{f_{ct,eff}}{E_s \rho_{p,eff}}(1 + \alpha_e \rho_{p,eff}) \geq 0.6\frac{\sigma_s}{E_s} \tag{5.200}$$

where σ_s is the stress in the tensile reinforcement assuming a cracked section. For a pretensioned member, σ_s may be replaced by the stress variation in the tendons $\Delta\sigma_p$ from the state of zero strain of the concrete at the same level; k_t is a factor that depends on the duration of load and equals 0.6 for short-term loading and 0.4 for long-term loading; α_e is the modular ratio E_s/E_{cm}; $f_{ct,eff}$ is the mean value of the axial tensile strength of concrete at the time cracking is expected; $\rho_{p,eff}$ is the reinforcement ratio given by

$(A_s + \xi_1^2 A_p')/A_{c,\text{eff}}$; A_p' and $A_{c,\text{eff}}$ are as defined under Equation 5.195; and ξ_1 is given by Equation 5.196.

For cross-sections with bonded reinforcement fixed at reasonably close centres, i.e. bar spacing $\leq 5(c + 0.5\phi)$, the maximum final crack width may be calculated from:

$$s_{r,\text{max}} = 3.4c + 0.425k_1k_2\phi/\rho_{p,\text{eff}} \tag{5.201}$$

in which ϕ is the bar diameter. Where a section contains different bar sizes in the tensile zone, an equivalent bar diameter ϕ_{eq} should be used in Equation 5.201. For a section containing i different bar diameters:

$$\phi_{eq} = \frac{n_1\phi_1^2 + \cdots + n_i\phi_i^2}{n_1\phi_1 + \cdots + n_i\phi_i} \tag{5.202}$$

and n_i is the number of bars of diameter ϕ_i; c is the clear cover to the longitudinal reinforcement; k_1 is a coefficient that accounts for the bond properties of the bonded reinforcement, with $k_1 = 0.8$ for high bond bars and $k_1 = 1.6$ for plain bars and prestressing tendons; and k_2 is a coefficient that accounts for the longitudinal strain distribution, with $k_2 = 0.5$ for bending and $k_2 = 1.0$ for pure tension. For cases in combined tension and bending, $k_2 = (\varepsilon_1 + \varepsilon_2)/(2\varepsilon_1)$ and ε_1 is the greater and ε_2 is the lesser of the tensile strains at the boundaries of the cross-section (assessed on the basis of a cracked section).

Where the spacing of the bonded reinforcement exceeds $5(c + 0.5\phi)$, or where there is no bonded reinforcement in the tensile zone, an upper bound to the crack width is obtained by assuming a maximum crack spacing of:

$$s_{r,\text{max}} = 1.3 (h - x) \tag{5.203}$$

where x is the depth to the neutral axis on the cracked section.

In a member that is reinforced in two orthogonal directions, where the angle between the axes of principal stress and the direction of the reinforcement is significant ($>15°$), the crack spacing may be taken as:

$$s_{r,\text{max}} = \frac{1}{\dfrac{\cos\theta}{s_{r,\text{max},y}} + \dfrac{\sin\theta}{s_{r,\text{max},z}}} \tag{5.204}$$

where θ is the angle between the reinforcement in the y direction and the direction of the axis of principal tension and $s_{r,max,y}$ and $s_{r,max,z}$ are the crack spacings in the y and z directions, respectively, calculated using Equation 5.201.

5.12.4 Crack control for restrained shrinkage and temperature effects

Direct tension cracks due to restrained shrinkage and temperature changes may lead to serviceability problems, particularly in regions of low moment and in directions with little or no prestress. Such cracks usually extend completely through the member and are more parallel-sided than flexural cracks. If uncontrolled, these cracks can become very wide and lead to waterproofing and corrosion problems. They can also disrupt the integrity and the structural action of the member.

Evidence of direct tension type cracks is common in concrete slab systems. For example, consider a typical one-way beam and slab floor system. The load is usually carried by the slab in the *primary direction* across the span to the supporting beams, while in the orthogonal direction (the *secondary direction*), the bending moment is small. Shrinkage is the same in both directions and restraint to shrinkage usually exists in both directions.

In the primary direction, prestress may eliminate flexural cracking, but if the level of prestress is such that flexural cracking does occur, shrinkage will cause small increases in the widths of flexural cracks and may cause additional flexure type cracks in the previously uncracked regions. However, in the secondary direction, which is in effect a direct tension situation, there may be little or no prestress and shrinkage may cause a few widely spaced cracks that penetrate completely through the slab. Frequently, more reinforcement is required in the secondary direction to control these direct tension cracks than is required for bending in the primary direction. As far as cracking is concerned, it is not unreasonable to say that shrinkage is a greater problem when it is not accompanied by flexure and when the level of prestress is low.

When determining the amount of reinforcement required in a slab to control shrinkage- and temperature-induced cracking, account should be taken of the influence of bending, the degree of restraint against in-plane movements and the exposure classification.

Where the ends of a slab are restrained and the slab is not free to expand or contract in the secondary direction, the minimum area of reinforcement in the restrained direction given by Equation 5.205 is recommended.

$$A_{s.min} = (6.0 - 2.5\sigma_{cp})b\,h \times 10^{-3} \qquad (5.205)$$

where σ_{cp} is the average prestress $P_{m,t}/A$. When a slab or wall is greater than 500 mm thick, the reinforcement required near each surface may be determined assuming that $h = 250$ mm in Equation 5.205.

5.12.5 Crack control at openings and discontinuities

Openings and discontinuities in slabs are the cause of stress concentrations that may result in diagonal cracks emanating from re-entrant corners. Additional reinforcing bars are generally required to trim the hole and control the propagation of these cracks. A suitable method of estimating the number and size of the trimming bars is to postulate a possible crack and provide reinforcement to carry a force at least equivalent to the area of the crack surface multiplied by the mean direct tensile strength of the concrete. For crack control, the maximum stress in the trimming bars should be limited to about 200 MPa.

While this additional reinforcement is required for serviceability to control cracking at re-entrant corners, it should not be assumed that this same steel is satisfactory for strength. For a small hole through a slab, it is generally sufficient for bending to place additional steel on either side of the hole equivalent to the steel that must be terminated at the face of the opening. The effects of a large hole or opening should be determined by appropriate analysis accounting for the size, shape and position of the opening. Plastic methods of design, such as the yield line method (see Section 12.9.7) or the simplified strip method, are convenient ways of designing such slabs to meet requirements for strength.

REFERENCES

1. EN 1992-1-1. 2004. Eurocode 2: Design of concrete structures. Part 1-1: General rules and rules for buildings. British Standards Institution, London, UK.
2. ACI 318-14M. 2014. Building code requirements for reinforced concrete. American Concrete Institute, Detroit, MI.
3. Magnel, G. 1954. *Prestressed Concrete*, 3rd edn. London, UK: Concrete Publications Ltd.
4. Lin, T.Y. 1963. *Prestressed Concrete Structures*. New York: Wiley.
5. Warner, R.F. and Faulkes, K.A. 1979. *Prestressed Concrete*. Melbourne, Victoria, Australia: Pitman Australia.
6. Gilbert, R.I. and Ranzi, G. 2011. *Time-Dependent Behaviour of Concrete Structures*. London, UK: Spon Press.
7. Gilbert, R.I. 1988. *Time Effects in Concrete Structures*. Amsterdam, the Netherlands: Elsevier.
8. Ghali, A., Favre, R. and Elbadry, M. 2002. *Concrete Structures: Stresses and Deformations*, 3rd edn. London, UK: Spon Press.
9. AS3600-2009. Australian standard concrete structures. Standards Australia, Sydney, New South Wales, Australia.
10. Branson, D.E. 1963. Instantaneous and time-dependent deflection of simple and continuous reinforced concrete beams. Alabama Highway Research Report, No. 7. Bureau of Public Roads, Montgomery, AL.

11. Bischoff, P.H. 2005. Reevaluation of deflection prediction for concrete beams reinforced with steel and FRP bars. *Journal of Structural Engineering, ASCE*, 131(5), 752–767.
12. Gilbert, R.I. 2001. Deflection calculation and control – Australian code amendments and improvements (Chapter 4). *ACI International SP 203, Code Provisions for Deflection Control in Concrete Structures*. Farmington Hills, MI: American Concrete Institute, pp. 45–78.

Chapter 6

Flexural resistance

6.1 INTRODUCTION

An essential design objective for a structure or a component of a structure is the provision of adequate strength. The consequences and costs of strength failures are high and therefore the probability of such failures must be very small.

The satisfaction of concrete and steel stress limits at service loads does not necessarily ensure adequate strength and does not provide a reliable indication of either the actual strength or the safety of a structural member. It is important to consider the non-linear behaviour of the member in the overloaded range to ensure that it has an adequate structural capacity. Only by calculating the design resistance of a member can a sufficient margin between the service load and the ultimate load be guaranteed.

The design resistance of a cross-section in bending M_{Rd} is calculated from a rational and well-established procedure involving consideration of the design strength of both the concrete and the steel in the compressive and tensile parts of the cross-section. The prediction of the design flexural strength is described and illustrated in this chapter. When M_{Rd} is determined, the design requirements for the strength limit state (as discussed in Section 2.4) may be checked and satisfied.

In addition to calculating the design strength of a section, a measure of the ductility of each section must also be established. Ductility is an important objective in structural design. Ductile members undergo large deformations prior to failure, thereby providing warning of failure and allowing indeterminate structures to establish alternative load paths. In fact, it is only with adequate ductility that the predicted strength of indeterminate members and structures can be achieved in practice.

6.2 FLEXURAL BEHAVIOUR AT OVERLOADS

The load at which collapse of a flexural member occurs is called the *ultimate load*. If the member has sustained large deformations prior to reaching the ultimate load, it is said to have ductile behaviour. If, on the

219

other hand, it has only undergone relatively small deformations prior to failure, the member is said to have brittle behaviour. There is no defined deformation or curvature that distinguishes ductile from brittle behaviour. Codes of practice, however, usually impose a ductility requirement by limiting the curvature of a beam or slab at the ultimate load to some minimum value, thereby ensuring that significant deformation occurs in a flexural member prior to failure.

Since beam failures that result from a breakdown of bond between the concrete and the steel reinforcement, or from excessive shear, or from failure of the anchorage zone tend to be brittle in nature, every attempt should be made to ensure that, if a beam is overloaded, a ductile flexural failure would initiate the collapse. Therefore, the design philosophy should ensure that a flexural member does not fail before the required design moment capacity of the critical section is attained.

Consider the prestressed concrete cross-section shown in Figure 6.1. The section contains non-prestressed reinforcement in the compressive and tensile zones and bonded prestressing steel. Typical strain and stress distributions for four different values of applied moment are also shown in Figure 6.1. As the applied moment M increases from typical in-service levels into the overload range, the neutral axis gradually rises and eventually material behaviour becomes non-linear. The non-prestressed tensile steel

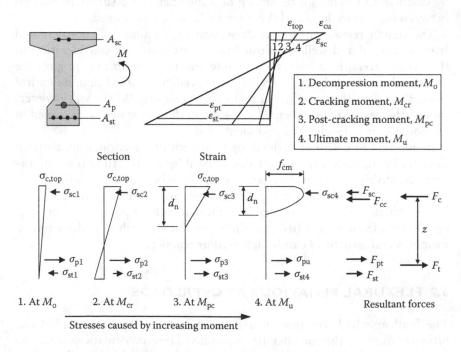

Figure 6.1 Stress and strain distributions caused by increasing moment.

may yield (if its strain ε_s exceeds the yield strain ε_{yk}, where $\varepsilon_{yk} = f_{yk}/E_s$), the prestressed steel may enter the non-linear part of its stress–strain curve as ε_{pt} increases, the concrete compressive stress distribution becomes non-linear when the extreme fibre stress exceeds about $0.5f_{ck}$, and the non-prestressed compressive steel may yield (if the magnitude of its strain exceeds the yield strain ε_{yk}).

A flexural member, which is designed to exhibit ductile behaviour, usually has failure of the critical section preceded by yielding of the bonded tensile steel, i.e. by effectively exhausting the capacity of the tensile steel to carry any additional force. Such a member is said to be under-reinforced.

Because the stress–strain curve for the prestressing steel has no distinct yield point and the stress increases monotonically as the strain increases (see Figure 4.11), the capacity of the prestressing steel to carry additional force is never entirely used up until the steel actually fractures. When the tendon strain exceeds about 0.01 (for wire or strand), the stress–strain curve becomes relatively flat and the rate of increase of stress with strain is small. After yielding of the steel, the resultant internal tensile force (i.e. $F_t = F_{st} + F_{pt}$ in Figure 6.1) remains approximately constant (as does the resultant internal compressive force F_c, which is equal and opposite to F_t). The moment capacity can be further increased slightly by an increase in the lever arm between F_c and F_t. Under increasing deformation, the neutral axis rises, the compressive zone becomes smaller and the maximum compressive concrete stress increases. Eventually, after considerable deformation, a compressive failure of the concrete above the neutral axis occurs and the section reaches its ultimate capacity. It is, however, the strengths of the prestressing tendons and the non-prestressed reinforcement in the tensile zone that control the strength of a ductile section. In fact, the difference between the moment at first yielding of the tensile steel and the ultimate moment is usually relatively small.

A flexural member, which is over-reinforced, on the other hand, does not have significant ductility at failure and fails by crushing of the compressive concrete without the prestressed or non-prestressed tensile reinforcement reaching yield or deforming significantly after yield. At the ultimate load condition, both the tensile strain at the steel level and the section curvature are relatively small and, consequently, there is little deformation or warning of failure.

Because it is the deformation at failure that defines ductility, it is both usual and reasonable in design to define a minimum ultimate curvature to ensure the ductility of a cross-section. This is often achieved by placing a maximum limit on the depth to the neutral axis at the ultimate load condition. Ductility can be increased by the inclusion of non-prestressed reinforcing steel in the compression zone of the beam. With compressive steel included, the internal compressive force F_c is shared between the concrete and the steel. The volume of the concrete stress block above the neutral axis is therefore reduced and, consequently, the depth to the neutral axis is decreased. Some compressive reinforcement is normally included in beams to provide anchorage for transverse shear reinforcement.

Ductility is desirable in prestressed (and reinforced) concrete flexural members. In continuous or statically indeterminate members, ductility is particularly necessary. Large curvatures are required at the peak moment regions in order to permit the inelastic moment redistribution that must occur if the moment diagram assumed in design is to be realised in practice. Consider the stress distribution caused by the ultimate moment on the section in Figure 6.1. The resultant compressive force of magnitude F_c equals the resultant tensile force F_t and the ultimate moment capacity M_u (also termed the resistance) is calculated from the internal couple:

$$M_u (= M_R) = F_c z = F_t z \tag{6.1}$$

The lever arm z between the internal compressive and tensile resultants (F_c and F_t) is usually about $0.9d$, where d is the *effective depth* of the section and may be defined as the distance from the extreme compressive fibre to the position of the resultant tensile force in all the steel on the tensile side of the neutral axis.

To find the lever arm z more accurately, the location of the resultant compressive force in the concrete F_c needs to be determined by considering the actual stress–strain relationship for concrete in the compression zone and locating the position of its centroid.

6.3 DESIGN FLEXURAL RESISTANCE

6.3.1 Assumptions

In the analysis of a cross-section to determine its design bending resistance M_{Rd}, the following assumptions are usually made:

1. the variation of strain on the cross-section is linear, i.e. strains in the concrete and the bonded steel are calculated on the assumption that plane sections remain plane;
2. perfect bond exists between the concrete and the bonded reinforcement or bonded tendons, i.e. the change in strain in the bonded reinforcement or bonded tendons is the same as that in the adjacent concrete;
3. concrete carries no tensile stress, i.e. the tensile strength of the concrete is ignored;
4. the stresses in the compressive concrete and in the steel reinforcement (both prestressed and non-prestressed) are obtained from actual or idealised stress–strain relationships for the respective materials; and
5. the initial strain in the prestressing tendons is taken into account when determining the stress in the tendon.

6.3.2 Idealised compressive stress blocks for concrete

In order to simplify numerical calculations for the design flexural resistance, EN 1992-1-1 [1] specifies the idealised stress blocks shown in Figure 6.2 for the concrete on the compressive side of the neutral axis. Compressive strains are shown as positive in these figures. The strain limits and the exponent n are given in Table 6.1 for the standard strength grades of concrete. As an alternative, to the stress blocks in Figure 6.2, an idealised rectangular stress block may be used to model the compressive stress distribution in the concrete.

In Figure 6.3a, an under-reinforced section at the ultimate moment is shown. The section has a single layer of bonded prestressing steel. The strain diagram and the actual concrete stress distribution used for the ultimate limit state design are also shown. In Figure 6.3b, the idealised rectangular stress block specified in EN 1992-1-1 [1] to model the design compressive stress distribution in the concrete above the neutral axis is shown. The dimensions of the rectangular stress block are calibrated such that the volume of the stress block and the position of its centroid are approximately the same as for the curvilinear design stress block.

At the design ultimate moment, the extreme fibre-compressive strain is taken to be ε_{cu3} and the depth to the neutral axis is x. In reality, the actual extreme fibre strain may vary depending on the degree of confinement, but for under-reinforced members, with the flexural resistance very much controlled by the strength of the tensile steel (both prestressed and non-prestressed), variation in the assumed value of ε_{cu3} does not have a significant effect on the design resistance M_{Rd}.

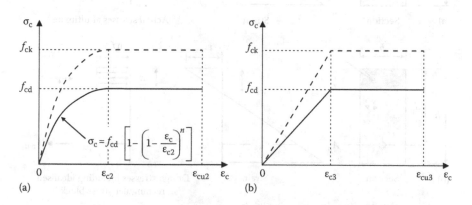

Figure 6.2 Idealised stress blocks for concrete in compression. (a) Parabola–rectangle diagram. (b) Bilinear diagram.

Table 6.1 Strain limits for idealised stress blocks [1]

f_{ck} MPa	Strength classes for concrete						Analytical relationship for $f_{ck} \geq 50$ MPa
	≤50	55	60	70	80	90	
f_{cm} MPa	–	63	68	78	88	98	$f_{cm} = f_{ck} + 8$ (MPa)
ε_{cu1}	0.0035	0.0032	0.003	0.0028	0.0028	0.0028	$\varepsilon_{cu1} = 0.0028 + 0.027$ $[(98 - f_{cm})/100]^4$
ε_{c2}	0.002	0.0022	0.0023	0.0024	0.0025	0.0026	$\varepsilon_{c2} = 0.002 + 0.085 \times 10^{-3}$ $(f_{ck} - 50)^{0.53}$
ε_{cu2}	0.0035	0.0031	0.0029	0.0027	0.0026	0.0026	$\varepsilon_{cu2} = 0.0026 + 0.035$ $[(90 - f_{ck})/100]^4$
n	2.0	1.74	1.6	1.45	1.4	1.4	$n = 1.4 + 23.4$ $[(90 - f_{ck})/100]^4$
ε_{c3}	0.00175	0.0018	0.0019	0.002	0.0022	0.0023	$\varepsilon_{c3} = 0.00175 + 0.00055$ $[(f_{ck} - 50)/40]$
ε_{cu3}	0.0035	0.0031	0.0029	0.0027	0.0026	0.0026	$\varepsilon_{cu3} = 0.0026 + 0.035$ $[(90 - f_{ck})/100]^4$

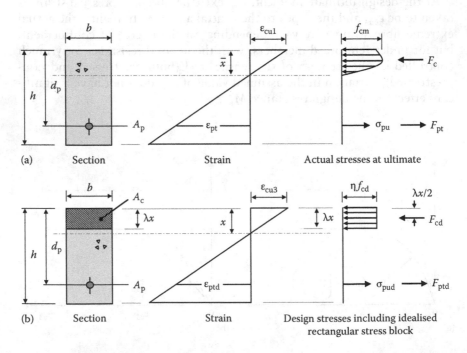

Figure 6.3 Flexural behaviour of a singly reinforced section at the ultimate limit state [1]. (a) Curvilinear stress block at ultimate moment. (b) Idealised rectangular stress block used to calculate the design resistance M_{Rd}.

The depth of the rectangular stress block (in Figure 6.3b) is λx and the uniform stress intensity is ηf_{cd}. For the rectangular section of Figure 6.3b, the hatched area A_c (= $\lambda x b$) is therefore assumed to be subjected to a uniform stress of ηf_{cd}. In EN 1992-1-1 [1], λ and η depend on the compressive strength of concrete and are given by:

$$\lambda = 0.8 \qquad\qquad\qquad \text{for } f_{ck} \leq 50 \text{ MPa} \qquad\qquad\qquad (6.2)$$

$$\lambda = 0.8 - (f_{ck} - 50)/400 \quad \text{for } 50 < f_{ck} \leq 90 \text{ MPa} \qquad (6.3)$$

$$\eta = 1.0 \qquad\qquad\qquad \text{for } f_{ck} \leq 50 \text{ MPa} \qquad\qquad\qquad (6.4)$$

$$\eta = 1.0 - (f_{ck} - 50)/200 \quad \text{for } 50 < f_{ck} \leq 90 \text{ MPa} \qquad (6.5)$$

EN 1992-1-1 [1] recommends that the value of ηf_{cd} should be reduced by 10% for cross-sections that reduce in width as the extreme compressive fibre is approached.

For the rectangular section of Figure 6.3b, the resultant compressive force F_{cd} is the volume of the rectangular stress block given by:

$$F_{cd} = \eta f_{cd} A_c = \eta f_{cd} \lambda x b \qquad\qquad\qquad (6.6)$$

and the line of action of F_{cd} passes through the centroid of the hatched area A_c, i.e. at a depth of $\lambda x/2$ below the extreme compressive fibre (provided, of course, that A_c is rectangular). The resultant tensile force F_{ptd} on the cross-section is the force in the tendon:

$$F_{ptd} = \sigma_{pud} A_p \qquad\qquad\qquad (6.7)$$

where σ_{pud} is the design stress in the tendon and is determined from considerations of equilibrium, strain compatibility and the design stress–strain relationship for the tendon (given in Figure 4.12).

Axial equilibrium requires that $F_{ptd} = F_{cd}$ and therefore:

$$\sigma_{pud} = \eta f_{cd} \lambda x b / A_p \qquad\qquad\qquad (6.8)$$

The flexural design resistance is obtained from Equation 6.1:

$$M_{Rd} = F_{ptd} z = \sigma_{pud} A_p \left(d_p - \frac{\lambda x}{2} \right) \qquad\qquad (6.9)$$

The ultimate design curvature κ_{ud} is an indicator of ductility and is the slope of the design strain diagram at failure (as shown in Figure 6.3b):

$$\kappa_{ud} = \frac{\varepsilon_{cu3}}{x} \qquad\qquad\qquad (6.10)$$

Ductile failures are associated with large deformations at the ultimate load condition. Ductility is generally acceptable if the depth of the neutral axis at the design resistance x is less than about 0.3 d, where d is the effective depth to the line of action of F_{ptd}. EN 1992-1-1 [1] requires that in regions of plastic hinges, x/d should not exceed 0.45 when $f_{ck} \leq 50$ MPa and 0.35 when $f_{ck} \geq 55$ MPa.

6.3.3 Prestressed steel strain components (for bonded tendons)

For reinforced concrete sections, the strains in the reinforcing steel and in the concrete at the steel level are the same at every stage of loading, while for the tendons on a prestressed concrete section, this is not the case. The strain in the bonded prestressing steel at any stage of loading is equal to the strain caused by the initial prestress plus the change in strain in the concrete at the steel level caused by the applied load. To calculate accurately the design flexural resistance of a section, an accurate estimate of the final strain in the prestressed and non-prestressed steel is required. The design tensile strain in the prestressing steel ε_{pud} is much larger than the tensile strain in the concrete at the steel level, owing to the large initial prestress. For a bonded tendon, ε_{pud} is usually considered to be the sum of several subcomponents. Figure 6.4 shows the instantaneous strain distributions on a prestressed section at three stages of loading.

Stage (a) shows the elastic instantaneous concrete strain caused by the effective prestress $P_{m,t}$, when the externally applied moment is zero. The instantaneous strain in the concrete at the steel level is compressive, with magnitude approximately equal to:

$$\varepsilon_{ce} = \frac{1}{E_{cm}} \left(\frac{P_{m,t}}{A} + \frac{P_{m,t}e^2}{I} \right)$$

(6.11)

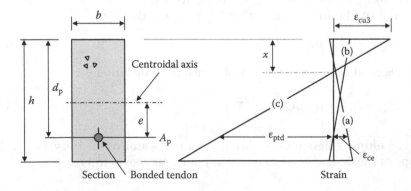

Figure 6.4 Instantaneous strain distributions at three stages of loading.

where A is the area of the section, I is the second moment of area of the section about its centroidal axis and e is the eccentricity of the prestressing force (as shown in Figure 6.4).

The stress and strain in the prestressing steel at stage (a) are:

$$\sigma_{pm,t} = \frac{P_{m,t}}{A_p} \tag{6.12}$$

and

$$\varepsilon_{pe} = \frac{\sigma_{pm,t}}{E_p} = \frac{P_{m,t}}{E_p A_p} \tag{6.13}$$

provided that the steel stress is within the elastic range.

Stage (b) is the concrete strain distribution when the applied moment is sufficient to decompress the concrete at the steel level. Provided that there is bond between the steel and the concrete, the change in strain in the pre-stressing steel is equal to the change in concrete strain at the steel level. The strain in the prestressing steel at stage (b) is therefore equal to the value at stage (a) plus a tensile increment of strain equal in magnitude to ε_{ce} (from Equation 6.11).

Strain diagram (c) in Figure 6.4 corresponds to the design ultimate load condition. The concrete strain at the steel level ε_{ptd} can be expressed in terms of the extreme compressive fibre strain ε_{cu3} and the depth to the neutral axis x and is given by:

$$\varepsilon_{ptd} = \varepsilon_{cu3} \frac{d_p - x}{x} \tag{6.14}$$

From the requirements of strain compatibility, the change in strain in the bonded prestressing steel between load stages (b) and (c) is also equal to ε_{ptd}. Therefore, the strain in the bonded tendon at the design ultimate load condition may be obtained from:

$$\varepsilon_{pud} = \varepsilon_{pe} + \varepsilon_{ce} + \varepsilon_{ptd} \tag{6.15}$$

and ε_{pud} can therefore be determined in terms of the position of the neutral axis at failure x and the extreme compressive fibre strain ε_{cu3}. If ε_{pud} is known, the design stress σ_{pud} in the prestressing steel at the design resistance can be determined from the design stress–strain diagram for the prestressing steel (Figure 4.12). With the area of prestressing steel known, the design tensile force F_{ptd} can be calculated. In general, however, the design steel stress is not known at failure, and it is necessary to equate the design tensile force in the steel tendon (plus the design tensile force in any non-prestressed tensile steel) with the design concrete compressive force (plus the design compressive force in any non-prestressed compressive steel) in order to locate the neutral axis depth and hence find ε_{pud}.

In general, the magnitude of ε_{ce} in Equation 6.15 is very much smaller than either ε_{pe} or ε_{ptd}, and may often be ignored without introducing serious errors.

6.3.4 Determination of M_{Rd} for a singly reinforced section with bonded tendons

Consider the section shown in Figure 6.3a and the idealised compressive stress block shown in Figure 6.3b. In order to calculate the design bending resistance using Equation 6.9, the depth to the neutral axis x and the final stress in the prestressing steel σ_{pud} must first be determined.

An iterative trial-and-error procedure is usually used to determine the value of x for a given section. The depth to the neutral axis is adjusted until horizontal equilibrium is satisfied, i.e. $F_{ptd} = F_{cd}$, in which both F_{ptd} and F_{cd} are functions of x. For this singly reinforced cross-section, F_{cd} is the volume of the compressive stress block given by Equation 6.6 and F_{ptd} depends on the strain in the prestressing steel ε_{pud}. For any value of x, the strain in the prestressing steel is calculated using Equation 6.15 (and Equations 6.11 through 6.14). The design steel stress σ_{pud}, which corresponds to the calculated value of strain ε_{pud}, can be obtained from the design stress–strain curve for the prestressing steel and the corresponding tensile force is given by Equation 6.7. When the correct value of x is found (i.e. when $F_{ptd} = F_{cd}$), the design flexural resistance M_{Rd} may be calculated from Equation 6.9.

A suitable iterative procedure is outlined and illustrated in the following example. About three iterations are usually required to determine a good estimate of x and hence M_{Rd}.

1. With ε_{cu3} taken from the bottom row in Table 6.1, select an *appropriate* trial value of x (= x_1) and determine the corresponding value of ε_{pud} (= ε_{pud1}) from Equation 6.15 and F_{cd} (= F_{cd1}) from Equation 6.6. By equating the tensile force in the steel to the compressive force in the concrete, the stress in the tendon σ_{pud} (= σ_{pud1}) may be determined from Equation 6.8.

2. Plot the points ε_{pud1} and σ_{pud1} on the graph containing the design stress–strain curve for the prestressing steel (as illustrated subsequently in Figure 6.6). If the point falls on the curve, then the value of x selected in step 1 is correct. If the point is not on the curve, then the stress–strain relationship for the prestressing steel is not satisfied and the value of x is not correct.

3. If the points ε_{pud1} and σ_{pud1} plotted in step 2 are not sufficiently close to the design stress–strain curve for the steel, repeat steps 1 and 2 with a new estimate of x (= x_2) to obtain revised estimates of tendon strain and stress (ε_{pud2} and σ_{pud2}). A larger value for x is required if the

point plotted in step 2 is below the design stress–strain curve and a smaller value is required if the point is above the curve. Plot the new points ε_{pud2} and σ_{pud2} on the curve.

4. Interpolate between the plots from steps 2 and 3 to obtain a close estimate for ε_{pud} and σ_{pud} and the corresponding value for x.

5. With the values of σ_{pud} and x determined in step 4, calculate the design moment resistance M_{Rd}. If the area above the neutral axis is rectangular, M_{Rd} is obtained from Equation 6.9. Non-rectangular-shaped cross-sections are considered in Section 6.5.

EXAMPLE 6.1

The design flexural resistance M_{Rd} of the rectangular section of Figure 6.5 is to be calculated.

The steel tendon consists of ten 12.9 mm strands (steel type Y1860S). From Table 4.8, A_p = 1,000 mm², f_{pk} = 1,860 MPa, f_{pd} = 1,391 MPa, E_p = 195,000 MPa, γ_s = 1.15 and ε_{uk} = 0.035. The effective prestress is $P_{m,t}$ = 1,200 kN. The design stress–strain relationship for prestressing steel is shown in Figure 6.6 (taken as Line 1 from Figure 4.12). The concrete properties are f_{ck} = 40 MPa and E_{cm} = 35,000 MPa.

With the partial safety factor for concrete γ_C = 1.5 and the coefficient α_{cc} = 1.0, the design strength of concrete is given by Equation 4.11:

$$f_{cd} = \frac{\alpha_{cc}f_{ck}}{\gamma_C} = \frac{1.0 \times 40}{1.5} = 26.67 \text{ MPa}$$

From Equation 6.2, λ = 0.8 and, from Equation 6.4, η = 1.0.

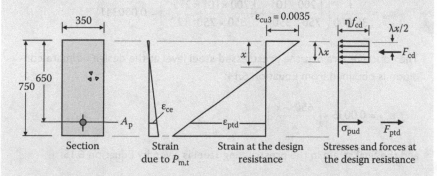

Section   Strain   Strain at the design   Stresses and forces at
due to $P_{m,t}$ resistance   the design resistance

Figure 6.5 Section details and stress and strain distributions used for the calculation of design flexural resistance M_{Rd} (Example 6.1).

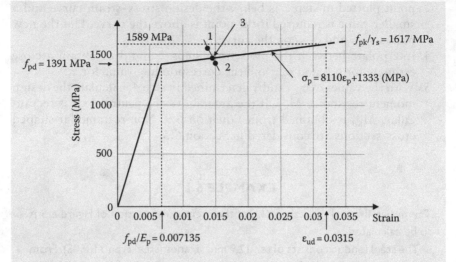

Figure 6.6 Stress–strain curve for strand (Example 6.1).

The initial strain in the tendons due to the effective prestress is given by Equation 6.13:

$$\varepsilon_{pe} = \frac{P_{m,t}}{E_p A_p} = \frac{1,200 \times 10^3}{195,000 \times 1,000} = 0.00615$$

The strain in the concrete caused by the effective prestress at the level of the prestressing steel (ε_{ce} in Figure 6.4) is calculated using Equation 6.11. Because ε_{ce} is small compared with ε_{pe}, it is usually acceptable to use the properties of the gross cross-section for its determination:

$$\varepsilon_{ce} = \frac{1}{35,000} \left(\frac{1,200 \times 10^3}{750 \times 350} + \frac{1,200 \times 10^3 \times 275^2}{350 \times 750^3 / 12} \right) = 0.000341$$

The concrete strain at the prestressed steel level at the design ultimate condition is obtained from Equation 6.14:

$$\varepsilon_{ptd} = 0.0035 \times \left(\frac{650 - x}{x} \right)$$

and the final strain in the prestressing steel is given by Equation 6.15:

$$\varepsilon_{pud} = 0.00615 + 0.000341 + 0.0035 \times \left(\frac{650 - x}{x} \right) \tag{6.1.1}$$

When ε_{pd} (= f_{pd}/E_p) $\leq \varepsilon_{pud} \leq \varepsilon_{ud}$, the stress–strain relationship for the tendon is obtained from Figure 4.12 as:

$$\frac{\sigma_{pud} - f_{pd}}{(f_{pk}/\gamma_s) - f_{pd}} = \frac{\varepsilon_{pud} - (f_{pd}/E_p)}{\varepsilon_{uk} - (f_{pd}/E_p)}$$

The magnitude of resultant compressive force F_{cd} carried by the concrete on the rectangular section is the volume of the idealised rectangular stress block in Figure 6.5 and is given by Equation 6.6:

$$F_{cd} = \eta f_{cd} \lambda x b = 1.0 \times 26.67 \times 0.8 \times 350x = 7467x$$

The resultant tensile force F_{ptd} is given by Equation 6.7:

$$F_{ptd} = 1000 \times \sigma_{pud}$$

Horizontal equilibrium requires that $F_{cd} = F_{ptd}$ and hence:

$$\sigma_{pud} = 7.467x \qquad (6.1.2)$$

Trial values of x are selected and the corresponding values of ε_{pud} and σ_{pud} (calculated from Equations 6.1.1 and 6.1.2 earlier) are tabulated in the following text and plotted on the stress–strain curve for the steel in Figure 6.6.

Trial x (mm)	ε_{pud} Equation 6.1.1	σ_{pud} (MPa) Equation 6.1.2	Point plotted on Figure 6.6
210	0.0138	1568	1
190	0.0150	1419	2
195	0.0147	1456	3

Point 3 lies sufficiently close to the stress–strain curve for the tendon and therefore the correct value for x is close to 195 mm. With $x/d = 0.300 < 0.45$, the ductility requirements of EN 1992-1-1 [1] are satisfied.

The design moment resistance is given by Equation 6.9:

$$M_{Rd} = 1456 \times 1000 \left(650 - \frac{0.8 \times 195}{2} \right) = 833 \times 10^6 \text{ Nmm} = 833 \text{ kNm}$$

and Equation 6.10 gives the design curvature corresponding to M_{Rd}:

$$\kappa_{ud} = \frac{0.0035}{195} = 18.0 \times 10^{-6} \text{ mm}^{-1}$$

6.3.5 Determination of M_{Rd} for sections containing non-prestressed reinforcement and bonded tendons

Frequently, in addition to the prestressing reinforcement, prestressed concrete beams contain non-prestressed longitudinal reinforcement in both the compressive and tensile zones. This reinforcement may be included for a variety of reasons. For example, non-prestressed reinforcement is included in the tensile zone to provide additional flexural strength when the strength provided by the prestressing steel is not adequate. Non-prestressed tensile steel is also included to improve crack control when cracking is anticipated at service loads. Non-prestressed compressive reinforcement may be used to strengthen the compressive zone in beams that might otherwise be over-reinforced. In such beams, the inclusion of compression reinforcement not only increases the design ultimate strength, but also increases the curvature at failure and, therefore, improves ductility.

The use of compressive reinforcement also reduces long-term deflections caused by creep and shrinkage and, therefore, improves serviceability. If for no other reason, compression reinforcement may be included to provide anchorage and bearing for the transverse reinforcement (stirrups) in beams.

When compressive reinforcement is included, closely spaced transverse ties should be used to laterally brace the highly stressed bars in compression and prevent them from buckling outward. In general, the spacing of these ties should not exceed about 16 times the diameter of the compressive bar.

Consider the *doubly reinforced* section shown in Figure 6.7a. The resultant design compressive force consists of a steel component $F_{sd(1)}$ (= $\sigma_{sd(1)} A_{s(1)}$) and a concrete component F_{cd} (= $\eta f_{cd} \lambda x b$). The magnitude of the strain in the compressive reinforcement is determined from the geometry of the linear strain diagram shown in Figure 6.7b and is given by:

$$\varepsilon_{sd(1)} = \frac{\varepsilon_{cu3}(x - d_{s(1)})}{x} \tag{6.16}$$

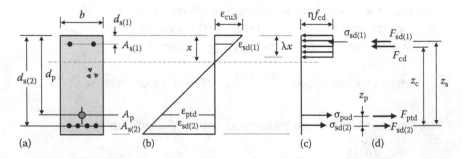

Figure 6.7 Doubly reinforced rectangular cross-section at the ultimate design moment. (a) Section. (b) Strain. (c) Stresses. (d) Forces.

If the idealised elastic–plastic design stress–strain relationship of Figure 4.8 is used, when $\varepsilon_{sd(1)}$ is less than or equal to the design yield strain of the non-prestressed steel ($\varepsilon_{yd} = f_{yd}/E_s = f_{yk}/(\gamma_S E_s)$) then the design stress in the compressive steel is $\sigma_{sd(1)} = \varepsilon_{sd(1)} E_s$. If $\varepsilon_{sd(1)}$ exceeds the design yield strain, then $\sigma_{sd(1)} = f_{yd} = f_{yk}/\gamma_S$.

The resultant tensile force in Figure 6.7d consists of a prestressed component F_{ptd} (= $\sigma_{pud} A_p$) and a non-prestressed steel component $F_{sd(2)}$ (= $\sigma_{sd(2)} A_{s(2)}$). The design stress in the non-prestressed tensile steel is determined from the strain $\varepsilon_{sd(2)}$ given by:

$$\varepsilon_{sd(2)} = \frac{\varepsilon_{cu3}(d_{s(2)} - x)}{x} \tag{6.17}$$

If $\varepsilon_{sd(2)} \leq \varepsilon_{yd}$, then $\sigma_{sd(2)} = \varepsilon_{sd(2)} E_s$. If $\varepsilon_{sd(2)} > \varepsilon_{yd}$, then $\sigma_{sd(2)} = f_{yd}$.

In order to calculate the depth to the neutral axis x at the ultimate design moment, a trial-and-error approach similar to that outlined in Section 6.3.4 can be employed. Successive values of x are tried until the value which satisfies the following horizontal equilibrium equation is determined:

$$F_{ptd} + F_{sd(2)} = F_{cd} + F_{sd(1)} \tag{6.18}$$

Since one of the reasons for the inclusion of compressive reinforcement is to improve ductility, most doubly reinforced beams are, or should be, under-reinforced, i.e. the non-prestressed tensile steel $A_{s(2)}$ is at yield at the ultimate design moment. Whether or not the compressive steel $A_{s(1)}$ has yielded depends on its depth $d_{s(1)}$ from the top compressive surface of the section and on the depth to the neutral axis x.

For any value of x, with the stresses in the compressive and tensile reinforcement determined from the strains $\varepsilon_{sd(1)}$ and $\varepsilon_{sd(2)}$ (given by Equations 6.16 and 6.17, respectively), Equation 6.18 can be expanded as:

$$F_{cd} = \eta f_{cd} \lambda x b$$
$$= F_{ptd} + F_{sd(2)} - F_{sd(1)} = \sigma_{pud} A_p + \sigma_{sd(2)} A_{s(2)} - \sigma_{sd(1)} A_{s(1)}$$

and this can be rearranged to give the following expressions for x and σ_{pud}:

$$x = \frac{\sigma_{pud} A_p - \sigma_{sd(2)} A_{s(2)} + \sigma_{sd(1)} A_{s(1)}}{\eta f_{cd} \lambda b} \tag{6.19}$$

$$\sigma_{pud} = \frac{\eta f_{cd} \lambda x b - \sigma_{sd(2)} A_{s(2)} + \sigma_{sd(1)} A_{s(1)}}{A_p} \tag{6.20}$$

When the value of σ_{pud} (calculated from Equation 6.20) and the value of ε_{pud} (calculated from Equation 6.15) together satisfy the stress–strain relationship of the prestressing steel, the correct value of x has been found.

If it has been assumed that the non-prestressed steel has yielded in the calculations, the corresponding steel strains should be checked to ensure that the steel has, in fact, yielded. If the compressive steel is not at yield, then the compressive force $F_{s(1)}$ has been overestimated and the correct value of x is slightly greater than the calculated value. The compressive steel stress $\sigma_{sd(1)}$ in Equations 6.19 and 6.20 should be taken as $\varepsilon_{sd(1)}E_s$ instead of f_{yd}. Further iteration may be required to determine the correct value of x and the corresponding internal forces F_{cd}, $F_{sd(1)}$, F_{ptd} and $F_{sd(2)}$.

With horizontal equilibrium satisfied, the design moment resistance of the section may be determined by taking moments of the internal forces about any convenient point on the cross-section. Taking moments about the non-prestressed tensile reinforcement level gives:

$$M_{Rd} = F_{cd}\, z_c + F_{sd(1)}\, z_s - F_{ptd}\, z_p \tag{6.21}$$

For the rectangular section shown in Figure 6.7, the lever arms from the non-prestressed tensile reinforcement to each of the internal forces in Equation 6.21 are:

$$z_c = d_{s(2)} - \frac{\lambda x}{2} \quad z_s = d_{s(2)} - d_{s(1)} \quad z_p = d_{s(2)} - d_p$$

In these equations, $F_{sd(1)}$ and F_{cd} are the *magnitudes* of the compressive forces in the steel and concrete, respectively, and are therefore considered to be positive.

The design ultimate curvature is obtained from Equation 6.10. For ductility to be acceptable, the depth of the neutral axis x should be less than about $0.3d$, where d is the effective depth to the line of action of the resultant of the tensile forces F_{ptd} and $F_{sd(2)}$. The minimum design curvature required for ductility is therefore:

$$(\kappa_{ud})_{min} = \frac{3.33\varepsilon_{cu3}}{d} \tag{6.22}$$

EXAMPLE 6.2

To the cross-section shown in Figure 6.5 and analysed in Example 6.1, non-prestressed reinforcing bars of area $A_s = 1350$ mm^2 are added in the tensile zone at a depth $d_s = 690$ mm. Calculate the design flexural resistance M_{Rd} of the section. From Table 4.6, for B500B type reinforcing steel, the design yield stress is $f_{yd} = 435$ MPa and the elastic modulus is $E_s = 200{,}000$ MPa. All other material properties and cross-sectional details are as specified in Example 6.1.

The design strain in the prestressing steel is as calculated in Example 6.1:

$$\varepsilon_{pud} = 0.0064 9 1 + 0.0035 \times \left(\frac{650 - x}{x} \right) \tag{6.2.1}$$

and the magnitude of the compressive force F_{cd} carried by the concrete above the neutral axis is:

$$F_{cd} = \eta f_{cd} \lambda x b = 1.0 \times 26.67 \times 0.8 \times 350x = 7467x$$

From Equation 6.17 and with $\varepsilon_{cu3} = 0.0035$, the non-prestressed tensile steel is at yield, i.e. $\varepsilon_{sd} \geq \varepsilon_{yd} (= f_{yd}/E_s = 0.002175)$, provided that the depth to the neutral axis x is less than or equal to $0.6167d_s (= 425.6$ mm). If σ_{sd} is assumed to equal f_{yd}, the resultant tensile force $F_{td} (= F_{ptd} + F_{sd})$ is given by:

$$F_{td} = \sigma_{pud} A_p + f_{yd} A_s = 1000\sigma_{pud} + (435 \times 1350)$$

$$= 1000 \times (\sigma_{pud} + 587.3)$$

and enforcing horizontal equilibrium (i.e. $F_{cd} = F_{td}$):

$$\sigma_{pud} = 7.467x - 587.3 \tag{6.2.2}$$

Trial values of x are now selected, and the respective values of ε_{pud} and σ_{pud} are tabulated here and plotted on the stress–strain curve in Figure 6.8:

Trial x (mm)	ε_{pud} Equation 6.2.1	σ_{pud} (MPa) Equation 6.2.2	Point plotted on Figure 6.8
280	0.0111	1503	4
260	0.0117	1354	5
269.5	0.0114	1425	6

Since point 6 lies sufficiently close to the stress–strain curve for the tendon, the value for x is taken as 269.5 mm, and the effective depth to the resultant tensile force is $d = 670$ mm and, therefore, $x/d = 0.402$.

It is apparent in Figure 6.8 that the strain in the prestressing steel is decreased by the introduction of tensile reinforcement (from point 3 to point 6) and the depth to the neutral axis is increased. From Equation 6.10, the design ultimate curvature is:

$$\kappa_{ud} = \frac{0.0035}{269.5} = 13.0 \times 10^{-6} \text{ mm}^{-1}$$

and this is 27.8% less than that obtained in Example 6.1 (where $A_s = 0$).

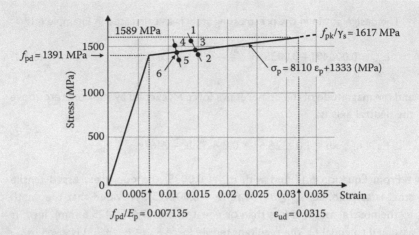

Figure 6.8 Stress–strain curve for strand (Example 6.2).

The strain in the tensile reinforcement at the strength limit state is given by Equation 6.17:

$$\varepsilon_{sd(2)} = \frac{0.0035(690 - 269.5)}{269.5} = 0.0055 > \varepsilon_{yd}$$

and therefore the non-prestressed steel has yielded, as previously assumed. The depth from the top surface to the resultant force in the tensile steel is:

$$d = \frac{\sigma_{pud}A_p d_p + f_{yd}A_s d_s}{\sigma_{pud}A_p + f_{yd}A_s} = 662 \text{ mm}$$

The minimum curvature required to ensure some measure of ductility is obtained from Equation 6.22:

$$(\kappa_{ud})_{min} = \frac{3.33 \times 0.0035}{662} = 17.6 \times 10^{-6} \text{ mm}^{-1}$$

and this is greater than κ_{ud}. The section is therefore non-ductile and, in design, it would be prudent to insert some non-prestressed compressive reinforcement to increase the design ultimate curvature and improve ductility (at least to the level required by Equation 6.22).

The design compressive force in the concrete is $F_{cd} = 7467 \times 269.5 \times 10^{-3}$ = 2012 kN and the tensile force in the tendon is $F_{ptd} = \sigma_{pud}A_p = 1425$ kN. The design moment resistance is calculated from Equation 6.21:

$$M_{Rd} = F_{cd}z_c - F_{ptd}z_p = F_{cd}\left(d_s - \frac{\lambda x}{2}\right) - F_{ptd}(d_s - d_p)$$

$$= 2012 \times \left(690 - \frac{0.8 \times 269.5}{2}\right) \times 10^{-3} - 1425 \times (690 - 650) \times 10^{-3}$$

$$= 1114\,kNm$$

EXAMPLE 6.3

Consider the effect on both strength and ductility of the cross-section of Example 6.2 if reinforcement of area $A_{s(1)} = 900$ mm^2 is included in the compression zone. Details of the cross-section are shown in Figure 6.9, together with the stress and strain distributions at the ultimate design moment. All data are as specified in Examples 6.1 and 6.2.

From Examples 6.1 and 6.2, the design ultimate strain in the tendons is:

$$\varepsilon_{pud} = 0.006491 + 0.0035 \times \left(\frac{650 - x}{x}\right) \tag{6.3.1}$$

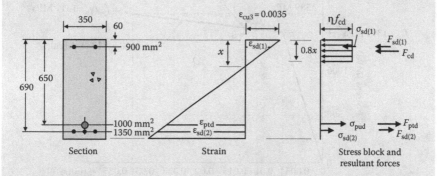

Figure 6.9 Section details and stress and strain distributions at the design ultimate moment condition (Example 6.3).

and the strain in the non-prestressed tensile reinforcement in Example 6.2 is greater than ε_{sd} and, hence, $\sigma_{sd(2)} = f_{yd}$.

The magnitude of the compressive steel strain is given by Equation 6.16:

$$\varepsilon_{sd(1)} = \frac{0.0035(x-60)}{x} \tag{6.3.2}$$

and the stress in the compression steel can be readily obtained from $\varepsilon_{sd(1)}$ for any value of x.

By equating $(F_{cd} + F_{sd(1)})$ with $(F_{ptd} + F_{sd(2)})$, the expression for σ_{pud} given by Equation 6.20 becomes:

$$\sigma_{pud} = \frac{1.0 \times 26.67 \times 0.8 \times x \times 350 - 435 \times 1350 + \sigma_{sd(1)} \times 900}{1000}$$

$$= 7.468x - 587 + 0.9\sigma_{sd(1)} \tag{6.3.3}$$

Values of ε_{pud}, $\varepsilon_{sd(1)}$, $\sigma_{sd(1)}$ and σ_{pud} for trial values of x are tabulated below and plotted as points 7–9 in Figure 6.10:

Trial x (mm)	ε_{pud} Equation 6.3.1	$\varepsilon_{sd(1)}$ Equation 6.3.2	$\sigma_{sd(1)}$ (MPa)	σ_{pud} (MPa) Equation 6.3.3	Point plotted on Figure 6.10
230	0.0129	0.00259	435	1522	7
210	0.0138	0.00250	435	1372	8
219	0.0134	0.00254	435	1440	9

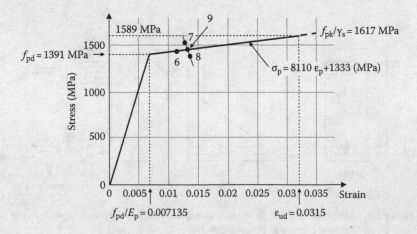

Figure 6.10 Stress–strain curve for strand (Example 6.3).

From Figure 6.10, point 9 lies very close to actual stress–strain curve and therefore the neutral axis depth is taken as $x = 219$ mm.

It is apparent from Figure 6.10 that the design strain in the prestressing steel is increased by the introduction of compressive reinforcement (from point 6 to point 9) and the depth to the neutral axis is decreased. The design ultimate curvature is obtained from Equation 6.10:

$$\kappa_{ud} = \frac{0.0035}{219} = 16.0 \times 10^{-6} \text{ mm}^{-1}$$

which represents a 23% increase in final curvature caused by the introduction of the compressive reinforcement and an improvement in the ductility of the cross-section.

The magnitudes of the resultant forces on the cross-section are:

$$F_{cd} = 1635 \text{ kN} \quad F_{sd(1)} = 392 \text{ kN} \quad F_{ptd} = 1440 \text{ kN} \quad F_{sd(2)} = 587 \text{ kN}$$

and the design moment resistance is calculated using Equation 6.21:

$$M_{Rd} = F_{cd}z_c + F_{sd(1)}z_s - F_{ptd}z_p = F_{cd}\left(d_{s(2)} - \frac{\lambda x}{2}\right) + F_{sd(1)}\left(d_{s(2)} - d_{s(1)}\right) - F_{ptd}\left(d_{s(2)} - d_p\right)$$

$$= \left[1635 \times \left(690 - \frac{0.8 \times 219}{2}\right) + 392 \times (690 - 60) - 1440 \times (690 - 650)\right] \times 10^{-3}$$

$$= 1174 \text{ kNm}$$

This represents a 5.4% increase in strength compared to the section without compressive steel that was analysed in Example 6.2. In general, for non-ductile sections, the addition of compressive reinforcement causes a significant increase in curvature (i.e. a significant increase in ductility) and a less significant, but nevertheless appreciable, increase in strength.

6.3.6 Members with unbonded tendons

In post-tensioned concrete, where the prestressing steel is not bonded to the concrete, the design stress in the tendon σ_{pud} is significantly less than that predicted for a bonded tendon, and the final strain in the tendon is more difficult to determine accurately. The design resistance of a section containing unbonded tendons may be as low as 75% of the strength of an equivalent section containing bonded tendons. Hence, from a strength point of view, bonded construction is to be preferred.

An unbonded tendon is not restrained by the concrete along its length, and slip between the tendon and the duct takes place as the external loads are applied and the member deforms. The tendon strain is more uniform along the length of the member and tends to be lower in regions of maximum moment than would be the case for a bonded tendon. The design resistance of the section may be reached before the stress in the unbonded tendon reaches its yield stress. For members not containing any bonded reinforcement, crack control may be a problem if cracking occurs in the member for any reason. If flexural cracking occurs, the number of cracks in the tensile zone is fewer than in a beam containing bonded reinforcement, but the cracks are wider and less serviceable.

To determine the increase in design stress in an unbonded tendon at the ultimate limit state $\Delta\sigma_p$, it is necessary to consider the deformation of the whole member using the mean values of the material properties. The design value of the stress increase is $\Delta\sigma_{pd} = \Delta\sigma_p\gamma_{\Delta P}$. When non-linear analysis is undertaken, the partial safety factor $\gamma_{\Delta P}$ is taken as 1.2 when the upper characteristic value of $\Delta\sigma_{pd}$ is required, and $\gamma_{\Delta P} = 0.8$ when the lower characteristic value of $\Delta\sigma_{pd}$ is required. If linear analysis is applied, with uncracked section properties, the calculated member deformation will generally underestimate the actual deformations and EN1992-1-1 [1] permits $\gamma_{\Delta P}$ to be taken as 1.0 and $\Delta\sigma_{pd} = \Delta\sigma_p$.

If no detailed calculation of the change in length of the tendon is made, EN 1992-1-1 [1] allows the stress in the tendon at the ultimate limit state to be assumed to equal the effective prestress (after all losses) plus $\Delta\sigma_{p,ULS} = 100$ MPa.

To ensure robustness and some measure of crack control, it is good practice to include non-prestressed bonded tensile reinforcement in members where the post-tensioned tendons are to remain unbonded for a significant period during and after construction.

EXAMPLE 6.4

The design flexural resistance of a simply-supported post-tensioned beam containing a single unbonded cable is to be calculated. The beam spans 12 m and its cross-section at mid-span is shown in Figure 6.5. Material properties and prestressing arrangement are as specified in Example 6.1.

The stress in the tendon caused by the effective prestressing force $P_{m,t} = 1200$ kN is:

$$\sigma_{pm,t} = P_{m,t}/A_p = 1200 \times 10^3/1000 = 1200 \text{ MPa}$$

EN1992-1-1 [1] permits the design stress in the tendon at the strength limit state to be taken as:

$$\sigma_{pud} = \sigma_{pm,t} + 100 = 1300\,MPa$$

and therefore the tensile force in the steel is $F_{ptd} = 1300$ kN $(= F_{cd})$. This is almost 10.7% lower than the value determined in Example 6.1 where the tendon was bonded to the concrete. The depth to the neutral axis is calculated as:

$$x = \frac{F_{ptd}}{\eta f_{cd} \lambda b} = \frac{1300 \times 10^3}{1.0 \times 26.67 \times 0.8 \times 350} = 174.1\,mm$$

and Equation 6.9 gives:

$$M_{Rd} = 1300 \times 1000 \times \left(650 - \frac{0.8 \times 174.1}{2}\right) \times 10^{-6} = 754\,kNm$$

In Example 6.1, the design bending resistance of the same cross-section with a bonded tendon was calculated to be 833 kNm. Clearly, the strength afforded by a post-tensioned tendon is significantly reduced if it remains unbonded.

6.4 DESIGN CALCULATIONS

6.4.1 Discussion

The magnitude of the effective prestressing force $P_{m,t}$ and the quantity of the prestressing steel A_p are usually selected to satisfy the serviceability requirements of the member, i.e. to control deflection or to reduce or eliminate cracking. With serviceability satisfied, the member is then checked for adequate strength. The design resistance M_{Rd} for the section containing the prestressing steel (plus any non-prestressed steel added for crack control or deflection control) is calculated and the design resistance is compared with the design action, in accordance with the design requirements outlined in Section 2.4. The design action M_{Ed} is the moment caused by the most severe factored load combination specified for the strength limit state (see Section 2.3.2). The design requirement is expressed by:

$$M_{Rd} \geq M_{Ed} \tag{6.23}$$

The prestressing steel needed for the satisfaction of serviceability requirements may not be enough to provide adequate strength. When this is the case, the design moment resistance can be increased by the inclusion of additional non-prestressed tensile reinforcement. Additional compressive reinforcement may also be required to improve ductility.

6.4.2 Calculation of additional non-prestressed tensile reinforcement

Consider the singly reinforced cross-section shown in Figure 6.11a. It is assumed that the effective prestress $P_{m,t}$, the area of the prestressing steel A_p and the cross-sectional dimensions have been designed to satisfy the serviceability requirements of the member. The idealised strain and stress distributions specified in EN1992-1-1 [1] for the ultimate limit state are also shown in Figure 6.11a. The design moment resistance of the section, denoted as M_{Rd1}, is calculated as follows:

$$M_{Rd1} = \sigma_{pud1} A_p \left(d_p - \frac{\lambda x_1}{2} \right) \tag{6.24}$$

where the tendon stress at the ultimate limit state σ_{pud1} can be calculated from the actual stress–strain curve for the steel (as illustrated in Example 6.1) or from the approximation illustrated in Example 6.4.

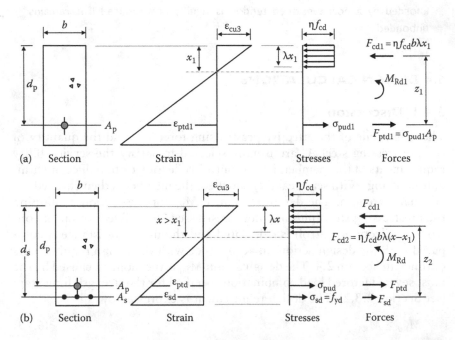

Figure 6.11 Cross-section containing tensile reinforcement – ultimate design condition. (a) Singly reinforced prestressed cross-section. (b) Cross-section containing both prestressed and non-prestressed tensile reinforcement.

If the design resistance M_{Rd1} is greater than or equal to M_{Ed}, then no additional tensile steel is necessary, and the cross-section has adequate strength. If M_{Rd1} is less than M_{Ed}, the section is not adequate and additional tensile reinforcement is required.

In addition to providing adequate strength, it is important also to ensure that the section is ductile. To ensure that the design ultimate curvature κ_{ud} is large enough to provide sufficient ductility, an upper limit for the depth to the neutral axis of about $0.3d_p$ should be enforced. If the value of x_1 in Figure 6.11a is greater than $0.3d_p$, some additional non-prestressed compressive reinforcement is required to relieve the concrete compressive zone and reduce the depth to the neutral axis. The design procedure outlined in Section 6.4.3 for doubly reinforced cross-sections is recommended in such a situation.

For the cross-section shown in Figure 6.11a, if M_{Rd1} is less than M_{Ed} and if x_1 is small so that ductility is not a problem, the aim in design is to calculate the minimum area of non-prestressed tensile reinforcement A_s that must be added to the section to satisfy strength requirements (i.e. the value of A_s such that $M_{Rd} = M_{Ed}$). In Figure 6.11b, the cross-section containing A_s is shown, together with the revised strain and stress distributions at the ultimate limit state design condition. With x small enough to ensure ductility, the tensile steel strain ε_{sd} is greater than the yield strain ε_{yd} ($= f_{yd}/E_s$), so that $\sigma_{sd} = f_{yd}$. The addition of A_s to the cross-section causes an increase in the resultant design tension ($F_{ptd} + F_{sd}$) and hence an increase in the resultant compression F_{cd} ($= F_{cd1} + F_{cd2}$). To accommodate this additional compression, the depth of the compressive stress block in Figure 6.11b must be greater than the depth of the stress block in Figure 6.11a (i.e. $\lambda x > \lambda x_1$). The increased value of x results in a reduction in the design ultimate curvature (i.e. a decrease in ductility), a reduction in the strain in the prestressing steel and a consequent decrease in σ_{pud}. While the decrease in σ_{pud} is relatively small, it needs to be verified that the modified cross-section possesses adequate ductility (i.e. that the value of x remains less than about $0.3d$).

If σ_{pud} is assumed to remain constant, a first estimate of the magnitude of the area of non-prestressed steel A_s required to increase the design resistance from M_{Rd1} (the strength of the section prior to the inclusion of the additional steel) to the design bending moment M_{Ed} (equal to the required minimum strength of the section) may be obtained from:

$$A_s \geq \frac{M_{Ed} - M_{Rd1}}{f_{yd}z_2} \tag{6.25}$$

where z_2 is the lever arm between the design tension force in the additional steel F_{sd} and the equal and opposite compressive force F_{cd2} which results from the increase in the depth of the compressive stress block. The lever arm z_2 may be approximated initially as:

$$z_2 = 0.9(d_s - \lambda x_1) \tag{6.26}$$

EXAMPLE 6.5

The design resistance of the singly reinforced cross-section shown in Figure 6.12 is M_{Rdl} = 931 kNm. The stress and strain distributions corresponding to M_{Rdl} are also shown in Figure 6.12 and the material properties are f_{ck} = 40 MPa (f_{cd} = 26.67 MPa), λ = 0.8, η = 1.0, f_{pk} = 1860 MPa, f_{pd} = 1391 MPa, E_p = 195000 MPa. Calculate the additional amount of non-prestressed tensile reinforcement located at d_s = 840 mm (f_{yd} = 435 MPa) if the design bending moment on the section is M_{Ed} = 1250 kNm.

For the section in Figure 6.12, x_1 = 159 mm = $0.212d_p$ and the section is ductile. If the additional tensile steel is to be added at d_s = 840 mm, then the lever arm z_2 in Equation 6.26 may be approximated by:

$$z_2 = 0.9(d_s - \lambda x_1) = 0.9 \times (840 - 0.8 \times 159) = 641.5 \text{ mm}$$

and the required area of non-prestressed steel is estimated using Equation 6.25:

$$A_s \geq \frac{(1250 - 931) \times 10^6}{435 \times 641.5} = 1143 \text{ mm}^2$$

Choose four 20 mm diameter bars (A_s = 1256 mm²) located at a depth d_s = 840 mm.

A check of this section to verify that $M_{Rd} \geq$ 1250 kNm, and also that the section is ductile, can now be made using the trial-and-error procedure illustrated in Example 6.2.

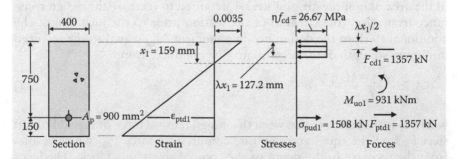

Figure 6.12 Singly reinforced cross-section of Example 6.5.

6.4.3 Design of a doubly reinforced cross-section

For a singly reinforced section (such as that shown in Figure 6.13a) in which x_1 is greater than about $0.3d_p$, the inclusion of additional tensile reinforcement may cause ductility problems. In such cases, the design resistance may be increased by the inclusion of suitable quantities of both tensile and compressive non-prestressed reinforcement without causing any reduction in curvature, i.e. without increasing x. If the depth to the neutral axis is held constant at x_1, the values of both F_{cd} (the compressive force carried by the concrete) and F_{ptd} (the tensile force in the prestressing steel) in Figures 6.13a and b are the same. In each figure, F_{cd} is equal and opposite to F_{ptd}. With the strain diagram in Figure 6.13b known, the strains at the levels of the top and bottom non-prestressed steel may be calculated using Equations 6.16 and 6.17, respectively, and hence the non-prestressed steel stresses $\sigma_{sd(1)}$ and $\sigma_{sd(2)}$ may be determined. The equal and opposite forces which result from the inclusion of the non-prestressed steel are:

$$F_{sd(1)} = A_{s(1)}\sigma_{sd(1)} \tag{6.27}$$

and

$$F_{sd(2)} = A_{s(2)}\sigma_{sd(2)} \tag{6.28}$$

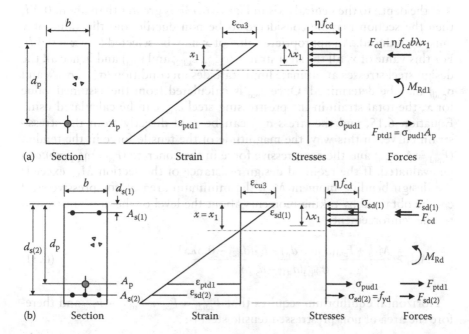

Figure 6.13 Doubly reinforced section at the ultimate limit state condition. (a) Cross-section containing prestressed steel only. (b) Cross-section containing top and bottom non-prestressed reinforcement.

When the depth to the compressive reinforcement is less than λx, the compressive force $F_{sd(1)}$ could be calculated as $F_{sd(1)} = A_{s(1)}(\sigma_{sd(1)} - \eta f_{cd})$, in order to account for the voids in the compressive concrete created by the compressive reinforcement.

If M_{Rd1} is the design resistance of the singly reinforced section in Figure 6.13a (calculated using Equation 6.9) and M_{Ed} is the design moment (equal to the minimum required strength of the doubly reinforced cross-section), the minimum area of the tensile reinforcement is given by:

$$A_{s(2)} = \frac{M_{Ed} - M_{Rd1}}{\sigma_{sd(2)}(d_{s(2)} - d_{s(1)})} \tag{6.29}$$

For conventional non-prestressed steel, $\sigma_{sd(2)}$ is usually at yield (i.e. $\sigma_{sd(2)} = f_{yd}$) provided that $\varepsilon_{sd(2)} \geq \varepsilon_{yd}$ and the depth to the neutral axis x satisfies the stated ductility requirements. For equilibrium, the forces in the top and bottom non-prestressed steel are equal and opposite, i.e. $F_{sd(1)} = F_{sd(2)}$, since $F_{cd} = F_{ptd}$. From Equations 6.27 and 6.28:

$$A_{s(1)} = \frac{A_{s(2)}\sigma_{sd(2)}}{\sigma_{sd(1)}} \tag{6.30}$$

If the depth to the neutral axis in Figure 6.13b is greater than about $0.3d$, then the section may be considered to be non-ductile and the value of x must be reduced. An appropriate value of x may be selected (say $x = 0.3d$). For this value of x, all the steel strains ($\varepsilon_{sd(1)}$, $\varepsilon_{sd(2)}$ and ε_{ptd}) and hence all the design steel stresses at ultimate limit state design condition ($\sigma_{sd(1)}$, $\sigma_{sd92)}$ and σ_{pud}) may be determined. Once ε_{ptd} is calculated from the assumed value for x, the total strain in the prestressing steel ε_{pud} can be calculated using Equation 6.15, and the stress σ_{pud} can be read directly from the stress–strain curve. In this way, the magnitude of the tensile force in the tendon ($F_{ptd} = A_p\sigma_{pud}$) and the compressive force in the concrete ($F_{cd} = \eta f_{cd}\lambda xb$) can be evaluated. If the required design resistance of the section M_{Rd} exceeds the design bending moment M_{Ed}, the minimum area of compressive steel can be obtained by taking moments about the level of the non-prestressed tensile reinforcement:

$$A_{s(1)} \geq \frac{M_{Ed} + F_{ptd}(d_{s(1)} - d_p) - F_{cd}(d_{s(2)} - 0.5\lambda x)}{\sigma_{sd(1)}(d_{s(1)} - d_{s(2)})} \tag{6.31}$$

Horizontal equilibrium requires that $F_{sd(2)} = F_{cd} + F_{sd(1)} - F_{ptd}$ and therefore the area of non-prestressed tensile steel is:

$$A_{st(2)} = \frac{\eta f_{cd}\lambda xb + A_{s(1)}\sigma_{sd(1)} - A_p\sigma_{pud}}{\sigma_{sd(2)}} \tag{6.32}$$

EXAMPLE 6.6

Determine the additional non-prestressed steel required to increase the design flexural resistance of the section in Figure 6.5 (and analysed in Example 6.1) if the design bending moment M_{Ed} is 1150 kNm. Take the depth to the additional tensile steel as $d_{s(2)}$ = 690 mm and the depth to the compressive steel (if required) as $d_{s(1)}$ = 60 mm. Assume for the reinforcement f_{yd} = 435 MPa and E_s = 200,000 MPa.

From Example 6.1, M_{Rd1} = 833 kNm and x_1 = 195 mm. If only non-prestressed tensile steel were to be added, the lever arm z in Equation 6.26 would be:

$$z = 0.9 \times (690 - 0.8 \times 195) = 481\,mm$$

and from Equation 6.25:

$$A_{s(2)} \geq \frac{(1150 - 833) \times 10^6}{435 \times 481} = 1515\,mm^2$$

This corresponds to the addition of five 20 mm diameter bars (1570 mm²) in the bottom of the section at a depth $d_{s(2)}$ = 690 mm.

A check of the section to verify that $M_{Rd} \geq$ 1150 kNm can next be made using the trial-and-error procedure illustrated in Example 6.2. In this example, however, the neutral axis depth increases above 0.3d, and the curvature at the ultimate limit state condition is less than the minimum value recommended in Equation 6.22. For this cross-section, it is appropriate to supply the additional moment capacity via both tensile and compressive non-prestressed reinforcement.

If the depth to the neutral axis is held constant at the value determined in Example 6.1, i.e. $x = x_1$ = 195 mm, then the stress and strain in the prestressed steel remain as previously calculated, i.e. ε_{pud} = 0.0147 and σ_{pud} = 1456 MPa.

With $d_{s(1)}$ = 60 mm, from Equation 6.16:

$$\varepsilon_{sd(1)} = \frac{0.0035(195 - 60)}{195} = 0.00242 > \varepsilon_{yd} \quad \text{and} \quad \sigma_{sd(1)} = f_{yd} = 435\,MPa$$

From Equation 6.17:

$$\varepsilon_{sd(2)} = \frac{0.0035(690 - 195)}{195} = 0.00888 > \varepsilon_{yd} \quad \text{and} \quad \sigma_{sd(2)} = f_{yd} = 435\,MPa$$

The minimum areas of additional tensile and compressive steel are obtained using Equations 6.29 and 6.30, respectively:

$$A_{s(2)} \geq \frac{(1150 - 833) \times 10^6}{435 \times (690 - 60)} = 1157 \text{ mm}^2$$

$$A_{s(1)} = \frac{1157 \times 435}{435} = 1157 \text{ mm}^2$$

A suitable solution is to include four 20 mm diameter reinforcing bars in the top and bottom of the section (at $d_{s(1)} = 60$ mm and $d_{s(2)} = 690$ mm).

6.5 FLANGED SECTIONS

Flanged sections such as those shown in Figure 6.14a are commonly used in prestressed concrete construction, where the bending efficiency of I-, T- and box-shaped sections can be effectively utilised. Frequently, in the construction of prestressed floor systems, beams or wide bands are poured monolithically with the slabs. In such cases, a portion of slab acts as either a top or a bottom flange of the beam, as shown in Figure 6.14b. The *effective flange width* (b_{eff} in Figure 6.14b) is selected such that the stresses across the width of the flanged beam may be assumed to be uniform and b_{eff} depends on the beam

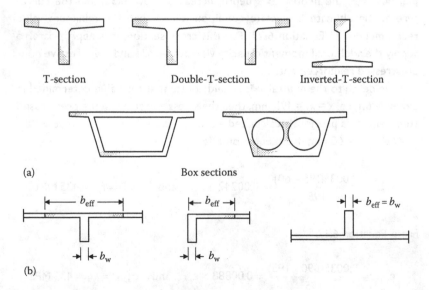

Figure 6.14 Typical flanged sections. (a) Precast. (b) Monolithic.

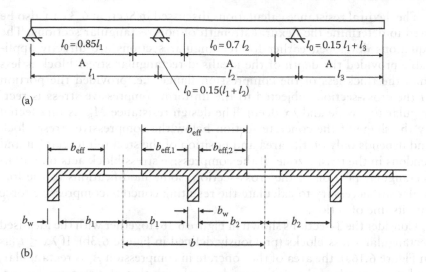

Figure 6.15 Effective flange width parameters [I]. (a) Elevation – definitions of l_0.
(b) Effective beam cross-sections.

and slab dimensions, the span and the support conditions, the type of loading and the amount and distribution of transverse reinforcement in the slab.

EN1992-1-1 [1] specifies the effective width in terms of the distance l_0 along the beam between the points of zero moment (as illustrated in Figure 6.15a) and the cross-sectional geometry (as defined in Figure 6.15b):

For T-sections: $\quad b_{eff} = b_{eff,1} + b_{eff,2} + b_w$ (6.33)

For L-sections: $\quad b_{eff} = b_{eff,1} + b_w$ (6.34)

where:

$$b_{eff,i} = 0.2b_i + 0.1l_0 \le 0.2l_0 \qquad (6.35)$$

except that the overhanging part of the effective flange should not exceed half the clear distance to the next parallel beam (i.e. $b_{eff,i} \le b_i$). In structural analysis, it is permissible to assume that the effective width is constant over the whole span, with the value of b_{eff} determined for the span section (marked region A in Figure 6.15a).

It is recommended in ACI 318M-14 [2] that the overhanging part of the effective flange on each side of the web of a T-beam should not exceed eight times the slab thickness. For L-beams with a slab on one side only, the overhanging part of the effective flange width should not exceed six times the slab thickness. Although these are not formal requirements of EN1992-1-1 [1], their satisfaction is recommended here.

The flexural resistance calculations discussed in Section 6.3 can also be used to determine the flexural strength of non-rectangular sections. The equations developed earlier for rectangular sections are directly applicable provided the depth of the idealised rectangular stress block is less than the thickness of the compression flange, i.e. provided the portion of the cross-section subjected to the uniform compressive stress is rectangular (b_{eff} wide and λx deep). The design resistance M_{Rd} is unaffected by the shape of the concrete section below the compressive stress block and depends only on the area and position of the steel reinforcement and tendons in the tensile zone. If the compressive stress block acts on a non-rectangular portion of the cross-section, some modifications to the formulae are necessary to calculate the resulting concrete compressive force and its line of action.

Consider the T-sections shown in Figure 6.16, together with the idealised rectangular stress blocks (previously defined in Figure 6.3b). If $\lambda x \leq t$ (as in Figure 6.16a), the area of the concrete in compression A_c is rectangular, and the strength of the section is identical with that of a rectangular section of width b_{eff} containing the same tensile steel at the same effective

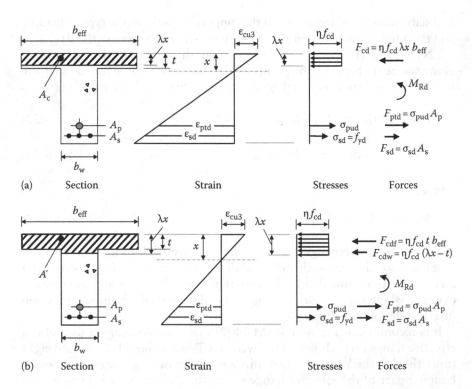

Figure 6.16 Flanged sections subjected to the design flexural resistance. (a) Compressive stress block in the flange. (b) Compressive stress block in the flange and web.

depth. Equation 6.21 may therefore be used to calculate the design resistance of such a section. The depth of the neutral axis x may be calculated using Equation 6.19, except that b_{eff} replaces b in the denominator.

If $\lambda x > t$, the area of concrete in compression A_c is T-shaped, as shown in Figure 6.16b. Although not strictly applicable, the idealised stress block may still be used on this non-rectangular compressive zone. A uniform stress of ηf_{cd} may therefore be considered to act over the area A_c.

It is convenient to separate the resultant compressive force in the concrete into a force in the flange F_{cdf} and a force in the web F_{cdw} as follows (and shown in Figure 6.16b):

$$F_{cdf} = \eta f_{cd} t b_{eff} \tag{6.36}$$

and

$$F_{cdw} = \eta f_{cd} (\lambda x - t) b_w \tag{6.37}$$

By equating the tensile and compressive forces, the depth to the neutral axis x can be determined by trial and error, and the design moment resistance M_{Rd} can be obtained by taking moments of the internal forces about any convenient point on the cross-section.

EXAMPLE 6.7

Evaluate the design flexural resistance of the double-tee section shown in Figure 6.17. The cross-section contains a total of twenty-six 12.9 mm diameter strands (13 in each cable) placed at an eccentricity of 408 mm to the centroidal axis. The effective prestressing force $P_{m,t}$ is 3250 kN. The stress–strain relationship for the prestressing steel is shown in Figure 6.18, and its elastic modulus and tensile strength are $E_p = 195{,}000$ MPa and $f_{pk} = 1{,}860$ MPa, respectively. The properties of the section and other relevant material data are as follows:

$A = 371 \times 10^3$ mm^2; $\quad I = 22.8 \times 10^9$ mm^4; $\quad Z_{btm} = 43.7 \times 10^6$ mm^3;
$Z_{top} = 82.5 \times 10^6$ mm^3; $\quad A_p = 26 \times 100 = 2{,}600$ mm^2;
$f_{pd} = 1{,}391$ MPa; $\quad E_{cm} = 35{,}000$ MPa; $\quad f_{ck} = 40$ MPa; $\quad f_{cd} = 26.67$ MPa;
$\lambda = 0.8 \quad$ and $\quad \eta = 1.0$.

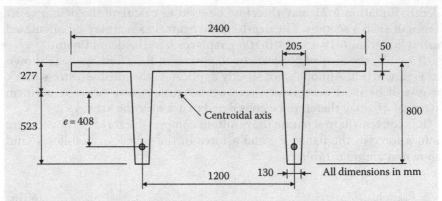

Figure 6.17 Double-tee cross-section (Example 6.7).

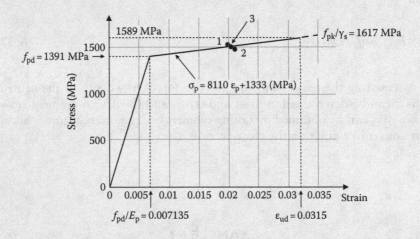

Figure 6.18 Stress–strain curve for strand (Example 6.7).

Using the same procedure as was illustrated in Example 6.1, the strain components in the prestressing steel are obtained from Equations 6.11 through 6.14:

$$\varepsilon_{ce} = \frac{1}{35,000}\left(\frac{3,250 \times 10^3}{371 \times 10^3} + \frac{3,250 \times 10^3 \times 408^2}{22.8 \times 10^9}\right) = 0.000928$$

$$\varepsilon_{pe} = \frac{3,250 \times 10^3}{2,600 \times 195,000} = 0.00641$$

$$\varepsilon_{ptd} = 0.0035\left(\frac{685 - x}{x}\right)$$

and from Equation 6.15:

$$\varepsilon_{pud} = 0.00734 + 0.0035\left(\frac{685 - x}{x}\right) \tag{6.7.1}$$

At this point, an assumption must be made regarding the depth of the equivalent stress block. If x is less than the flange thickness, the calculation would proceed as in the previous examples. However, a simple check of horizontal equilibrium indicates that λx is significantly greater than the flange thickness of 50 mm. This means that the entire top flange and part of the top of each web is in compression. From Equation 6.36:

$$F_{cdf} = 1.0 \times 26.67 \times 50 \times 2400 = 3200 \times 10^3 \text{ N}$$

In this example, the web is tapering and b_w varies with the depth. The width of the web at a depth of λx is given by:

$$b_{w.\lambda x} = 210 - \frac{\lambda x}{10} = 210 - 0.08x$$

The compressive force in the web is therefore:

$$F_{cdw} = 26.67 \times (0.8x - 50) \times \left(\frac{205 + b_{w.\lambda x}}{2}\right) \times 2$$

$$= -1.7069x^2 + 8961.1x - 553,400$$

The resultant compression force is the sum of the flange and web compressive forces:

$$F_{cd} = F_{cdf} + F_{cdw} = -1.7069x^2 + 8961.1x + 2,647,000$$

and the resultant tensile force in the tendons is:

$$F_{ptd} = 2600\, \sigma_{pud}$$

Equating F_{cd} and F_{ptd} gives:

$$\sigma_{pud} = -0.0006565x^2 + 3.4466x + 1018.1 \tag{6.7.2}$$

Trial values of x may now be used to determine ε_{pud} and σ_{pud} from the previous expressions, and the resulting points are tabulated here and plotted on the stress–strain curves in Figure 6.18:

Trial x (mm)	ε_{pud} Equation 6.7.1	σ_{pud} (MPa) Equation 6.7.2	Point plotted on Figure 6.18
150	0.0198	1520	1
140	0.0210	1488	2
144	0.0205	1500	3

Since point 3 lies sufficiently close to the stress–strain curve for the tendon, the value taken for x is 144 mm.

The depth of the stress block is $\lambda x = 115.2$ mm, which is greater than the flange thickness (as was earlier assumed). The resultant forces on the cross-section are:

$$F_{cd} = F_{ptd} = 2600 \times 1500 \times 10^{-3} = 3900 \text{ kN}$$

For this section, $x = 0.210d_p < 0.3d_p$ and therefore the failure may be considered to be ductile. The compressive force in the flange $F_{cdf} = 3200$ kN acts 25 mm below the top surface, and the compressive force in the web $F_{cdw} = 700$ kN acts at the centroid of the trapezoidal areas of the webs above λx, i.e. 82.4 mm below the top surface.

By taking moments of these internal compressive forces about the level of the tendons, we get:

$$M_{Rd} = 3200 \times (685 - 25) \times 10^{-3} + 700 \times (685 - 82.4) \times 10^{-3} = 2534 \text{ kNm}$$

6.6 DUCTILITY AND ROBUSTNESS OF PRESTRESSED CONCRETE BEAMS

6.6.1 Introductory remarks

Ductility is the ability of a structure or structural member to undergo large plastic deformations without significant loss of load carrying capacity. Ductility is important for many reasons. It provides indeterminate structures with alternative load paths and the ability to redistribute internal actions as the collapse load is approached. After the onset of cracking, concrete structures are non-linear and inelastic. The stiffness varies from location to location depending on the extent of cracking and the reinforcement/tendon layout. In addition, the stiffness of a particular cross-section or region is time-dependent, with the distribution of internal actions changing under service loads due to creep and shrinkage, as well as other imposed deformations such as support settlements and temperature changes and gradients. All these factors cause the actual distribution of internal actions in an indeterminate structure to deviate from that assumed in an elastic analysis. Despite these difficulties, codes of practice, including EN1992-1-1 [1], permit the design of concrete structures based on elastic analysis. This is quite reasonable provided the critical regions possess sufficient ductility (plastic rotational capacity) to enable the actions to redistribute towards the calculated elastic distribution as the collapse load is approached. If critical regions have little ductility (such as in over-reinforced elements), the member may not be able to undergo the necessary plastic deformation and the safety of the structure could be compromised.

Ductility is also important to resist impact and seismic loading, and to provide robustness. With proper detailing, ductile structures can absorb the energy associated with sudden impact (as may occur in an accident or a blast or a seismic event) without collapse of the structure. With proper detailing, ductile structures can also be designed to resist progressive collapse.

Figure 6.19 shows the load–deflection curves for two prestressed concrete beams, one under-reinforced (Curve A) and one over-reinforced (Curve B). Curve A indicates ductile behaviour with large plastic deformations developing as the peak load is approached. The relatively flat post-yield plateau (1–2) in Curve A, where the structure deforms while maintaining its full load carrying capacity (or close to it) is characteristic of ductile behaviour. Curve B indicates non-ductile or brittle behaviour, with relatively little plastic deformation before the peak load. There is little or no evidence of a flat *plastic plateau,* and the beam immediately begins to unload when the peak load is reached.

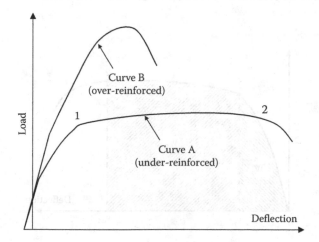

Figure 6.19 Ductile and non-ductile load–deflection curves.

Structures with load–deflection relationships similar to Curve B in Figure 6.19 are simply too brittle to perform adequately under significant impact or seismic loading and they cannot resist progressive collapse. Prestressed concrete beams can be designed to be robust and not to suddenly collapse when overloaded, but ductility is the key and a ductile load-displacement relationship such as that shown as Curve A in Figure 6.19 is an essential requirement. Structures should be designed to be robust, but in most codes of practice qualitative statements rather than quantitative recommendations are made and relatively little guidance is available. To design a prestressed structure for robustness, some quantitative measure of robustness is required.

Beeby [3] stated that a structure is robust if it is able to absorb damage resulting from unforeseen events without collapse. He also argued that this could form the basis of a design approach to quantify robustness. Explosions or impacts are clearly inputs of energy. Beeby [3] suggests that accidents or even design mistakes could also be considered as inputs of energy and that robustness requirements could be quantified in terms of a structure's ability to absorb energy.

The area under the load–deflection response of a member or structure is a measure of the energy absorbed by the structure in undergoing that deformation. Consider the load–deflection response of a simply-supported under-reinforced prestressed beam shown in Figure 6.20. The area under the curve up to point 1 (before the tensile steel yields) is W_1 and represents the elastic energy. The area under the curve between points 1 and 2 (when the plastic hinge develops and the peak load is reached) is W_2, which represents the plastic energy. A minimum value of the ratio W_2/W_1 could be specified to ensure an acceptable level of ductility and, if all members and connections were similarly ductile and appropriately detailed, an acceptable level of robustness (or resistance to collapse) could be achieved.

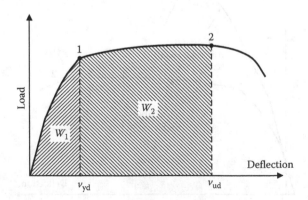

Figure 6.20 Typical under-reinforced load–deflection curve.

A ductile simply-supported member is one for which W_2/W_1 exceeds about 2.0, but for statically indeterminate structures where significant redistribution of internal actions may be required as the peak load is approached, satisfaction of the following is recommended:

$$W_2/W_1 \geq 3.0 \tag{6.38}$$

6.6.2 Calculation of hinge rotations

A typical moment curvature relationship for an under-reinforced prestressed concrete cross-section was shown in Figure 1.15, and an idealised elastic–plastic design moment–curvature curve is shown in Figure 6.21. A plastic hinge is assumed to develop at a point in a beam or slab when the peak (design ultimate) moment is reached at a curvature of κ_{yd} and rotation of the plastic hinge occurs as the curvature increases from κ_{yd} to κ_{ud}. The rotation at the plastic hinge θ_s is the change in curvature multiplied by the length of the plastic hinge l_h in the direction of the member axis. For under-reinforced cross-sections with ductile tensile reinforcement and tendons, the length of the plastic hinge l_h may be taken as 1.2 times the depth of the member [1], i.e. $l_h = 1.2 h$. The maximum rotation available at a plastic hinge may therefore be approximated by:

$$\theta_s = l_h(\kappa_{ud} - \kappa_{yd}) \approx 1.2h(\kappa_{ud} - \kappa_{yd}) \tag{6.39}$$

6.6.3 Quantifying ductility and robustness of beams and slabs

To investigate the ductility of a prestressed concrete beam, it is convenient to assume that the load–deflection curve is elastic perfectly plastic. As an example, consider the idealised load–deflection response shown in Figure 6.22 of a simply-supported prestressed concrete beam of span l and subjected to a point load P applied at mid-span. The deflection is the downward

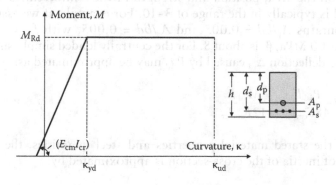

Figure 6.21 Idealised elastic–plastic moment–curvature relationship.

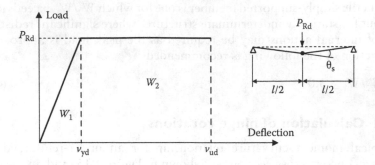

Figure 6.22 Idealised load–deflection curve.

deflection at mid-span caused by the applied load. A plastic hinge is assumed to develop at mid-span when the applied load first reaches P_{Rd} and the mid-span deflection caused by P_{Rd} is v_{yd}. The design moment resistance at the plastic hinge is $M_{Rd} = P_{Rd}l/4$. After the formation of the plastic hinge, it is assumed that the deflection at mid-span increased from v_{yd} to v_{ud} by rotation of the plastic hinge through an angle θ_s, and this can be described as follows:

$$(v_{ud} - v_{yd}) = \frac{l\theta_s}{4} \tag{6.40}$$

The design moment resistance at the plastic hinge M_{Rd} may be determined from Equation 6.21 and may be expressed as:

$$M_{Rd} = \sigma_{pud}A_p\left(d_p - \frac{\lambda x}{2}\right) + f_{yd}A_s\left(d_s - \frac{\lambda x}{2}\right) \approx \beta_1 bd^2 \tag{6.41}$$

where b is the section width and d is the effective depth to the resultant of the tensile forces in the prestressed and non-prestressed steel. The term β_1 depends on the area, position and strength of the reinforcement and tendons and is typically in the range of 3–10. For example, if we assume the beam contains $A_p/bd = 0.005$ and $A_s/bd = 0.005$, with $f_{ck} = 40$ MPa, $E_{cm} = 35,000$ MPa, β_1 is about 8. For the centrally loaded simply-supported beam, the deflection Δ_{yd} caused by P_{Rd} may be approximated as:

$$v_{yd} = \frac{P_{Rd}l^3}{48E_{cm}I_{cr}} = \frac{M_{Rd}l^2}{12E_{cm}I_{cr}} \tag{6.42}$$

and, for the stated material properties and steel quantities, the cracked moment of inertia of the cross-section is approximated by:

$$I_{cr} = 0.037bd^3 \tag{6.43}$$

Substituting Equations 6.41 and 6.43 into Equation 6.42 gives:

$$v_{yd} = \frac{l^2}{1940d} \qquad (6.44)$$

and the elastic energy W_1 (shown in Figure 6.22) may be approximated as:

$$W_1 = \frac{P_{Rd}v_{yd}}{2} = 0.00825bdl \qquad (6.45)$$

If we assume that satisfaction of Equation 6.38 is required for robustness, the minimum internal plastic energy W_2 that must be absorbed during the hinge rotation is $3W_1 = 0.0247bdl$ and this must equal the external work:

$$P_{Rd}(v_{ud} - v_{yd}) = W_2 \geq 0.0247bdl \qquad (6.46)$$

and substituting Equations 6.40 and 6.41 into Equation 6.46, we get:

$$(4M_u/l)(l\theta_s/4) = 8bd^2\theta_s \geq 0.0247bdl$$

$$\therefore \theta_s \geq 3.09 \times 10^{-3} \frac{l}{d} \qquad (6.47)$$

It is evident that the plastic rotation required at the hinge at mid-span depends on the span to effective depth ratio. To achieve a ductility corresponding to $W_2/W_1 = 3.0$, the minimum rotation required at the hinge at mid-span and the span to final deflection ratio (l/v_{ud}) are determined from Equations 6.47, 6.40 and 6.44 and given in the following table.

l/d	Minimum θ_s (rad)	(l/v_{ud})
10	0.031	77
14	0.043	56
18	0.056	43
22	0.068	35
26	0.080	30

In a simplified procedure, EN 1992-1-1 [1] suggests that the rotation θ_s should be less than an allowable rotation given by the product of $\theta_{pl.d}$ and k_λ, where $\theta_{pl.d}$ is the basic value of allowable rotation given in Figure 6.23 for steel Classes B and C and tendons, and k_λ is a factor that depends on the shear slenderness. The values of $\theta_{pl.d}$ in Figure 6.23 apply for a shear slenderness $\lambda = 3$. For different values of shear slenderness, $k_\lambda = (\lambda/3)^{0.5}$. The shear slenderness λ is the ratio of the distance along the beam from the

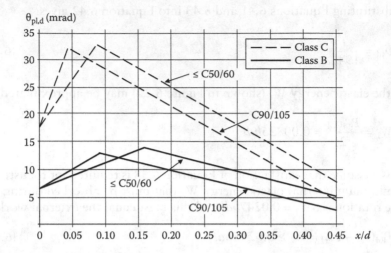

Figure 6.23 Basic values of allowable rotation $\theta_{pl,d}$ when $\lambda = 3$ [I].

plastic hinge to the nearest point of zero moment (after redistribution) and the effective depth d.

REFERENCES

1. EN 1992-1-1. 2004. Eurocode 2: Design of concrete structures – Part 1-1: General rules and rules for buildings. British Standards Institution, London, UK.
2. ACI 318M-14. 2014. Building code requirements for structural concrete. Detroit, MI: American Concrete Institute.
3. Beeby, A.W. 1999. Safety of structures, a new approach to robustness. *The Structural Engineer*, 77(4), 16–21.

Chapter 7

Design resistance in shear and torsion

7.1 INTRODUCTION

In Chapter 5, methods were presented for the determination of the strains and stresses normal to a cross-section caused by the longitudinal prestress and the bending moment acting at the cross-section. Procedures for calculating the flexural resistance of beams were discussed in Chapter 6. In structural design, shear failure must also be guarded against. Shear failure is sudden and difficult to predict with accuracy. It results from diagonal tension in the web of a concrete member produced by shear stress in combination with the longitudinal normal stress. Torsion, or twisting of the member about its longitudinal axis, also causes shear stresses which lead to diagonal tension in the concrete and consequential inclined cracking.

Conventional reinforcement in the form of transverse stirrups is used to carry the tensile forces in the webs of prestressed concrete beams after the formation of diagonal cracks. This reinforcement should be provided in sufficient quantities to ensure that flexural failure, which can be predicted accurately and is usually preceded by extensive cracking and large deformation, will occur before diagonal tension failure.

In slabs and footings, a local shear failure at columns or under concentrated loads may also occur. This so-called *punching shear* type of failure often controls the thickness of flat slabs and plates in the regions above the supporting columns. In this chapter, the design for adequate strength of prestressed concrete beams in shear and in combined shear and torsion is described. Procedures for determining the punching shear strength of slabs and footings are also presented.

7.2 SHEAR IN BEAMS

7.2.1 Inclined cracking

Cracking in prestressed concrete beams subjected to overloads, as shown in Figure 7.1, depends on the local magnitudes of moment and shear. In regions where the moment is large and the shear is small, vertical

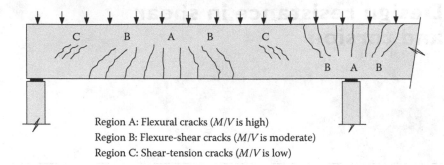

Region A: Flexural cracks (M/V is high)
Region B: Flexure-shear cracks (M/V is moderate)
Region C: Shear-tension cracks (M/V is low)

Figure 7.1 Types of cracking at overload.

flexural cracks appear after the normal tensile stress in the extreme concrete fibres exceeds the tensile strength of concrete. These are the cracks referred to in Sections 5.8.1 and 5.12.1 and are shown in Figure 7.1 as crack type A.

Where both the moment and shear force are relatively large, flexural cracks which are vertical at the extreme fibres become inclined as they extend deeper into the beam owing to the presence of shear stresses in the beam web. These inclined cracks, which are often quite flat in a prestressed beam, are called *flexure-shear* cracks and are designated crack type B in Figure 7.1. If adequate shear reinforcement is not provided, a flexure-shear crack may lead to a so-called shear-compression failure, in which the area of concrete in compression above the advancing inclined crack is so reduced as to be no longer adequate to carry the compression force resulting from flexure.

A second type of inclined crack sometimes occurs in the web of a prestressed beam in the regions where moment is small and shear is large, such as the cracks designated type C adjacent to the discontinuous support and near the point of contraflexure in Figure 7.1. In such locations, high principal tensile stress may cause inclined cracking in the mid-depth region of the beam before flexural cracking occurs in the extreme fibres. These shear-tension cracks (also known as *web-shear* cracks) occur most often in beams with relatively thin webs.

7.2.2 Effect of prestress

The longitudinal compression introduced by prestress delays the formation of each of the crack types shown in Figure 7.1. The effect of prestress on the formation and direction of inclined cracks can be seen by examining the stresses acting on a small element located at the centroidal axis of the uncracked beam shown in Figure 7.2. Using a simple Mohr's circle construction, the principal stresses and their directions are readily found.

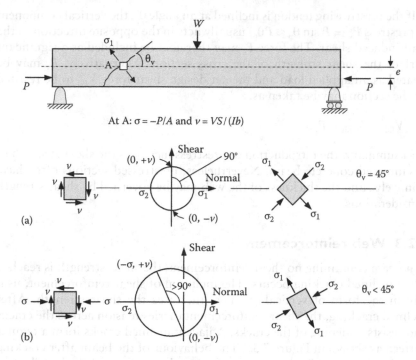

At A: $\sigma = -P/A$ and $v = VS/(Ib)$

Figure 7.2 **Effect of prestress on the principal stresses in a beam web. (a) At P = 0. (b) At P > 0.**

When the principal tensile stress σ_1 reaches the tensile strength of concrete, cracking occurs and the cracks form in the direction perpendicular to the direction of σ_1.

When the prestress is zero, σ_1 is equal to the shear stress v and acts at 45° to the beam axis, as shown in Figure 7.2a. If diagonal cracking occurs, it will be perpendicular to the principal tensile stress, i.e. at 45° to the beam axis. When the prestress is not zero, the normal compressive stress σ (= P/A) reduces the principal tension σ_1, as illustrated in Figure 7.2b. The angle between the principal stress direction and the beam axis increases, and consequently, if cracking occurs, the inclined crack is flatter. Prestress therefore improves the effectiveness of any transverse reinforcement (stirrups) that may be used to increase the shear strength of a beam. With prestress causing the inclined crack to be flatter, a larger number of vertical stirrup legs are crossed by the crack and, consequently, a larger tensile force can be carried across the crack.

In the case of I-beams, the maximum principal tension may not occur at the centroidal axis of the uncracked beam where the shear stress is greatest, but may occur at the flange–web junction where shear stresses are still high and the longitudinal compression is reduced by external bending.

If the prestressing tendon is inclined at an angle θ_p, the vertical component of prestress P_v $(= P \sin \theta_p \approx P\theta_p)$ usually acts in the opposite direction to the load-induced shear. The force P_v may therefore be included as a significant part of the shear strength of the cross-section. Alternatively, P_v may be treated as an applied load and the net design shear force V_{Ed} to be resisted by the section may be taken as:

$$V_{Ed} = V_{loads} - P_v \tag{7.1}$$

In summary, the introduction of prestress increases the shear strength of a reinforced concrete beam. Nevertheless, prestressed sections often have thin webs, and the thickness of the web may be governed by shear strength considerations.

7.2.3 Web reinforcement

In a beam containing no shear reinforcement, the shear strength is reached when inclined cracking occurs. The inclusion of shear reinforcement, usually in the form of vertical stirrups, increases the shear strength. After inclined cracking, the shear reinforcement carries tension across the cracks and resists widening of the cracks. Adjacent inclined cracks form a regular pattern as shown in Figure 7.3a. The behaviour of the beam after cracking is explained conveniently in terms of an analogous truss, first described by Ritter [1] and shown in Figure 7.3b.

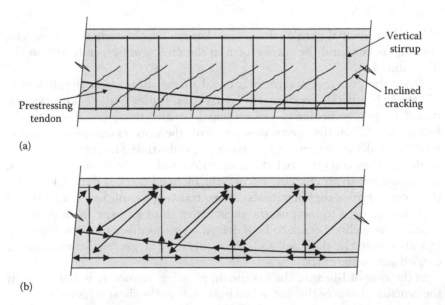

Figure 7.3 The analogous truss used to model a beam with shear reinforcement. (a) Beam elevation after inclined cracking. (b) The truss analogy.

The web members of the analogous truss resist the applied shear and consist of vertical tension members (which represent the vertical legs of the closely spaced steel stirrups) and inclined compression members (which model the concrete segments between the inclined cracks). In reality, there exists a continuous field of diagonal compression in the concrete between the diagonal cracks. This is idealised in the analogous truss by the discrete diagonal compression struts. In a similar manner, the vertical members of the analogous truss may represent a number of more closely spaced vertical stirrups. The top compressive chord of the analogous truss represents the concrete compressive zone plus any longitudinal compressive reinforcement, and the bottom chord models the longitudinal prestressed and non-prestressed reinforcement in the tensile zone. At each panel point along the bottom chord of the analogous truss, the vertical component of the compressive force in the inclined concrete strut must equal the tension in the vertical steel member, and the horizontal component must equal the change in the tensile force in the bottom chord (i.e. the change in force in the prestressing tendon and any other longitudinal non-prestressed reinforcement).

The analogous truss can be used to visualise the flow of forces in a beam after inclined cracking, but it is at best a simple model of a rather complex situation. The angle of the inclined compressive strut θ has traditionally been taken as 45°, although in practical beams it is usually less. The stirrup stresses predicted by a 45° analogous truss are considerably higher than those measured in real beams [2] because the truss is based on the assumption that the entire shear force is carried by the vertical stirrups. In fact, part of the shear is carried by dowel action of the longitudinal tensile steel and part by friction on the mating surfaces of the inclined cracks (known as *aggregate interlock*). Some shear is also carried by the uncracked concrete compressive zone. In addition, the truss model neglects the tension carried by the concrete between the inclined cracks. The stress in the vertical leg of a stirrup in a real concrete beam is therefore maximum at the inclined crack and is significantly lower away from the crack.

At the ultimate limit state, shear failure may be initiated by yielding of the stirrups or, if large amounts of web reinforcement are present, crushing of the concrete compressive strut. The latter is known as *web-crushing* and is usually avoided by placing upper limits on the quantity of web reinforcement. Not infrequently, premature shear failure occurs because of inadequately anchored stirrups. The truss analogy shows that the stirrup needs to be able to carry the full tensile force from the bottom panel point (where the inclined compressive force is resolved both vertically and horizontally) to the top panel point. To achieve this, care must be taken to detail the stirrup anchorages adequately to ensure that the full tensile capacity of the stirrup can be developed at any point along the vertical leg. After all, an inclined crack may cross the vertical leg of the stirrup at any point.

Larger diameter longitudinal bars should be included in the corners of the stirrup to form a rigid cage and to improve the resistance to pull out of the hooks at the stirrup anchorage. These longitudinal bars also disperse the concentrated force from the stirrup and reduce the likelihood of splitting in the plane of the stirrup anchorage. Stirrup hooks should be located on the compression side of the beam where anchorage conditions are most favourable and the clamping action of the transverse compression greatly increases the resistance to pull out. If the stirrup hooks are located on the tensile side of the beam, anchorage may be lost if flexural cracks form in the plane of the stirrup. In current practice, stirrup anchorages are most often located at the top of a beam. In the negative moment regions of such beams, adjacent to the internal supports, for example where shear and moment are relatively large, the shear capacity may be significantly reduced owing to loss of stirrup anchorages after flexural cracking.

It is good practice for the shear reinforcement calculated as being necessary at any cross-section to be provided for a distance h from that cross-section in the direction of decreasing shear, where h is the overall depth of the member. The first stirrup at each end of a span should be located within 50 mm from the face of the support. Shear reinforcement should extend as close to the compression face and the tension face of the member as cover requirements and the proximity of other reinforcement and tendons permit. The bends in bars used as stirrups should also enclose, and be in contact with, a longitudinal bar with a diameter not less than the diameter of the fitment (stirrup) bar.

In Figure 7.4, some satisfactory and some unsatisfactory stirrup arrangements are shown. Generally, stirrup hooks should be bent through an angle of at least 135°. A 90° bend (a cog) will become ineffective should the cover be lost, for any reason, and will not provide adequate anchorage. Fitment cogs of 90° should not be used when the cog is located within 50 mm of any concrete surface.

In addition to carrying diagonal tension produced by shear, and controlling inclined web cracks, closed stirrups also provide increased ductility to a beam by confining the compressive concrete. The *open stirrups*

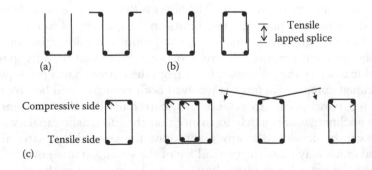

Figure 7.4 Stirrup shapes. (a) Incorrect. (b) Satisfactory in some situations. (c) Desirable.

shown in Figure 7.4b are commonly used, particularly in post-tensioned beams where the opening at the top of the stirrup facilitates the placement and positioning of the post-tensioning duct along the member. This form of stirrup does not provide confinement for the concrete in the compression zone and is undesirable in heavily reinforced beams where confinement of the compressive concrete may be required to improve ductility of the member.

It is good practice to use adequately anchored stirrups, even in areas of low shear, particularly when tensile steel quantities are relatively high and cross-sectional ductility is an issue. EN 1992-1-1 [3] suggests that when calculations indicate that no shear reinforcement is required (see Section 7.2.4), minimum shear reinforcement should nevertheless be provided (as given by Equation 7.17). EN 1992-1-1 [3] also specifies that minimum shear reinforcement may be omitted in slabs where transverse distribution of loads is possible and in members of minor importance that do not contribute to the overall strength and stability of the structure.

7.2.4 Design strength of beams without shear reinforcement

The design shear strength of a beam without shear reinforcement $V_{Rd,c}$ is usually considered to be the load required to cause the first inclined crack. In regions of a beam where the design shear force $V_{Ed} \leq V_{Rd,c}$, no calculated shear reinforcement is necessary.

According to EN 1992-1-1 [3], the design value of the shear resistance $V_{Rd,c}$ (in N) in a prestressed concrete member cracked in bending is given by:

$$V_{Rd,c} = \left[C_{Rd,c}k(100\rho_l f_{ck})^{1/3} + k_1\sigma_{cp} \right]b_w d \geq (v_{min} + k_1\sigma_{cp})b_w d \qquad (7.2)$$

where f_{ck} is in MPa; $k = 1 + \sqrt{200/d} \leq 2.0$ (with d in mm); $\rho_l = A_{sl}/(b_w d) \leq 0.02$; A_{sl} is the area of the longitudinal tensile reinforcement that extends at least $(l_{bd} + d)$ beyond the section considered; b_w is the smallest width of the cross-section in the tensile area (in mm); $\sigma_{cp} = N_{Ed}/A_c < 0.2f_{cd}$ (in MPa); N_{Ed} is the magnitude of the axial compressive force on the cross-section due to the design loading and prestress (in N); A_c is the area of the concrete cross-section (in mm²); $C_{Rd,c} = 0.18/\gamma_c = 0.12$ for persistent and transient loads; $k_1 = 0.15$ and $v_{min} = 0.035k^{1.5}f_{ck}^{0.5}$. Equation 7.2 may be considered as the shear force required to develop a flexure-shear crack (refer Figure 7.1).

For members with loads applied to the top surface relatively close to the support (i.e. when $a_v \leq 2d$), the contribution of such loads to the design shear force V_{Ed} may be multiplied by $\beta = a_v/2d \geq 0.25$, where a_v is defined in Figure 7.5 and is measured from the edge of a rigid support (as shown) or

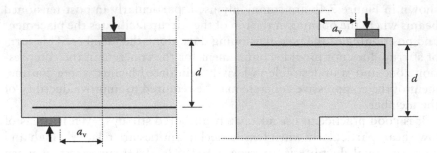

Figure 7.5 Loads near a support [3].

the centre of bearing for a flexible bearing. The factored design shear force calculated without reduction by β should always satisfy:

$$V_{Ed} \leq 0.3 b_w d f_{cd} \left[1 - \frac{f_{ck}}{250} \right] \tag{7.3}$$

In regions of a beam that are uncracked in bending (i.e. when the extreme fibre flexural tensile stress is less than $f_{ctk,0.05}/\gamma_c$), the design shear resistance is limited by the tensile strength of the concrete in the web and is given by:

$$V_{Rd,c} = \frac{I b_w}{S} \sqrt{(f_{ctd})^2 + \alpha_1 \sigma_{cp} f_{ctd}} \tag{7.4}$$

where I is the second moment of area; b_w is the width of the cross-section at the centroidal axis (allowing for the presence of ducts in accordance with Equations 7.19 and 7.20); S is the first moment of the area above and about the centroidal axis; $\alpha_1 = l_x/l_{pt2} \leq 1.0$ for pretensioned members and $\alpha_1 = 1.0$ for other types of prestressed members; l_x is the distance of the cross-section under consideration from the start of the transmission length; σ_{cp} is the magnitude of the concrete compressive stress at the centroidal axis due to axial loading and/or prestressing ($\sigma_{cp} = N_{Ed}/A_c$ in MPa); l_{pt2} is the upper bound of the transmission length ($l_{pt2} = 1.2 l_{pt}$) and l_{pt} is given by Equation 8.2.

On cross-sections, where the width varies over the height, the maximum principal tension may occur on an axis other than the centroidal. In such cases, Equation 7.4 should be used to locate the axis for which $V_{Rd,c}$ is at its minimum value with b_w and S referring to the axis under consideration and σ_{cp} replaced by the concrete stress caused by prestress and bending at the axis under consideration. Equation 7.4 should be considered as the shear force required to cause a shear-tension crack (refer Figure 7.1)

7.2.5 Design resistance of beams with shear reinforcement

In Figure 7.6, the transfer of shear force across a diagonal crack in a beam with vertical shear reinforcement is shown. The part of the shear force

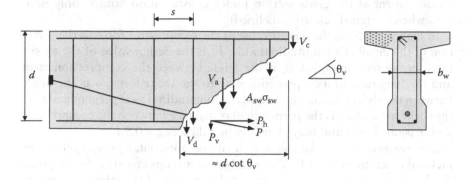

Figure 7.6 Transfer of shear at an inclined crack.

carried by shear stresses in the uncracked concrete compression zone is V_c, the part carried by bearing and friction between the two surfaces of the inclined crack is V_a, the part carried by dowel action in the longitudinal steel crossing the crack is V_d and the contribution of stirrups to the shear strength of the beam depends on the area of the vertical legs of each stirrup A_{sw} and the stress in each stirrup crossed by the inclined crack.

At the ultimate limit state, EN 1992-1-1 [3] conservatively assumes that the design strength in shear V_{Rd} is calculated based on the truss model shown in Figure 7.7 and is given by:

$$V_{Rd} = V_{Rd,s} + V_{ccd} + V_{td} \tag{7.5}$$

where $V_{Rd,s}$ is the design value of the maximum shear force that can be resisted by the yielding shear reinforcement and is given by Equation 7.7; V_{ccd} is the design value of the shear component of the compressive force in the compression chord of the truss at the cross-section under consideration (and is only non-zero when the compression chord is inclined) and V_{td} is the design value of the shear component of the tensile force in the tensile

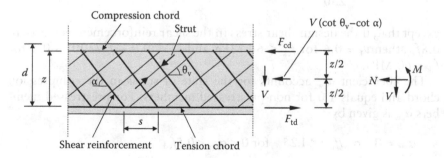

Figure 7.7 Truss model for shear in a beam with shear reinforcement [3].

reinforcement at the cross-section under consideration (and is only non-zero when the tensile chord is inclined).

In Figure 7.7, F_{cd} is the design value of the compressive force in the direction of the member's longitudinal axis, F_{td} is the design value of the tensile force in the tension chord, θ_v is the angle between the compression strut and the longitudinal axis perpendicular to the shear force, α is the angle between the shear reinforcement and the longitudinal axis perpendicular to the shear force and z is the perpendicular distance between the compression and tension chords and may generally be taken as $z = 0.9d$.

It is reasonable to take the length of the horizontal projection of the inclined crack to be $z \cot \theta_v$. The number of stirrups crossing the diagonal crack is therefore $z \cot \theta_v / s$, where s is the spacing of the stirrups required for shear in the direction of the member axis. EN 1992-1-1 [3] imposes the limits on the angle θ_v given by:

$$1 \leq \cot \theta_v \leq 2.5 \tag{7.6}$$

For members with shear reinforcement perpendicular to the longitudinal axis of the member (i.e. $\alpha = 90°$), the design shear resistance V_{Rd} is given by either Equation 7.7 or 7.8, whichever is the smaller value:

$$V_{Rd,s} = A_{sw} f_{ywd} \frac{z \cot \theta_v}{s} \tag{7.7}$$

$$V_{Rd,max} = \frac{\alpha_{cw} b_w z \nu_1 f_{cd}}{(\cot \theta_v + \tan \theta_v)} \tag{7.8}$$

where A_{sw} is the cross-sectional area of each stirrup, b_w is the minimum width of the cross-section between the tension and compression chords (allowing for any post-tensioning ducts in the web – see Equations 7.19 and 7.20), f_{ywd} is the design yield strength of the shear reinforcement, ν_1 is a strength reduction factor for concrete cracked in shear given by:

$$\nu_1 = 0.6 \left[1 - \frac{f_{ck}}{250} \right] \tag{7.9}$$

except that, if the design shear stress in the shear reinforcement is less than $0.8 f_{yk}$, then $\nu_1 = 0.6$ for $f_{ck} \leq 60$ MPa and $\nu_1 = (0.9 - f_{ck}/200) > 0.5$ for $f_{ck} > 60$ MPa.

The coefficient α_{cw} accounts for the state of stress in the compression chord and equals 1.0 for non-prestressed members. For prestressed members α_{cw} is given by:

$$\alpha_{cw} = (1 + \sigma_{cp}/f_{cd}) \leq 1.25 \quad \text{for } 0 < \sigma_{cp} < 0.5 f_{cd}$$

$$\alpha_{cw} = 2.5(1 - \sigma_{cp}/f_{cd}) \qquad \text{for } 0.5 f_{cd} < \sigma_{cp} < 1.0 f_{cd} \tag{7.10}$$

where σ_{cp} is the magnitude of the mean compressive stress in the concrete due to the design axial force.

For members with inclined shear reinforcement (i.e. $45° \leq \alpha < 90°$), the design shear resistance V_{Rd} is given by either Equation 7.11 or 7.12, whichever is the smaller value:

$$V_{Rd,s} = A_{sw}f_{ywd}\frac{z(\cot\theta_v + \cot\alpha)\sin\alpha}{s} \tag{7.11}$$

$$V_{Rd,max} = \frac{\alpha_{cw}b_w z v_1 f_{cd}(\cot\theta_v + \cot\alpha)}{(1 + \cot^2\theta_v)} \tag{7.12}$$

The design shear resistance $V_{Rd,s}$ in Equations 7.7 and 7.11 is equal to the vertical force carried by the web reinforcement when the design stress in every stirrup crossed by the inclined crack is f_{ywd}. The resistance $V_{Rd,max}$ in Equations 7.8 and 7.12 is the maximum design shear resistance and is governed by crushing of the concrete in the inclined compression strut between the shear cracks.

The maximum area of vertical shear reinforcement (i.e. $\alpha = 90°$) that can be included in a particular cross-section is when $\cot\theta_v = 1$ and is obtained by equating Equations 7.7 and 7.8:

$$\frac{A_{sw,max}}{s} = \frac{\alpha_{cw}b_w v_1 f_{cd}}{2f_{ywd}} \tag{7.13}$$

For inclined stirrups (i.e. $45° \leq \alpha < 90°$), equating Equations 7.11 and 7.12 with $\cot\theta_v = 1$ gives:

$$\frac{A_{sw,max}}{s} = \frac{\alpha_{cw}b_w v_1 f_{cd}}{2f_{ywd}\sin\alpha} \tag{7.14}$$

In regions of a beam, where the design shear force diagram is continuous, the shear reinforcement required in any length $l = z(\cot\theta_v + \cot\alpha)$ may be calculated using the smallest value of the design shear force (V_{Ed}) in that length.

The shear reinforcement ratio is given by:

$$\rho_w = \frac{A_{sw}}{(sb_w\sin\alpha)} \geq \rho_{min} \tag{7.15}$$

where:

$$\rho_{min} = \frac{0.08\sqrt{f_{ck}}}{f_{yk}} \tag{7.16}$$

and from Equation 7.15, the minimum quantity of shear reinforcement is therefore:

$$\frac{A_{sw,min}}{s} = \frac{0.08 b_w \sin\alpha \sqrt{f_{ck}}}{f_{yk}} \tag{7.17}$$

EN 1992-1-1 [3] specifies that the maximum longitudinal spacing between stirrups is:

$$s_{l,max} = 0.75d \,(1 + \cot\alpha) \tag{7.18}$$

For webs containing grouted metal ducts of outer diameter $\phi > b_w/8$, the maximum design shear resistance $V_{Rd.max}$ should be calculated by substituting $b_{w,nom}$ for b_w in either Equation 7.8 or 7.12 as appropriate, where:

$$b_{w,nom} = b_w - 0.5\Sigma\phi \tag{7.19}$$

and $\Sigma\phi$ is determined for the most unfavourable level on the cross-section. For grouted metal ducts with $\phi \leq b_w/8$, $b_{w,nom} = b_w$. For non-grouted ducts, grouted plastic ducts and unbonded tendons:

$$b_{w,nom} = b_w - \Sigma\phi \tag{7.20}$$

The additional tensile force ΔF_{td} in the longitudinal reinforcement due to the design shear force V_{Ed} is given by:

$$\Delta F_{td} = 0.5 V_{Ed} \,(\cot\theta_v - \sin\alpha) \tag{7.21}$$

except that the force in the tensile reinforcement $(M_{Ed}/z) + \Delta F_{td}$ should not be taken greater than $(M_{Ed,max}/z)$, where $M_{Ed,max}$ is the maximum moment along the beam.

For members with loads applied to the top surface relatively close to the support (i.e. when $a_v \leq 2d$), as shown in Figure 7.8, the contribution of such loads to the design shear force V_{Ed} may be reduced by $\beta = a_v/2d \geq 0.25$. In addition, the design shear force calculated this way should satisfy:

$$V_{Ed} \leq A_{sw} f_{ywd} \sin\alpha \tag{7.22}$$

where $A_{sw} f_{ywd}$ is the design resistance of the shear reinforcement located within the length $0.75a_v$ centrally located within the shear span a_v, as shown in Figure 7.8. The reduction factor β should only be applied when calculating the required shear reinforcement and only when the longitudinal reinforcement is fully anchored at the support. In addition, the design shear force V_{Ed} calculated without reduction by β should always satisfy Equation 7.3.

In EN 1992-1-1 [3], θ_v may be varied between the limits specified in Equation 7.6, i.e. $45° \geq \theta_v \geq 21.8°$. It is evident from Equations 7.7 and 7.11 that the contribution of stirrups to the shear strength of a beam

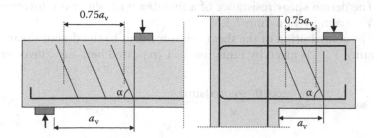

Figure 7.8 Shear reinforcement in short shear spans with direct strut action.

depends on θ_v. The flatter the inclined crack (i.e. the smaller the value of θ_v), the greater is the number of effective stirrups and the greater is the value of $V_{Rd,s}$. In order to achieve the desired shear strength in design, fewer stirrups are required as θ_v is reduced. The choice of $\theta_v = 21.8°$ leads to the least amount of shear reinforcement. However, if the slope of the diagonal compression member in the analogous truss of Figure 7.3 is small, the change in force in the longitudinal tensile steel is relatively large (see Equation 7.21). More longitudinal steel is required in the shear span near the support than would otherwise be the case, and greater demand is placed on the anchorage requirements of these bars. It should also be noted that a choice of $\theta_v = 45°$ will give the largest value of $V_{Rd,max}$ in Equation 7.12 and the smallest possible web width b_w.

For a given amount of shear reinforcement (A_{sw}/s), the optimum value of θ_v is when the design shear resistance provided by the stirrups $V_{Rd,s}$ is equal to the maximum design shear resistance provided by the compressive strut $V_{Rd,max}$. In design, $V_{Rd,s} \geq V_{Ed}$. Therefore, if $V_{Rd,max}$ is set equal to V_{Ed}, the optimum value of θ_v can be determined.

EN 1992-1-1 [3] also requires that the longitudinal steel necessary for flexure at any particular section must be provided and developed at a section a distance a_1 along the beam in the direction of increasing shear. For beams without shear reinforcement, $a_1 = d$. For members with shear reinforcement:

$$a_1 = z(\cot \theta_v - \cot \alpha)/2 \tag{7.23}$$

7.2.6 Summary of design requirements for shear

The design requirements for shear in EN 1992-1-1 [3] are summarised in the following:

1. The design shear resistance of a member without shear reinforcement is $V_{Rd,c}$ given by Equation 7.2 for a member cracked in bending or given by Equation 7.4 for a member uncracked in bending.

2. The design shear resistance of a member with shear reinforcement is V_{Rd} given by Equation 7.5.
3. The contribution of the shear reinforcement to the design shear resistance $V_{Rd,s}$ is given by Equation 7.11 (repeated here for convenience):

$$V_{Rd,s} = A_{sw} f_{ywd} \frac{z(\cot\theta_v + \cot\alpha)\sin\alpha}{s} \qquad (7.11)$$

where s is the centre-to-centre spacing of the shear reinforcement measured parallel to the axis of the member and θ_v is the angle of the concrete compression strut to the beam axis and may be taken as any value between 21.8° and 45°.

4. The maximum design shear resistance of a member with shear reinforcement is governed by crushing of the inclined concrete struts in the web of the member. In no case should the design shear strength $V_{Rd,s}$ exceed $V_{Rd,max}$ (as defined in Equation 7.12). The corresponding upper limit on the amount of shear reinforcement permitted in the member is given by Equation 7.14. The minimum quantity of shear reinforcement permitted in a beam is given by Equation 7.17.

5. For members subject to predominantly uniformly distributed load, the design shear force need not be checked at a distance less than d from the face of the support. Any shear reinforcement required at the section d from the face of the support should continue to the face of the support. Notwithstanding this requirement, the shear at the support should not exceed $V_{Rd,max}$ given by Equation 7.12. In addition, the longitudinal tensile reinforcement required at d from the face of the support shall be continued into the support and shall be fully anchored past that face.

   Where diagonal cracking can take place at the support or extend into the support, such as when the support is above the beam, the design shear force should be checked at the face of the support.

6. Where the $V_{Ed} \leq V_{Rd,c}$, no shear reinforcement is theoretically required, but minimum shear reinforcement as given by Equation 7.17 should be included in all members, with the exception of slabs and members of minor importance that do not contribute to the overall strength and stability of the structure.

   Where $V_{Ed} > V_{Rd,c}$, shear reinforcement should be provided in accordance with Equation 7.7 (for vertical stirrups) or Equation 7.11 (for inclined stirrups).

7. The maximum spacing between stirrups $s_{l,max}$, measured in the direction of the beam axis, is given by Equation 7.18. The maximum transverse spacing between the vertical legs of a stirrup measured across the web of a beam should not exceed the lesser of $0.75\,d$ or 600 mm.

The first stirrup at each end of a span should be positioned no more than 50 mm from the face of the adjacent support.

8. Stirrups should be anchored on the compression side of the beam using standard hooks bent through an angle of at least 135° around a larger diameter longitudinal bar. It is important that the stirrup anchorage be located as close to the compression face of the beam as is permitted by concrete cover requirements and the proximity of other reinforcement and tendons.

7.2.7 The design procedure for shear

An appropriate procedure for the design of a beam for shear resistance is outlined below.

1. Calculate the design shear force V_{Ed} along the span. The maximum value need not be determined closer than d from the face of the support.

2. Calculate the unreinforced concrete capacity $V_{Rd,c}$ using Equation 7.2 for a section that has been cracked in flexure or Equation 7.4 for a section that is uncracked in flexure.

3. If $V_{Ed} \leq V_{Rd,c}$, then minimum reinforcement given by Equation 7.17 is required.
   Go to Step 6.

4. If $V_{Ed} > V_{Rd,c}$, shear reinforcement is required to resist the design shear force V_{Ed}.
   Calculate $V_{Rd,max,1}$ from Equation 7.12 (or Equation 7.8 for vertical stirrups) for cot $\theta_v = 1.0$.
   Calculate $V_{Rd,max,2.5}$ from Equation 7.12 for cot $\theta_v = 2.5$.
   If $V_{Ed} > V_{Rd,max,1}$, the section is inadequate and requires a redesign involving an increase in section dimensions and/or an increase in concrete strength.
   If $V_{Ed} \leq V_{Rd,max,2.5}$, take cot $\theta_v = 2.5$, i.e. $\theta_v = 21.8°$.
   If $V_{Ed} > V_{Rd,max,2.5}$, θ_v may be estimated using:

$$\theta_v = 21.8 + 23.2 \times \frac{V_{Ed} - V_{Rd.max,2.5}}{V_{Rd,max,1} - V_{Rd.max,2.5}} \tag{7.24}$$

5. Determine the amount of shear reinforcement using the value of θ_v determined in Step 4 and Equation 7.11, by satisfying the strength requirement $V_{Ed} \leq V_{Rd,s}$. If vertical stirrups are used, Equations 7.5 and 7.7 give:

$$V_{Ed} \leq A_{sw} f_{ywd} \frac{z \cot \theta_v}{s} + V_{ccd} + V_{td}$$

and therefore:

$$\frac{A_{sw}}{s} \geq \frac{V_{Ed} - V_{ccd} - V_{td}}{f_{ywd}z\cot\theta_v} \tag{7.25}$$

6. The maximum longitudinal spacing of shear reinforcement is given by Equation 7.18. That is:

$$s_{l,max} = 0.75d(1 + \cot\alpha).$$

EXAMPLE 7.1

The shear reinforcement for the post-tensioned beam shown in Figure 7.9 is to be designed (assume 12 mm diameter stirrups each with two vertical legs and f_{yk} = 500 MPa). The beam is simply-supported over a span of 30 m and carries a uniformly distributed load, consisting of an imposed load w_Q = 25 kN/m and a permanent load w_G = 40 kN/m (which includes the beam self-weight). The beam is prestressed by a bonded parabolic cable with an eccentricity of 700 mm at mid-span and zero at each support. The area of the prestressing steel is A_p = 3800 mm^2 and the metal duct diameter is 120 mm. The prestressing force at each support is 4500 kN and at mid-span is 4200 kN and is assumed to vary linearly along the beam length. The concrete

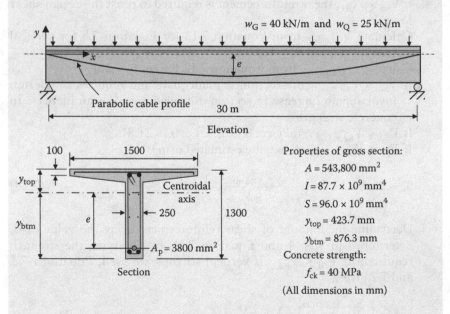

Figure 7.9 Beam details (Example 7.1).

cover to the reinforcement is taken to be 40 mm. The strength class for concrete is C40/50.

The factored design load combination for the strength limit state (see Section 2.3.2) is given by Equation 2.1:

$$w_{Ed} = 1.35w_G + 1.5w_Q = 1.35 \times 40 + 1.5 \times 25 = 91.5 \text{ kN/m}$$

At x m from support A, the design shear force V_{Ed} is the shear caused by w_{Ed} minus the vertical component of prestress at this point. Accordingly:

$$V_{Ed} = 1372.5 - 91.5x - P_v(x) \quad \text{and} \quad M_{Ed} = 1372.5x - 45.75x^2 \qquad (7.1.1)$$

Using Equations 1.3 and 1.4, the distance of the parabolic prestressing cable below the centroidal axis of the section at x m from A and the slope of the cable at that point are:

$$y = -2.8\left[\frac{x}{30} - \left(\frac{x}{30}\right)^2\right] \quad \text{and} \quad y' = -\frac{2.8}{30}\left(1 - \frac{x}{15}\right) \qquad (7.1.2)$$

With 40 mm cover and assuming 12 mm diameter stirrups and 28 mm diameter bottom reinforcement in the tension chord, the depth to the centroid of the tensile reinforcement is $d = 1300 - 40 - 12 - 14 = 1234$ mm and the lever arm between the compression and tension chords is approximated by:

$$z = 0.9d = 1111 \text{ mm}$$

In Table 7.1, a summary of the calculations and reinforcement requirements at a number of sections along the beam is presented. In the following, sample calculations are provided for the sections at 1 m from the support and at 2 m from the support. In this example, $b_{w,nom} = b_w - 0.5 \Sigma\phi = 190$ mm, and from Equations 4.11 and 4.12, when $f_{ck} = 40$ MPa, $f_{cd} = 26.67$ MPa and $f_{ctd} = 1.67$ MPa. From Table 4.6, $f_{ywd} = 435$ MPa.

From Equation 7.10, $\alpha_{cw} = 1.25$, and from Equation 7.9, $\nu_1 = 0.6 \times (1 - 40/250) = 0.504$. The maximum and minimum quantities of shear reinforcement are obtained from Equations 7.14 and 7.17, respectively:

$$\frac{A_{sw,max}}{s} = \frac{1.25 \times 190 \times 0.504 \times 26.67}{2 \times 435 \times 1.0} = 3.67$$

$$\frac{A_{sw,min}}{s} = \frac{0.08 \times 190 \times 1.0 \times \sqrt{40}}{500} = 0.192$$

Using 12 mm diameter vertical stirrups with 2 vertical legs, i.e. A_{sw} = 220 mm², the maximum stirrup spacing to satisfy the minimum steel requirement is s_{max} = 220/0.192 = 1146 mm. However, the maximum spacing is limited to 0.75d (Equation 7.18).

At x = 1.0 m
From Equation 7.1.2:

$$y = -0.0902 \text{ m (and therefore } e = 90.2 \text{ mm)} \quad \text{and} \quad y' = \theta_p = -0.0871 \text{ rad}$$

For this cross-section, $P_{m,t}$ = 4480 kN and the concrete stress at the centroidal axis and the vertical component of prestress are:

$$\sigma_{cp} = P_{m,t}/A = 8.24 \text{ MPa} \quad \text{and} \quad P_v = -P_{m,t}\theta_p = 390 \text{ kN}$$

From Equation 7.1.1:

$$V_{Ed} = 891 \text{ kN} \quad \text{and} \quad M_{Ed} = 1327 \text{ kNm}$$

Check whether the section has cracked in bending under the full design bending moment:

$$\sigma_{btm} = -\frac{P_{m,t}}{A} - \frac{P_{m,t}ey_{btm}}{I} + \frac{M_{Ed}y_{btm}}{I} = -8.24 - 4.04 + 13.26 = 0.98 \text{ MPa} < f_{ctd}$$

and therefore, the cross-section has not cracked.
From Equation 7.4:

$$V_{Rd,c} = \frac{87.7 \times 10^9 \times 190}{96.0 \times 10^6}\sqrt{(1.67)^2 + 1.0 \times 8.24 \times 1.67} = 706.2 \text{ kN} < V_{Ed}$$

Therefore, shear reinforcement is required.
For cot θ_v = 1, Equation 7.8 gives:

$$V_{Rd,max.1} = \frac{1.25 \times 190 \times 1111 \times 0.504 \times 26.67}{(1+1)} = 1773 \text{ kN}$$

For cot θ_v = 2.5, Equation 7.8 gives:

$$V_{Rd,max.2.5} = \frac{1.25 \times 190 \times 1111 \times 0.504 \times 26.67}{(2.5 + 0.4)} = 1223 \text{ kN}$$

With $V_{Ed} < V_{Rd,max,2.5}$, the slope of the inclined compressive strut may be taken as $\theta_v = 21.8°$ (i.e. cot $\theta_v = 2.5$). Equation 7.25 gives the required amount of shear reinforcement:

$$\frac{A_{sw}}{s} \geq \frac{(891 - 0 - 0) \times 10^3}{435 \times 1111 \times 2.5} = 0.738$$

and with $A_{sw} = 220$ mm², the stirrup spacing must satisfy $s \leq 220/0.738 = 298$ mm.

Using 12 mm diameter steel stirrups (with two vertical legs) at 290 mm centres satisfies both the minimum and maximum transverse steel requirements and the maximum spacing requirement (Equation 7.18) of EN 1992-1-1 [3].

From Equation 7.21, the additional tensile force ΔF_{td} in the longitudinal reinforcement due to the design shear force V_{Ed} is $\Delta F_{td} = 0.5 V_{Ed}$ cot $\theta_v = 1114$ kN, and the area of anchored tensile reinforcement in the bottom chord to carry this force is $A_s = \Delta F_{td}/f_{yd} = 2561$ mm².

Use five 28 mm diameter bottom bars. Therefore $\rho_l = A_s/(b_w d) = 0.0131$.

At x = 2.0 m

From Equation 7.1.2:

$$y = -0.174 \text{ m (and therefore } e = 174 \text{ mm)} \quad \text{and} \quad y' = \theta_p = -0.0809 \text{ rad}$$

For this cross-section, $P_{m,t} = 4460$ kN. The concrete stress at the centroidal axis and the vertical component of prestress are:

$$\sigma_{cp} = P_{m,t}/A = 8.20 \text{ MPa} \quad \text{and} \quad P_v = -P_{m,t}\theta_p = 360.8 \text{ kN}$$

From Equation 7.1.1:

$$V_{Ed} = 829 \text{ kN} \quad \text{and} \quad M_{Ed} = 2562 \text{ kNm}$$

Check whether the section has cracked in bending under the full design bending moment:

$$\sigma_{btm} = -\frac{P_{m,t}}{A} - \frac{P_{m,t}ey_{btm}}{I} + \frac{M_{Ed}y_{btm}}{I} = -8.20 - 7.75 + 25.60 = +9.65 \text{ MPa} > f_{ctd}$$

and therefore, the cross-section has cracked.

In Equation 7.2, $C_{Rd,c} = 0.12$, $k = 1 + \sqrt{200/1234} = 1.40$, $v_{min} = 0.035 \times 1.40^{1.5} \times 40^{0.5} = 0.37$, $k_1 = 0.15$ and $\rho_l = 0.0131$. From Equation 7.2:

$$V_{Rd,c} = \left[0.12 \times 1.40 \times (100 \times 0.0131 \times 40)^{1/3} + 0.15 \times 8.20 \right] \times 190 \times 1234 \times 10^{-3}$$

$$= 435.8 \text{ kN}$$

This is less than V_{Ed}, and therefore, shear reinforcement is required.

With $V_{Ed} < V_{Rd,max,2.5}$, the slope of the inclined compressive strut may be taken as $\theta_v = 21.8°$ (i.e. cot $\theta_v = 2.5$) and the required amount of shear reinforcement is obtained from Equation 7.25:

$$\frac{A_{sw}}{s} \geq \frac{(829-0-0)\times10^3}{435\times1111\times2.5} = 0.686$$

With $A_{sw} = 220$ mm², the stirrup spacing must satisfy $s \leq 220/0.686 = 321$ mm.

Using 12 mm diameter steel stirrups (with two vertical legs) at 320 mm centres satisfies both the minimum and maximum transverse steel requirements and the maximum spacing requirement (Equation 7.18) of EN 1992-1-1 [3].

For other cross-sections, results are shown in Table 7.1. When x exceeds about 8.3 m, the design shear V_{Ed} is less than $V_{Rd,c}$ and the concrete alone can carry the shear. The minimum amount of shear reinforcement is required, therefore, in the middle portion of the span (from $x = 9$ to 21 m).

Table 7.1 Summary of results – Example 7.1

x (m)	V_{Ed} (kN)	M_{Ed} (kNm)	$P_{m,t}$ (kN)	Cracked in bending	Equation 7.4 $V_{Rd,c}$ (kN)	Equation 7.2 $V_{Rd,c}$ (kN)	Specified spacing of 12 mm vertical steel stirrups (2 legs) (mm)
1	890.7	1327	4480	No	706.2	–	298
2	828.7	2562	4460	Yes	–	435.8	321
3	766.5	3706	4440	Yes	–	434.5	347
4	704.0	4758	4420	Yes	–	433.3	378
5	641.2	5719	4400	Yes	–	432.0	415
6	578.2	6588	4380	Yes	–	430.7	460
7	515.0	7366	4360	Yes	–	429.4	516
8	451.5	8052	4340	Yes	–	428.1	588
9	397.7	8647	4320	Yes	–	426.8	833*
10	323.7	9150	4300	Yes	–	425.5	833*

* Minimum steel required ($V_{Ed} < V_{Rd,c}$) and $s = 0.75d$.

7.2.8 Shear between the web and flange of a T-section

As the moment varies along a T-beam, the change in force in the flange is transmitted to the web of the beam by longitudinal shear stresses that develop at the junction between the web of the beam and the flange on each side of the web. Figure 7.10 shows a segment of a T-beam where the longitudinal force in the flange on each side of the web due to bending at the ultimate limit state varies from F_d at Section A–A to $F_d + \Delta F_d$ at Section B–B. EN 1992-1-1 [3] specifies that the shear strength of the flange may be calculated by treating the flange as a system of compressive struts and tensile ties in the direction perpendicular to the beam axis. The average longitudinal shear stress v_{Ed} at the junction between one side of the flange and the web is given by:

$$v_{Ed} = \frac{\Delta F_d}{h_f \Delta x} \tag{7.26}$$

where h_f is the flange thickness and Δx is the length of the beam over which ΔF_d develops (as shown in Figure 7.10).

The length Δx should not be taken greater than half the distance from the section where the moment is zero to the section where the moment is the maximum. For a beam subjected to point loads, Δx should not be greater than the distance between point loads.

The area per unit length of transverse reinforcement in the slab must be sufficient to carry the transverse tension generated by the inclined compression struts in the flange shown in Figure 7.10 and may be determined from:

$$A_{sf} \geq \frac{v_{Ed} h_f s_f}{f_{yd} \cot \theta_f} \tag{7.27}$$

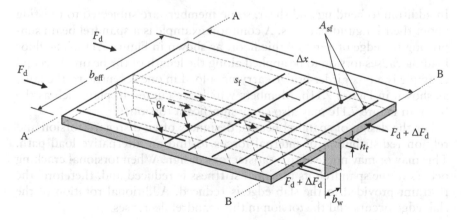

Figure 7.10 Notation for shear at a flange–web junction [3].

where the angle between the compression strut and the longitudinal axis of the beam may be taken within the ranges $45° \geq \theta_f \geq 26.5°$ (i.e. $1.0 \leq \cot \theta_f \leq 2.0$) for compressive flanges and $45° \geq \theta_f \geq 38.6°$ (i.e. $1.0 \leq \cot \theta_f \leq 1.25$) for tensile flanges.

To prevent crushing of the compression struts in the flange, v_{Ed} should satisfy:

$$v_{Ed} \leq v f_{cd} \sin \theta_f \cos \theta_f \tag{7.28}$$

where v is the strength reduction factor for concrete cracked in shear and is equal to v_1 given by Equation 7.9.

In beam and slab floor systems, the slab spans between the beams resulting in transverse bending in the flange of the T-beam. The reinforcement required to resist this bending will often exceed the reinforcement required to satisfy Equation 7.27 (i.e. required to resist the shear between the flange and the web).

When the flange does resist transverse bending, as well as shear between the flange and the web, EN 1992-1-1 [3] requires that the area of transverse steel should be greater than the value given by Equation 7.27 or half that given by Equation 7.27 plus that required for transverse bending.

If v_{Ed} is less than $0.4f_{ctd}$, no additional reinforcement needs to be provided other than that required for flexure.

If longitudinal tensile reinforcement is required in the flange at a particular section, it should be anchored beyond the strut required to transmit the bar force back to the web at the section where this reinforcement is required (see Figure 7.10).

7.3 TORSION IN BEAMS

7.3.1 Compatibility torsion and equilibrium torsion

In addition to bending and shear, some members are subjected to twisting about their longitudinal axes. A common example is a spandrel beam supporting the edge of a monolithic floor, as shown in Figure 7.11a. The floor loading causes torsion to be applied along the length of the beam. A second example is a box girder bridge carrying a load in one eccentric traffic lane, as shown in Figure 7.11b. Members which are curved in plan such as the beam in Figure 7.11c may also carry significant torsion.

For the design of spandrel beams, designers often disregard torsion and rely on redistribution of internal forces to find an alternative load path. This may or may not lead to a satisfactory design. When torsional cracking occurs in the spandrel, its torsional stiffness is reduced and, therefore, the restraint provided to the slab edge is reduced. Additional rotation of the slab edge occurs and the torsion in the spandrel decreases.

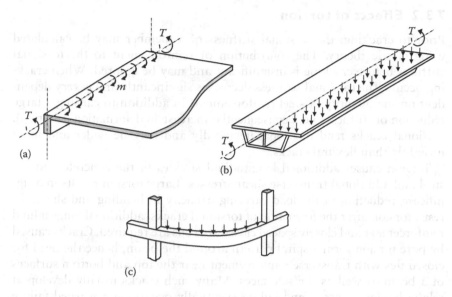

Figure 7.11 Members subjected to torsion. (a) Compatibility torsion. (b) Equilibrium torsion. (c) Equilibrium and compatibility torsion.

Torsion that may be reduced by redistribution, such as the torsion in the spandrel beam, is often called *compatibility torsion*. Whereas indeterminate structures generally tend to behave in accordance with the design assumptions, full redistribution will occur only if the structure possesses adequate ductility and may be accompanied by excessive cracking and large local deformations. Ductile reinforcement is essential. For some statically indeterminate members (and for statically determinate members) twisted about their longitudinal axes, some torsion is required for equilibrium and cannot be ignored. In the case of the box girder bridge of Figure 7.11b, for example, torsion cannot be disregarded and will not be redistributed, as there is no alternative load path. This is *equilibrium torsion* and must be considered in design. EN 1992-1-1 [3] states that where equilibrium of the structure depends on the torsional resistance of its elements, a full torsional design covering both ultimate and serviceability limit states is required. However, in statically indeterminate structures, where torsion arises from considerations of compatibility only, it is usually not necessary to consider torsion at the ultimate limit state, but to provide sufficient reinforcement to control cracking at service loads.

The behaviour of beams carrying combined bending, shear and torsion is complex. Most current design recommendations rely heavily on gross simplifications and empirical estimates derived from experimental observations and the provisions of EN 1992-1-1 [3] are no exception.

7.3.2 Effects of torsion

Prior to cracking, the torsional stiffness of a member may be calculated using elastic theory. The contribution of reinforcement to the torsional stiffness before cracking is insignificant and may be ignored. When cracking occurs, the torsional stiffness decreases significantly and is very dependent on the quantity of steel reinforcement. In addition to causing a large reduction of stiffness and a consequential increase in deformation (twisting), torsional cracks tend to propagate rapidly and may be wider and more unsightly than flexural cracks.

Torsion causes additional longitudinal stresses in the concrete and the steel and additional transverse shear stresses. Large torsion results in a significant reduction in the load carrying capacity in bending and shear. To resist torsion after the formation of torsional cracks, additional longitudinal reinforcement and closely spaced closed stirrups are required. Cracks caused by pure torsion form a spiral pattern around the beam, hence the need for closed ties with transverse reinforcement near the top and bottom surfaces of a beam as well as the side faces. Many such cracks usually develop at relatively close centres, and failure eventually occurs on a warped failure surface. The angles between the crack and the beam axis on each face of the beam are approximately the same. EN 1992-1-1 [3] allows the effects of torsion and shear to be superimposed, assuming the same value for the strut angle θ_v, with the limits on θ_v as specified in Equation 7.6.

Tests show that prestress increases the torsional stiffness of a member significantly but does not greatly affect the strength in torsion. The introduction of prestress delays the onset of torsional cracking, thereby improving the member stiffness and increasing the cracking torque. The strength contribution of the concrete after cracking, however, is only marginally increased by prestress, and the contribution of the transverse reinforcement is unchanged by prestress.

For a beam in pure torsion, the behaviour after cracking can be described in terms of the three-dimensional analogous truss shown in Figure 7.12. The closed stirrups act as transverse tensile web members (both vertical and horizontal); the longitudinal reinforcement in each corner of the stirrups acts as the longitudinal chords of the truss and the compressive web members inclined at an angle θ_v on each face of the truss carry the inclined compressive forces and represent the concrete between the inclined cracks on each face of the beam.

The three-dimensional analogous truss ignores the contribution of the interior concrete to the post-cracking torsional strength of the member. The diagonal compressive struts are located on each face of the truss and, in the actual beam, diagonal compressive stress is assumed to be located close to each surface of the member. The beam is therefore assumed to behave similarly to a hollow thin-walled section. Tests of members in pure torsion tend to support these assumptions.

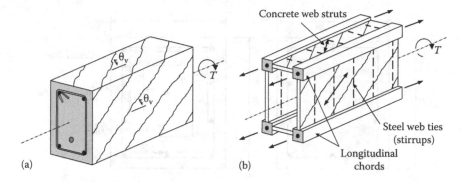

Figure 7.12 Three-dimensional truss analogy for a beam in pure torsion. (a) Beam segment. (b) Analogous truss.

Design models for reinforced and prestressed concrete beams in torsion are usually based on a simple model such as that of Figure 7.12.

7.3.3 Design provisions for torsion

When designing for torsion in accordance with EN 1992-1-1 [3], a solid cross-section may be idealised as a thin-walled closed section with shear flow in each of the section walls resisting the applied torque. Non-rectangular cross-sections, such as T-sections, are divided into subsections, each modelled as an equivalent thin-walled section. The torsional strength of the section is taken as the sum of the strengths of the individual subsections. The distribution of the applied design torsion to the various subsections is in proportion to their uncracked torsional stiffness and each subsection may be designed separately.

Consider the cross-section shown in Figure 7.13a and the idealised thin-walled section shown in Figure 7.13b. If the total area of the cross-section inside the outer circumference is A (in Figure 7.13a, $A = bh$) and the outer circumference is u, the effective thickness of the thin walls of Figure 7.13b is $t_{ef} = A/u$, but not less than twice the distance between the outside edge and the centre of the longitudinal reinforcement.

For a cross-section subjected to an applied design torsion T_{Ed}, the shear flow $q_{t,i}$ in the i-th wall of the idealised section is the product of the shear stress $\tau_{t,i}$ and the effective wall thickness $t_{ef,i}$ and may be taken as:

$$q_{t,i} = \tau_{t,i}t_{ef,i} = \frac{T_{Ed}}{2A_k} \tag{7.29}$$

and the shear force in the ith wall due to torsion is:

$$V_{Ed,i} = q_{t,i}z_i \tag{7.30}$$

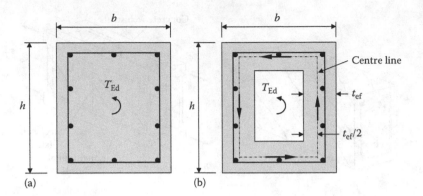

Figure 7.13 Idealisation of cross-section in torsion. (a) Solid cross-section. (b) Idealised thin-walled cross-section.

where z_i is the side length of wall i defined by the distance between the intersection points with the centre line of adjacent walls and A_k is the area enclosed by the centre lines of the connecting walls, including inner hollow areas.

The additional transverse reinforcement required in each wall of the idealised thin-walled cross-section is determined inserting Equations 7.29 and 7.30 into Equation 7.25:

$$\frac{A_{sw}}{s} \geq \frac{V_{Ed,i}}{f_{ywd}z_i \cot\theta_v} = \frac{T_{Ed}}{2A_k f_{ywd} \cot\theta_v} \tag{7.31}$$

where A_{sw} is the area of the single leg of transverse steel in each wall of the idealised thin-walled section.

The additional cross-sectional area of longitudinal reinforcement ΣA_{sl} required to resist torsion may be determined from:

$$\frac{\Sigma A_{sl} f_{yd}}{u_k} = \frac{T_{Ed}}{2A_k} \cot\theta_v \tag{7.32}$$

where u_k is the perimeter of the area A_k, θ_v is the angle of the compressive strut and f_{yd} is the design yield stress of the longitudinal steel.

In the compression side of the cross-section, the longitudinal reinforcement may be reduced in proportion to the available compressive force caused by bending. In the tension side, the amount of longitudinal reinforcement required for torsion should be added to that required for bending and axial tension. The longitudinal steel required for torsion should generally be distributed over the length of the side z_i but for small cross-sections it can be lumped at the ends of this length, i.e. in the corners of the closed ties.

To avoid crushing of the concrete struts, for a section in combined shear and torsion, the following interaction equation must be satisfied:

$$\frac{T_{Ed}}{T_{Rd,max}} + \frac{V_{Ed}}{V_{Rd,max}} \leq 1.0 \qquad (7.33)$$

where V_{Ed} is the design transverse force and $T_{Rd,max}$ is the maximum design torsional resistance given by:

$$T_{Rd,max} = 2 v \alpha_{cw} f_{cd} A_k t_{ef,i} \sin\theta_v \cos\theta_v \qquad (7.34)$$

v is the same as v_1 in Equation 7.9 and α_{cw} is given in Equation 7.10. In Equation 7.33, $V_{Rd,max}$ is the maximum design shear resistance given by either Equation 7.8 or 7.12.

For solid rectangular sections that satisfy Equation 7.35, only minimum reinforcement is required for shear and torsion:

$$\frac{T_{Ed}}{T_{Rd,c}} + \frac{V_{Ed}}{V_{Rd,c}} \leq 1.0 \qquad (7.35)$$

where $T_{Rd,c}$ is the torsion required to cause first cracking in an otherwise unloaded beam, determined by setting $\tau_{t,i} = f_{ctd}$, and may be taken as:

$$T_{Rd,c} = J_t f_{ctd} \sqrt{(1 + 10\sigma_{cp}/f_{ck})} \qquad (7.36)$$

The torsional constant J_t may be taken as:

$J_t = 0.33x^2y$ for solid rectangular sections;

$= 0.33\Sigma x^2y$ for solid T-shaped, L-shaped or I-shaped sections; and

$= 2A_k t_w$ for thin-walled hollow sections, where A_k is the area enclosed by the centre lines of the walls of a single closed cell and t_w is the minimum thickness of the wall of the hollow section.

The terms x and y are, respectively, the shorter and longer overall dimensions of the rectangular part(s) of the solid section. The beneficial effect of the prestress on $T_{Rd,c}$ is accounted for by the term $\sqrt{(1 + 10\sigma_{cp}/f_{ck})}$ and σ_{cp} is the average effective prestress $P_{m,t}/A$.

EXAMPLE 7.2

A prestressed concrete beam has a rectangular cross-section 400 mm wide and 550 mm deep. At a particular cross-section, the beam must resist the following factored design actions: M_{Ed} = 300 kNm, V_{Ed} = 150 kN and T_{Ed} = 60 kNm.

The dimensions and properties of the cross-section are shown in Figure 7.14. An effective prestress of 700 kN is applied at a depth of 375 mm by a single cable consisting of seven 12.9 mm diameter strands in a grouted duct of 60 mm diameter. The area of the strands is A_p = 700 mm² and the characteristic tensile strength is f_{pk} = 1860 MPa. The vertical component of the prestressing force at the section under consideration is 50 kN, and the concrete strength is f_{ck} = 40 MPa (and therefore f_{cd} = 26.67 MPa and f_{ctd} = 1.67 MPa).

Determine the longitudinal and transverse reinforcement requirements.

(1) Initially, the cross-section is checked for web-crushing. The effective width of the web for shear is obtained from Equation 7.19:

$$b_{w,nom} = 400 - 0.5 \times 60 = 370 \text{ mm}$$

With $\sigma_{cp} = P_{m,t}/A$ = 3.18 MPa, Equation 7.10 gives α_{cw} = 1.12 and Equation 7.9 gives ν_1 = 0.504.

$A = 220 \times 10^3$ mm²;

$I = 5546 \times 10^6$ mm⁴;

$u = 2 \times (400 + 550) = 1900$ mm;

$t_{ef} = A/u = 115.8$ mm;

$A_k = (400 - 115.8) \times (550 - 115.8)$

$\quad = 284.2 \times 434.2 = 123.4 \times 10^3$ mm²;

$u_k = 2 \times (284.2 + 434.2) = 1436.8$ mm;

$P_{m,t} = 700$ kN; $P_v = 50$ kN

$A_p = 700$ mm²; $\phi_{duct} = 60$ mm;

$f_{yk} = f_{ywk} = 500$ MPa.

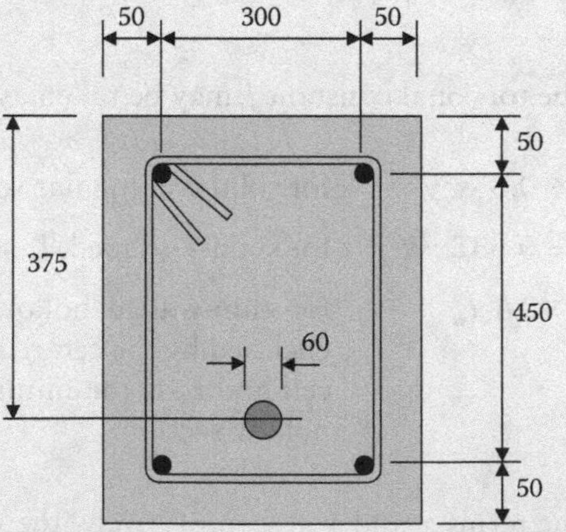

Figure 7.14 Cross-section details (Example 7.2).

Initially ignoring any non-prestressed tensile steel, $d = d_p = 375$ mm (and therefore, $z = 0.9d = 337.5$ mm), $A_{sl} = A_p = 700$ mm, $\rho_l = 700/(370 \times 375) = 0.005$ and, with $\cot \theta_v = 2.5$ (i.e. $\theta_v = 21.8°$), Equations 7.8 and 7.34 give, respectively:

$$V_{Rd,max} = \frac{1.12 \times 370 \times 337.5 \times 0.504 \times 26.67}{(2.5 + 0.4)} \times 10^{-3} = 648.3 \text{ kN}$$

$$T_{Rd,max} = 2 \times 0.504 \times 1.12 \times 26.67 \times 123.4 \times 10^3 \times 115.8$$

$$\times \sin 21.8 \cos 21.8 \times 10^{-6}$$

$$= 148.4 \text{ kNm}$$

Checking Equation 7.33:

$$\frac{T_{Ed}}{T_{Rd,max}} + \frac{V_{Ed}}{V_{Rd,max}} = \frac{60}{148.4} + \frac{150}{648.3} = 0.64 \le 1.0$$

Therefore, web-crushing will not occur and the size of the cross-section is acceptable.

(2) Check whether transverse reinforcement is required for shear and torsion. The torsional constant J_t is:

$J_t = 0.33x^2y = 0.33 \times 400^2 \times 550 = 29.0 \times 10^6$ mm^3

and Equation 7.36 gives:

$$T_{Rd,c} = 29 \times 10^6 \times 1.67\sqrt{(1 + 10 \times 3.18 / 40)} = 64.9 \text{ kNm}$$

From Equation 7.2:

$$V_{Rd,c} = \left[0.12 \times (1 + \sqrt{200/375}) \times (100 \times 0.0050 \times 40)^{1/3} + 0.15 \times 3.18\right]$$

$$\times 370 \times 375 \times 10^{-3}$$

$$= 144.6 \text{ kN}$$

Equation 7.35 gives:

$$\frac{60}{64.9} + \frac{150}{144.6} = 1.96 > 1.0$$

Therefore, transverse reinforcement in the form of closed stirrups is required.

(3) The longitudinal reinforcement required for bending must next be calculated. From the procedures outlined in Chapter 6, the design bending resistance provided by the seven prestressing strands (ignoring the longitudinal non-prestressed reinforcement) is $M_{Rd} = 332$ kNm (refer Example 6.1) and the distance z between F_{cd} and F_{td} is 315 mm and the neutral axis depth $x = 119$ mm. With $M_{Rd} > M_{Ed}$, no non-prestressed steel is required for flexural strength.

(4) Determine the quantity of transverse reinforcement required. Assume 12 mm closed stirrups are to be used. Assuming $\theta_v = 21.8°$ (i.e. $\cot \theta_v = 2.5$) and two vertical legs per stirrup, the stirrup spacing ($s = s_v$) required to carry V_{Ed} is obtained from Equation 7.25:

$$\frac{2 \times 110}{s_v} \geq \frac{150 \times 10^3 - 50 \times 10^3}{(500/1.15) \times 315 \times 2.5} = 0.292$$

Therefore the spacing of stirrups required for shear $s_v \leq 753$ mm, i.e. at least 1.328 stirrups are required for shear per metre length along the beam.

The spacing ($s = s_t$) of the additional 12 mm closed stirrups required for torsion ($A_{sw} = 110$ mm^2) is obtained from Equation 7.31:

$$\frac{110}{s_t} \geq \frac{60 \times 10^6}{2 \times 123.4 \times 10^3 \times (500/1.15) \times 2.5} = 0.224 \quad \therefore s_t \leq 491.8 \text{ mm}$$

i.e. an additional 2.033 stirrups are required for torsion per metre length of beam.

Summing the transverse steel requirements for shear and torsion, we have at least 3.361 stirrups required per metre, i.e. $s \leq 1000/3.361 = 297.5$ mm. This is greater than the maximum permitted spacing of $0.75d = 281$ mm.

Use 12 mm diameter stirrups at 280 mm centres ($f_{ywk} = 500$ MPa).

Note: The required maximum spacing of transverse stirrups could have been calculated directly from s_v and s_t using:

$$\frac{1}{s} \geq \frac{1}{s_v} + \frac{1}{s_t} \quad \text{i.e. } s \leq \frac{s_v s_t}{s_v + s_t}$$

(5) The additional longitudinal tensile force caused by torsion is obtained from Equation 7.32:

$$\Sigma A_{sl} f_{yd} = \frac{T_{Ed}}{2A_k} u_k \cot \theta_v = \frac{60 \times 10^6}{2 \times 123.4 \times 10^3} \times 1436.8 \times 2.5 \times 10^{-3} = 873.3 \text{ kN}$$

The depth to the neutral axis in bending is 119 mm (23.8% of the cross-section) and bending causes a design compression force in this region $F_{cd} = M_{Rd}/z = 1050$ kN, so that no additional steel is required for torsion in the compressive zone. Two 20 mm bars will be included in the top corners of the stirrups. The remaining 76.2% of the cross-section (i.e. that part of the cross-section in tension due to bending) will be subjected to an additional tensile force due to torsion of $0.762 \times 873.3 = 665.5$ kN. The additional non-prestressed reinforcement required to resist torsion in the tensile zone (with $f_{yd} = 500/1.15 = 435$ MPa) is therefore:

$$\Sigma A_{sl} = \frac{665.5 \times 10^3}{435} = 1530 \text{ mm}^2$$

This is equivalent to 5–20 mm diameter deformed longitudinal bars. Use 1–20 mm bar in each corner of the stirrup and one additional bar in the middle of the bottom and the two side legs of the stirrup as shown in Figure 7.15.

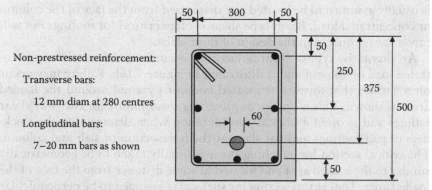

Non-prestressed reinforcement:

Transverse bars:

12 mm diam at 280 centres

Longitudinal bars:

7–20 mm bars as shown

Figure 7.15 Reinforcement details (Example 7.2).

7.4 SHEAR IN SLABS AND FOOTINGS

7.4.1 Punching shear

In the design of slabs and footings, strength in shear frequently controls the thickness of the member, particularly in the vicinity of a concentrated load or a column. Consider the pad footing shown in Figure 7.16. Shear failure may occur on one of two different types of failure surface. The footing may act essentially as a wide beam and shear failure may occur across the entire width of the member, as illustrated in Figure 7.16a. This is *beam-type shear* (or one-way shear) and the shear strength of the critical section

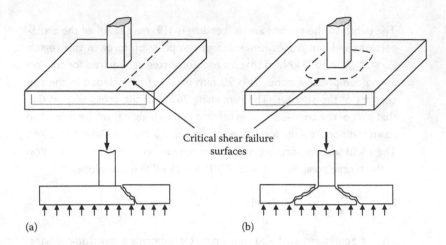

Figure 7.16 Shear failure surfaces in a footing or slab. (a) Beam-type shear. (b) Punching shear.

is calculated as for a beam. The critical section for this type of shear failure is usually assumed to be located at a distance *d* from the face of the column or concentrated load. Beam-type shear is often critical for footings but will rarely cause concern in the design of floor slabs.

An alternative type of shear failure may occur in the vicinity of a concentrated load or column and is illustrated in Figure 7.16b. Failure may occur on a surface that forms a truncated cone or pyramid around the loaded area, as shown. This is known as *punching shear failure* (or two-way shear failure) and is often a critical consideration when determining the thickness of pad footings and flat slabs at the intersection of slab and column. The critical section for punching shear is usually taken to be geometrically similar to the loaded area and located at some distance from the face of the loaded area. The critical section (or surface) is assumed to be perpendicular to the plane of the footing or slab.

Ideally in design, the column support should be large enough for the concrete to carry satisfactorily the moments and shears being transferred across the critical surface without the need for any shear reinforcement. If this is not possible, procedures for the design of an adequate quantity of properly detailed reinforcement must be established. The remainder of this chapter is concerned with this type of shear failure.

7.4.2 The basic control perimeter

Where shear failure can occur locally around a column support or concentrated load, the design shear strength of the slab must be greater than or equal to the design shear force V_{Ed} acting on the *critical shear perimeter*. EN 1992-1-1 [3] requires that the shear resistance must be checked at

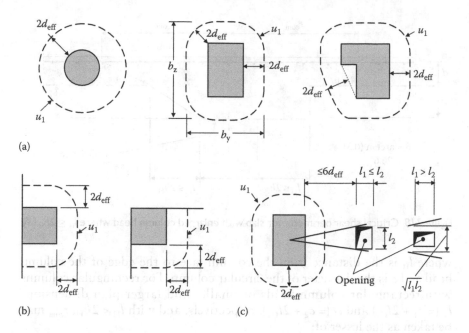

Figure 7.17 Basic control perimeters. (a) Around loaded areas. (b) Around loaded areas at an edge or corner. (c) Near an opening.

the face of the column and around the *basic control perimeter* of length u_1. The basic control perimeter is taken to be at a distance $2d_{\text{eff}}$ from the loaded area A_{load} (i.e. the shaded areas shown in Figure 7.17) and the basic control perimeter is so constructed that its length is minimised, as shown in Figure 7.17a. The distance d_{eff} is the average of the effective depths of the tensile reinforcement in two orthogonal directions in the slab or footing, i.e. $d_{\text{eff}} = (d_y + d_z)/2$. The basic control area A_{cont} is the area between the basic control perimeter and the loaded area.

For loaded areas, at or near the edge or a corner of the slab or footing, the control perimeter is as shown in Figure 7.17b. When the loaded area is less than $6d_{\text{eff}}$ from an opening, that part of the control perimeter contained between two tangents drawn to the opening from the centroid of the loaded area (as shown in Figure 7.17c) is considered to be ineffective.

The control section is perpendicular to the mid-plane of the slab or footing, it follows the control perimeter and has an effective depth d_{eff}.

For slabs with column capitals or drop panels, with $l_H \leq 2h_H$ as shown in Figure 7.18, the punching shear stresses need only be checked on the critical shear perimeter outside the column head located at a distance r_{cont} from the centroid of the column. For circular column heads:

$$r_{\text{cont}} = 2d_{\text{eff}} + l_H + 0.5c \qquad (7.37)$$

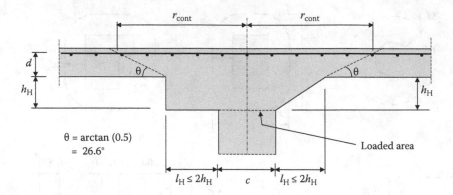

Figure 7.18 Critical shear perimeter for slab with enlarged column head where $l_H \le 2h_H$ [3].

where l_H is the distance from the column face to the edge of the column head and c is the diameter of the circular column. For rectangular columns with rectangular column heads of smaller and larger plan dimensions l_1 ($= c_1 + 2l_{H1}$) and l_2 ($= c_2 + 2l_{H2}$), respectively, and with $l_H < 2h_H$, r_{cont} may be taken as the lesser of:

$$r_{cont} = 2d_{eff} + 0.56\sqrt{l_1 l_2} \qquad (7.38)$$

and

$$r_{cont} = 2d_{eff} + 0.69l_1 \qquad (7.39)$$

For slabs column connections with drop panels or column capitals where $l_H > 2h_H$, checks must be made on shear perimeters both within the slab beyond the column head and within the column head, where d_{eff} is taken as d_H, as defined in Figure 7.19. For circular columns, the distances from the centroid of the column (loaded area) to the control perimeters in Figure 7.19 are:

$$r_{cont.ext} = l_H + 2d_{eff} + 0.5c \qquad (7.40)$$

and

$$r_{cont.int} = 2(d_{eff} + h_H) + 0.5c \qquad (7.41)$$

7.4.3 Shear resistance of critical shear perimeters

The design values of the punching shear resistance of a control section with or without shear reinforcement are denoted in EN 1992-1-1 [3] as $v_{Rd,cs}$ and $v_{Rd,c}$, respectively.

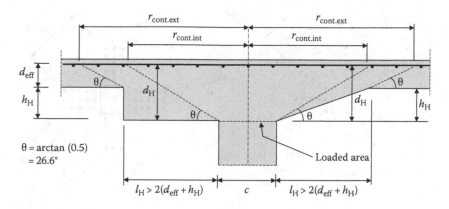

Figure 7.19 Critical shear perimeter for slab with enlarged column head ($l_H > 2(d + h_H)$) [3].

For a slab or column base without shear reinforcement:

$$v_{Rd,c} = C_{Rd,c}k(100\rho_1 f_{ck})^{1/3} + k_1\sigma_{cp} \geq (v_{min} + k_1\sigma_{cp}) \qquad (7.42)$$

which is similar to the design stress contained within Equation 7.2. In Equation 7.42, f_{ck} is in MPa; $k = 1 + \sqrt{200/d} \leq 2.0$ (with d in mm); $C_{Rd,c} = 0.18/\gamma_c = 0.12$ for persistent and transient loads; $k_1 = 0.1$; $v_{min} = 0.035k^{1.5}f_{ck}^{0.5}$; $\rho_1 = \sqrt{\rho_{1y}\rho_{1z}} \leq 0.2$; ρ_{1y} and ρ_{1z} are the tensile steel reinforcement ratios in the y and z directions, respectively, related to a slab width equal to the column width plus $3d$ on each side of the column; $\sigma_{cp} = (\sigma_{cy} + \sigma_{cz})/2$; σ_{cy} and σ_{cz} are the normal compressive stresses on the control section in the orthogonal y and z directions, respectively (in MPa, compression is positive), i.e. $\sigma_{cy} = N_{Ed,y}/A_{cy}$ and $\sigma_{cz} = N_{Ed,z}/A_{cz}$ and $N_{Ed,y}$ and $N_{Ed,z}$ are the longitudinal forces in each direction (due to both applied load and prestressing acting on the concrete areas A_{cy} and A_{cz}) across the full bay for internal columns and across the control perimeter for edge columns.

For a slab or column base with shear reinforcement placed as shown in Figure 7.20a or 7.20b, the design punching shear resistance of the shear perimeter is:

$$v_{Rd,cs} = 0.75v_{Rd,c} + 1.5\frac{d_{eff}}{s_r}A_{sw}f_{ywd,ef}\frac{1}{(u_1 d_{eff})}\sin\alpha \qquad (7.43)$$

where A_{sw} is the area of steel in one perimeter of shear reinforcement around the column (mm²); s_r is the radial spacing of perimeters of shear reinforcement (mm); $f_{ywd,ef}$ is the effective design strength of the shear reinforcement given by $f_{ywd,ef} = 250 + 0.25d_{eff} \leq f_{ywd}$ (MPa) and α is the angle between the shear reinforcement and the plane of the slab.

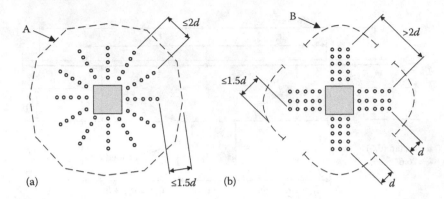

Figure 7.20 Outer control perimeters beyond the punching shear reinforcement [3]. (a) Perimeter A, u_{out}. (b) Perimeter B, $u_{out,ef}$.

An outer shear perimeter (of length u_{out} or $u_{out,ef}$) should be found where shear reinforcement is no longer required (marked as perimeter A in Figure 7.20a and perimeter B in Figure 7.20b). The shear reinforcement should extend radially from the column at a radial spacing not exceeding $0.75d_{eff}$ to a minimum distance of $1.5d_{eff}$ from the first outer shear perimeter where shear reinforcement is not required. The spacing of shear bars around a perimeter within a distance $2d_{eff}$ from the column face should not exceed $1.5d_{eff}$ and, for perimeters outside the basic control perimeter, the spacing should not exceed $2d_{eff}$. The first bar in each radial line of shear reinforcement should be located about $0.5d_{eff}$ from the column face.

7.4.4 Design for punching shear

According to EN 1992-1-1 [3], checks for punching shear must be made at the face of the column and at the basic control perimeter u_1. At the face of the column (i.e. at the perimeter of the loaded area, u_0), the maximum punching shear stress v_{Ed} should not exceed the maximum design shear stress $v_{Rd,max}$:

$$v_{Ed} \leq v_{Rd,max} \tag{7.44}$$

where:

$$v_{Rd,max} = 0.3 \times \left[1 - \frac{f_{ck}}{250}\right] f_{cd} \tag{7.45}$$

Punching shear reinforcement is not necessary if, at the basic control perimeter, the punching shear stress v_{Ed} is less than $v_{Rd,c}$. If v_{Ed} is greater than

$v_{Rd,c}$ on the basic control perimeter, punching shear reinforcement is required and may be calculated using Equation 7.43.

Where the support reaction is eccentric to the control perimeter and the control perimeter must carry a shear force V_{Ed} and a moment M_{Ed}, the maximum shear stress is given by:

$$v_{Ed} = \beta \frac{V_{Ed}}{u_i d_{eff}} \qquad (7.46)$$

where u_i is the length of the control perimeter being considered; β accounts for the eccentricity of the support reaction and is given by:

$$\beta = 1 + k \frac{M_{Ed}}{V_{Ed}} \frac{u_1}{W_1} \qquad (7.47)$$

k is a coefficient that depends on the ratio of the plan dimensions of the column c_1/c_2 and is given in Table 7.2, where c_1 is the column dimension parallel to the eccentricity of load, c_2 is the column dimension perpendicular to the eccentricity of load, and u_1 is the length of the basic control perimeter.

W_1 corresponds to the distribution of shear stress illustrated in Figure 7.21 and is akin to the plastic modulus of the control perimeter about the axis of bending acting through the centroid of the control perimeter and is given by:

$$W_1 = \int_0^{u_i} |e| \, dl \qquad (7.48)$$

where dl is the length increment around the shear perimeter and e is the distance of dl from the axis about which the moment M_{Ed} acts.

For an internal rectangular column:

$$W_1 = 0.5c_1^2 + c_1 c_2 + 4c_2 d_{eff} + 16 d_{eff}^2 + 2\pi d_{eff} c_1 \qquad (7.49)$$

For an internal rectangular column where the loading is eccentric about both axes, β may be approximated by:

$$\beta = 1 + 1.8 \sqrt{\left(\frac{e_y}{b_z}\right)^2 + \left(\frac{e_z}{b_y}\right)^2} \qquad (7.50)$$

Table 7.2 Value of k for rectangular loaded areas [3]

c_1/c_2	≤0.5	1.0	2.0	≥3.0
k	0.45	0.6	0.7	0.8

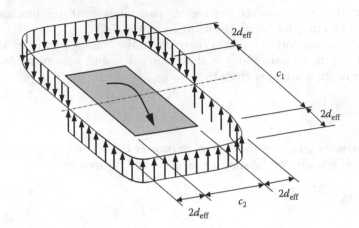

Figure 7.21 Shear distribution due to an unbalanced moment at slab–internal column connection [3].

where $e_y = M_{Ed,z}/V_{Ed}$ and $e_z = M_{Ed,y}/V_{Ed}$ are the eccentricities along the y- and z-axes, respectively, and b_z and b_y are the dimensions of the control perimeter (see Figure 7.17a).

For an internal circular column of diameter D, β may be approximated by:

$$\beta = 1 + 0.6\pi \frac{e}{D + 4d_{eff}} \tag{7.51}$$

where e is the eccentricity M_{Ed}/V_{Ed}.

Figure 7.22a shows the control perimeter for an edge column, where the unbalanced moment acts about an axis parallel to the slab edge and the eccentricity is towards the interior in the direction of c_1. For the rectangular column of Figure 7.22a:

$$W_1 = 0.25c_2^2 + c_1c_2 + 4c_1d_{eff} + 8d_{eff}^2 + \pi d_{eff}c_2 \tag{7.52}$$

When the eccentricity perpendicular to the slab edge is not towards the interior, β should be calculated using Equation 7.47.

Where there are eccentricities in both orthogonal directions, β may be calculated from:

$$\beta = \frac{u_1}{u_{1*}} + k \frac{u_1}{W_1} e_{par} \tag{7.53}$$

where u_1 is the perimeter of the basic control perimeter (see Figure 7.17b); u_{1*} is the reduced basic control perimeter (see Figure 7.22a); k may be

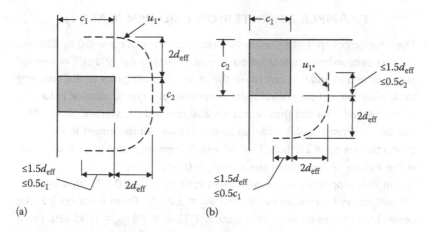

Figure 7.22 Reduced basic control perimeter u_{1*} [3]. (a) Edge column. (b) Corner column.

obtained from Table 7.2 but replacing the ratio c_1/c_2 with $c_1/2c_2$; W_1 is cal-culated for the basic control perimeter u_1 (see Figure 7.17b) and e_{par} is the eccentricity parallel to the slab edge resulting from a moment about an axis perpendicular to the slab edge.

For corner columns, where the eccentricity is towards the interior of the slab, the punching shear may be assumed to be uniformly distributed along the reduced control perimeter u_{1*} shown in Figure 7.22b. In this case, β may be taken as:

$$\beta = \frac{u_1}{u_{1*}} \tag{7.54}$$

If the eccentricity is towards the exterior, β should be calculated using Equation 7.47.

In structures that are laterally braced and do not rely on frame action between the slabs and the columns for lateral stability, the following approximate values of β may be used: $\beta = 1.15$ for internal columns; $\beta = 1.4$ for edge columns and $\beta = 1.5$ for corner columns.

Control perimeters at a distance less than $2d$ from the loaded area should be considered where the concentrated force is opposed by high pressure, such as may occur under a column in a pad footing. For foot-ings, the soil pressure within the control perimeter may be considered when determining the punching shear force V_{Ed} that must be resisted on any control perimeter.

EXAMPLE 7.3   INTERIOR COLUMN (CASE I)

The interior columns of a prestressed concrete flat plate are 450 by 450 mm in cross-section and are located on a regular rectangular grid at 9 m centres in one direction and 6 m centres in the other. The adequacy of the punching shear resistance of the critical shear perimeter of a typical interior column is to be checked. The slab thickness is $h = 250$ mm and the average of the effective depths from the slab soffit up to the tensile reinforcement in the y and z directions is $d_{eff} = 210$ mm. The slab reinforcement ratios in each direction in the vicinity of the column are $\rho_{ly} = \rho_{lz} = 0.008$.

The slab supports a permanent dead load of $w_G = 7.0$ kPa (that includes self-weight) and a variable live load of $w_Q = 3.0$ kPa. From Equation 2.2, the design load for the strength limit state is $1.35 w_G + 1.5 w_Q = 13.95$ kPa. From a frame analysis, the design load transferred from the slab to the column is 800 kN and the moment transferred to the column is 60 kNm.

Assume $\sigma_{cp} = (\sigma_{cy} + \sigma_{cz})/2 = 2.5$ MPa, $f_{ck} = 30$ MPa and $f_{cd} = 20$ MPa.

The perimeter of the loaded area is $u_0 = 4 \times 450 = 1800$ mm. The basic control perimeter is $u_1 = 4 \times 450 + 2\pi \times 2d_{eff} = 1800 + 4\pi \times 210 = 4439$ mm. The area of the slab subjected to load inside the basic control perimeter is $(4 \times 450 \times 420 + \pi \times 420^2) \times 10^{-6} = 1.31$ m². Therefore, the design actions transmitted on the basic control perimeter are:

$$V_{Ed} = 800 - 1.31 \times 13.95 = 782 \text{ kN} \quad \text{and} \quad M_{Ed} = 60 \text{ kNm}$$

The design value of the maximum punching shear resistance on any control section is obtained from Equation 7.45:

$$v_{Rd,max} = 0.3 \times \left[1 - \frac{30}{250}\right] \times 20 = 5.28 \text{ MPa}$$

The design punching shear resistance for a slab without shear reinforcement is obtained from Equation 7.42. With $k = 1.976$, $\rho_l = 0.008$ and $v_{min} = 0.035 \times k^{1.5} \times f_{ck}^{0.5} = 0.532$, Equation 7.42 gives:

$$v_{Rd,c} = 0.12 \times 1.976 \times (100 \times 0.008 \times 30)^{1/3} + 0.1 \times 2.5$$
$$= 0.934 \text{ MPa} > (v_{min} + k_1\sigma_{cp})$$

From Equation 7.49:

$$W_1 = 0.5 \times 450^2 + 450 \times 450 + 4 \times 450 \times 210 + 16 \times 210^2 + 2\pi \times 210 \times 450$$

$$= 1981 \times 10^3 \text{ mm}^2$$

and from Equation 7.47 and Table 7.2:

$$\beta = 1 + 0.6 \times \frac{60 \times 10^6}{782 \times 10^3} \times \frac{4439}{1981 \times 10^3} = 1.103$$

The maximum shear stress at the face of the column, i.e. on the perimeter u_0 of the loaded area, is given by Equation 7.46:

$$v_{Ed} = 1.103 \times \frac{782 \times 10^3}{1800 \times 210} = 2.28 \text{ MPa}$$

which is less than $v_{Rd,max}$ and, therefore, the section at the column perimeter is adequate.

From Equation 7.46, the maximum shear stress on the basic control perimeter is:

$$v_{Ed} = 1.103 \times \frac{782 \times 10^3}{4439 \times 210} = 0.925 \text{ MPa}$$

which is just less than $v_{Rd,c}$ and, therefore, shear reinforcement is not required. The punching shear resistance of the critical section of the slab is adequate.

EXAMPLE 7.4   INTERIOR COLUMN (CASE 2)

The slab–column connection analysed in Example 7.3 is to be rechecked for the case when $V_{Ed} = 1200$ kN and $M_{Ed} = 80$ kNm.

As in Example 7.3, $u_0 = 1800$ mm, $u_1 = 4439$ mm, $v_{Rd,max} = 5.28$ MPa, $v_{Rd,c} = 0.934$ MPa and $W_1 = 1981 \times 10^3$ mm^2. From Equation 7.47 and Table 7.2:

$$\beta = 1 + 0.6 \times \frac{80 \times 10^6}{1200 \times 10^3} \times \frac{4439}{1981 \times 10^3} = 1.090$$

The maximum shear stress at the face of the column, i.e. on the perimeter u_0 of the loaded area, is given by Equation 7.46:

$$v_{Ed} = 1.090 \times \frac{1200 \times 10^3}{1800 \times 210} = 3.46 \text{ MPa}$$

which is less than $v_{Rd,max}$ and, therefore, the section at the column perimeter is adequate.

From Equation 7.46, the maximum shear stress on the basic control perimeter is:

$$v_{Ed} = 1.090 \times \frac{1200 \times 10^3}{4439 \times 210} = 1.40 \text{ MPa}$$

which is greater than $v_{Rd,c}$ and, therefore, shear reinforcement is required. We will adopt a shear reinforcement layout similar to that shown in Figure 7.20a and shown again in Figure 7.23.

It is first necessary to determine the outer perimeter u_{out} at which shear reinforcement is no longer required. Let the radial distance from the centroid of the column to the outer perimeter be r_{out}. The outer perimeter may be approximated by $u_{out} = 2\pi r_{out}$.

Assuming conservatively that the shear force V_{Ed} acting on the outer shear perimeter remains at 1200 kN, we have:

$$v_{Ed} = 1.090 \times \frac{1200 \times 10^3}{2\pi r_{out} \times 210} = v_{Rd,c}$$

and solving gives $r_{out} = 1061$ mm and therefore $u_{out} = 6669$ mm.

From Figure 7.20a, the outermost perimeter of shear reinforcement must be located at a distance not greater than $1.5d_{eff} = 315$ mm from u_{out}, i.e. at a distance greater than $1061 - 315 - 225 = 521$ mm from the face of the column.

Assume vertical shear reinforcement (i.e. $\alpha = 90°$) and $f_{ywd.ef} = 250 + 0.25 \times 210 = 302.5$ MPa. From Table 4.7, the area of a 10 mm diameter bar is 78.5 mm² and with 12 radial lines of reinforcement, $A_{sw} = 12 \times 78.5 = 942$ mm².

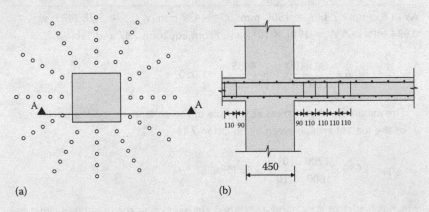

(a)   (b)

Figure 7.23 Shear reinforcement layout (Example 7.4). (a) Plan. (b) Section A–A.

The required radial spacing of the bars can be obtained from Equation 7.43. At the basic control perimeter $2d_{eff}$ from the face of the column:

$$v_{Rd,cs} = 0.75 \times 0.934 + 1.5 \times \frac{210}{s_r} \times 942 \times 302.5 \times \frac{1}{(4439 \times 210)} \sin 90 \geq 1.40 \, \text{MPa}$$

and therefore $s_r \leq 137$ mm (which is less than $0.75d_{eff}$ and is therefore satisfactory).

In each radial line of shear reinforcement, the first vertical bar is located at about $0.5d_{eff}$ from the column face, the outermost bar must be at least 521 mm from the loaded face and the spacing of the vertical bars must be less than 137 mm. The layout of the shear reinforcement shown in Figure 7.23 meets these requirements. In addition, the circumferential spacing between bars in each bar perimeter does not exceed $1.5d_{eff}$ inside the basic control perimeter and $2d_{eff}$ outside the basic control perimeter.

Successfully anchoring and locating stirrups within a 250 mm thick slab can be a difficult process. An alternative solution is to use fabricated steel shear heads to improve resistance to punching shear. Proprietary punching shear reinforcement consisting of headed bars and prefabricated units similar to those shown in Figure 7.24 are available. Usually the most economical and structurally efficient solution, however, is to increase the size of the critical section. The slab thickness can often be increased locally by the introduction of a drop panel, or alternatively the critical shear perimeter may be increased by introducing a column capital or simply by increasing the column dimensions. In general, provided such dimensional changes are architecturally acceptable, they represent the best structural solution.

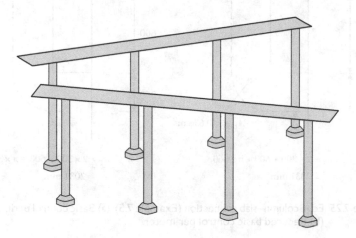

Figure 7.24 Typical prefabricated units used as punching shear reinforcement.

EXAMPLE 7.5 EDGE COLUMN

Consider the edge column–slab connection with basic control perimeter u_1 as shown in Figure 7.25a. The factored design load on the slab is 14 kPa, the design reaction force in the column is 320 kN and the moment transferred from the slab to the column acting about the centroidal axis of the basic control perimeter parallel to the free edge of the slab is 160 kNm. The basic control area is $A_{cont} = 0.636$ m², and the reduced basic control perimeter u_{1*} is shown in Figure 7.25b. The slab thickness is 220 mm, and the average effective depth for the top tensile steel in the slab is $d_{eff} = 180$ mm. Take $f_{ck} = 40$ MPa and $f_{cd} = 26.67$ MPa.

When designing a slab for punching shear at an edge (or corner) column, the average prestress σ_{cy} perpendicular to the free edge across part of the critical section should be taken as zero, unless care is taken to ensure that the slab tendons are positioned so that this part of the critical section is subjected to prestress. Often this is not physically possible, as discussed in Section 12.2 and illustrated in Figure 12.6. In this example, however, it is assumed that $\sigma_{cy} = \sigma_{cz} = 2.0$ MPa. It is also assumed that $\rho_{ly} = \rho_{lz} = 0.008$.

On the basic control perimeter, $V_{Ed} = 320 - 0.636 \times 14 = 311.1$ kN and $M_{Ed} = 160$ kNm.

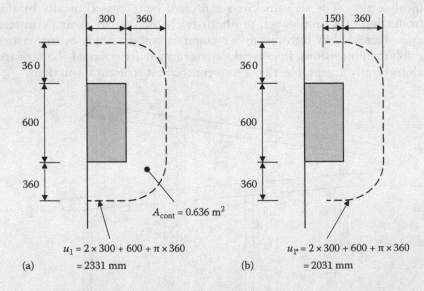

$u_1 = 2 \times 300 + 600 + \pi \times 360$
$= 2331$ mm

(a)

$u_{1*} = 2 \times 300 + 600 + \pi \times 360$
$= 2031$ mm

(b)

Figure 7.25 Edge column–slab connection (Example 7.5). (a) Basic control perimeter. (b) Reduced basic control perimeter.

The design value of the maximum punching shear resistance on any control section is obtained from Equation 7.45:

$$v_{Rd,max} = 0.3 \times \left[1 - \frac{40}{250}\right] \times 26.67 = 6.72 \text{ MPa}$$

The design punching shear resistance for a slab without shear reinforcement is obtained from Equation 7.42. With $k = 2.0$, $\rho_l = 0.008$ and $v_{min} = 0.035 \times k^{1.5} \times f_{ck}^{0.5} = 0.626$, Equation 7.42 gives:

$$v_{Rd,c} = 0.12 \times 2.0 \times (100 \times 0.008 \times 40)^{1/3} + 0.1 \times 2.0 = 0.962 \text{ MPa} \quad (>v_{min} + k_1\sigma_{cp})$$

From Equation 7.52:

$$W_l = 0.2 \times 600^2 + 300 \times 600 + 4 \times 300 \times 180 + 8 \times 180^2 + \pi \times 180 \times 600$$

$$= 1084.5 \times 10^3 \text{ mm}^2$$

and from Equation 7.47 and Table 7.2:

$$\beta = 1 + 0.45 \times \frac{160 \times 10^6}{311.1 \times 10^3} \times \frac{2331}{1084.5 \times 10^3} = 1.497$$

The maximum shear stress at the face of the column, i.e. on the perimeter u_0 of the loaded area, is given by Equation 7.46:

$$v_{Ed} = 1.497 \times \frac{311.1 \times 10^3}{1200 \times 180} = 2.16 \text{ MPa}$$

which is less than $v_{Rd,max}$ and, therefore, the section at the column perimeter is adequate.

From Equation 7.46, the maximum shear stress on the reduced basic control perimeter is:

$$v_{Ed} = 1.497 \times \frac{311.1 \times 10^3}{2031 \times 180} = 1.274 \text{ MPa}$$

which is greater than $v_{Rd,c}$ and, therefore, shear reinforcement is required. A shear layout similar to that shown in Figure 7.26 will be adopted here.

The effective outer perimeter $u_{out,ef}$ at which shear reinforcement is no longer required and the area inside the outer perimeter are calculated in Figure 7.26 as $u_{out,ef} = 2468$ mm and $A_{out} = 2.571 r_{link}^2 + 1.582 r_{link} + 0.324$ m^2, where r_{link} is in metres.

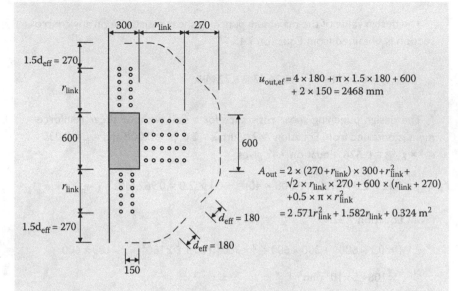

Figure 7.26 Effective outer shear perimeter (Example 7.5).

The shear force V_{Ed} acting on the outer shear perimeter (in kN) is $320 - 14.0 \times A_{out} = 320 - 14.0 \times (2.57 \, lr_{link}^2 + 1.582 r_{link} + 0.324)$, and the corresponding shear stress is:

$$v_{Ed} = 1.497 \times \frac{[320 - 14 \times (2.57 \, lr_{link}^2 + 1.582 r_{link}^2 + 0.324)] \times 10^3}{2468 \times 180} = v_{Rd,c}$$

and solving gives $r_{link} = 0.656$ m.

From Figure 7.26, the outermost perimeter of shear reinforcement must be located at a distance of not less than $r_{link} = 656$ mm from the face of the column.

Assume vertical shear reinforcement (i.e. $\alpha = 90°$) and $f_{ywd.ef} = 250 + 0.25 \times 180 = 295$ MPa. From Table 4.7, the area of a 10 mm diameter bar is 78.5 mm^2 and with 8 lines of reinforcement perpendicular to the column faces, $A_{sw} = 8 \times 78.5 = 628$ mm^2 and the spacing of the bars in each line can be obtained from Equation 7.43. At the basic control perimeter $2d_{eff}$ from the face of the column:

$$v_{Rd,cs} = 0.75 \times 0.962 + 1.5 \times \frac{180}{s_r} \times 628 \times 295 \times \frac{1}{(2331 \times 180)} \sin 90 \geq 1.274 \text{ MPa}$$

and therefore $s_r \leq 217$ mm. The maximum spacing of $0.75d_{eff} = 135$ mm governs and is therefore adopted, as shown in Figure 7.27.

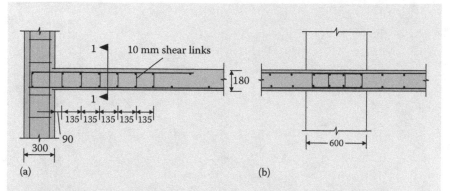

Figure 7.27 Sections through slab–column connection (Example 7.5). (a) Section perpendicular to slab edge. (b) Section I–I.

In each of the eight lines of shear reinforcement perpendicular to the column face (as shown in Figure 7.26), the first vertical bar is located at about $0.5d_{eff}$ = 90 mm from the column face, the outermost bar must be at least 656 mm from the column face. The shear reinforcement layout is shown in Figure 7.27.

REFERENCES

1. Ritter, W. 1899. *Die Bauweise Hennebique (Construction Methods of Hennebique)*. Zurich, Switzerland: Schweizerische Bayzeitung.
2. Hognestad, E. 1952. What do we know about diagonal tension and web reinforcement in concrete? University of Illinois Engineering Experiment Station, Circular Series No. 64, Urbana, IL.
3. EN 1992-1-1. 2004. Eurocode 2: Design of concrete structures – Part 1-1: General rules and rules for buildings. British Standards Institution, London, UK.

Chapter 8

Anchorage zones

8.1 INTRODUCTION

In prestressed concrete structural members, the prestressing force is usually transferred from the prestressing steel to the concrete in one of two different ways. In post-tensioned construction, relatively small anchorage plates transfer the force from the tendon to the concrete immediately behind the anchorage by bearing at each end of the tendon. For pretensioned members, the force is transferred by bond between the tendon and the concrete. In either case, the prestressing force is transferred in a relatively concentrated fashion, usually at the end of the member, and involves high local pressures and forces. A finite length of the member is required for the concentrated forces to disperse to form the linear compressive stress distribution usually assumed in design.

The length of member over which this dispersion of stress takes place is called the *dispersion length* (in the case of pretensioned members) and the *anchorage length* (for post-tensioned members). Within these so-called *anchorage zones,* a complex stress condition exists. Transverse tension is produced by the dispersion of the longitudinal compressive stress trajectories and may lead to longitudinal cracking within the anchorage zone. Similar zones of stress exist in the immediate vicinity of any concentrated force, including the concentrated reaction forces at the supports of a member.

The anchorage length in a post-tensioned member and the magnitude of the transverse forces (both tensile and compressive), which act perpendicular to the longitudinal prestressing force, depend on the magnitude of the prestressing force and on the size and position of the anchorage plate or plates. Both single and multiple anchorages are commonly used in post-tensioned construction. A careful selection of the number, size and location of the anchorage plates can often minimise the transverse tension and hence minimise the transverse reinforcement requirements within the anchorage zone.

The stress concentrations within the anchorage zone in a pretensioned member are not usually as severe as in a post-tensioned anchorage zone. There is a more gradual transfer of prestress. The full prestress is transmitted to the concrete by bond over a significant length of the tendon (called the

309

transmission length l_{pt}), and there are usually numerous individual tendons that are well distributed throughout the anchorage zone. In addition, the high concrete bearing stresses behind the anchorage plates in post-tensioned members do not occur in pretensioned construction.

8.2 PRETENSIONED CONCRETE: FORCE TRANSFER BY BOND

In pretensioned concrete, the tendons are usually tensioned within a casting bed. The concrete is cast around the tendons and, after the concrete has gained sufficient strength, the pretensioning force is released. The extent of the anchorage zone and the distribution of stresses within that zone depend on the quality of bond between the tendon and the concrete. The transfer of prestress usually occurs only at the end of the member, with the steel stress varying from zero at the end of the tendon, to the prescribed amount (full prestress P_{m0}) at some distance (the *transmission length* l_{pt}) in from the end. Over the transmission length bond stresses are high. The better the quality of the steel–concrete bond, the more efficient is the force transfer and the shorter is the transmission length. Outside the transmission length, bond stresses at transfer are small and the prestressing force in the tendon is approximately constant. Bond stresses and localised bond failures may occur outside the transfer length after the development of flexural cracks and under overloads, but a bond failure of the entire member involves failure of the anchorage zone at the ends of the tendons.

The main mechanisms that contribute to the strength of the steel–concrete bond are chemical adhesion of steel to concrete, friction at the steel–concrete interface and mechanical interlocking of concrete and steel (associated primarily with deformed or twisted strands). When the tendon is released from its anchorage within the casting bed and the force is transferred to the concrete, there is a small amount of tendon slip at the end of the member. This slippage destroys the bond for a short distance into the member at the released end, after which adhesion, friction and mechanical interlock combine to transfer the tendon force to the concrete.

During the stressing operation, there is a reduction in the diameter of the tendon due to Poisson's ratio effect. The concrete is then cast around the highly tensioned tendon. When the tendon is released, the unstressed portion of the tendon at the end of the member returns to its original diameter, whilst at some distance into the member, where the tensile stress in the tendon is still high, the tendon remains at its reduced diameter. Within the transmission length, the tendon diameter varies as shown in Figure 8.1 and there is a radial pressure exerted on the surrounding concrete. This pressure produces a frictional component which assists in the transferring of force from the steel to the concrete. The wedging action due to this radial strain is known as the Hoyer effect [1].

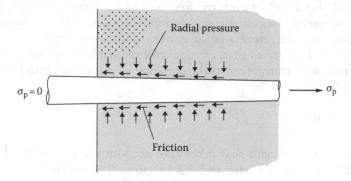

Figure 8.1 The Hoyer effect [1].

The transmission length and the rate of development of the steel stress along the tendon depend on many factors, including the size of the strand (i.e. the surface area in contact with the concrete), the surface conditions of the tendon, the type of tendon, the degree of concrete compaction within the anchorage zone, the degree of cracking in the concrete within the anchorage zone, the method of release of the prestressing force into the member and, to a minor degree, the compressive strength of the concrete.

The factors of size and surface condition of a tendon affect bond capacity in the same way as they do for non-prestressed reinforcement. A light coating of rust on a tendon will provide greater bond than for steel that is clean and bright. The surface profile has a marked effect on transfer length. Stranded cables have a shorter transfer length than crimped, indented or plain steel wires of equivalent area owing to the interlocking between the helices forming the strand. The strength of concrete, within the range of strengths used in prestressed concrete members, does not greatly affect the transmission length. However, with increased concrete strength, there is greater shear strength of the concrete embedded between the individual wires in the strand.

An important factor in force transfer is the quality and degree of concrete compaction. The transmission length in poorly compacted concrete is significantly longer than in well-compacted concrete. A prestressing tendon anchored at the top of a member generally has a greater transmission length than a tendon located near the bottom of the member. This is because the concrete at the top of a member is subject to increased sedimentation and is generally less well compacted than the concrete at the bottom of a member. When the tendon is released suddenly and the force is transferred to the concrete with impact, the transmission length is greater than for the case when the force in the steel is gradually imparted to the concrete.

Depending on these factors, transmission lengths are generally within the range 50–150 times the tendon diameter. The force transfer is not linear, with about 50% of the force transferred in the first quarter of the transfer length and about 80% within the first half of the length. For design purposes, however, it is reasonable and generally conservative to assume a linear variation of steel stress over the entire transmission length.

EN 1992-1-1 [2] states that, at release of the tendons, the prestress may be assumed to be transferred to the concrete by a constant bond stress f_{bpt} given by:

$$f_{bpt} = \eta_{p1}\eta_1 f_{ctd}(t) \tag{8.1}$$

where η_{p1} is a coefficient that takes into account the type of tendon and equals 2.7 for indented wire and 3.2 for strand; η_1 depends on the bond conditions and equals 1.0 for good bond conditions and 0.7 otherwise and $f_{ctd}(t)$ is the design tensile strength of concrete at the time of release, i.e. the lower characteristic tensile strength $(0.7f_{ctm}(t)$, see Equation 4.9) divided by the partial material factor $\gamma_c = 1.5$.

The *basic* value of the transmission length l_{pt} is specified in EN 1992-1-1 [2] as:

$$l_{pt} = \alpha_1\alpha_2\phi\sigma_{pm0}/f_{bpt} \tag{8.2}$$

where α_1 depends on the method of release of the tendon and equals 1.0 for gradual release and 1.25 for sudden release; α_2 depends on the type of tendon and equals 0.25 for round wire and 0.19 for seven-wire strand; ϕ is the nominal diameter of the tendon and σ_{pm0} is the tendon stress (P_{m0}/A_p) just after release. The *design* value of the transmission length is taken as the least favourable of two alternative values $l_{pt1} = 0.8\ l_{pt}$ or $l_{pt2} = 1.2\ l_{pt}$ depending on the design situation. When local stresses are being checked at release, l_{pt1} is appropriate. When the ultimate limit state of the anchorage and the anchorage zone is being checked, l_{pt2} is appropriate.

More information about the transmission length may be obtained from specialist literature, including References [3] to [6].

Eurocode 2 [2] states that concrete stresses caused by prestress in a pretensioned element may be assumed to be linear across any cross-section outside the dispersion length l_{disp}, where:

$$l_{disp} = \sqrt{l_{pt}^2 + d^2} \tag{8.3}$$

as shown in Figure 8.2.

The value of stress in the tendon, in regions outside the transmission length, remains approximately constant under service loads or whilst the member remains uncracked. After cracking in a flexural member, however, the behaviour becomes more like that of a reinforced concrete member and the stress in the tendon increases with increasing moment. If the critical moment location occurs at or near the end of a member, such as may occur

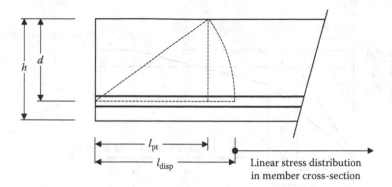

Figure 8.2 Length parameters at transfer of prestress in a pretensioned member.

in a short-span beam or a cantilever, the required development length for the tendon is significantly greater than the transmission length. In such cases, the bond capacity of the tendons needs to be carefully considered.

EN 1992-1-1 [2] requires that anchorage of tendons should be checked in regions where cracking is likely, i.e. in regions where the concrete tensile stress exceeds $f_{ctk,0,05}$. For this check, the tendon force on a cracked section should be used. The bond strength of the tendon at the ultimate limit state is specified as:

$$f_{bpd} = \eta_{p2}\eta_1 f_{ctd} \tag{8.4}$$

where η_{p2} depends on the type of tendon (η_{p2} = 1.4 for indented wires and η_{p2} = 1.2 for seven-wire strand); η_1 is as defined for Equation 8.1 and f_{ctd} is obtained from Equation 4.12, but due to the increasing brittleness of high-strength concrete, $f_{ctk,0,05}$ should not be taken greater than the value for C60/75 concrete (i.e. 3.1 MPa from Table 4.2).

The anchorage length required to develop a stress of σ_{pd} at the ultimate limit state is given by:

$$l_{bpd} = l_{pt2} + \alpha_2\phi(\sigma_{pd} - \sigma_{pm\infty})/f_{bpd} \tag{8.5}$$

where l_{pt2} = 1.2l_{pt}; α_2 is as defined for Equation 8.2 and $\sigma_{pm\infty}$ is the tendon stress after all losses. The development length l_{bpd} is illustrated in Figure 8.3 and is the sum of the upper design value of the transmission length l_{pt2} and the additional bonded length necessary to develop the increase of steel stress from $\sigma_{pm\infty}$ to σ_{pd}.

Where debonding of a strand is specified near the end of a pretensioned member, and the design allows for tension at service loads within the development length, a minimum development length of the debonded strand of $2l_{bpd}$ is recommended here.

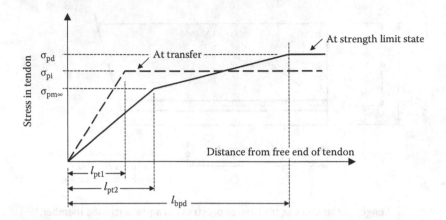

Figure 8.3 Variation of steel stress near the free end of a pretensioned tendon [2].

Where a prestressing tendon is not initially stressed, i.e. it is used in a member as non-prestressed reinforcement, and the tendon is required to develop its full characteristic breaking strength f_{pk}, the development length required on either side of the critical cross-section should be at least 2.5 times the transmission length specified in Equation 8.2. Care should be taken in situations where a sudden change in the effective depth of the tendon occurs due to an abrupt change in the member depth. In these locations, it may not be possible to develop the full strength of the initially untensioned tendon. Local bond failure may occur in the vicinity of the step, limiting the stress that can be developed in the tendon. Such a situation may develop if the calculated stress change in the strand required in the region of high local bond stresses exceeds about 500 MPa [7].

From their test results, Marshall and Mattock [8] proposed the following simple equation for determining the amount of transverse reinforcement A_{sb} (in the form of stirrups) in the end zone of a pretensioned member:

$$A_{sb} = 0.021 \frac{h}{l_{pt}} \frac{P_{m0}}{\sigma_{sb}} \qquad (8.6)$$

where h is the overall depth of the member, P_{m0} is the prestressing force immediately after transfer and σ_{sb} is the permissible steel stress required for crack control. According to EN 1992-1-1 [2], no check on crack widths is necessary if the stress in the reinforcement is limited to 300 MPa. The transverse steel A_{sb} should be equally spaced within $0.2h$ from the end face of the member.

8.3 POST-TENSIONED CONCRETE ANCHORAGE ZONES

8.3.1 Introduction

In post-tensioned concrete structures, failure of the anchorage zone is perhaps the most common cause of problems arising during construction. Such failures are difficult and expensive to repair, and usually necessitate replacement of the entire structural member. Anchorage zones may fail owing to uncontrolled cracking or splitting of the concrete resulting from insufficient well-anchored transverse reinforcement. Bearing failures immediately behind the anchorage plate are also relatively common and may be caused by inadequately dimensioned bearing plates or poor workmanship resulting in poorly compacted concrete in the heavily reinforced region behind the bearing plate. Great care should therefore be taken in both the design and construction of post-tensioned anchorage zones.

Consider the case of a single square anchorage plate (h_p by h_p) centrally positioned at the end of a prismatic member of depth h and width b, as shown in Figure 8.4. In the disturbed region of length l_a immediately behind the anchorage plate (i.e. the anchorage zone), plane sections do not remain plane and simple beam theory does not apply. High bearing stresses at the anchorage plate disperse throughout the anchorage zone, creating high transverse stresses, until at a distance l_a from the anchorage plate the linear stress and strain distributions predicted by simple beam theory are produced. The dispersion of stress that occurs within the anchorage zone is illustrated in Figure 8.4b.

The stress trajectories directly behind the anchorage are convex to the centre line of the member, as shown, and therefore produce a transverse

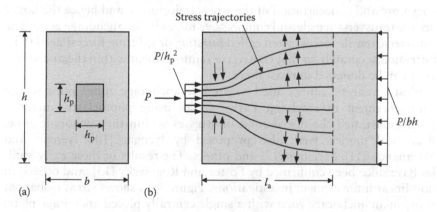

Figure 8.4 Stress trajectories for a centrally placed anchorage plate. (a) End elevation. (b) Side elevation.

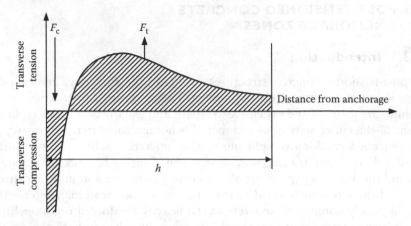

Figure 8.5 Distribution of transverse stress behind a single central anchorage.

component of compressive stress normal to the member axis. Further from the anchorage, the compressive stress trajectories become concave to the member axis and, as a consequence, produce transverse tensile stress components. The stress trajectories are closely spaced directly behind the bearing plate where compressive stress is high, and become more widely spaced as the distance from the anchorage plate increases. St Venant's principle suggests that the length of the disturbed region, for the single centrally located anchorage shown in Figure 8.4, is approximately equal to the depth of the member h. The variation of the transverse stresses along the centre line of the member, and normal to it, is represented in Figure 8.5.

The degree of curvature of the stress trajectories is dependent on the size of the bearing plate. The smaller the bearing plate, the larger are both the curvature and concentration of the stress trajectories, and hence the larger are the transverse tensile and compressive forces in the anchorage zone. The transverse tensile forces (often called *bursting* or *splitting* forces) need to be estimated accurately so that transverse reinforcement within the anchorage zone can be designed to resist them.

Elastic analysis can be used to analyse anchorage zones prior to the commencement of cracking. Early studies using photoelastic methods [9] demonstrated the distribution of stresses within the anchorage zone. Analytical models were also proposed by Iyengar [10], Iyengar and Yogananda [11], Sargious [12] and others. The results of these early studies have since been confirmed by Foster and Rogowsky [13] (and others) in non-linear finite element investigations. Figure 8.6a shows stress isobars of $|\sigma_y/\sigma_x|$ in an anchorage zone with a single centrally placed anchorage plate. Results are presented for three different anchorage plate sizes: $h_p/h = 0$, $h_p/h = 0.25$ and $h_p/h = 0.5$. These isobars are similar to those obtained in

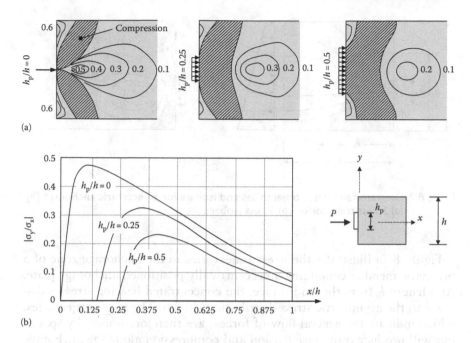

Figure 8.6 Transverse stress distributions for central anchorage [9]. (a) Stress isobars $|\sigma_y/\sigma_x|$. (b) Transverse stress along member axis.

photoelastic studies reported by Guyon [9]. σ_y is the transverse stress and σ_x is the average longitudinal compressive stress $P/(bh)$. The transverse compressive stress region in Figure 8.6a is hatched.

The effect of varying the size of the anchor plate on both the magnitude and position of the transverse stress along the axis of the member can be more clearly seen in Figure 8.6b. As the plate size increases, the magnitude of the maximum transverse tensile stress on the member axis decreases and its position moves further along the member (i.e. away from the anchorage plate). Tensile stresses also exist at the end surface of the anchorage zone in the corners adjacent to the bearing plate. Although these stresses are relatively high, they act over a small area and the resulting tensile force is small. Guyon [9] suggested that a tensile force of about 3% of the longitudinal prestressing force is located near the end surface of a centrally loaded anchorage zone when h_p/h is greater than 0.10.

The position of the line of action of the prestressing force with respect to the member axis has a considerable influence on the magnitude and distribution of stress within the anchorage zone. As the distance of the applied force from the axis of the member increases, the tensile stress at the loaded face adjacent to the anchorage also increases.

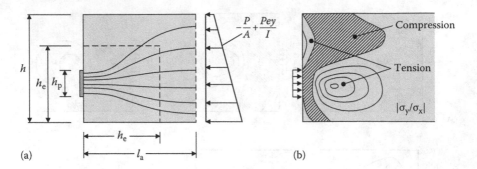

Figure 8.7 Diagrammatic stress trajectories and isobars for an eccentric anchorage [9]. (a) Stress trajectories. (b) Stress isobars.

Figure 8.7a illustrates the stress trajectories in the anchorage zone of a prismatic member containing an eccentrically positioned anchorage plate. At a length l_a from the loaded face, the concentrated bearing stresses disperse to the asymmetric stress distribution shown. The stress trajectories, which indicate the general flow of forces, are therefore unequally spaced but will produce transverse tension and compression along the anchorage axis in a manner similar to that for the single centrally placed anchorage. Isobars of $|\sigma_y/\sigma_x|$ are shown in Figure 8.7b. High bursting forces exist along the axis of the anchorage plate and, away from the axis of the anchorage, tensile stresses are induced on the end surface. These end tensile stresses, or spalling stresses, are typical of an eccentrically loaded anchorage zone.

Transverse stress isobars in the anchorage zones of members containing multiple anchorage plates are shown in Figure 8.8. The length of the member over which significant transverse stress exists (l_a) reduces with the number of symmetrically placed anchorages. The zone directly behind each anchorage contains bursting stresses and the stress isobars that resemble those in a single anchorage centrally placed in a much smaller end zone, as indicated in Figure 8.8. Tension also exists at the end face between adjacent anchorage plates. Guyon [9] suggested that the tensile force near the end face between any two adjacent bearing plates is about 4% of the sum of the longitudinal prestressing forces at the two anchorages.

The isobars presented in this section are intended only as a means of visualising the structural behaviour. Concrete is not a linear-elastic material, and a cracked prestressed concrete anchorage zone does not behave as depicted by the isobars in Figures 8.6 through 8.8. However, the linear-elastic analyses indicate the areas of high tension, both behind each anchorage plate and on the end face of the member, where cracking of the concrete can be expected during the stressing operation. The formation of such cracks reduces the stiffness in the transverse direction and leads to a significant redistribution of forces within the anchorage zone.

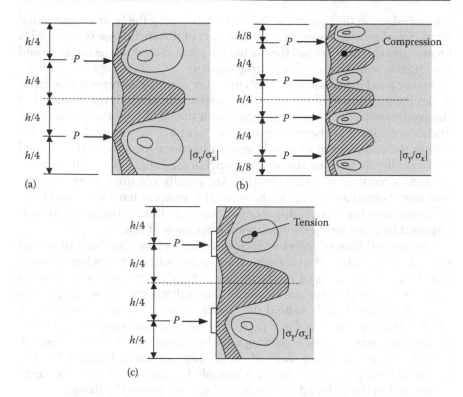

Figure 8.8 Transverse stress isobars for end zones with multiple anchorages [9]. (a) Two symmetrically placed anchorages – $h_p/h = 0$. (b) Four symmetrically placed anchorages – $h_p/h = 0$. (c) Two symmetrically placed anchorage plates.

8.3.2 Methods of analysis

The design of the anchorage zone of a post-tensioned member involves both the arrangement of the anchorage plates to minimise transverse stresses and the determination of the amount and distribution of reinforcement to carry the transverse tension after cracking of the concrete. Relatively large amounts of transverse reinforcement, usually in the form of stirrups, are often required within the anchorage zone and careful detailing of the steel is essential to permit the satisfactory placement and compaction of the concrete. In thin-webbed members, the anchorage zone is often enlarged to form an *end block* which is sufficient to accommodate the anchorage devices. This also facilitates the detailing and fixing of the reinforcement and the subsequent placement of concrete.

The anchorages usually used in post-tensioned concrete are patented by the manufacturer and prestressing companies for each of the types and arrangements of tendons. Typical anchorages are shown in Figures 3.6e, 3.8 and 3.9. The units are usually recessed into the end of the member and have bearing areas which are sufficient to prevent bearing problems in well-compacted concrete. Often the anchorages are manufactured with *fins* that are embedded in the concrete to assist in distributing the large concentrated force. Spiral reinforcement often forms part of the anchorage system and is located immediately behind the anchorage plate to confine the concrete and thus significantly improve its bearing capacity.

As discussed in Section 8.3.1, the curvature of the stress trajectories determines the magnitude of the transverse stresses. In general, the dispersal of the prestressing forces occurs through both the depth and the width of the anchorage zone and therefore transverse reinforcement must be provided within the end zone in two orthogonal directions (usually, vertically and horizontally on sections through the anchorage zone). The reinforcement quantities required in each direction are usually obtained from separate two-dimensional analyses, i.e. the vertical transverse tension is calculated by considering the vertical dispersion of forces and the horizontal tension is obtained by considering the horizontal dispersion of forces.

The internal flow of forces in each direction can be visualised in several ways. A simple model is to consider truss action within the anchorage zone. For the anchorage zone of the beam of rectangular cross-section shown in Figure 8.9, a simple strut-and-tie model shows that a transverse compressive force (F_{bc}) exists directly behind the bearing plate, with transverse tension, often called the bursting force (F_{bt}), at some distance along the member. Design using strut-and-tie modelling is outlined in more detail in Section 8.4.

Consider the anchorage zone of the T-beam shown in Figure 8.10. The strut-and-tie arrangement shown is suitable for calculating both the vertical tension in the web and the horizontal tension across the flange.

An alternative model for estimating the internal tensile forces is to consider the anchorage zone as a deep beam loaded on one side by the bearing stresses immediately under the anchorage plate and resisted on the other side by the statically equivalent, linearly distributed stresses in the beam.

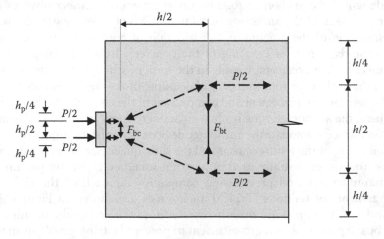

Figure 8.9 Strut-and-tie model of an anchorage zone.

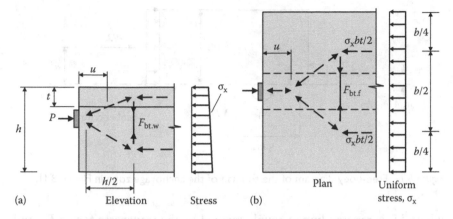

Figure 8.10 Vertical and horizontal tension in the anchorage zone of a post-tensioned T-beam. (a) Vertical tension in web. (b) Horizontal tension across flange.

The depth of the deep beam is taken as the anchorage length l_a. This approach was proposed by Magnel [14] and was further developed by Gergely and Sozen [15] and Warner and Faulkes [16].

8.3.2.1 Single central anchorage

The beam analogy model is illustrated in Figure 8.11 for a single central anchorage, together with the bending moment diagram for the idealised beam. Since the maximum moment tends to cause bursting along the axis of the anchorage, it is usually denoted by M_b and called the *bursting moment*.

By considering one half of the end block as a free-body diagram, as shown in Figure 8.12, the bursting moment M_b required for rotational equilibrium is obtained from statics. Taking moments about any point on the member axis gives:

$$M_b = \frac{P}{2}\left(\frac{h}{4} - \frac{h_p}{4}\right) = \frac{P}{8}(h - h_p) \tag{8.7}$$

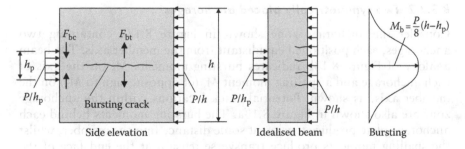

Figure 8.11 Beam analogy for a single centrally placed anchorage.

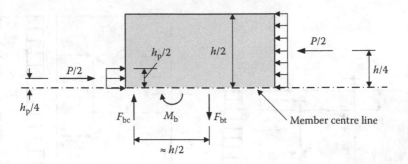

Figure 8.12 Free-body diagram of the top half of the anchorage zone in Figure 8.11.

where M_b is resisted by the couple formed by the transverse forces F_{bc} and F_{bt}, as shown.

As has already been established, the position of the resulting transverse (vertical) tensile force F_{bt} in Figures 8.11 and 8.12 is located at some distance from the anchorage plate. For a linear-elastic anchorage zone, the exact position of F_{bt} is the centroid of the area under the appropriate transverse tensile stress curve in Figure 8.6b. For the single, centrally placed anchorage of Figures 8.6, 8.11 and 8.12, the lever arm between F_{bc} and F_{bt} is approximately equal to $h/2$. This approximation also proves to be a reasonable one for a cracked concrete anchorage zone. Therefore, using Equation 8.7, we get:

$$F_{bt} \approx \frac{M_b}{h/2} = \frac{P}{4}\left(1 - \frac{h_p}{h}\right) \tag{8.8}$$

Expressions for the bursting moment and the horizontal transverse tension resulting from the lateral dispersion of bearing stresses across the width b of the section are obtained by replacing the depth h in Equations 8.7 and 8.8 with the width b.

8.3.2.2 Two symmetrically placed anchorages

Consider the anchorage zone shown in Figure 8.13a containing two anchorages, each positioned equidistant from the member axis. The beam analogy of Figure 8.13b indicates bursting moments M_b on the axis of each anchorage and a spalling moment M_s (of opposite sign to M_b) on the member axis, as shown. Potential crack locations within the anchorage zone are also shown in Figure 8.13a. The bursting moments behind each anchorage plate produce tension at some distance into the member, whilst the spalling moments produce transverse tension at the end face of the member. This simple analysis is consistent with the stress isobars for the

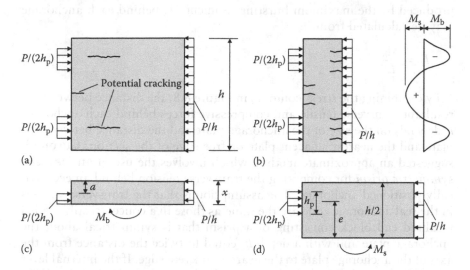

Figure 8.13 Beam analogy for two symmetrically placed anchorages. (a) End block-side elevation. (b) Idealised beam with bursting and spalling moments. (c) Free-body diagram through top anchorage. (d) Free-body diagram of top half of end block.

linear-elastic end block of Figure 8.8c. Consider the free-body diagram shown in Figure 8.13c. The maximum bursting moment behind the top anchorage occurs at the distance x below the top fibre, where the shear force at the bottom edge of the free-body is zero. That is:

$$\frac{P}{h}x = \frac{P}{2h_p}(x-a) \quad \text{or} \quad x = \frac{ah}{(h-2h_p)} \tag{8.9}$$

Summing moments about any point in Figure 8.13c gives:

$$M_b = \frac{Px^2}{2h} - \frac{P(x-a)^2}{4h_p} \tag{8.10}$$

The maximum spalling moment M_s occurs at the member axis, where the shear is also zero, and may be obtained by taking moments about any point on the member axis in the free-body diagram of Figure 8.13d:

$$M_s = \frac{P}{2}\left(e - \frac{h}{4}\right) \tag{8.11}$$

After the maximum bursting and spalling moments have been determined, the resultant internal compressive and tensile forces can be estimated provided that the lever arm between them is known. The internal tension F_{bt}

produced by the maximum bursting moment M_b behind each anchorage may be calculated from:

$$F_{bt} = \frac{M_b}{l_b} \qquad (8.12)$$

By examining the stress contours in Figure 8.8, the distance between the resultant transverse tensile and compressive forces behind each anchorage l_b depends on the size of the anchorage plate and the distance between the plate and the nearest adjacent plate or free edge of the section. Guyon [9] suggested an approximate method which involves the use of an *idealised symmetric prism* for computing the transverse tension behind an eccentrically positioned anchorage. The assumption is that the transverse stresses in the real anchorage zone are the same as those in a concentrically loaded idealised end block consisting of a prism that is symmetrical about the anchorage plate and with a depth h_e equal to twice the distance from the axis of the anchorage plate to the nearest concrete edge. If the internal lever arm l_b is assumed to be half the depth of the symmetrical prism (i.e. $h_e/2$), then the resultant transverse tension induced along the line of action of the anchorage is obtained from an equation that is identical to Equation 8.8, except that the depth of the symmetric prism h_e replaces h:

$$F_{bt} = \frac{P}{4}\left(1 - \frac{h_p}{h_e}\right) \qquad (8.13)$$

where h_p and h_e are, respectively, the dimensions of the anchorage plate and the symmetric prism in the direction of the transverse tension F_{bt}. For a single concentrically located anchorage plate $h_e = h$ (for vertical tension) and Equations 8.8 and 8.13 are identical.

Alternatively, the tension F_{bt} can be calculated from the bursting moment obtained from the statics of the real anchorage zone using a lever arm $l_b = h_e/2$. It should be noted that Equation 8.13 is an approximation that underestimates the transverse tension. Guyon [9] suggested that a conservative estimate of F_{bt} will always result if the bursting tension calculated by Equation 8.13 is multiplied by h/h_e, but this may be very conservative.

For anchorage zones containing multiple bearing plates, the bursting tension behind each anchorage, for the case where all anchorages are stressed, may be calculated using Guyon's symmetric prism. The depth of the symmetric prism h_e associated with a particular anchorage may be taken as the smaller of the following:

1. the distance in the direction of the transverse tension from the centre of the anchorage to the centre of the nearest adjacent anchorage; and
2. twice the distance in the direction of the transverse tension from the centre of the anchorage to the nearest edge of the anchorage zone.

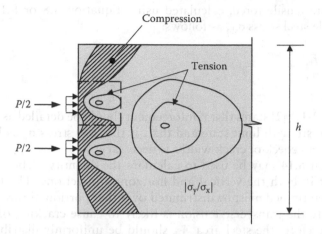

Figure 8.14 Two closely spaced symmetric anchorage plates.

For each symmetric prism, the lever arm l_b between the resultant transverse tension and compression may be taken as $h_e/2$.

The anchorage zone shown in Figure 8.14 contains two symmetrically placed anchorage plates located close together near the axis of the member. The stress contours show the bulb of tension immediately behind each anchorage plate. Also shown in Figure 8.14 is the symmetric prism to be used to calculate the resultant tension and the transverse reinforcement required in this region. Tension also exists further along the axis of the member in a similar location to the tension that occurs behind a single concentrically placed anchorage. Where the distance between two anchorages is less than 0.3 times the total depth of a member, consideration must also be given to the effects of the pair of anchorages acting in a manner similar to a single anchorage subject to the combined forces.

The loading cases to be considered in the design of a post-tensioned anchorage zone with multiple anchorage plates are:

1. all anchorages loaded; and
2. critical loading cases during the stressing operation.

8.3.3 Reinforcement requirements

In general, reinforcement should be provided to carry all the transverse tension in an anchorage zone. It is unwise to assume that the concrete will be able to carry any tension or that the concrete in the anchorage zone will not crack. The quantity of transverse reinforcement A_{sb} required to carry the transverse tension caused by bursting can be obtained by dividing the appropriate tensile force, calculated using Equation 8.8 or 8.13, by the permissible steel stress σ_{sb}, as follows:

$$A_{sb} = \frac{F_{bt}}{\sigma_{sb}} \qquad\qquad (8.14)$$

EN 1992-1-1 [2] states that reinforcement should be detailed as appropriate for the strength limit state and that, if the steel stress σ_{sb} is limited to 300 MPa, no check of crack widths is necessary.

Equation 8.14 may be used to calculate the quantity of bursting reinforcement in both the vertical and horizontal directions. The transverse steel so determined must be distributed over that portion of the anchorage zone where the transverse tension is likely to cause cracking of the concrete. Therefore, the steel area A_{sb} should be uniformly distributed over the portion of beam located from about $0.20h_e$ to $1.0h_e$ from the loaded end face. For the particular bursting moment being considered, h_e is the depth of the symmetric prism in the direction of the transverse tension and equals h for a single concentric anchorage. The size and spacing of the transverse reinforcement required in this region should also be provided in the portion of the beam from $0.20h_e$ to as near as practicable to the loaded face.

For spalling moments, the lever arm l_{sp} between the resultant transverse tension F_{spt} and compression F_{spc} is usually larger than for bursting, as can be seen from the isobars in Figure 8.7. For a single eccentric anchorage, the transverse tension at the loaded face remote from the anchorage may be calculated by assuming that l_{sp} is half the overall depth of the member. Between two widely spaced anchorages, the transverse tension at the loaded face may be obtained by taking l_{sp} equal to 0.6 times the spacing of the anchorages. The reinforcement required to resist the transverse tension at the loaded face A_{ssp} is obtained from:

$$A_{ssp} = \frac{F_{spt}}{\sigma_{ssp}} = \frac{M_s}{\sigma_{ssp}l_{sp}} \qquad\qquad (8.15)$$

According to EN 1992-1-1 [2], if $\sigma_{ssp} \leq 300$ MPa, it is not necessary to check for crack control. The steel area A_{ssp} should be located as close to the loaded face as is permitted by concrete cover and compaction requirements.

8.3.4 Bearing stresses behind anchorages

Local concrete bearing failures can occur in post-tensioned members immediately behind the anchorage plates if the bearing area is inadequate

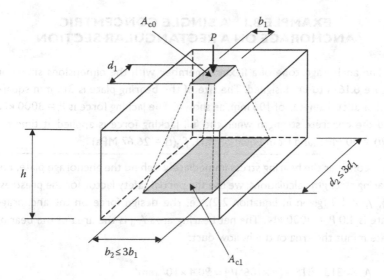

Figure 8.15 Design distribution areas for determination of bearing resistance force [2].

or the concrete strength is too low. The design resistance F_{Rdu} that can be supported on a bearing area A_{c0} is specified in EN 1992-1-1 [2] as:

$$F_{Rdu} = A_{c0}f_{cd}\sqrt{A_{c1}/A_{c0}} \leq 3.0f_{cd}A_{c0} \tag{8.16}$$

where f_{cd} is the design compressive strength of the concrete at the time of first loading (i.e. Equation 4.11 at transfer), A_{c0} is the bearing area, A_{c1} is the largest area of the concrete supporting surface that is geometrically similar to and concentric with A_{c0}, with maximum dimensions as indicated in Figure 8.15 (taken from EN 1992-1-1). The centre of the design distribution area A_{c1} is on the line of action of the force P passing through the centre of the bearing area A_{c0}. If there is more than one bearing plate at the end of the member, the design distribution areas should not overlap.

In commercial post-tensioned anchorages, the concrete immediately behind the anchorage is confined by spiral reinforcement (see Figure 3.6e), in addition to the transverse bursting and spalling reinforcement (often in the form of closed stirrups). In addition, the transverse compression at the loaded face immediately behind the anchorage plate significantly improves the bearing capacity of such anchorages. Therefore, provided the concrete behind the anchorage is well compacted, the bearing stress given by Equation 8.16 is usually conservative. Commercial anchorages are typically designed for bearing stresses of about 40 MPa, and bearing strength is specified by the manufacturer and is usually based on satisfactory test performance.

EXAMPLE 8.1 A SINGLE CONCENTRIC ANCHORAGE ON A RECTANGULAR SECTION

The anchorage zone of a flexural member with the dimensions shown in Figure 8.16 is to be designed. The size of the bearing plate is 315 mm square with a duct diameter of 106 mm, as shown. The jacking force is $P_j = 3000$ kN, and the concrete strength when the full jacking force is applied at time t is $f_{ck}(t) = 40$ MPa (and from Equation 4.11, $f_{cd}(t) = 26.67$ MPa).

First consider the bearing stress immediately behind the anchorage plate. For bearing strength calculations, we use the partial safety factor for the prestress (P), $\gamma_P = 1.0$ (given in Equation 2.2), i.e. the design force on the anchorage plate is 1.0 $P_j = 3000$ kN. The net bearing area A_{c0} is the area of the bearing plate minus the area of the hollow duct:

$$A_{c0} = 315 \times 315 - (\pi \times 106^2)/4 = 90.4 \times 10^3 \text{ mm}^2$$

For this anchorage, the dimensions of the design distribution area A_{c1} are $b_2 = d_2 = 480$ mm (since A_{c1} must be geometrically similar to A_{c0}). Therefore:

$$A_{c1} = 480 \times 480 = 230.4 \times 10^3 \text{ mm}^2$$

The design resistance force is obtained from Equation 8.16:

$$F_{Rdu} = 90.4 \times 10^3 \times 26.67 \times \sqrt{230.4 \times 10^3/(90.4 \times 10^3)} = 3849 \text{ kN}$$

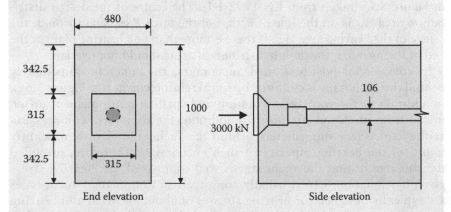

Figure 8.16 Details of anchorage zone (Example 8.1).

which is greater than the design force, $P_i = 3000$ kN and therefore acceptable. In practice, confinement reinforcement included behind the anchorage will significantly increase the design bearing strength.

Consider moments in the vertical plane:

The forces and bursting moments in the vertical plane are illustrated in Figure 8.17a. From Equation 8.7:

$$M_b = \frac{3000 \times 10^3}{8}(1000 - 315) = 256.9 \text{ kNm}$$

and the vertical bursting tension is obtained from Equation 8.8:

$$F_{bt} = \frac{256.9 \times 10^6}{1000/2} \times 10^{-3} = 513.8 \text{ kN}$$

With σ_{sb} taken equal to 300 MPa, the amount of vertical transverse reinforcement required to resist bursting is calculated from Equation 8.14:

$$A_{sb} = \frac{513.8 \times 10^3}{300} = 1713 \text{ mm}^2$$

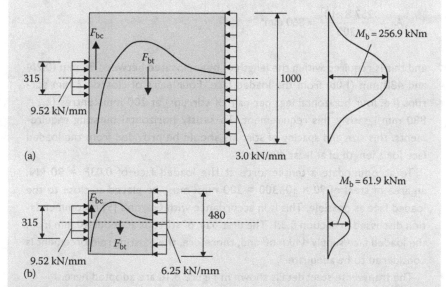

Figure 8.17 Bursting force and moment diagrams (Example 8.1). (a) Bursting in the vertical plane. (b) Bursting in the horizontal plane.

This area of transverse steel must be provided within the length of beam located from 0.20h to 1.0h from the loaded end face, i.e. over a length of 0.8h = 800 mm. Two 12 mm diameter stirrups (four vertical legs) are required every 200 mm along the 800 mm length (i.e. 4 sets of stirrups give A_{sb} = 4 × 4 × 110 = 1760 mm^2 within the 800 mm length). This size and spacing of stirrups must be provided over the entire anchorage zone, i.e. for a distance of 1000 mm from the loaded face.

Consider moments in the horizontal plane:

The forces and bursting moments in the horizontal plane are illustrated in Figure 8.17b. With b = 480 mm replacing h in Equations 8.7 and 8.8, the bursting moment and horizontal tension are:

$$M_b = \frac{3000 \times 10^3}{8}(480 - 315) = 61.9 \text{ kNm}$$

$$F_{bt} = \frac{61.9 \times 10^6}{480/2} \times 10^{-3} = 257.8 \text{ kN}$$

The amount of horizontal transverse steel is obtained from Equation 8.14 as:

$$A_{sb} = \frac{257.8 \times 10^3}{300} = 860 \text{ mm}^2$$

and this is required within the length of beam located between 96 mm (0.2b) and 480 mm (1.0b) from the loaded face. Four pairs of closed 12 mm stirrups (i.e. four horizontal legs per pair of stirrups) at 200 mm centres (A_{sb} = 880 mm^2) satisfy this requirement. To satisfy horizontal bursting requirements, this size and spacing of stirrups should be provided from the loaded face for a length of at least 480 mm.

To accommodate a tensile force at the loaded face of 0.03P_i = 90 kN, an area of steel of 90 × 10^3/300 = 300 mm^2 must be placed as close to the loaded face as possible. This is in accordance with Guyon's [9] recommendation discussed in Section 8.3.1. The first pair of stirrups at about 40 mm from the loaded face supply 440 mm^2 and, therefore, the existing reinforcement is considered to be adequate.

The transverse steel details shown in Figure 8.18 are adopted here.

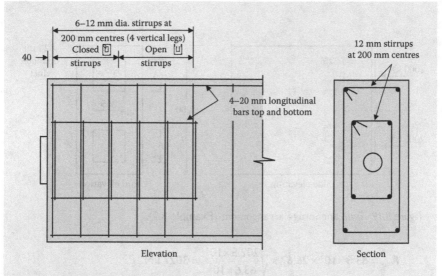

Figure 8.18 Reinforcement details (Example 8.1).

Within the first 480 mm, where horizontal transverse steel is required, the stirrups are closed at the top, as indicated, but for the remainder of the anchorage zone, between 480 and 1000 mm from the loaded face, open stirrups may be used to facilitate placement of the concrete. The first stirrup is placed as close as possible to the loaded face, as shown.

EXAMPLE 8.2 TWIN ECCENTRIC ANCHORAGES ON A RECTANGULAR SECTION

The anchorage shown in Figure 8.19 is to be designed. The jacking force at each of the two anchorage plates is $P_j = 2000$ kN, and the concrete strength at the time of transfer is $f_{ck}(t) = 40$ MPa (and from Equation 4.11, $f_{cd}(t) = 26.67$ MPa).

Check bearing stresses behind each anchorage:

As in Example 8.1, the design resistance force in bearing F_{Rdu} is calculated using Equation 8.16:

$$A_{c0} = 265^2 - \frac{\pi \times 92^2}{4} = 63.6 \times 10^3 \text{ mm}^2; \quad A_{c1} = 450^2 = 202.5 \times 10^3 \text{ mm}^2$$

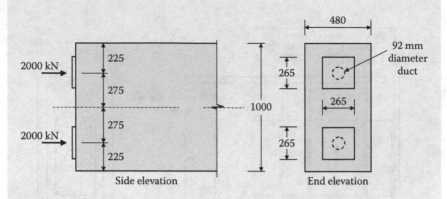

Figure 8.19 Twin anchorage arrangement (Example 8.2).

$$F_{Rdu} = 63.6 \times 10^3 \times 26.67 \times \sqrt{\frac{202.5 \times 10^3}{63.6 \times 10^3}} = 3027 \text{ kN}$$

which is greater than the design force less P_j = 2000 kN and is therefore satisfactory.

Case (a) – Consider the lower cable only stressed:

It is necessary first to examine the anchorage zone after just one of the tendons has been stressed. The stresses, forces and corresponding moments acting on the eccentrically loaded anchorage zone are shown in Figure 8.20a through c.

The maximum bursting moment M_b occurs at a distance x from the bottom surface at the point of zero shear in the free-body diagram of Figure 8.20d. From statics:

$$7.55 \times (x - 92.5) = \frac{5.3 + (5.3 - 0.0066x)}{2} x$$

$$\therefore x = 231.8 \text{ mm} \quad \text{and} \quad w_x = 3.77 \text{ kN/mm}$$

and

$$M_b = \left[5.30 \times \frac{231.8^2}{2} - (5.30 - 3.77) \times \frac{231.8^2}{6} - 7.55 \times \frac{(231.8 - 92.5)^2}{2} \right] \times 10^{-3}$$

$$= 55.5 \text{ kNm}$$

The maximum spalling moment M_s occurs at 394 mm below the top surface where the shear is also zero, as shown in Figure 8.20e, and from equilibrium:

$$M_s = 1.3 \times \frac{394^2}{6} \times 10^{-3} = 33.6 \text{ kNm}$$

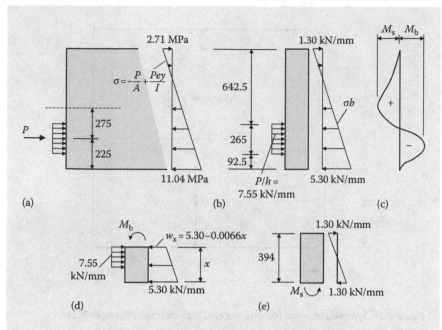

Figure 8.20 Actions on anchorage zone when the lower cable only is stressed (Example 8.2). (a) Side elevation and stresses. (b) Forces. (c) Moments. (d) Free-body of analogous beam at M_b. (e) Free-body of analogous beam at M_s.

Design for M_b:

The *symmetric prism* which is concentric with and directly behind the lower anchorage plate has a depth of $h_e = 450$ mm and is shown in Figure 8.21. From Equation 8.12:

$$F_{bt} = \frac{M_b}{l_b} = \frac{55.5 \times 10^3}{450/2} = 246.5 \text{ kN}$$

By contrast, Equation 8.13 gives:

$$F_{bt} = \frac{2000}{4}\left(1 - \frac{265}{450}\right) = 206 \text{ kN}$$

and this is considerably less conservative. Adopting the value of F_{bt} obtained from the actual bursting moment, Equation 8.14 gives:

$$A_{sb} = \frac{246.5 \times 10^3}{300} = 822 \text{ mm}^2$$

This area of steel must be distributed over a distance of $0.8h_e = 360$ mm.

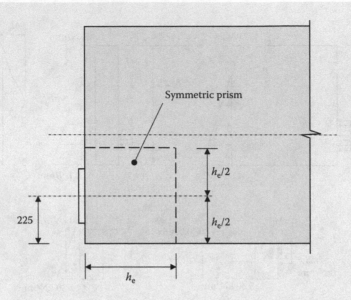

Figure 8.21 Symmetric prism for one eccentric anchorage (Example 8.2).

For the steel arrangement illustrated in Figure 8.23, 12 mm diameter stirrups are used at the spacings indicated, i.e. a total of four vertical legs of area 440 mm² per stirrup location are used behind each anchorage. The number of such stirrups required in the 360 mm length of the anchorage zone is 822/440 = 1.87 and therefore the maximum spacing of the stirrups is 360/1.87 = 192 mm. This size and spacing of stirrups are required from the loaded face to 450 mm therefrom. The spacing of the stirrups in Figure 8.23 is less than that calculated here because the horizontal bursting moment and spalling moment requirements are more severe. These are examined subsequently.

Design for M_s:

The lever arm l_s between the resultant transverse compression and tension forces that resist M_s is taken as $0.5h = 500$ mm. The area of transverse steel required within $0.2h = 200$ mm from the front face is given by Equation 8.15:

$$A_{ss} = \frac{33.6 \times 10^6}{300 \times 500} = 224 \text{ mm}^2$$

The equivalent of about two vertical 12 mm diameter steel legs is required close to the loaded face of the member to carry the resultant tension caused by spalling. This requirement is easily met by the three full depth 12 mm diameter stirrups (six vertical legs) located within 0.2h of the loaded face, as shown in Figure 8.23.

Case (b) – Consider both cables stressed:

Figure 8.22 shows the force and moment distribution for the end block when both cables are stressed.

Design for M_b:

The maximum bursting moment behind the anchorage occurs at the level of zero shear, x mm below the top surface and x mm above the bottom surface. From Equation 8.9:

$$x = \frac{92.5 \times 1000}{(1000 - 2 \times 265)} = 196.8 \text{ mm}$$

and Equation 8.10 gives:

$$M_b = \left[\frac{4000 \times 196.8^2}{2 \times 1000} - \frac{4000 \times (196.8 - 92.5)^2}{4 \times 265}\right] \times 10^{-6} = 36.4 \text{ kNm}$$

This is less than the value for M_b when only the single anchorage was stressed. Since the same symmetric prism is applicable here, the reinforcement requirements for bursting determined in case (a) are more than sufficient.

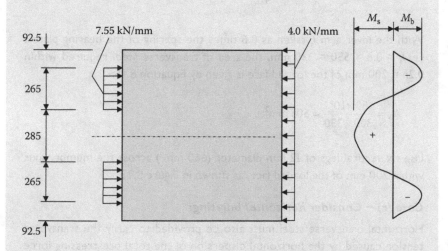

Figure 8.22 Forces and moments when both cables are stressed.

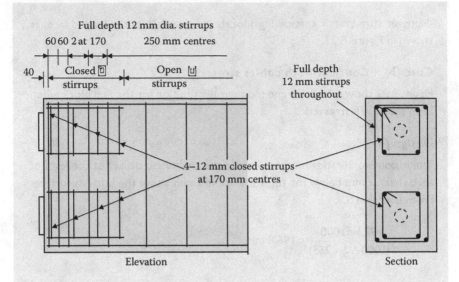

Figure 8.23 Reinforcement details (Example 8.2).

Design for M_s:

The spalling moment at the mid-depth of the anchorage zone (on the member axis) is obtained from Equation 8.11:

$$M_s = \frac{4000 \times 10^3}{2}\left(275 - \frac{1000}{4}\right) \times 10^{-6} = 50 \text{ kNm}$$

With the lever arm l_s taken as 0.6 times the spacing of the bearing plates, i.e. $l_s = 0.6 \times 550 = 330$ mm, the area of transverse steel required within $0.2h = 200$ mm of the loaded face is given by Equation 8.15:

$$A_{ss} = \frac{50 \times 10^6}{300 \times 330} = 505 \text{ mm}^2$$

Use six vertical legs of 12 mm diameter (660 mm²) across the member axis within 200 mm of the loaded face, as shown in Figure 8.23.

Case (c) – Consider horizontal bursting:

Horizontal transverse steel must also be provided to carry the transverse tension caused by the horizontal dispersion of the total prestressing force

(P = 4000 kN) from a 265 mm wide anchorage plate into a 480 mm wide section. With b = 480 mm used instead of h, Equations 8.7 and 8.8 give:

$$M_b = 107.5 \text{ kNm} \quad \text{and} \quad F_{bt} = 448 \text{ kN}$$

and the amount of horizontal steel is obtained from Equation 8.14:

$$A_{sb} = 1493 \text{ mm}^2$$

With the steel arrangement shown in Figure 8.23, six horizontal 12 mm diameter bars exist at each stirrup location, i.e. 660 mm² at each stirrup location. The required stirrup spacing within the length $0.8b$ (= 384 mm) is 170 mm. Therefore, within 480 mm from the end face of the beam, all available horizontal stirrup legs are required and therefore all stirrups in this region must be closed.

The reinforcement details shown in Figure 8.21 are adopted.

EXAMPLE 8.3 SINGLE CONCENTRIC ANCHORAGE IN A T-BEAM

The anchorage zone of the T-beam shown in Figure 8.24a is to be designed. The member is prestressed by strands located within a single 92 mm diameter duct, with a 265 mm square anchorage plate located at the centroidal axis of the cross-section. The jacking force is P_j = 2000 kN, and the characteristic concrete strength at transfer is 40 MPa, i.e. $f_{cd}(t)$ = 26.67 MPa. The distributions of forces on the anchorage zone in elevation and in plan are shown in Figure 8.24b and c, respectively.

The bearing area and the design distribution area are A_{c0} = 63.6 × 10^3 mm² and A_{c1} = 122.5 × 10^3 mm², respectively, and the design resistance force in bearing F_{Rdu} is calculated using Equation 8.16:

$$F_{Rdu} = 63.6 \times 10^3 \times 26.67 \times \sqrt{\frac{122.5 \times 10^3}{63.6 \times 10^3}} = 2354 \text{ kN}$$

which is greater than the design force less P_j = 2000 kN and is therefore satisfactory.

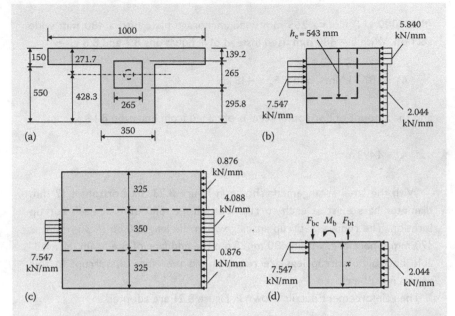

Figure 8.24 Details of the anchorage zone of the T-beam (Example 8.3). (a) End elevation. (b) Side elevation. (c) Plan. (d) Part side elevation.

Consider moments in the vertical plane:

The maximum bursting moment occurs at the level of zero shear at x mm above the bottom of the cross-section. From Figure 8.24d:

$$2.044 \times x = 7.547 \times (x - 295.8) \quad \therefore x = 405.7 \text{ mm}$$

and

$$M_b = \left[\frac{2.044 \times 405.7^2}{2} - \frac{7.547 \times (405.7 - 295.8)^2}{2} \right] \times 10^{-3} = 122.6 \text{ kNm}$$

As indicated in Figure 8.24b, the depth of the symmetric prism associated with M_b is $h_e = 2 \times 139.2 + 265 = 543$ mm and the vertical tension is:

$$F_{bt} = \frac{M_b}{h_e/2} = 451 \text{ kN}$$

The vertical transverse reinforcement required in the web is obtained from Equation 8.14:

$$A_{sb} = \frac{451 \times 10^3}{300} = 1503 \text{ mm}^2$$

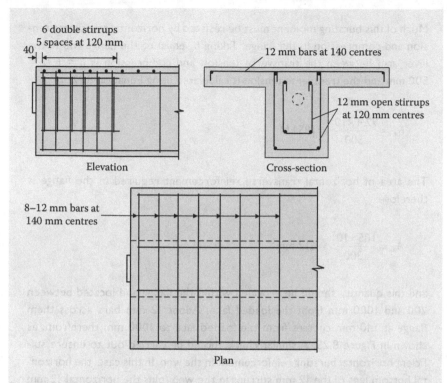

Figure 8.25 Reinforcement details for anchorage zone of T-beam (Example 8.3).

This area of steel must be located within the length of the beam between $0.2h_e = 109$ mm and $h_e = 543$ mm from the loaded face.

By using 12 mm stirrups over the full depth of the web and additional 12 mm stirrups immediately behind the anchorage, as shown in Figure 8.25 (i.e. $A_{sb} = (4 \times 110) = 440$ mm^2 per stirrup location), the number of double stirrups required is 1503/440 = 3.42 and the required spacing is (543 − 109)/3.42 = 127 mm. With a full depth stirrup located 40 mm from the loaded face, Guyon's recommendation that steel be provided near the loaded face to carry $0.03P_i$ is satisfied.

Consider moments in the horizontal plane:

Significant lateral dispersion of prestress in plan occurs in the anchorage zone as the concentrated prestressing force finds its way out into the flange of the T-section. By taking moments of the forces shown in Figure 8.24c about a point on the axis of the anchorage, the horizontal bursting moment is:

$$M_b = (0.876 \times 325 \times 337.5 + 4.088 \times 175 \times 87.5 - 7.547 \times 132.5^2/2) \times 10^{-3}$$

$$= 92.4 \text{ kNm}$$

Much of this bursting moment must be resisted by horizontal transverse tension and compression in the flange. Taking h_e equal to the flange width, the lever arm between the transverse tension and compression is $l_b = h_e/2 = 500$ mm and the transverse tension is calculated using Equation 8.12:

$$F_{bt} = \frac{92.4 \times 10^3}{500} = 185 \text{ kN}$$

The area of horizontal transverse reinforcement required in the flange is therefore:

$$A_{sb} = \frac{185 \times 10^3}{300} = 617 \text{ mm}^2$$

and this quantity should be provided within the flange and located between 200 and 1000 mm from the loaded face. Adopt 12 mm bars across them flange at 140 mm centres from the loaded face to 1000 mm therefrom, as shown in Figure 8.25. A similar check should be carried out to ensure sufficient horizontal bursting reinforcement in the web. In this case, the horizontal bottom legs of the 12 mm stirrups in the web (plus the horizontal 12 mm bars in the flange) are more than sufficient.

Alternative design using the strut-and-tie method:

An alternative approach to the design of the anchorage zone in a flanged member, and perhaps a more satisfactory approach, involves the use of strut-and-tie modelling, as illustrated in Figure 8.26.

The vertical dispersion of the prestress in the anchorage zone of Example 8.3 may be visualised using the simple strut-and-tie model illustrated in Figure 8.26a. The strut-and-tie model extends from the bearing plate into the beam for a length of about half the depth of the symmetric prism (i.e. $h_e/2 = 272$ mm in this case). The total prestressing force carried in the flange is 876 kN and this force is assumed to be applied to the analogous truss at A and B, as shown. The total prestressing force in the web of the beam is 1124 kN, and this is assumed to be applied to the analogous truss at the quarter points of the web depth, i.e. at D and F. From statics, the tension force in the vertical tie DF is 405 kN, which is in reasonable agreement with the bursting tension (451 kN) calculated previously using the deep beam analogy.

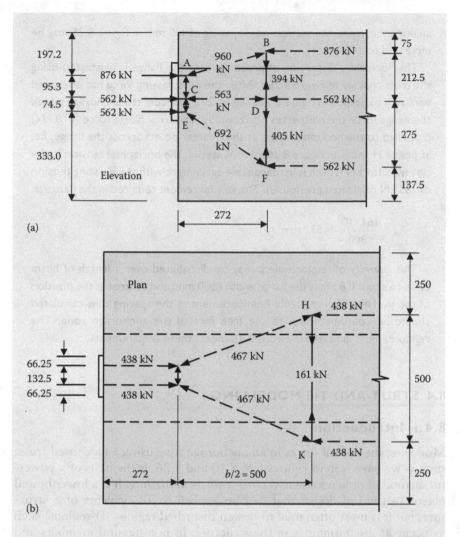

Figure 8.26 Truss analogy of the anchorage zone of T-beam (Example 8.3). (a) Vertical dispersion of prestress. (b) Horizontal dispersion of prestress.

The area of steel required to carry the vertical tension in the analogous truss is:

$$A_{sb} = \frac{405 \times 10^3}{300} = 1350 \text{ mm}^2$$

and this may be distributed over a length of the anchorage zone of about $0.8h_e$ (= 435 mm) centred on the tie BF in Figure 8.26a. According to the truss analogy, therefore, the vertical steel spacing of 120 mm in Figure 8.25 may be increased to 140 mm.

The horizontal dispersion of prestress into the flange is illustrated using the truss analogy of Figure 8.26b. After the prestressing force has dispersed vertically to point B in Figure 8.26a (i.e. at 272 mm from the anchorage plate), the flange force then disperses horizontally. The total flange force (876 kN) is applied to the horizontal truss at the quarter points across the flange, i.e. at points H and K in Figure 8.26b. From statics, the horizontal tension in the tie HK is 161 kN (which is in reasonable agreement with the bursting tension of 185 kN calculated previously). The reinforcement required in the flange is:

$$A_{sb} = \frac{161 \times 10^3}{300} = 537 \text{ mm}^2$$

This quantity of reinforcement may be distributed over a length of beam equal to about 0.8 times the flange width (800 mm) and centred at the position of the tie HK in Figure 8.26b. Reinforcement at the spacing thus calculated should be continued back to the free face of the anchorage zone. The reinforcement indicated in Figure 8.25 meets these requirements.

8.4 STRUT-AND-TIE MODELLING

8.4.1 Introduction

Modelling the flow of forces in an anchorage zone using an idealised truss, such as we have seen in Figures 8.9, 8.10 and 8.26, is the basis of a powerful method of design known as *strut-and-tie modelling*. It is a lower bound plastic method of design that can be applied to all elements of a structure, but it is most often used to design disturbed regions (D-regions) such as occur at discontinuities in the structure, in non-flexural members and within supports and connections. Anchorage zones in prestressed concrete structures are such regions.

Strut-and-tie modelling became popular in the 1980s with Marti [17,18] and Schlaich et al. [19] making important contributions. The designer selects a load path consisting of internal concrete struts and steel ties connected at nodes. The internal forces carried by the struts and ties must be in equilibrium with the external loads. Each element of the strut-and-tie model (i.e. the concrete struts representing compressive stress fields, the steel ties representing the reinforcement and the nodes connecting them) must then be designed and detailed so that the load path is everywhere sufficiently strong enough to carry the applied loads through the structure and into the supports. Care must be taken to ensure that strut-and-tie model selected is compatible with the applied loads and the supports and that both the struts and the ties possess sufficient ductility to accommodate the redistribution of internal forces necessary to achieve the desired load path.

EN 1992-1-1 [2] permits the use of strut-and-tie modelling as a basis for strength design (and for evaluating strength) in non-flexural regions of members. It may also be used for the design of members where linear distribution of strain exists on the cross-section. The following requirements are generally applicable when designing using strut-and-tie modelling:

1. loads are applied only at nodes with struts and ties carrying only axial force;
2. the model must be in equilibrium;
3. when determining the geometry of the model, the dimensions of the struts, ties and nodes must be accounted for;
4. if required, ties may cross struts;
5. struts are permitted to cross or intersect only at nodes; and
6. the angle between the axis of any strut and any tie at a node point shall not be less than about 30° for a reinforced concrete tie or 20° in a prestressed concrete member when a tendon is acting as the tie.

8.4.2 Concrete struts

8.4.2.1 Types of struts

Depending on the geometry of the member and its supports and loading points, the struts in a strut-and-tie model can be fan shaped, bottle shaped or prismatic, as shown in Figure 8.27. If unimpeded by the edges of a member or any penetrations through the member, compressive stress fields diverge. A prismatic strut, such as shown in Figure 8.27c, can only develop if the stress field is physically unable to diverge because of the geometry of

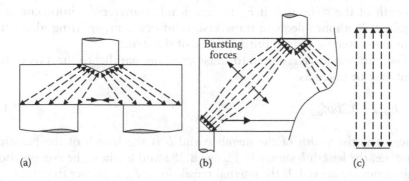

Bursting forces

(a)   (b)   (c)

Figure 8.27 Types of concrete struts. (a) Fan-shaped struts. (b) Bottle-shaped strut. (c) Prismatic strut.

the structure. When the compressive stress field can diverge without interruption and is not constrained at its ends, so that the stress trajectories remain straight, a fan-shaped strut results, as shown in Figure 8.27a. When the compressive stress field is free to diverge laterally along its length, but is constrained at either end, a bottle-shaped strut develops with curved stress trajectories similar to that shown in Figure 8.27b. Such curved compressive stress trajectories create bursting forces F_{bt} at right angles to the strut axis.

8.4.2.2 Strength of struts

According to EN 1992-1-1 [2], the design strength of a strut in a region with transverse compressive stress (or no transverse stress) is the product of the smallest cross-sectional area of the concrete strut at any point along its length and the design compressive stress f_{cd}. For struts in cracked compressive zones, with transverse tension, the design strength may be calculated using a reduced design compressive stress $\sigma_{Rd.max}$ given by:

$$\sigma_{Rd.max} = 0.6\left(1 - \frac{f_{ck}}{250}\right)f_{cd} \tag{8.17}$$

Properly detailed longitudinal reinforcement placed parallel to the axis of the strut and located within the strut may be used to increase the strength of a strut. The longitudinal reinforcement should be enclosed by suitably detailed ties or spiral reinforcement (see Section 14.6.2). The strength of a strut containing longitudinal reinforcement may be calculated as for a prismatic, pin-ended short column of cross-sectional area A_c and the same length as the strut (see Section 13.3).

8.4.2.3 Bursting reinforcement in bottle-shaped struts

The bursting tension in a bottle-shaped strut reduces the compressive strength of the strut and, if F_{bt} is significant, transverse reinforcement is required. Without adequate transverse reinforcement, splitting along the strut can initiate a sudden brittle failure of the strut.

The bursting force required to cause cracking parallel to the axis of the strut may be taken as:

$$F_{bt.cr} = 0.7bhf_{ctd} \tag{8.18}$$

where b is the width of the member and h is the length of the bursting zone, i.e. the length h shown in Figure 8.28a and b where the compression trajectories are curved. If the internal tensile force F_{bt} is greater than $0.5F_{bt.cr}$, it is recommended that adequate reinforcement be included to carry the entire bursting tension F_{bt}. EN 1992-1-1 [2] suggests that this should be

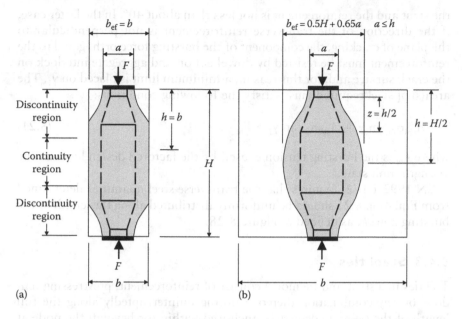

Figure 8.28 Dimensions for the determination of transverse tensile forces in a compression field with smeared reinforcement [2]. (a) Partially constrained ($b \leq H/2$). (b) Unconstrained ($b > H/2$).

distributed over the length h shown in Figure 8.28a and b. The bursting force F_{bt} may be determined from either Equation 8.19 for a partially constrained strut (as shown in Figure 8.28a) or Equation 8.20 for an unconstrained bottle-shaped strut (as shown in Figure 8.28b):

$$F_{bt} = \frac{1}{4}\frac{b-a}{b}F \qquad (8.19)$$

$$F_{bt} = \frac{1}{4}\left(1 - 0.7\frac{a}{h}\right)F \qquad (8.20)$$

where F is the compressive force carried by the strut and caused by the factored design loads at the strength limit state.

For the control of cracking along the strut at service loads, the maximum stress in the transverse reinforcement σ_{sb} should be limited to 300 MPa (perhaps even lower in regions where a strong degree of crack control is required for appearance or where cracks may reflect through finishes).

The transverse reinforcement requirements can be met by including reinforcement of areas A_{sb1} and A_{sb2} in two orthogonal directions γ_1 and γ_2 to the axis of the strut. Alternatively, transverse reinforcement of area A_{sb1} in one direction only can be used, provided the angle γ_1 between the axis of the strut and the reinforcement is not less than about 40°. In the latter case, if the direction of the transverse reinforcement is not perpendicular to the plane of cracking, the component of the bursting force orthogonal to the reinforcement must be resisted by dowel action and aggregate interlock on the crack surface and, for this reason, a minimum limit is placed on γ_1. The area(s) of steel required must satisfy the following conditions:

$$A_{sb1}\sigma_{sb} \sin \gamma_1 + A_{sb2}\sigma_{sb} \sin \gamma_2 \geq F_{bt} \tag{8.21}$$

where F_{bt} is the bursting tension caused by the factored design loads at the strength limit state.

EN 1992-1-1 [2] requires that the transverse steel quantities determined from Equation 8.21 should be uniformly distributed along the length of the bursting zone h, as defined in Figure 8.28.

8.4.3 Steel ties

The ties in a strut-and-tie model consist of reinforcement, prestressing tendons or any combination thereof running uninterruptedly along the full length of the tie and adequately anchored within (or beyond) the node at each end of the tie. The reinforcement and/or tendons should be evenly distributed across the end nodes and arranged so that the resultant tension in the steel coincides with the axis of the tie in the strut-and-tie model.

The design strength of the tie is given by:

$$T_{ud} = A_s f_{yd} + A_p(\sigma_{pm,t} + \Delta\sigma_p) \tag{8.22}$$

where $\sigma_{pm,t}$ is the effective prestress in the tendons after all the losses (see Equation 6.12) and $\Delta\sigma_p$ is the incremental force in the tendons due to the design external loads. The sum $\sigma_{pm,t} + \Delta\sigma_p$ should not be taken to be greater than the design strength of the tendons f_{pd}.

EN 1992-1-1 [2] requires that for adequate anchorage at each end of the tie, all reinforcement shall be fully anchored in accordance with the procedures outlined in Section 14.3.2. Alternatively, anchorage can be provided by a welded or mechanical anchorage entirely located beyond the node.

8.4.4 Nodes

At a node connecting struts and ties, at least three forces must be acting to satisfy equilibrium. The strength of the concrete within a node must also be checked. EN 1992-1-1 [2] identifies three types of nodes depending on the arrangement of the struts and ties entering the node.

    1. A compression node or a CCC node is one with only struts (or compressive loading points or reactions) entering the node, as shown in

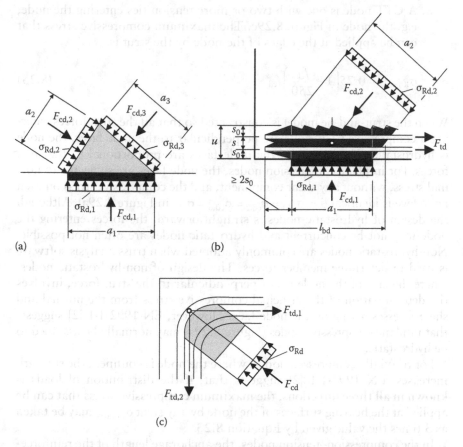

Figure 8.29 Node types in strut-and-tie models. (a) CCC (compression) node with-out ties. (b) CCT (compression–tension) node. (c) CTT (compression–tension) node.

Figure 8.29a. The maximum compressive stress that can be applied on the bearing surfaces at the edges of the node is:

$$\sigma_{Rd.max} = \left(1 - \frac{f_{ck}}{250}\right) f_{cd} \qquad (8.23)$$

2. A CCT node is one with two or more struts and single tension tie entering the node, e.g. the node in Figure 8.29b. The maximum com-pressive stress that can be applied at the bearing surfaces of the node by either of the two struts is:

$$\sigma_{Rd.max} = 0.85\left(1 - \frac{f_{ck}}{250}\right) f_{cd} \qquad (8.24)$$

3. A CTT node is one with two or more tension ties entering the node, e.g. the node in Figure 8.29c. The maximum compressive stress that can be applied at the edges of the node by the strut is:

$$\sigma_{Rd.max} = 0.75\left(1 - \frac{f_{ck}}{250}\right)f_{cd} \tag{8.25}$$

When the strut-and-tie model is constructed so that all the strut forces entering a compressive CCC node are perpendicular to the node faces, the node is hydrostatic. The lengths of the node faces are proportional to the strut forces. For in-plane compression nodes, the node faces are subjected to normal stress, without any shear component, and the compressive stress on each node face is identical (i.e. $\sigma_{cd,1} = \sigma_{cd,2} = \sigma_{cd,3} = \sigma_{cd,0}$ in Figure 8.29a). Although the design of hydrostatic nodes is straightforward, the forces entering the node may not be concurrent and hydrostatic nodes are often not possible. Non-hydrostatic nodes are commonly adopted when truss analysis software is used to determine member forces. The design of non-hydrostatic nodes, where the face of the node is not perpendicular to the strut force, involves the determination of the principal compressive stress from the normal and shear stresses acting on the node face. However, EN 1992-1-1 [2] suggests that in-plane compressive nodes (Figure 8.29a) may normally be assumed to be hydrostatic.

For triaxially compressed nodes, where the node is confined, the strength increases. EN 1992-1-1 [2] suggests that, if the distribution of loads is known in all three directions, the maximum compressive stress that can be applied at the bearing surfaces of the node by any strut $\sigma_{Rd,max}$ may be taken as 3 times the value given by Equation 8.23.

In the compression–tension nodes, the anchorage length of the reinforcement starts at the beginning of the node and should extend through the entire node and if necessary beyond the node (as shown in Figure 8.29b). The anchorage and bending requirements for reinforcement are discussed in Section 14.3.2.

REFERENCES

1. Hoyer, E. 1939. *Der Stahlsaitenbeton.* Berlin, Germany: Elsner.
2. BS EN 1992-1-1. 2004. Eurocode 2: Design of concrete structures – Part 1-1: General rules and rules for buildings. British Standards Institution, London, UK.
3. Logan, D.R. 1997. Acceptance criteria for bond quality of strand for pretensioned concrete applications. *PCI Journal,* 42(2), 52–90.
4. Rose, D.R. and Russell, B.W. 1997. Investigation of standardized tests to measure the bond performance of prestressing strands. *PCI Journal,* 42(4), 56–80.

5. Martin, L. and Korkosz, W. 1995. Strength of prestressed members at sections where strands are not fully developed. *PCI Journal*, 40(5), 58–66.
6. MNL-120-4, Precast/Prestressed Concrete Institute. 2005. *PCI Design Handbook: Precast And Prestressed Concrete*, 6th edn. Chicago, IL: Precast/Prestressed Concrete Institute, pp. 4-27–4-29.
7. Gilbert, R.I. 2012. Unanticipated bond failure over supporting band beams in grouted post-tensioned slab tendons with little or no prestress. *Bond in Concrete, Fourth International Symposium*, 17–20 June, Brescia, Italy.
8. Marshall, W.T. and Mattock, A.H. 1962. Control of horizontal cracking in the ends of pretensioned prestressed concrete girders. *Journal of the Prestressed Concrete Institute*, 7(5), 56–74.
9. Guyon, Y. 1953. *Prestressed Concrete*, English edn. London, UK: Contractors Record and Municipal Engineering.
10. Iyengar, K.T.S.R. 1962. Two-dimensional theories of anchorage zone stresses in post-tensioned concrete beams. *Journal of the American Concrete Institute*, 59, 1443–1446.
11. Iyengar, K.T.S.R. and Yogananda, C.V. 1966. A three dimensional stress distribution problem in the end zones of prestressed beams. *Magazine of Concrete Research*, 18, 75–84.
12. Sargious, M. 1960. Beitrag zur Ermittlung der Hauptzugspannungen am Endauflager vorgespannter Betonbalken. Dissertation. Stuttgart, Germany: Technische Hochschule.
13. Foster, S.J. and Rogowsky, D.M. 1997. Bursting forces in concrete members resulting from in-plane concentrated loads. *Magazine of Concrete Research*, 49(180), 231–240.
14. Magnel, G. 1954. *Prestressed Concrete*, 3rd edn. New York: McGraw-Hill.
15. Gergely, P. and Sozen, M.A. 1967. Design of anchorage zone reinforcement in prestressed concrete beams. *Journal of the Prestressed Concrete Institute*, 12(2), 63–75.
16. Warner, R.F. and Faulkes, K.A. 1979. *Prestressed Concrete*, 1st edn. Melbourne, Victoria, Australia: Pitman Australia.
17. Marti, P. 1985. Truss models in detailing. *Concrete International – American Concrete Institute*, 7(1), 46–56.
18. Marti, P. 1985. Basic tools of reinforced concrete beam design. *Concrete International – American Concrete Institute*, 7(12), 66–73.
19. Schlaich J., Schäfer, K. and Jennewein, M. 1987. Towards a consistent design of structural concrete. Special Report. *PCI Journal*, 32(3), 74–150.

Chapter 9

Composite members

9.1 TYPES AND ADVANTAGES OF COMPOSITE CONSTRUCTION

Composite construction in prestressed concrete usually consists of precast prestressed members acting in combination with a cast in-situ concrete component. The composite member is formed in at least two separate stages with some or all of the prestressing normally applied before the completion of the final stage. The precast and the cast in-situ elements are mechanically bonded to each other to ensure that the separate components act together as a single composite member.

Composite members can take a variety of forms. In building construction, the precast elements are often pretensioned slabs (which may be either solid or voided), or single- or double-tee beams. The cast in-situ element is a thin, lightly reinforced topping slab placed on top of the precast units after the units have been erected to their final position in the structure. Single- or double-tee precast units are used extensively in building and bridge structures because of the economies afforded by this type of construction.

Composite prestressed concrete beams are widely used in the construction of highway bridges. For short- and medium-span bridges, standardised I-shaped or trough-shaped girders (which may be either pretensioned or post-tensioned) are erected between the piers and a reinforced concrete slab is cast onto and across the top flange of the girders. The precast girders and the in-situ slab are bonded together to form a stiff and strong composite bridge deck.

The two concrete elements, which together form the composite structure, may have different concrete strengths, different elastic moduli and different creep and shrinkage characteristics. The concrete in the precast element is generally of better quality than the concrete in the cast in-situ element because it usually has a higher specified target strength and experiences better quality control during construction and better curing conditions. With the concrete in the precast element being older and of better quality than the in-situ concrete, restraining actions will develop in the composite

351

structure with time owing to differential creep and shrinkage movements. These effects should be carefully considered in design.

Prestressed concrete composite construction has many advantages over non-composite construction. In many situations, a significant reduction in construction costs can be achieved. The use of precast elements can greatly speed up construction time. When the precast elements are standardised and factory produced, the cost of long-line pretensioning may be considerably less than the cost of post-tensioning on site. Of course, the cost of transporting precast elements to the site must be included in these comparisons and it is often transportation difficulties that limit the size of the precast elements and the range of application of this type of construction. In addition, it is easier and more economical to manufacture concrete elements with high mechanical properties in a controlled prestressing plant rather than on a building or bridge site.

During construction, the precast elements can support the formwork for the cast in-situ concrete, thereby reducing falsework and shoring costs. The elimination of scaffolding and falsework is often a major advantage over other forms of construction, and permits the construction to proceed without interruption to the work or traffic beneath. Apart from providing significant increases to both the strength and stiffness of the precast girders, the in-situ concrete can perform other useful structural functions. It can provide continuity at the ends of precast elements over adjacent spans. In addition, it provides lateral stability to the girders and also provides a means for carrying lateral loads back to the supports. Stage stressing can be used to advantage in some composite structures. A composite member consisting of a pretensioned, precast element and an in-situ slab may be subsequently post-tensioned to achieve additional economies of section. This situation may arise, for example, when a relatively large load is to be applied at some time after composite action has been achieved.

Cross-sections of some typical composite prestressed concrete members commonly used in buildings and bridges are shown in Figure 9.1.

9.2 BEHAVIOUR OF COMPOSITE MEMBERS

The essential requirement for a composite member is that the precast and cast in-situ elements act together as one unit. To achieve this, it is necessary to have good bond between the two elements.

When a composite member is subjected to bending, a horizontal shear force develops at the interface between the precast and the in-situ elements. This results in a tendency for horizontal slip on the mating surfaces, if the bond is inadequate. Resistance to slip is provided by the naturally achieved adhesion and friction that occurs between the two elements. Often the top surface of the precast element is deliberately roughened during manufacture to improve its bonding characteristics and facilitate the transfer of

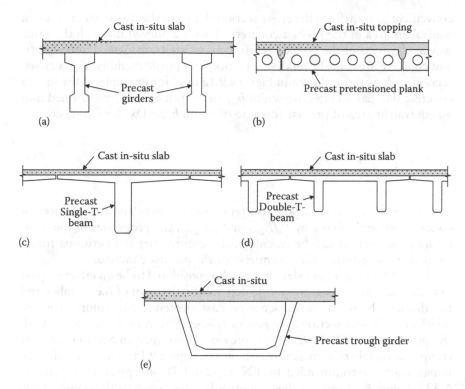

Figure 9.1 Typical composite prestressed concrete cross-sections. (a) Slab and girder. (b) Pretensioned plank plus topping. (c) Single-T-sections. (d) Double-T-sections. (e) Trough girder.

horizontal shear through mechanical interlock. Where the contact surface between the two elements is broad (such as in Figure 9.1b through d), natural adhesion and friction are usually sufficient to resist the horizontal shear. Where the contact area is small (such as between the slab and girders in Figure 9.1a and e), other provisions are necessary. Frequently, the web reinforcement in the precast girder is continued through the contact surface and anchored in the cast in-situ slab. This reinforcement resists horizontal shear primarily by dowel action, but assistance is also gained by clamping the mating surfaces together and increasing the frictional resistance.

If the horizontal shear on the element interface is resisted without slip (or with small slip only), the response of the composite member can be determined in a similar manner to that of a monolithic member. Stresses and strains on the composite cross-section due to service loads applied after the in-situ slab has been placed (and has hardened) may be calculated using the properties of the combined cross-section calculated using the transformed area method. If the elastic modulus of the concrete in the in-situ part of the cross-section E_{cm2} is different from that in the precast element E_{cm1}, it is convenient to transform the cross-sectional area of the in-situ element to an equivalent area of the precast concrete. This is achieved in much the same way as the areas of the bonded reinforcement are transformed into equivalent concrete areas in the analysis of a non-composite member. For a cross-section such as those shown in Figures 9.1a or e, for example, if the in-situ concrete slab has an effective width b_{eff} and depth h_s, it is transformed into an equivalent area of precast concrete of depth h_s and width b_{tr}, where:

$$b_{tr} = \frac{E_{cm2}}{E_{cm1}} b_{eff} = \alpha_c b_{ef} \tag{9.1}$$

If the bonded steel areas are also replaced by equivalent areas of precast concrete (by multiplying by E_s/E_{cm1} or E_p/E_{cm1}), the properties of the composite cross-section can be calculated by considering the fictitious transformed cross-section made up entirely of the precast concrete.

The width of the in-situ slab that can be considered to be an effective part of the composite cross-section (b_{eff}) depends on the span of the member and the distance between the adjacent precast elements. Maximum effective widths for flanged sections are generally specified in building codes, with the provisions of EN 1992-1-1 [1] previously outlined in Section 6.5. For composite members such as those shown in Figure 9.1a and e, the effective flange widths recommended by EN 1992-1-1 [1] are given in Equations 6.33 through 6.35, except that the term b_w now refers to the width of the slab–girder interface.

The design of prestressed concrete composite members is essentially the same as that of non-composite members, provided that certain behavioural differences are recognised and taken into account. It is important to appreciate that part of the applied load is resisted by the precast element(s) prior to the establishment of composite action. Care must be taken, therefore, when designing for serviceability to ensure that behaviour of the cross-section and its response to various load stages are accurately modelled. It is also necessary in design to ensure adequate horizontal shear capacity at the element interface. With these issues taken into account, the design procedures for flexural, shear and torsional strengths are similar to that of a non-composite member.

9.3 STAGES OF LOADING

As mentioned in the previous section, the precast part of a composite member may be required to carry loads prior to the establishment of composite action. When loads are applied during construction, before the cast in-situ slab has set, flexural stresses are produced on the precast element. After the in-situ concrete has been placed and cured, the properties of the cross-section are altered for all subsequent loadings. Moments due to service live loads, for example, modify the stress distribution in the precast element and introduce stresses into the cast in-situ slab. Creep and shrinkage of the concrete also cause a substantial redistribution of stress with time between the precast and the in-situ elements, and between the concrete and the bonded reinforcement in each element.

In the design of a prestressed concrete composite member, the following load stages usually need to be considered.

1. *Initial prestress at transfer in the precast element*: This normally involves calculation of elastic stresses due to both the initial prestress P_{m0} and the self-weight of the precast member. This load stage frequently occurs off-site in a precast plant.
2. *Period before casting the in-situ slab*: This involves a time analysis to determine the stress redistribution and change in curvature caused by creep and shrinkage of the concrete in the precast element during the period after the precast element is prestressed and prior to casting the in-situ concrete. The only loads acting are the prestress (after initial losses) and the self-weight of the precast element. A reasonably accurate time analysis can be performed using the analysis described in Section 5.7.3. Typical concrete stresses at load stages 1 and 2 at the mid-span of the precast element are illustrated in Figure 9.2a.
3. *Immediately after casting the in-situ concrete and before composite action*: This load stage involves a short-term analysis of the precast element to calculate the instantaneous effects of the additional superimposed dead loads prior to composite action. If the precast element is unshored (i.e. not temporarily supported by props during construction), the superimposed dead load mentioned here includes the weight of the wet in-situ concrete. The additional increments of stress and instantaneous strain in the precast element are added to the stresses and strains obtained at the end of load stage 2. Typical concrete stresses at the critical section of an unshored member at load stage 3 are shown in Figure 9.2b.

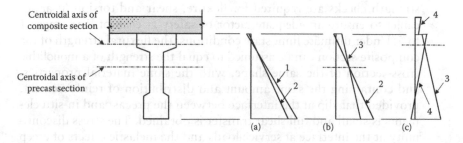

Figure 9.2 Concrete stresses at the various load stages. (a) Load stages 1 and 2. (b) Load stages 2 and 3. (c) Load stages 3 and 4.

If the precast member is shored prior to placement of the cast in-situ slab, the applied loads do not produce internal actions or deformations in the member and the imposed loads are carried by the shoring. Therefore, no additional stresses or strains occur in a fully shored precast element at this load stage. When curing of the cast in-situ component has been completed, the shoring is removed and the self-weight of the cast in-situ concrete, together with any other loading applied at this time, produce deformations and flexural stresses and are considered in load stage 4.

4. *Immediately after the establishment of composite action*: This involves a short-term analysis of the composite cross-section (see Section 9.5.2) to determine the change of stresses and deformations on the composite cross-section as all the remaining loads are applied. The instantaneous effect of any dead load or service live load and any additional prestressing not previously considered (i.e. not applied previously to the non-composite precast element) are considered here. If cracking occurs, a cracked section analysis is required. Additional prestress may be applied to the composite member by re-stressing existing post-tensioned tendons or tensioning previously unstressed tendons. If the composite section remains uncracked, the increments of stress and strain calculated at this load stage on the precast part of the composite cross-section are added to the stresses and strains calculated in stage 3 prior to the establishment of composite action. Typical concrete stresses at the end of load stage 4 are shown in Figure 9.2c.

5. *Period after the establishment of composite action*: A time analysis of the composite cross-section is required (see Section 9.5.3) for the period beginning at the time the sustained load is first applied (usually soon after the in-situ concrete is poured) and ending after all creep and shrinkage deformations have taken place. The long-term effects of creep and shrinkage of concrete and relaxation of the prestressing steel on the behaviour of the composite section subjected to the sustained service loads are determined.

6. *The ultimate limit state condition for the composite section*: Ultimate strength checks are required for flexure, shear and torsion (if applicable) to ensure an adequate factor of safety at each of load stages 1 to 5. Under ultimate limit state conditions, the flexural strength of the composite section can be assumed to equal the strength of a monolithic cross-section of the same shape, with the same material properties, and containing the same amount and distribution of reinforcement, provided that slip at the interface between the precast and in-situ elements is small and full shear transfer is obtained. The stress discontinuity at the interface at service loads and the inelastic effects of creep and shrinkage have an insignificant effect on the design strength and can be ignored at the ultimate limit state condition.

9.4 DETERMINATION OF PRESTRESS

In practice, the initial prestress and the eccentricity of prestress at the critical section in the precast element (P_{m0} and e_{pc}, respectively) are calculated to satisfy preselected stress limits at transfer. In general, cracking is avoided at transfer by limiting the tensile stress to about $f_{ct,0} = 0.15 \times [f_{cm}(t_0)]^{2/3}$. In addition, in order to avoid unnecessarily large creep deformations, it is prudent to ensure that the initial compressive stresses do not exceed $f_{cc,0} = -0.45f_{cm}(t_0)$. In the case of trough girders, as shown in Figure 9.1e, the centroidal axis of the precast element is often not far above the bottom flange, so that loads applied to the precast element prior to or during placement of the in-situ slab may cause unacceptably large compressive stresses in the top fibres of the precast girder.

Satisfaction of stress limits in the precast element at transfer and immediately prior to the establishment of composite action (at the end of load stage 3) can be achieved using the procedure discussed in Section 5.4.1. For the case of a precast girder, Equations 5.3 through 5.6 become:

$$P_{m0} \le \frac{A_{pc}f_{ct,0} + \alpha_{top.pc}M_1}{\alpha_{top.pc}e_{pc} - 1} \tag{9.2}$$

$$P_{m,0} \le \frac{-A_{pc}f_{cc,0} + \alpha_{btm.pc}M_1}{\alpha_{btm.pc}e_{pc} + 1} \tag{9.3}$$

$$P_{m,0} \ge \frac{-A_{pc}f_{ct,t} + \alpha_{btm.pc}M_3}{\Omega_3(\alpha_{btm.pc}e_{pc} + 1)} \tag{9.4}$$

$$P_{m,0} \ge \frac{A_{pc}f_{cc,t} + \alpha_{top.pc}M_3}{\Omega_3(\alpha_{top.pc}e_{pc} - 1)} \tag{9.5}$$

where e_{pc} is the eccentricity of prestress from the centroidal axis of the precast section; $\alpha_{top.pc} = A_{pc}/Z_{top.pc}$; $\alpha_{btm.pc} = A_{pc}/Z_{btm.pc}$; A_{pc} is the cross-sectional area of the precast member and $Z_{top.pc}$ and $Z_{btm.pc}$ are the top and bottom section moduli of the precast element, respectively. The moment M_1 is the moment applied at load stage 1 (usually resulting from the self-weight of the precast member); M_3 is the maximum in-service moment applied to the precast element prior to composite action (in load stage 3) and $\Omega_3 P_{m,0}$ is the prestressing force at load stage 3. An estimate of the losses of prestress between the transfer and the placement of the in-situ slab deck is required for the determination of Ω_3.

Equations 9.2 and 9.3 provide an upper limit to $P_{m,0}$, and Equations 9.4 and 9.5 establish a minimum level of prestress in the precast element.

After the in-situ slab has set, the composite cross-section resists all subsequent loading. There is a change both in the size and the properties of the cross-section, and a stress discontinuity exists at the element interface. If cracking is to be avoided under the service loads, a limit $f_{ct,t}$ (say $0.5f_{ctm}$) is placed on the magnitude of the extreme fibre tensile stress at the end of load stage 5, i.e. after all prestress losses and under full service loads. This requirement places another, perhaps more severe limit on the minimum amount of prestress compared to that imposed by Equation 9.4. Alternatively, this requirement may suggest that an additional prestressing force is required on the composite member, i.e. the member may need to be further post-tensioned after the in-situ slab has developed its target strength.

The bottom fibre tensile stress immediately before the establishment of composite action may be approximated by:

$$\sigma_{btm,3} = -\frac{\Omega_3 P_{m,0}}{A_{pc}}\left(1 + \frac{A_{pc}e_{pc}}{Z_{btm.pc}}\right) + \frac{M_3}{Z_{btm.pc}} \tag{9.6}$$

If the maximum additional moment applied to the composite cross-section in load stage 4 is M_4 and the prestressing force reduces to $\Omega P_{m,0}$ with time, then the final maximum bottom fibre stress at the end of load stage 5 may be approximated by:

$$\sigma_{btm,5} = -\frac{\Omega P_{m,0}}{A_{pc}}\left(1 + \frac{A_{pc}e_{pc}}{Z_{btm.pc}}\right) + \frac{M_3}{Z_{btm.pc}} + \frac{M_4}{Z_{btm.comp}} \tag{9.7}$$

where $Z_{btm,comp}$ is the section modulus for the bottom fibre of the composite cross-section. If the bottom fibre stress in load stage 5 is to remain less than the stress limit $f_{ct,t}$, then Equation 9.7 can be rearranged to give:

$$P_{m,0} \geq \frac{A_{pc}\left(\dfrac{M_3}{Z_{btm.pc}} + \dfrac{M_4}{Z_{btm.comp}} - f_{ct,t}\right)}{\Omega(\alpha_{btm.pc}e_{pc} + 1)} \tag{9.8}$$

Equation 9.8, together with Equations 9.2, 9.3 and 9.5, can be used to establish a suitable combination of $P_{m,0}$ and e_{pc}. In some cases, the precast section may be proportioned so that the prestress and eccentricity satisfy all stress limits prior to composite action (i.e. Equations 9.2 through 9.5). However, if when the additional requirement of Equation 9.8 is included, no combination of $P_{m,0}$ and e_{pc} can be found to satisfy all the stress limits, additional prestress may be applied to the composite member after the in-situ slab is in place.

If cracking can be tolerated in the composite member under full service loads, a cracked section analysis may be required to check for crack control and to determine the reduction of stiffness and its effect on deflection.

Care must be taken in such an analysis to accurately model stresses in the various parts of the cross-section and the stress discontinuity at the slab–girder interface.

In many situations, cracking may be permitted under the full live load but not under the permanent sustained load. In such a case, M_4 in Equation 9.8 can be replaced by the sustained part of the moment applied at load stage 4 ($M_{4.sus}$) and the so-modified Equation 9.8 can be used to determine the minimum level of prestress on a *partially-prestressed* composite section.

9.5 METHODS OF ANALYSIS AT SERVICE LOADS

9.5.1 Introductory remarks

After the size of the concrete elements and the quantity and disposition of prestressing steel have been determined, the behaviour of the composite member at service loads should be investigated to determine the deflection (and shortening) at the various load stages (and times) and also to check for the possibility of cracking. The short-term and time-dependent analyses of uncracked composite cross-sections can be carried out conveniently using procedures similar to those described in Sections 5.6.2 and 5.7.3 for non-composite cross-sections. The approaches described here were also presented by Gilbert and Ranzi [2].

Consider a cross-section made up of a precast, pretensioned girder (element 1) and a cast in-situ reinforced concrete slab (element 2), as shown in Figure 9.3. The concrete in each element has different deformation characteristics. This particular cross-section contains four layers of non-prestressed reinforcement and two layers of prestressing steel, although any number of steel layers can be handled without added difficulty. As was demonstrated in Tables 5.1 and 5.2, the presence of non-prestressed

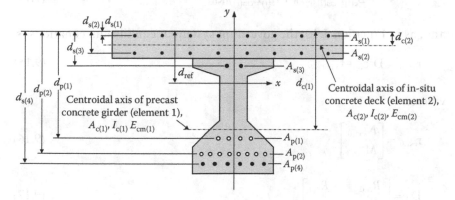

Figure 9.3 Typical prestressed concrete composite cross-section.

reinforcement may affect the time-dependent deformation of the section significantly and cause a reduction of the compressive stresses in the concrete. In the following analyses, no slip is assumed to occur between the two concrete elements or between the steel reinforcement and the concrete.

9.5.2 Short-term analysis

As outlined in Section 5.6.2, the constitutive relationships for each material for use in the instantaneous analysis are (Equations 5.23 through 5.26):

$$\sigma_{c(i),0} = E_{cm(i),0}\varepsilon_0 \tag{9.9}$$

$$\sigma_{s(i),0} = E_{s(i)}\varepsilon_{s(i),0} \tag{9.10}$$

$$\sigma_{p(i),0} = E_{p(i)}\left(\varepsilon_{cp(i),0} + \varepsilon_{p(i),init}\right) \quad \text{if } A_{p(i)} \text{ is bonded} \tag{9.11}$$

$$\sigma_{p(i),0} = E_{p(i)}\varepsilon_{p(i),init} \quad \text{if } A_{p(i)} \text{ is unbonded} \tag{9.12}$$

Similar to Equation 5.29, the internal actions carried by the i-th concrete element (for inclusion in the equilibrium equations) can be expressed as:

$$N_{c(i),0} = \int_{A_{c(i)}} \sigma_{c(i),0}\, dA = \int_{A_{c(i)}} E_{cm(i),0}\left(\varepsilon_{r,0} - y\kappa_0\right)dA$$

$$= A_{c(i)}E_{cm(i),0}\varepsilon_{r,0} - B_{c(i)}E_{cm(i),0}\kappa_0 \tag{9.13}$$

$$M_{c(i),0} = \int_{A_{c(i)}} -y\sigma_{c(i),0}\, dA = \int_{A_{c(i)}} -E_{cm(i),0}y\left(\varepsilon_{r,0} - y\kappa_0\right)dA$$

$$= -B_{c(i)}E_{cm(i),0}\varepsilon_{r,0} + I_{c(i)}E_{cm(i),0}\kappa_0 \tag{9.14}$$

and, as in Equation 5.40, the governing system of equilibrium equations is:

$$r_{ext,0} = D_0\varepsilon_0 + f_{p,init} \tag{9.15}$$

where:

$$r_{ext,0} = \begin{bmatrix} N_{ext,0} \\ M_{ext,0} \end{bmatrix} \tag{9.16}$$

$$D_0 = \begin{bmatrix} R_{A,0} & -R_{B,0} \\ -R_{B,0} & R_{I,0} \end{bmatrix} \tag{9.17}$$

$$\varepsilon_0 = \begin{bmatrix} \varepsilon_{r,0} \\ \kappa_0 \end{bmatrix} \tag{9.18}$$

$$f_{p,init} = \sum_{i=1}^{m_p} \begin{bmatrix} A_{p(i)}E_{p(i)}\varepsilon_{p(i),init} \\ y_{p(i)}A_{p(i)}E_{p(i)}\varepsilon_{p(i),init} \end{bmatrix} = \sum_{i=1}^{m_p} \begin{bmatrix} P_{init(i)} \\ y_{p(i)}P_{init(i)} \end{bmatrix} \tag{9.19}$$

Solving for the unknown strain variables gives (Equation 5.45):

$$\varepsilon_0 = D_0^{-1}\left(r_{ext,0} - f_{p,init}\right) = F_0\left(r_{ext,0} - f_{p,init}\right) \tag{9.20}$$

where:

$$F_0 = \frac{1}{R_{A,0}R_{I,0} - R_{B0}^2} \begin{bmatrix} R_{I,0} & R_{B,0} \\ R_{B,0} & R_{A,0} \end{bmatrix} \tag{9.21}$$

The cross-sectional rigidities forming the D_0 and F_0 matrices are:

$$R_{A,0} = \sum_{i=1}^{m_c} A_{c(i)}E_{cm(i),0} + \sum_{i=1}^{m_s} A_{s(i)}E_{s(i)} + \sum_{i=1}^{m_p} A_{p(i)}E_{p(i)}$$

$$= \sum_{i=1}^{m_c} A_{c(i)}E_{cm(i),0} + R_{A,s} + R_{A,p} \tag{9.22}$$

$$R_{B,0} = \sum_{i=1}^{m_c} B_{c(i)}E_{c(i),0} + \sum_{i=1}^{m_s} y_{s(i)}A_{s(i)}E_{s(i)} + \sum_{i=1}^{m_p} y_{p(i)}A_{p(i)}E_{p(i)}$$

$$= \sum_{i=1}^{m_c} B_{c(i)}E_{c(i),0} + R_{B,s} + R_{B,p} \tag{9.23}$$

$$R_{I,0} = \sum_{i=1}^{m_c} I_{c(i)}E_{c(i),0} + \sum_{i=1}^{m_s} y_{s(i)}^2 A_{s(i)}E_{s(i)} + \sum_{i=1}^{m_p} y_{p(i)}^2 A_{p(i)}E_{p(i)}$$

$$= \sum_{i=1}^{m_c} I_{c(i)}E_{c(i),0} + R_{I,s} + R_{I,p} \tag{9.24}$$

where for convenience the following notation has been introduced for the rigidities of the reinforcement and tendons:

$$R_{A,s} = \sum_{i=1}^{m_s} A_{s(i)}E_{s(i)} \quad R_{B,s} = \sum_{i=1}^{m_s} y_{s(i)}A_{s(i)}E_{s(i)} \quad R_{I,s} = \sum_{i=1}^{m_s} y_{s(i)}^2 A_{s(i)}E_{s(i)} \tag{9.25}$$

$$R_{A,p} = \sum_{i=1}^{m_p} A_{p(i)}E_{p(i)} \quad R_{B,p} = \sum_{i=1}^{m_p} y_{p(i)}A_{p(i)}E_{p(i)} \quad R_{I,p} = \sum_{i=1}^{m_p} y_{p(i)}^2 A_{p(i)}E_{p(i)} \quad (9.26)$$

The stress distribution is calculated from Equations 9.9 through 9.12:

$$\sigma_{c(i),0} = E_{cm(i),0}\varepsilon_0 = E_{cm(i),0}[1-y]\varepsilon_0 \tag{9.27}$$

$$\sigma_{s(i),0} = E_{s(i)}\varepsilon_{s(i),0} = E_{s(i)}[1-y_{s(i)}]\varepsilon_0 \tag{9.28}$$

$$\sigma_{p(i),0} = E_{p(i)}\left(\varepsilon_{cp(i),0} + \varepsilon_{p(i),init}\right)$$

$$= E_{p(i)}[1-y_{p(i)}]\varepsilon_0 + E_{p(i)}\varepsilon_{p(i),init} \quad \text{if } A_{p(i)} \text{ is bonded} \tag{9.29}$$

$$\sigma_{p(i),0} = E_{p(i)}\varepsilon_{p(i),init} \quad\quad\quad \text{if } A_{p(i)} \text{ is unbonded} \tag{9.30}$$

where $\varepsilon_0 = \varepsilon_{r,0} - y\kappa_0 = [1-y]\varepsilon_0$.

9.5.3 Time-dependent analysis

For the analysis of stresses and deformations on a composite concrete–concrete cross-section at time t_k after a period of sustained loading, the *age-adjusted effective modulus method* may be used, as outlined in Sections 5.7.2 and 5.7.3. The stress–strain relationships for each concrete element and for each layer of reinforcement and tendons at t_k are as follows (Equations 5.68 through 5.72 renumbered here for convenience):

$$\text{Concrete:} \quad \sigma_{c(i),k} = \bar{E}_{c,eff(i),k}\left(\varepsilon_k - \varepsilon_{cs(i),k}\right) + \bar{F}_{e(i),0}\sigma_{c(i),0} \tag{9.31}$$

$$\text{Steel reinforcement:} \quad \sigma_{s(i),k} = E_{s(i)}\varepsilon_{s(i),k} \tag{9.32}$$

Pretensioned tendons or post-tensioned tendons bonded at t_0:

$$\sigma_{p(i),k} = E_{p(i)}\left(\varepsilon_{cp(i),k} + \varepsilon_{p(i),init} - \varepsilon_{p.rel(i),k}\right) \tag{9.33}$$

Post-tensioned tendons unbonded at t_0:

$$\text{Bonded:} \quad \sigma_{p(i),k} = E_{p(i)}\left(\varepsilon_{cp(i),k} - \varepsilon_{cp(i),0} + \varepsilon_{p(i),init} - \varepsilon_{p.rel(i),k}\right) \tag{9.34}$$

$$\text{Unbonded:} \quad \sigma_{p(i),k} = E_{p(i)}\left(\varepsilon_{p(i),init} - \varepsilon_{p.rel(i),k}\right) \tag{9.35}$$

In this case, the contribution of the i-th concrete component to the internal axial force and moment can be determined as (similar to Equation 5.78):

$$N_{c(i),k} = \int_{A_{c(i)}} \sigma_{c(i),k} \, dA = \int_{A_{c(i)}} \left[\bar{E}_{c,\text{eff}(i),k} \left(\varepsilon_{r,k} - y\kappa_k - \varepsilon_{cs(i),k} \right) + \bar{F}_{e(i),0} \sigma_{c(i),0} \right] dA$$

$$= A_{c(i)} \bar{E}_{c,\text{eff}(i),k} \varepsilon_{r,k} - B_{c(i)} \bar{E}_{c,\text{eff}(i),k} \kappa_k - A_{c(i)} \bar{E}_{c,\text{eff}(i),k} \varepsilon_{cs(i),k} + \bar{F}_{e(i),0} N_{c(i),0}$$

$$(9.36)$$

$$M_{c(i),k} = \int_{A_{c(i)}} -y\sigma_{c(i),k} \, dA = \int_{A_{c(i)}} -y \left[\bar{E}_{c,\text{eff}(i),k} \left(\varepsilon_{r,k} - y\kappa_k - \varepsilon_{cs(i),k} \right) + \bar{F}_{e(i),0} \sigma_{c(i),0} \right] dA$$

$$= -B_{c(i)} \bar{E}_{c,\text{eff}(i),k} \varepsilon_{r,k} + I_{c(i)} \bar{E}_{c,\text{eff}(i),k} \kappa_k - B_{c(i)} \bar{E}_{c,\text{eff}(i),k} \varepsilon_{cs(i),k} + \bar{F}_{e(i),0} M_{c(i),0}$$

$$(9.37)$$

The equilibrium equations are (Equation 5.93):

$$r_{\text{ext},k} = D_k \varepsilon_k + f_{cr,k} - f_{cs,k} + f_{p,\text{init}} - f_{p,\text{rel},k} - f_{cp,0} \qquad (9.38)$$

where:

$$r_{\text{ext},k} = \begin{bmatrix} N_{\text{ext},k} \\ M_{\text{ext},k} \end{bmatrix} \qquad (9.39)$$

$$D_k = \begin{bmatrix} R_{A,k} & -R_{B,k} \\ -R_{B,k} & R_{I,k} \end{bmatrix} \qquad (9.40)$$

$$\varepsilon_k = \begin{bmatrix} \varepsilon_{r,k} \\ \kappa_k \end{bmatrix} \qquad (9.41)$$

$$f_{cr,k} = \sum_{i=1}^{m_c} \bar{F}_{e(i),0} \begin{bmatrix} N_{c(i),0} \\ M_{c(i),0} \end{bmatrix} = \sum_{i=1}^{m_c} \bar{F}_{e(i),0} E_{c(i),0} \begin{bmatrix} A_{c(i)}\varepsilon_{r,0} - B_{c(i)}\kappa_0 \\ -B_{c(i)}\varepsilon_{r,0} + I_{c(i)}\kappa_0 \end{bmatrix} \qquad (9.42)$$

$$f_{cs,k} = \sum_{i=1}^{m_c} \begin{bmatrix} A_{c(i)} \\ -B_{c(i)} \end{bmatrix} \bar{E}_{c,\text{eff}(i),k} \varepsilon_{cs(i),k} \qquad (9.43)$$

$$f_{p,init} = \sum_{i=1}^{m_p} \begin{bmatrix} P_{init(i)} \\ -y_{p(i)}P_{init(i)} \end{bmatrix} \tag{9.44}$$

$$f_{p.rel,k} = \sum_{i=1}^{m_p} \begin{bmatrix} P_{init(i)}\varphi_{p(i)} \\ -y_{p(i)}P_{init(i)}\varphi_{p(i)} \end{bmatrix} \tag{9.45}$$

and $f_{cp,0}$ is given by either Equation 9.46 or 9.47:

$$f_{cp,0} = \sum_{i=1}^{m_p} \begin{bmatrix} A_{p(i)}E_{p(i)}\varepsilon_{cp(i),0} \\ -y_{p(i)}A_{p(i)}E_{p(i)}\varepsilon_{cp(i),0} \end{bmatrix} \tag{9.46}$$

$$f_{cp,0} = \begin{bmatrix} 0 \\ 0 \end{bmatrix} \tag{9.47}$$

As discussed in Section 5.7.3, for post-tensioned tendons that were unbonded at t_0 and then bonded (grouted) soon after, Equation 9.46 applies, while for pretensioned tendons or post-tensioned tendons that were bonded at the time of the short-term analysis (t_0), Equation 9.47 applies. For a post-tensioned member with all tendons unbonded throughout the time period t_0 to t_k, Equation 9.47 also applies.

Solving Equation 9.38 gives the strain at time t_k:

$$\varepsilon_k = D_k^{-1}\left(r_{ext,k} - f_{cr,k} + f_{cs,k} - f_{p,init} + f_{p.rel,k} + f_{cp,0}\right)$$
$$= F_k\left(r_{ext,k} - f_{cr,k} + f_{cs,k} - f_{p,init} + f_{p.rel,k} + f_{cp,0}\right) \tag{9.48}$$

where:

$$F_k = \frac{1}{R_{A,k}R_{I,k} - R_{B,k}^2}\begin{bmatrix} R_{I,k} & R_{B,k} \\ R_{B,k} & R_{A,k} \end{bmatrix} \tag{9.49}$$

and the cross-sectional rigidities at t_k are:

$$R_{A,k} = \sum_{i=1}^{m_c} A_{c(i)}\bar{E}_{c,eff(i),k} + \sum_{i=1}^{m_s} A_{s(i)}E_{s(i)} + \sum_{i=1}^{m_p} A_{p(i)}E_{p(i)}$$
$$= \sum_{i=1}^{m_c} A_{c(i)}\bar{E}_{c,eff(i),k} + R_{A,s} + R_{A,p} \tag{9.50}$$

$$R_{B,k} = \sum_{i=1}^{m_c} B_{c(i)}\bar{E}_{c,\text{eff}(i),k} + \sum_{i=1}^{m_s} y_{s(i)}A_{s(i)}E_{s(i)} + \sum_{i=1}^{m_p} y_{p(i)}A_{p(i)}E_{p(i)}$$

$$= \sum_{i=1}^{m_c} B_{c(i)}\bar{E}_{c,\text{eff}(i),k} + R_{B,s} + R_{B,p} \tag{9.51}$$

$$R_{I,k} = \sum_{i=1}^{m_c} I_{c(i)}\bar{E}_{c,\text{eff}(i),k} + \sum_{i=1}^{m_s} y_{s(i)}^2 A_{s(i)}E_{s(i)} + \sum_{i=1}^{m_p} y_{p(i)}^2 A_{p(i)}E_{p(i)}$$

$$= \sum_{i=1}^{m_c} I_{c(i)}\bar{E}_{c,\text{eff}(i),k} + R_{I,s} + R_{I,p} \tag{9.52}$$

The stress distribution at time t_k in each concrete element is given by (Equation 5.103):

$$\sigma_{c(i),k} = \bar{E}_{c,\text{eff}(i),k}\left(\varepsilon_k - \varepsilon_{cs(i),k}\right) + \bar{F}_{e(i),0}\sigma_{c(i),0}$$

$$= \bar{E}_{c,\text{eff}(i),k}\left\{\begin{bmatrix}1 & -y\end{bmatrix}\varepsilon_k - \varepsilon_{cs(i),k}\right\} + \bar{F}_{e(i),0}\sigma_{c(i),0} \tag{9.53}$$

where $\varepsilon_k = \varepsilon_{r,k} - y\kappa_k = \begin{bmatrix}1 & -y\end{bmatrix}\varepsilon_k$. The stress in the non-prestressed reinforcement at time t_k is given by (Equation 5.104):

$$\sigma_{s(i),k} = E_{s(i)}\varepsilon_{s(i),k} = E_{s(i)}\begin{bmatrix}1 & -y_{s(i)}\end{bmatrix}\varepsilon_k \tag{9.54}$$

and the stress in any pretensioned tendons, or any post-tensioned tendons that were bonded to the concrete at the time of the short-term analysis (t_0), is:

$$\sigma_{p(i),k} = E_{p(i)}\left(\varepsilon_{p(i),k} + \varepsilon_{p(i),\text{init}} - \varepsilon_{p.\text{rel}(i),k}\right)$$

$$= E_{p(i)}\begin{bmatrix}1 - y_{p(i)}\end{bmatrix}\varepsilon_k + E_{p(i)}\varepsilon_{p(i),\text{init}} - E_{p(i)}\varepsilon_{p.\text{rel}(i),k} \tag{9.55}$$

For bonded post-tensioned tendons (initially unbonded at t_0), the stress at time t_k is:

$$\sigma_{p(i),k} = E_{p(i)}\left(\varepsilon_{cp(i),k} - \varepsilon_{cp(i),0} + \varepsilon_{p(i),\text{init}} - \varepsilon_{p.\text{rel}(i),k}\right)$$

$$= E_{p(i)}\begin{bmatrix}1 - y_{p(i)}\end{bmatrix}\varepsilon_k - E_{p(i)}\begin{bmatrix}1 - y_{p(i)}\end{bmatrix}\varepsilon_0 + E_{p(i)}\varepsilon_{p(i),\text{init}} - E_{p(i)}\varepsilon_{p.\text{rel}(i),k} \tag{9.56}$$

while for unbonded tendons:

$$\sigma_{p(i),k} = E_{p(i)}\left(\varepsilon_{p(i),\text{init}} - \varepsilon_{p.\text{rel}(i),k}\right) \tag{9.57}$$

EXAMPLE 9.1

The cross-section of a composite footbridge consists of a precast, pre-tensioned trough girder and a cast in-situ slab, as shown in Figure 9.4. The precast section is cast and moist cured for 4 days prior to transfer. The cross-section is subjected to the following load history.

At t = 4 days: The total prestressing force of 2000 kN is transferred to the trough girder. The centroid of all the pretensioned strands is located 100 mm above the bottom fibre, as shown. The moment on the section caused by the self-weight of the girder $M_4 = 320$ kNm is introduced at transfer. Shrinkage of the concrete also begins to develop at this time.

At t = 40 days: The in-situ slab deck is cast and cured, and the moment caused by the weight of the deck is applied to the precast section, $M_{40} = 300$ kNm.

At t = 60 days: A wearing surface is placed, and all other superimposed dead loads are applied to the bridge, thereby introducing an additional moment $M_{60} = 150$ kNm.

At t > 60 days: The moment remains constant from 60 days to time infinity.

Composite action gradually begins to develop as soon as the concrete in the deck sets. Full composite action may not be achieved for several days. However, it is assumed here that the in-situ deck and the precast section act

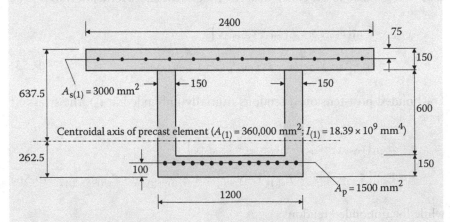

Figure 9.4 Details of composite cross-section of Example 9.1.

compositely at all times after $t = 40$ days. The stress and strain distributions on the composite cross-section are to be calculated:

(i) immediately after the application of the prestress at $t = 4$ days;
(ii) just before the slab deck is cast at $t = 40$ days;
(iii) immediately after the slab deck is cast at $t = 40$ days;
(iv) just before the road surface is placed at $t = 60$ days;
(v) immediately after the road surface is placed at $t = 60$ days; and
(vi) at time infinity after all creep and shrinkage strains have developed.

For the precast section (element 1): $f_{ck} = 40$ MPa

$$E_{cm(1),4} = E_{cm(1)}(4) = 25{,}000 \text{ MPa}; \quad E_{cm(1),40} = 31{,}500 \text{ MPa};$$

$$E_{cm(1),60} = 33{,}000 \text{ MPa}$$

$$\varepsilon_{cs(1),40} = \varepsilon_{cs(1)}(40) = -150 \times 10^{-6}; \quad \varepsilon_{cs(1),60} = -200 \times 10^{-6};$$

$$\varepsilon_{cs(1),\infty} = -500 \times 10^{-6}$$

$$\varphi_{(1)}(40, 4) = 0.9; \quad \varphi_{(1)}(60, 4) = 1.2; \quad \varphi_{(1)}(\infty, 4) = 2.4$$

$$\chi_{(1)}(40, 4) = 0.8; \quad \chi_{(1)}(60, 4) = 0.7; \quad \chi_{(1)}(\infty, 4) = 0.65$$

$$\varphi_{(1)}(60, 40) = 0.5; \quad \varphi_{(1)}(\infty, 40) = 1.6; \quad \varphi_{(1)}(\infty, 60) = 1.2$$

$$\chi_{(1)}(60, 40) = 0.8; \quad \chi_{(1)}(\infty, 40) = 0.65; \quad \chi_{(1)}(\infty, 60) = 0.65$$

For the in-situ slab (element 2): $f_{ck} = 25$ MPa

$$E_{cm(2),40} = 18{,}000 \text{ MPa}; \quad E_{cm(2),60} = 25{,}000 \text{ MPa}$$

$$\varepsilon_{cs(2),60} = -120 \times 10^{-6}; \quad \varepsilon_{cs(2),\infty} = -600 \times 10^{-6}$$

$$\varphi_{(2)}(60, 40) = 0.8; \quad \varphi_{(2)}(\infty, 40) = 3.0; \quad \varphi_{(2)}(\infty, 60) = 2.0$$

$$\chi_{(2)}(60, 40) = 0.8; \quad \chi_{(2)}(\infty, 40) = 0.65; \quad \chi_{(2)}(\infty, 60) = 0.65$$

To account for relaxation in the prestressing tendons, we take the creep coefficient to be:

$$\varphi_p(40) = 0.01; \quad \varphi_p(60) = 0.015; \quad \varphi_p(\infty) = 0.025$$

and the elastic moduli for the reinforcement and tendons are $E_s = E_p = 200{,}000$ MPa.

(i) At $t = 4$ days

The reference x-axis is here taken as the centroidal axis of the precast cross-section. The properties of the concrete part of the cross-section (with respect to the x-axis) are:

$$A_{c(1)} = A_{(1)} - A_p = 360,000 - 1,500 = 358,500 \text{ mm}^2$$

$$B_{c(1)} = A_{(1)}y_c - A_py_p = 360,000 \times 0 - 1,500 \times (-162.5) = 243,750 \text{ mm}^3$$

$$I_{c(1)} = I_{(1)} - A_py_p^2 = 18.39 \times 10^9 - 1500 \times (-162.5)^2 = 18.35 \times 10^9 \text{ mm}^4$$

For this member with bonded prestressing tendons, Equations 9.22 through 9.24 give the rigidities of the cross-section at first loading (age 4 days):

$$R_{A,4} = A_{c(1)}E_{c(1),4} + A_pE_p$$

$$= 358,500 \times 25,000 + 1,500 \times 200,000 = 9,263 \times 10^6 \text{ N}$$

$$R_{B,4} = B_{c(1)}E_{c(1),4} + y_pA_pE_p = 243,750 \times 25,000 + 1,500 \times (-162.5) \times 200,000$$

$$= -42,660 \times 10^6 \text{ Nmm}$$

$$R_{I,4} = I_{c(1)}E_{c(1),4} + y_p^2A_pE_p$$

$$= 18.35 \times 10^9 \times 25,000 + (-162.5)^2 \times 1,500 \times 200,000$$

$$= 466.8 \times 10^{12} \text{ Nmm}^2$$

and from Equation 9.21:

$$\mathbf{F}_4 = \frac{1}{R_{A,4}R_{I,4} - R_{B4}^2}\begin{bmatrix} R_{I,4} & R_{B,4} \\ R_{B,4} & R_{A,4} \end{bmatrix}$$

$$= \frac{1}{9,263 \times 10^6 \times 466.8 \times 10^{12} - (-42,660 \times 10^6)^2}\begin{bmatrix} 466.8 \times 10^{12} & -42,660 \times 10^6 \\ -42,660 \times 10^6 & 9,263 \times 10^6 \end{bmatrix}$$

$$= \begin{bmatrix} 108.0 \times 10^{-12} & -9.870 \times 10^{-15} \\ -9.870 \times 10^{-15} & 2.143 \times 10^{-15} \end{bmatrix}$$

The vector of internal actions at first loading is (Equation 9.16):

$$\mathbf{r}_{ext,4} = \begin{bmatrix} N_{ext,4} \\ M_{ext,4} \end{bmatrix} = \begin{bmatrix} 0 \\ M_4 \end{bmatrix} = \begin{bmatrix} 0 \\ 320 \times 10^6 \text{ Nmm} \end{bmatrix}$$

and the initial strain in the prestressing steel due to the initial prestressing force is obtained as follows (Equation 5.27):

$$\varepsilon_{p,init} = \frac{P_{init}}{A_p E_p} = \frac{2,000 \times 10^3}{1,500 \times 200,000} = 0.00667$$

The vector of internal actions caused by the initial prestress $\mathbf{f}_{p,init}$ is given by Equation 9.19:

$$\mathbf{f}_{p,init} = \begin{bmatrix} P_{init} \\ -y_p P_{init} \end{bmatrix} = \begin{bmatrix} 2000 \times 10^3 \\ -(-162.5) \times 2000 \times 10^3 \end{bmatrix} = \begin{bmatrix} 2000 \times 10^3 \text{ N} \\ 325 \times 10^6 \text{ Nmm} \end{bmatrix}$$

and the strain vector at first loading ε_4 containing the unknown strain variables is determined from Equation 9.20:

$$\varepsilon_4 = \mathbf{F}_4 \left(\mathbf{r}_{ext,4} - \mathbf{f}_{p,init} \right)$$

$$= \begin{bmatrix} 108.0 \times 10^{-12} & -9.870 \times 10^{-15} \\ -9.870 \times 10^{-15} & 2.143 \times 10^{-15} \end{bmatrix} \left(\begin{bmatrix} 0 \\ 320 \times 10^6 \end{bmatrix} - \begin{bmatrix} 2000 \times 10^3 \\ 325 \times 10^6 \end{bmatrix} \right)$$

$$= \begin{bmatrix} -216.0 \times 10^{-6} \\ 0.00902 \times 10^{-6} \end{bmatrix}$$

The strain at the reference axis and the curvature at first loading are therefore :

$$\varepsilon_{r(1),4} = -216.0 \times 10^{-6} \quad \text{and} \quad \kappa_{(1),4} = +0.00902 \times 10^{-6} \text{ mm}^{-1}$$

and the strains at the top fibre of the precast section (at $y = +487.5$ mm) and at the bottom fibre (at $y = -262.5$ mm) are:

$$\varepsilon_{(1),4(top)} = \varepsilon_{r,4} - 487.5 \times \kappa_4 = (-216.0 - 487.5 \times 0.00902) \times 10^{-6}$$

$$= -220.4 \times 10^{-6}$$

$$\varepsilon_{(1),4(btm)} = \varepsilon_{r,4} - (-262.5) \times \kappa_4 = (-216.0 + 262.5 \times 0.00902) \times 10^{-6}$$

$$= -213.6 \times 10^{-6}$$

The strain in the bonded prestressing steel is:

$$\varepsilon_{p,4} = \varepsilon_{p,init} + (\varepsilon_{r,4} - y_p \kappa_4) = 0.00667$$

$$+(-216.0 - (-162.5) \times 0.00902) \times 10^{-6} = 0.00645$$

The top and bottom fibre stresses in the concrete and the stress in the pre-stressing steel are obtained from Equations 9.9 and 9.11, respectively:

$$\sigma_{c(1),4(top)} = E_{c(1),4}\varepsilon_{(1),4(top)} = 25,000 \times (-220.4 \times 10^{-6}) = -5.51\,\text{MPa}$$

$$\sigma_{c(1),4(btm)} = E_{c(1),4}\varepsilon_{(1),4(btm)} = 25,000 \times (-213.6 \times 10^{-6}) = -5.34\,\text{MPa}$$

$$\sigma_{p,4} = E_p\varepsilon_{p,4} = 200,000 \times 0.00645 = +1,290\,\text{MPa}$$

The stress and strain distributions immediately after transfer at time $t = $ 4 days are shown in Figure 9.5b.

(ii) At t = 40 days: (prior to casting the in-situ slab)

The age-adjusted effective modulus at this time is (Equation 5.57):

$$\overline{E}_{c,eff(1),40} = \frac{E_{c,0}}{1 + \chi_{(1)}(40,4)\varphi_{(1)}(40,4)} = \frac{25,000}{1 + 0.8 \times 0.9} = 14,535\,\text{MPa}$$

and from Equation 5.60:

$$\overline{F}_{e(1),40} = \frac{\varphi_{(1)}(40,4)\left[\chi_{(1)}(40,4)-1\right]}{1 + \chi_{(1)}(40,4)\varphi_{(1)}(40,4)} = \frac{0.9 \times (0.8-1.0)}{1.0 + 0.8 \times 0.9} = -0.1047$$

With the properties of the concrete part of the section determined in part (i) as $A_c = 358,500$ mm^2, $B_c = 243,750$ mm^3 and $I_c = 18.35 \times 10^9$ mm^4, the cross-sectional rigidities are obtained from Equations 9.50 through 9.52:

$$R_{A,40} = A_{c(1)}\overline{E}_{c,eff(1),40} + A_pE_p$$

$$= 358,500 \times 14,535 + 1,500 \times 200,000 = 5,511 \times 10^6\,\text{N}$$

$$R_{B,40} = B_{c(1)}\overline{E}_{c,eff(1),40} + y_pA_pE_p$$

$$= 243,750 \times 14,535 + 1,500 \times (-162.5) \times 200,000$$

$$= -45,210 \times 10^6\,\text{Nmm}$$

$$R_{I,40} = I_{c(1)}\overline{E}_{c,eff(1),40} + y_p^2A_pE_p$$

$$= 18.35 \times 10^9 \times 14,535 + (-162.5)^2 \times 1,500 \times 200,000$$

$$= 274.7 \times 10^{12}\,\text{Nmm}^2$$

With $\varepsilon_{r,4} = -216.0 \times 10^{-6}$ and $\kappa_4 = +0.00902 \times 10^{-6}$ mm^{-1}, Equation 9.42 gives:

$$\mathbf{f}_{cr(1),40} = \bar{F}_{e(1),4}E_{c(1),4}\begin{bmatrix} A_c\varepsilon_{r,0} - B_c\kappa_0 \\ -B_c\varepsilon_{r,0} + I_c\kappa_0 \end{bmatrix}$$

$$= -0.1047 \times 25,000$$

$$\times \begin{bmatrix} 358,500 \times \left(-216.0 \times 10^{-6}\right) - 243,750 \times 0.00902 \times 10^{-6} \\ -243,750 \times \left(-216.0 \times 10^{-6}\right) + 18.35 \times 10^9 \times 0.00902 \times 10^{-6} \end{bmatrix}$$

$$= \begin{bmatrix} +202.6 \times 10^3 \text{ N} \\ -0.5711 \times 10^6 \text{ Nm} \end{bmatrix}$$

and from Equation 9.43:

$$\mathbf{f}_{cs(1),40} = \begin{bmatrix} A_c \\ -B_c \end{bmatrix}\bar{E}_{c,eff(1),40}\varepsilon_{cs(1),40} = \begin{bmatrix} 358,500 \times 14,535 \times \left(-150 \times 10^{-6}\right) \\ -243,750 \times 14,535 \times \left(-150 \times 10^{-6}\right) \end{bmatrix}$$

$$= \begin{bmatrix} -781.6 \times 10^3 \text{ N} \\ 0.5314 \times 10^6 \text{ Nmm} \end{bmatrix}$$

The vectors of initial prestressing actions and relaxation actions are given by Equations 9.44 and 9.45, respectively:

$$\mathbf{f}_{p,init} = \begin{bmatrix} 2000 \times 10^3 \text{ N} \\ 325 \times 10^6 \text{ Nmm} \end{bmatrix}$$

$$\mathbf{f}_{p,rel,40} = \begin{bmatrix} \sigma_{p4}A_p\varphi_p(40) \\ -y_p\sigma_{p4}A_p\varphi_p(40) \end{bmatrix} = \begin{bmatrix} 1290 \times 1500 \times 0.01 \\ -(-162.5) \times 1290 \times 1500 \times 0.01 \end{bmatrix}$$

$$= \begin{bmatrix} 19.4 \times 10^3 \text{ N} \\ 3.14 \times 10^6 \text{ Nmm} \end{bmatrix}$$

and Equation 9.49 gives:

$$\mathbf{F}_{40} = \frac{1}{R_{A,40}R_{I,40} - R_{B,40}^2}\begin{bmatrix} R_{I,40} & R_{B,40} \\ R_{B,40} & R_{A,40} \end{bmatrix}$$

$$= \frac{1}{5,511 \times 10^6 \times 274.7 \times 10^{12} - (-45,210 \times 10^6)^2}\begin{bmatrix} 274.7 \times 10^{12} & -45,210 \times 10^6 \\ -45,210 \times 10^6 & 5,511 \times 10^6 \end{bmatrix}$$

$$= \begin{bmatrix} 181.7 \times 10^{-12} \text{ N}^{-1} & -29.90 \times 10^{-15} \text{ N}^{-1}\text{mm}^{-1} \\ -29.90 \times 10^{-15} \text{ N}^{-1}\text{mm}^{-1} & 3.645 \times 10^{-15} \text{ N}^{-1}\text{mm}^{-2} \end{bmatrix}$$

The vector of internal actions at 40 days before the casting of the slab is the same as at age 4 days. That is:

$$\mathbf{r}_{ext,40} = \begin{bmatrix} N_{ext,40} \\ M_{ext,40} \end{bmatrix} = \begin{bmatrix} 0 \\ M_4 \end{bmatrix} = \begin{bmatrix} 0 \\ 320 \times 10^6 \text{ Nmm} \end{bmatrix}$$

and the strain ε_{40} at time $t_k = 40$ days is determined using Equation 9.48:

$$\varepsilon_{(1),40} = \mathbf{F}_{40} \left(\mathbf{r}_{ext,40} - \mathbf{f}_{cr(1),40} + \mathbf{f}_{cs(1),40} - \mathbf{f}_{p,init} + \mathbf{f}_{p.rel,40} + \mathbf{f}_{cp,0} \right)$$

$$= \begin{bmatrix} 181.7 \times 10^{-12} & -29.90 \times 10^{-15} \\ -29.90 \times 10^{-15} & 3.645 \times 10^{-15} \end{bmatrix}$$

$$\times \begin{bmatrix} (0 - 202.6 - 781.6 - 2000 + 19.40 + 0) \times 10^3 \\ (320 + 0.5711 + 0.5314 - 325 + 3.14 + 0) \times 10^6 \end{bmatrix}$$

$$= \begin{bmatrix} -538.7 \times 10^{-6} \\ +0.0859 \times 10^{-6} \end{bmatrix}$$

The strain at the reference axis and the curvature at time $t_k = 40$ days are therefore $\varepsilon_{r(1),40} = -538.7 \times 10^{-6}$ and $\kappa_{(1),40} = +0.0859 \times 10^{-6}$ mm^{-1}, respectively, and the strains at the top fibre of the precast section (at $y = +487.5$ mm) and at the bottom fibre (at $y = -262.5$ mm) are:

$$\varepsilon_{(1)40-(top)} = \varepsilon_{r(1),40} - 487.5 \times \kappa_{(1),40} = (-538.7 - (487.5 \times 0.0859)) \times 10^{-6}$$

$$= -580.6 \times 10^{-6}$$

$$\varepsilon_{(1)40-(btm)} = \varepsilon_{r,(1)40} - (-262.5) \times \kappa_{(1),40} = (-536.7 + 262.5 \times 0.0859) \times 10^{-6}$$

$$= -516.2 \times 10^{-6}$$

The concrete stress distribution at time $t_k = 40$ days is calculated using Equation 9.53:

$$\sigma_{c(1),40-(top)} = \bar{E}_{c,eff(1),40} \left(\varepsilon_{(1)40-(top)} - \varepsilon_{cs(1),40} \right) + \bar{F}_{e(1),40} \sigma_{c(1),4(top)}$$

$$= 14,535 \times [-580.6 - (-150)] \times 10^{-6} + (-0.1047) \times (-5.51)$$

$$= -5.68 \text{ MPa}$$

$$\sigma_{c(1),40-(btm)} = \bar{E}_{c,eff(1),40} \left(\varepsilon_{(1)40-(btm)} - \varepsilon_{cs(1),40} \right) + \bar{F}_{e(1),40} \sigma_{c(1),4(top)}$$

$$= 14,535 \times [-516.2 - (-150)] \times 10^{-6} + (-0.1047) \times (-5.34)$$

$$= -4.76 \text{ MPa}$$

The final stress in the prestressing steel at time $t_k = 40$ days is obtained from Equation 9.55:

$$\sigma_{p,40-} = E_p \left[(\varepsilon_{r,40} - y_p \kappa_{40}) + \varepsilon_{p,\text{init}} - \varepsilon_{p,\text{rel},40} \right]$$

$$= 200,000$$

$$\times \left[(-538.7 - (-162.5) \times 0.0859) + 0.00667 - 0.000065 \right] \times 10^{-6}$$

$$= 1,216 \text{ MPa}$$

where the relaxation strain $\varepsilon_{p,\text{rel},40}$ is calculated as $\varepsilon_{p,\text{rel},40} = \sigma_{p4} \varphi_p(40)/E_p$.

The stress and strain distributions at $t = 40$ days before the in-situ slab is cast are shown in Figure 9.5c. The increments of strain and stress that have developed in the precast girder during the sustained load period from 4 to 40 days are therefore:

$$\Delta\varepsilon_{r(1),(40-4)} = \varepsilon_{r(1),40} - \varepsilon_{r(1),4} = (-538.7 - (-216.0)) \times 10^{-6} = -322.7 \times 10^{-6}$$

$$\Delta\kappa_{(1),(40-4)} = \kappa_{(1),40} - \kappa_{(1),4} = (0.0859 - 0.00902) \times 10^{-6} \text{ mm}^{-1}$$

$$= 0.0769 \times 10^{-6} \text{ mm}^{-1}$$

$$\Delta\varepsilon_{(1),(40-4)(\text{top})} = \varepsilon_{(1),40(\text{top})} - \varepsilon_{(1),4(\text{top})} = (-580.6 - (-220.4)) \times 10^{-6}$$

$$= -360.2 \times 10^{-6}$$

$$\Delta\varepsilon_{(1),(40-4)(\text{btm})} = \varepsilon_{(1),40(\text{btm})} - \varepsilon_{(1),4(\text{btm})} = (-516.2 - (-213.6)) \times 10^{-6}$$

$$= -302.6 \times 10^{-6}$$

$$\Delta\varepsilon_{p,(40-4)} = \varepsilon_{p,40} - \varepsilon_{p,4} = 0.00608 - 0.00645 = -0.00037$$

$$\Delta\sigma_{c(1),(40-4)(\text{top})} = \sigma_{c(1),40(\text{top})} - \sigma_{c(1),4(\text{top})} = -5.68 - (-5.51) = -0.17 \text{ MPa}$$

$$\Delta\sigma_{c(1),(40-4)(\text{btm})} = \sigma_{c(1),40(\text{btm})} - \sigma_{c(1),4(\text{btm})} = -4.76 - (-5.34) = +0.58 \text{ MPa}$$

$$\Delta\sigma_{p,(40-4)} = \sigma_{p,40} - \sigma_{p,4} = 1216 - 1290 = -74 \text{ MPa}.$$

(iii) At $t = 40$ days (after casting the in-situ slab)

The increments of stress and strain caused by $M_{40} = 300$ kNm applied to the precast section at age 40 days are calculated using the same procedure as was outlined in part (i) of this example, except that the elastic modulus of the precast concrete has now increased.

The short-term rigidities of the cross-section at age 40 days are:

$$R_{A,40} = A_{c(1)} E_{c(1),40} + A_p E_p$$

$$= 358,500 \times 3,1500 + 1,500 \times 200,000 = 11,593 \times 10^6 \text{ N}$$

$$R_{B,40} = B_{c(1)}E_{c(1),4} + y_p A_p E_p = 243,750 \times 31,500 + 1,500 \times (-162.5) \times 200,000$$

$$= -41,072 \times 10^6 \text{ Nmm}$$

$$R_{I,40} = I_{c(1)}E_{c(1),40} + y_p^2 A_p E_p$$

$$= 18.35 \times 10^9 \times 31,500 + (-162.5)^2 \times 1,500 \times 200,000$$

$$= 586.1 \times 10^{12} \text{ Nmm}^2$$

and from Equation 9.21:

$$\mathbf{F}_{40} = \frac{1}{R_{A,40}R_{I,40} - R_{B,40}^2} \begin{bmatrix} R_{I,40} & R_{B,40} \\ R_{B,40} & R_{A,40} \end{bmatrix} = \begin{bmatrix} 86.28 \times 10^{-12} & -6.047 \times 10^{-15} \\ -6.047 \times 10^{-15} & 1.707 \times 10^{-15} \end{bmatrix}$$

For the load increment applied at 40 days, the vector of internal actions is:

$$\mathbf{r}_{ext,40} = \begin{bmatrix} 0 \\ M_{40} \end{bmatrix} = \begin{bmatrix} 0 \\ 300 \times 10^6 \text{ Nmm} \end{bmatrix}$$

and the vector of instantaneous strain caused by the application of M_{40} at age 40 days is:

$$\Delta \varepsilon_{40} = \mathbf{F}_{40}\mathbf{r}_{ext,40} = \begin{bmatrix} 86.28 \times 10^{-12} & -6.047 \times 10^{-15} \\ -6.047 \times 10^{-15} & 1.707 \times 10^{-15} \end{bmatrix} \begin{bmatrix} 0 \\ 300 \times 10^6 \end{bmatrix}$$

$$= \begin{bmatrix} -1.8 \times 10^{-6} \\ 0.5120 \times 10^{-6} \end{bmatrix}$$

The increment of instantaneous strain at the reference axis and the increment of curvature caused by the application of M_{40} at age 40 days are therefore:

$$\Delta \varepsilon_{r,40} = -1.8 \times 10^{-6} \quad \text{and} \quad \Delta \kappa_{40} = +0.512 \times 10^{-6} \text{ mm}^{-1}$$

and the increments of instantaneous strain at the top fibre of the precast section (at $y = +487.5$ mm) and at the bottom fibre (at $y = -262.5$ mm) are:

$$\Delta \varepsilon_{(1),40(top)} = \Delta \varepsilon_{r,40} - 487.5 \times \Delta \kappa_{40} = (-1.8 - 487.5 \times 0.512) \times 10^{-6}$$

$$= -251.4 \times 10^{-6}$$

$$\Delta\varepsilon_{(1),40(btm)} = \Delta\varepsilon_{r,40} - (-262.5) \times \Delta\kappa_{40} = (-1.8 + 262.5 \times 0.512) \times 10^{-6}$$

$$= +132.6 \times 10^{-6}$$

The increment of top and bottom fibre stresses in the concrete and the increment of stress in the prestressing steel are obtained from Equations 9.9 and 9.11, respectively:

$$\Delta\sigma_{c(1),40(top)} = E_{c(1),40}\Delta\varepsilon_{(1),40(top)} = 31,500 \times (-251.4 \times 10^{-6}) = -7.92\,\text{MPa}$$

$$\Delta\sigma_{c(1),40(btm)} = E_{c(1),40}\Delta\varepsilon_{(1),40(btm)} = 31,500 \times (+132.6 \times 10^{-6}) = +4.18\,\text{MPa}$$

$$\Delta\sigma_{p,40} = E_p\Delta\varepsilon_{p,40} = E_p(\Delta\varepsilon_{r,40} - y_p\Delta\kappa_{40})$$

$$= 200,000 \times \left[-1.8 - (-162.5) \times 0.512\right] \times 10^{-6}$$

$$= +16\,\text{MPa}$$

The extreme fibre concrete strains and stresses and the strain and stress in the tendons in the precast girder immediately after placing the in-situ slab at $t = 40$ days are obtained by summing the respective increments calculated in parts (i), (ii) and (iii):

$$\varepsilon_{(1),40+(top)} = \varepsilon_{(1),4,(top)} + \Delta\varepsilon_{(1),(40-4)(top)} + \Delta\varepsilon_{(1),40(top)}$$

$$= (-220.4 - 360.2 - 251.4) \times 10^{-6} = -832.0 \times 10^{-6}$$

$$\varepsilon_{(1),40+(btm)} = \varepsilon_{(1),4,(btm)} + \Delta\varepsilon_{(1),(40-4)(btm)} + \Delta\varepsilon_{(1),40(btm)}$$

$$= (-213.6 - 302.6 + 132.6) \times 10^{-6} = -383.6 \times 10^{-6}$$

$$\sigma_{c(1),40+(top)} = \sigma_{c(1),4(top)} + \Delta\sigma_{c(1),(40-4)(top)} + \Delta\sigma_{c(1),40(top)}$$

$$= -5.51 - 0.17 - 7.92 = -13.60\,\text{MPa}$$

$$\sigma_{c(1),40+(btm)} = \sigma_{c(1),4(btm)} + \Delta\sigma_{c(1),(40-4)(btm)} + \Delta\sigma_{c(1),40(btm)}$$

$$= -5.34 + 0.58 + 4.18 = -0.58\,\text{MPa}$$

$$\sigma_{p,40+} = \sigma_{p,4} + \Delta\sigma_{p,(40-4)} + \Delta\sigma_{p,40} = 1290 - 74 + 16 = 1232\,\text{MPa}$$

The stress and strain distributions at age 40 days immediately after the in-situ slab is cast are shown in Figure 9.5d. Stress levels in the precast girder are satisfactory at all stages prior to and immediately after placing the in-situ slab.

Cracking will not occur and compressive stress in the top fibre is not excessive. However, with a sustained compressive stress of -13.60 MPa in the top fibre, a relatively large subsequent creep differential will exist between the precast and the in-situ elements.

(iv) At t = 60 days (prior to placement of the wearing surface)

The change of stress and strain during the time interval from $t = 40$ to 60 days is to be calculated here. During this period, the precast section and the in-situ slab are assumed to act compositely. The concrete stress increments in the precast section, calculated in parts (i), (ii) and (iii) earlier, are applied at different times and are therefore associated with different creep coefficients.

For the stresses applied at $t = 4$ days in part (i), the creep coefficient for this time interval is $\Delta\varphi_{(1)}(60-40,4) = \varphi_{(1)}(60,4) - \varphi_{(1)}(40,4) = 0.30$, and from Equations 5.57 and 5.60:

$$\bar{E}_{c,eff(1),60} = \frac{E_{c(1),4}}{1 + \chi_{(1)}(60,40)\Delta\varphi_{(1)}(60-40,4)} = \frac{25,000}{1 + 0.8 \times 0.30} = 20,161\,\text{MPa}$$

$$\bar{F}_{e(1),60} = \frac{\Delta\varphi_{(1)}(60-40,4)\left[\chi_{(1)}(60,40)-1\right]}{1 + \chi_{(1)}(60,40)\Delta\varphi_{(1)}(60-40,4)} = \frac{0.30 \times (0.8-1.0)}{1.0 + 0.8 \times 0.30} = -0.0484$$

The stress increment calculated in part (ii), which is in fact gradually applied between $t = 4$ and 40 days, may be accounted for by assuming that it is suddenly applied at $t = 4$ days and using the reduced creep coefficient given by $\chi_{(1)}(40,4)[\varphi_{(1)}(60,4) - \varphi_{(1)}(40,4)] = 0.24$, and from Equations 5.57 and 5.60:

$$\bar{E}_{c,eff(1),60} = \frac{25,000}{1 + 0.8 \times 0.24} = 20,973\,\text{MPa} \quad \text{and}$$

$$\bar{F}_{e(1),60} = \frac{0.24 \times (0.8-1.0)}{1.0 + 0.8 \times 0.24} = -0.0403$$

For the stress increment calculated in part (iii) (and caused by M_{40}), the appropriate creep coefficient for the precast girder is $\varphi_{(1)}(60,40) = 0.5$, and from Equations 5.57 and 5.60:

$$\bar{E}_{c,eff(1),60} = \frac{31,500}{1 + 0.8 \times 0.5} = 22,500\,\text{MPa} \quad \text{and}$$

$$\bar{F}_{e(1),60} = \frac{0.5 \times (0.8-1.0)}{1.0 + 0.8 \times 0.5} = -0.0714$$

For the in-situ slab, the creep coefficient used in this time interval is $\varphi_{(2)}(60,40) = 0.8$, and from Equations 5.57 and 5.60:

$$\bar{E}_{c,eff(2),60} = \frac{18,000}{1 + 0.8 \times 0.8} = 10,976 \text{ MPa} \quad \text{and}$$

$$\bar{F}_{e(2),60} = \frac{0.8 \times (0.8 - 1.0)}{1.0 + 0.8 \times 0.8} = -0.0976$$

The shrinkage strains that develop in the precast section and the in-situ slab during this time interval are, respectively, $\varepsilon_{cs(1),60} - \varepsilon_{cs(1),40} = -50 \times 10^{-6}$ and $\varepsilon_{cs(2),60} = -120 \times 10^{-6}$.

The creep coefficient associated with this time interval for the prestressing steel is $\varphi_{p,60} - \varphi_{p,40} = 0.005$.

The section properties of the concrete part of the precast girder (element 1) and the in-situ slab (element 2) with respect to the centroidal axis of the precast girder are:

$$A_{c(1)} = 358,500 \text{ mm}^2; \quad B_{c(1)} = 243,750 \text{ mm}^3; \quad I_{c(1)} = 18.35 \times 10^9 \text{ mm}^4$$

$$A_{c(2)} = 357,000 \text{ mm}^2; \quad B_{c(2)} = 200.8 \times 10^6 \text{ mm}^3; \quad I_{c(2)} = 113.6 \times 10^9 \text{ mm}^4$$

To determine the internal actions required to restrain creep, shrinkage and relaxation, the initial elastic strain distribution caused by each of the previously calculated stress increments in each concrete element must be determined.

In the in-situ slab:

$$\varepsilon_{i,r(2),40} = 0 \quad \text{and} \quad \kappa_{i,(2),40} = 0 \quad \text{since the slab at } t = 40 \text{ days is unloaded.}$$

In the precast section:

For the stresses applied at 4 days, calculated in part (i): $\varepsilon_{r(1),4} = -216.0 \times 10^{-6}$; $\kappa_{(1),4} = +0.00902 \times 10^{-6} \text{ mm}^{-1}$. For the stress increment calculated in part (ii) and assumed to be applied at 4 days, the increments of instantaneous elastic strain at the top and bottom of the precast section are $\Delta\varepsilon_{i,(1),4(top)} = \Delta\sigma_{c(1),(40-4)(top)}/E_{c(1),4} = -0.17/25,000 = -6.92 \times 10^{-6}$ and $\Delta\varepsilon_{i,(1),4(btm)} = \Delta\sigma_{c(1),(40-4)(btm)}/E_{c(1),4} = +0.58/25,000 = 23.1 \times 10^{-6}$. Therefore, the increment of elastic curvature is $\Delta\kappa_{i,(1)} = [-(-6.92 \times 10^{-6}) + 23.1 \times 10^{-6}]/750 = +0.040 \times 10^{-6} \text{ mm}^{-1}$ and the instantaneous strain at the reference axis is $\Delta\varepsilon_{i,r,(1)} = +12.6 \times 10^{-6}$. For the stress increment applied to the precast element at 40 days [part (iii)], $\Delta\varepsilon_{r(1),40} = -1.8 \times 10^{-6}$ and $\Delta\kappa_{(1),40} = +0.512 \times 10^{-6} \text{ mm}^{-1}$.

The actions required to restrain creep due to these initial stresses are obtained from Equation 9.42:

$$f_{cr,60} = -0.0484 \times 25,000 \times \begin{bmatrix} 358,500 \times (-216.0 \times 10^{-6}) - 243,750 \times 0.00902 \times 10^{-6} \\ -243,750 \times (-216.0 \times 10^{-6}) + 18.35 \times 10^9 \times 0.00902 \times 10^{-6} \end{bmatrix}$$

$$- 0.0403 \times 25,000 \times \begin{bmatrix} 358,500 \times (+12.6 \times 10^{-6}) - 243,750 \times 0.040 \times 10^{-6} \\ -243,750 \times (+12.6 \times 10^{-6}) + 18.35 \times 10^9 \times 0.040 \times 10^{-6} \end{bmatrix}$$

$$- 0.0714 \times 31,500 \times \begin{bmatrix} 358,500 \times (-1.8 \times 10^{-6}) - 243,750 \times 0.512 \times 10^{-6} \\ -243,750 \times (-1.8 \times 10^{-6}) + 18.35 \times 10^9 \times 0.512 \times 10^{-6} \end{bmatrix}$$

$$= \begin{bmatrix} 90.88 \times 10^3 \ N \\ -22.14 \times 10^6 \ Nmm \end{bmatrix}$$

and the actions required to restrain shrinkage in each concrete element and relaxation of the tendons during this time period are obtained from Equations 9.43 and 9.45, respectively:

$$f_{cs,60} = \begin{bmatrix} 358,500 \times 22,500 \times (-50 \times 10^{-6}) + 357,000 \times 10,976 \times (-120 \times 10^{-6}) \\ -243,750 \times 22,500 \times (-50 \times 10^{-6}) - 200.8 \times 10^6 \times 10,976 \times (-120 \times 10^{-6}) \end{bmatrix}$$

$$= \begin{bmatrix} -873.5 \times 10^3 \ N \\ 264.5 \times 10^6 \ Nmm \end{bmatrix}$$

$$f_{p.rel,60} = \begin{bmatrix} 9.7 \times 10^3 \ N \\ 1.573 \times 10^6 \ Nmm \end{bmatrix}$$

The cross-sectional rigidities of the composite cross-section for the period 40 to 60 days are obtained from Equations 9.50 through 9.52:

$$R_{A,60} = 358,500 \times 22,500 + 357,000 \times 10,975 + 3,000 \times 200,000$$

$$+ 1,500 \times 200,000$$

$$= 12,885 \times 10^6 \ N$$

$$R_{B,60} = 243,750 \times 22,500 + 200.8 \times 10^6 \times 10,975 + 562.5 \times 3,000 \times 200,000$$

$$+ (-162.5) \times 1,500 \times 200,000$$

$$= 2.498 \times 10^{12} \ Nmm$$

$$R_{1,60} = 18.35 \times 10^9 \times 22{,}500 + 113.6 \times 10^9 \times 10{,}975 + 562.5^2 \times 3{,}000$$

$$\times 200{,}000 + (-162.5)^2 \times 1{,}500 \times 200{,}000$$

$$= 1{,}857.9 \times 10^{12} \text{ Nmm}^2$$

and from Equation 9.49:

$$\mathbf{F}_{60} = \frac{1}{R_{A,60}R_{1,60} - R_{B,60}^2}\begin{bmatrix} R_{1,60} & R_{B,60} \\ R_{B,60} & R_{A,60} \end{bmatrix}$$

$$= \begin{bmatrix} 105.0 \times 10^{-12} \text{ N}^{-1} & 141.2 \times 10^{-15} \text{ N}^{-1}\text{mm}^{-1} \\ 141.2 \times 10^{-15} \text{ N}^{-1}\text{mm}^{-1} & 0.728 \times 10^{-15} \text{ N}^{-1}\text{mm}^{-2} \end{bmatrix}$$

The vector of the change in strain that occurs between 40 and 60 days on the composite section is obtained from Equation 9.48:

$$\Delta\varepsilon_{(60-40)} = \mathbf{F}_{60}\left(-\mathbf{f}_{cr,60} + \mathbf{f}_{cs,60} + \mathbf{f}_{p.rel,60}\right)$$

$$= \begin{bmatrix} 105.0 \times 10^{-12} & 141.2 \times 10^{-15} \\ 141.2 \times 10^{-15} & 0.728 \times 10^{-15} \end{bmatrix}\begin{bmatrix} (-90.88 - 873.5 + 9.7) \times 10^3 \\ (+22.14 + 264.5 + 1.573) \times 10^6 \end{bmatrix}$$

$$= \begin{bmatrix} -59.5 \times 10^{-6} \\ 0.0753 \times 10^{-6} \end{bmatrix}$$

The increment of strain at the reference axis and the increment of curvature that occur between $t_k = 40$ and 60 days are therefore $\Delta\varepsilon_{r,(60-40)} = -59.5 \times 10^{-6}$ and $\Delta\kappa_{(60-40)} = +0.0753 \times 10^{-6} \text{ mm}^{-1}$, respectively, and the increments of strains at the top fibre of the precast section (at $y = +487.5$ mm) and at the bottom fibre (at $y = -262.5$ mm) are:

$$\Delta\varepsilon_{(1)60-(top)} = [-59.5 - (487.5 \times 0.0753)] \times 10^{-6} = -96.2 \times 10^{-6}$$

$$\Delta\varepsilon_{(1)60-(btm)} = [-59.5 + 262.5 \times 0.0753] \times 10^{-6} = -39.7 \times 10^{-6}$$

The increments of strain at the top fibre of the in-situ slab (at $y = +637.5$ mm) and at the bottom fibre (at $y = +487.5$ mm) are:

$$\Delta\varepsilon_{(2)60-(top)} = [-59.5 - (637.5 \times 0.0753)] \times 10^{-6} = -107.5 \times 10^{-6}$$

$$\Delta\varepsilon_{(2)60-(btm)} = [-59.5 - (487.5 \times 0.0753)] \times 10^{-6} = -96.2 \times 10^{-6}$$

The increments of concrete stress that develop between $t_k = 40$ and 60 days are calculated using Equation 9.53. In the precast element:

$$\Delta\sigma_{c(1),60-(top)} = \bar{E}_{c,eff(1),60}\left(\Delta\varepsilon_{(1),60-(top)} - \Delta\varepsilon_{cs(1),60-40}\right) + \bar{F}_{e(1),60}\sigma_{c(1),40+(top)}$$

$$= 22,500 \times [-96.2 - (-50)] \times 10^{-6} + (-0.0714) \times (-13.60)$$

$$= -0.07 \text{ MPa}$$

$$\Delta\sigma_{c(1),60-(btm)} = \bar{E}_{c,eff(1),60}\left(\Delta\varepsilon_{(1),60-(btm)} - \Delta\varepsilon_{cs(1),60-40}\right) + \bar{F}_{e(1),60}\sigma_{c(1),40+(btm)}$$

$$= 22,500 \times [-39.7 - (-50)] \times 10^{-6} + (-0.0714) \times (-0.58)$$

$$= +0.27 \text{ MPa}$$

and in the in-situ slab:

$$\Delta\sigma_{c(2),60-(top)} = \bar{E}_{c,eff(2),60}\left(\Delta\varepsilon_{(2),60-(top)} - \Delta\varepsilon_{cs(2),60}\right) + \bar{F}_{e(2),60}\sigma_{c(2),40+(top)}$$

$$= 10,976 \times [-107.5 - (-120)] \times 10^{-6} - 0.0976 \times 0 = +0.14 \text{ MPa}$$

$$\Delta\sigma_{c(2),60-(btm)} = \bar{E}_{c,eff(2),60}\left(\Delta\varepsilon_{(2),60-(btm)} - \Delta\varepsilon_{cs(2),60}\right) + \bar{F}_{e(2),60}\sigma_{c(2),40+(btm)}$$

$$= 10,976 \times [-96.2 - (-120)] \times 10^{-6} - 0.0976 \times 0 = +0.26 \text{ MPa}$$

The increment of stress in the bonded prestressing steel that develops between 40 and 60 days is:

$$\Delta\sigma_{p,60-} = E_p\left(\Delta\varepsilon_{r,(60-40)} - y_p\Delta\kappa_{(60-40)} - \Delta\varepsilon_{p,rel}\right) = -15.9 \text{ MPa}$$

and in the reinforcing steel in the in-situ slab:

$$\Delta\sigma_{s,60-} = E_s\left(\Delta\varepsilon_{r,(60-40)} - y_s\Delta\kappa_{(60-40)}\right) = -20.4 \text{ MPa}$$

The total stresses and strains at age 60 days before the application of the wearing surface are as follows.

In the precast girder:

$$\varepsilon_{(1),60-(top)} = \varepsilon_{(1),40+(top)} + \Delta\varepsilon_{(1)60-(top)} = -832.0 \times 10^{-6} - 96.2 \times 10^{-6}$$

$$= -928.2 \times 10^{-6}$$

$$\varepsilon_{(1),60-(btm)} = \varepsilon_{(1),40+(btm)} + \Delta\varepsilon_{(1)60-(btm)} = -423.3 \times 10^{-6}$$

$$\sigma_{c(1),60-(top)} = \sigma_{c(1),40+(top)} + \Delta\sigma_{c(1),60-(top)} = -13.60 - 0.07 = -13.67 \text{ MPa}$$

$$\sigma_{c(1),60-(btm)} = \sigma_{c(1),40+(btm)} + \Delta\sigma_{c(1),60-(btm)} = -0.31 \text{ MPa}$$

$$\sigma_{p,60-} = \sigma_{p,40+} + \Delta\sigma_{p,60-} = 1216 \text{ MPa}$$

In the in-situ slab:

$$\varepsilon_{(2),60-(top)} = \Delta\varepsilon_{(2)60-(top)} = -107.5 \times 10^{-6}$$

$$\varepsilon_{(2),60-(btm)} = \Delta\varepsilon_{(2)60-(btm)} = -96.2 \times 10^{-6}$$

$$\sigma_{c(2),60-(top)} = \Delta\sigma_{c(2),60-(top)} = +0.14 \text{ MPa}$$

$$\sigma_{c(2),60-(btm)} = \Delta\sigma_{c(2),60-(btm)} = +0.26 \text{ MPa}$$

$$\sigma_{s,60-} = \Delta\sigma_{s,60-} = -20.4 \text{ MPa}$$

The stress and strain distributions at age 60 days immediately before the wearing surface is placed are shown in Figure 9.5e.

There is a complex interaction taking place between the two concrete elements. The in-situ slab is shrinking at a faster rate than the precast element and, if this were the only effect, the in-situ slab would suffer a tensile restraining force and an equal and opposite compressive force would be imposed on the precast girder. Because of the high initial compressive stresses in the top fibres of the precast section, however, the precast concrete at the element interface is creeping more than the in-situ concrete and, as a result, the in-situ slab is being compressed by the creep deformations in the precast girder, resulting in a decrease in the tension caused by restrained shrinkage. In this example, the magnitude of the tensile force on the in-situ slab as a result of restrained shrinkage is a little larger than the magnitude of the compressive force due to the creep differential. The result is that relatively small tensile stresses develop in the in-situ slab with time and the magnitude of the compressive stress in the top fibre of the precast girder increases slightly.

(v) At t = 60 days (immediately after placement of the wearing surface)

The instantaneous increments of stress and strain caused by $M_{60} = 150$ kNm applied to the composite section at age 60 days are calculated here. The short-term rigidities of the composite cross-section at age 60 days (with respect to the reference axis at the centroid of the precast element) are:

$$R_{A,60} = A_{c(1)}E_{c(1),60} + A_{c(2)}E_{c(2),60} + A_pE_p + A_sE_s$$

$$= 358,500 \times 33,000 + 357,000 \times 25,000$$

$$+ 1,500 \times 200,000 + 3,000 \times 200,000$$

$$= 21,660 \times 10^6 \text{ N}$$

$$R_{B,60} = B_{c(1)}E_{c(1),60} + B_{c(2)}E_{c(2),60} + y_pA_pE_p + y_sA_sE_s = 5.317 \times 10^{12} \text{ Nmm}$$

$$R_{I,60} = I_{c(1)}E_{c(1),60} + I_{c(2)}E_{c(2),60} + y_p^2A_pE_p + y_s^2A_sE_s = 3644 \times 10^{12} \text{ Nmm}^2$$

and from Equation 9.21:

$$\mathbf{F}_{60} = \frac{1}{R_{A,60}R_{I,60} - R_{B,60}^2}\begin{bmatrix} R_{I,60} & R_{B,60} \\ R_{B,60} & R_{A,60} \end{bmatrix} = \begin{bmatrix} 71.96 \times 10^{-12} & 10.50 \times 10^{-12} \\ 10.50 \times 10^{-12} & 42.76 \times 10^{-15} \end{bmatrix}$$

For the load increment applied at 60 days, the vector of internal actions is:

$$\mathbf{r}_{ext,60} = \begin{bmatrix} 0 \\ M_{60} \end{bmatrix} = \begin{bmatrix} 0 \\ 150 \times 10^6 \text{ Nmm} \end{bmatrix}$$

and the vector of instantaneous strain caused by the application of M_{60} at age 60 days is:

$$\Delta\varepsilon_{60} = \mathbf{F}_{60}\mathbf{r}_{ext,60} = \begin{bmatrix} 71.96 \times 10^{-12} & 10.50 \times 10^{-12} \\ 10.50 \times 10^{-12} & 42.76 \times 10^{-15} \end{bmatrix}\begin{bmatrix} 0 \\ 150 \times 10^6 \end{bmatrix}$$

$$= \begin{bmatrix} +15.7 \times 10^{-6} \\ 0.0641 \times 10^{-6} \end{bmatrix}$$

The increment of instantaneous strain at the reference axis and the increment of curvature caused by the application of M_{60} at age 60 days are therefore:

$$\Delta\varepsilon_{r,60} = +15.7 \times 10^{-6} \quad \text{and} \quad \Delta\kappa_{60} = +0.0641 \times 10^{-6} \text{ mm}^{-1}$$

and the increment of strains at the top fibre of the precast section (at $y = +487.5$ mm) and at the bottom fibre (at $y = -262.5$ mm) are:

$$\Delta\varepsilon_{(1)60+(top)} = (+15.7 - (487.5 \times 0.0641)) \times 10^{-6} = -15.5 \times 10^{-6}$$

and

$$\Delta\varepsilon_{(1)60+(btm)} = (+15.7 + 262.5 \times 0.0641) \times 10^{-6} = +32.6 \times 10^{-6}$$

The increment of strains at the top fibre of the in-situ slab (at $y = +637.5$ mm) and at the bottom fibre (at $y = +487.5$ mm) are:

$$\Delta\varepsilon_{(2)60+(top)} = (+15.7 - (637.5 \times 0.0641)) \times 10^{-6} = -25.1 \times 10^{-6}$$

and

$$\Delta\varepsilon_{(2)60+(btm)} = (+15.7 - (487.5 \times 0.0641) \times 10^{-6} = -15.5 \times 10^{-6}$$

The increments of concrete stress that develop at $t_k = 60$ days due to the wearing surface are as follows.

In the precast girder:

$$\Delta\sigma_{c(1),60+(top)} = E_{c(1),60}\Delta\varepsilon_{(1)60+(top)} = -0.51 \text{ MPa}$$

$$\Delta\sigma_{c(1),60+(btm)} = E_{c(1),60}\Delta\varepsilon_{(1)60+(btm)} = +1.08 \text{ MPa}$$

In the in-situ slab:

$$\Delta\sigma_{c(2),60+(top)} = E_{c(2),60}\Delta\varepsilon_{(2)60+(top)} = -0.63 \text{ MPa}$$

$$\Delta\sigma_{c(2),60+(btm)} = E_{c(2),60}\Delta\varepsilon_{(2)60+(btm)} = -0.39 \text{ MPa}$$

The increment of stress in the bonded prestressing steel at $t_k = 60$ days due to the wearing surface is:

$$\Delta\sigma_{p,60+} = E_p(\Delta\varepsilon_{r,60} - y_p\Delta\kappa_{60}) = 5.2 \text{ MPa}$$

and in the reinforcing steel in the in-situ slab:

$$\Delta\sigma_{s,60+} = E_s(\Delta\varepsilon_{r,60} - y_s\Delta\kappa_{60}) = -4.1 \text{ MPa}$$

The total stresses and strains at age 60 days after the application of the wearing surface are as follows.

In the precast girder:

$$\varepsilon_{(1),60+(top)} = \varepsilon_{(1),60-(top)} + \Delta\varepsilon_{(1)60+(top)} = -943.7 \times 10^{-6}$$

$$\varepsilon_{(1),60+(btm)} = \varepsilon_{(1),60-(btm)} + \Delta\varepsilon_{(1)60+(btm)} = -390.7 \times 10^{-6}$$

$$\sigma_{c(1),60+(top)} = \sigma_{c(1),60-(top)} + \Delta\sigma_{c(1),60+(top)} = -14.18 \text{ MPa}$$

$$\sigma_{c(1),60+(btm)} = \sigma_{c(1),60-(btm)} + \Delta\sigma_{c(1),60+(btm)} = +0.77 \text{ MPa}$$

$$\sigma_{p,60+} = \sigma_{p,60-} + \Delta\sigma_{p,60+} = 1241 \text{ MPa}$$

In the in-situ slab:

$$\varepsilon_{(2),60+(top)} = \varepsilon_{(2),60-(top)} + \Delta\varepsilon_{(2)60+(top)} = -132.6 \times 10^{-6}$$

$$\varepsilon_{(2),60+(btm)} = \varepsilon_{(2),60-(btm)} + \Delta\varepsilon_{(2)60+(btm)} = -111.7 \times 10^{-6}$$

$$\sigma_{c(2),60-(top)} = \sigma_{c(2),60-(top)} + \Delta\sigma_{c(2),60-(top)} = -0.49 \text{ MPa}$$

$$\sigma_{c(2),60+(btm)} = \sigma_{c(2),60-(btm)} + \Delta\sigma_{c(2),60+(btm)} = -0.13 \text{ MPa}$$

$$\sigma_{s,60+} = \sigma_{s,60-} + \Delta\sigma_{s,60+} = -24.5 \text{ MPa}$$

and the stress and strain distributions at age 60 days immediately after the wearing surface has been placed are shown in Figure 9.5f.

(vi) At t = ∞

The change of stress and strain on the composite cross-section during the time interval from $t = 60$ days to $t = \infty$ is to be calculated here.

The relevant creep coefficients for each of the previously calculated stress increments determined in parts (i) to (v), together with the corresponding age-adjusted effective moduli and creep factors (from Equations 5.57 and 5.60) are as follows.

For the precast girder:

$$\text{Part}(i): \quad \varphi_{(1)}(\infty - 60, 4) = \varphi_{(1)}(\infty, 4) - \varphi_{(1)}(60, 4) = 1.2$$

$$E_{c(1),4} = 25,000 \text{ MPa}$$

$$\overline{E}_{c,eff(1),\infty} = \frac{E_{c(1),4}}{1 + \chi_{(1)}(\infty,60)\Delta\varphi_{(1)}(\infty - 60, 4)} = \frac{25,000}{1 + 0.65 \times 1.2}$$

$$= 14,045 \text{ MPa}$$

$$\overline{F}_{e(1),\infty} = \frac{1.2 \times (0.65 - 1.0)}{1.0 + 0.65 \times 1.2} = -0.236$$

Part (ii): $\chi_{(1)}(40,4)[\varphi_{(1)}(\infty,4) - \varphi_{(1)}(60,4)] = 0.96$

$E_{c(1),4} = 25,000 \text{ MPa} \quad \bar{E}_{c,\text{eff}(1),\infty} = 15,394 \text{ MPa}$

$\bar{F}_{e(1),\infty} = -0.2069$

Part (iii): $\varphi_{(1)}(\infty,40) - \varphi_{(1)}(60,40) = 1.1$

$E_{c(1),40} = 31,500 \text{ MPa} \quad \bar{E}_{c,\text{eff}(1),\infty} = 18,367 \text{ MPa}$

$\bar{F}_{e(1),\infty} = -0.2245$

Part (iv): $\chi_{(1)}(60,40)[\varphi_{(1)}(\infty,40) - \varphi_{(1)}(60,40)] = 0.88$

$E_{c(1),40} = 31,500 \text{ MPa} \quad \bar{E}_{c,\text{eff}(1),\infty} = 20,038 \text{ MPa}$

$\bar{F}_{e(1),\infty} = -0.1958$

Part (v): $\varphi_{(1)}(\infty,60) = 1.2$

$E_{c(1),60} = 33,000 \text{ MPa} \quad \bar{E}_{c,\text{eff}(1),\infty} = 18,539 \text{ MPa}$

$\bar{F}_{e(1),\infty} = -0.236$

For the in-situ slab:

Part (iv): $\chi_{(2)}(60,40)[\varphi_{(2)}(\infty,40) - \varphi_{(2)}(60,40)] = 1.76$

$E_{c(2),40} = 18,000 \text{ MPa} \quad \bar{E}_{c,\text{eff}(2),\infty} = 8,396 \text{ MPa}$

$\bar{F}_{e(2),\infty} = -0.2873$

Part (v): $\varphi_{(2)}(\infty,60) = 2.0$

$E_{c(2),60} = 25,000 \text{ MPa} \quad \bar{E}_{c,\text{eff}(2),\infty} = 10,870 \text{ MPa}$

$\bar{F}_{e(2),\infty} = -0.3043$

Note that the stress increments calculated in parts (ii) and (iv) are accounted for by assuming that they are suddenly applied at $t = 40$ and 60 days, respectively, using the appropriate reduced creep coefficients.

The shrinkage strains that develop during the time period after $t = 60$ days are:

$$\varepsilon_{cs(1),\infty} - \varepsilon_{cs(1),40} = -300 \times 10^{-6} \quad \text{and} \quad \varepsilon_{cs(2),\infty} - \varepsilon_{cs(2),40} = -480 \times 10^{-6}$$

and the creep coefficient associated with this time interval for the prestressing steel is $\varphi_{p,\infty} - \varphi_{p,60} = 0.01$.

To determine the internal actions required to restrain creep, shrinkage and relaxation, the initial elastic strain distribution caused by each of the previously calculated stress increments in each concrete element must be determined. The elastic strains due to the stress increments applied in parts (i) to (v) are:

$$\text{Part(i):} \quad \varepsilon_{i,r(1),4} = -216.0 \times 10^{-6}; \quad \kappa_{i(1),4} = +0.00902 \times 10^{-6} \text{ mm}^{-1}$$

$$\text{Part(ii):} \quad \Delta\varepsilon_{i,r,(1),4} = +12.6 \times 10^{-6}; \quad \Delta\kappa_{i(1)} = +0.0402 \times 10^{-6} \text{ mm}^{-1}$$

$$\text{Part(iii):} \quad \Delta\varepsilon_{i,r(1),40} = -1.8 \times 10^{-6} \quad \text{and} \quad \Delta\kappa_{i(1),40} = +0.512 \times 10^{-6} \text{ mm}^{-1}$$

$$\text{Part(iv):} \quad \Delta\varepsilon_{i,r(1),40} = +4.9 \times 10^{-6} \quad \text{and} \quad \Delta\kappa_{i(1),40} = +0.0145 \times 10^{-6} \text{ mm}^{-1}$$

$$\Delta\varepsilon_{i,r(2),40} = +36.9 \times 10^{-6} \quad \text{and}$$

$$\Delta\kappa_{i(2),40} = +0.0459 \times 10^{-6} \text{ mm}^{-1}$$

$$\text{Part(v):} \quad \Delta\varepsilon_{i,r(1),60} = +15.7 \times 10^{-6} \quad \text{and} \quad \Delta\kappa_{i(1),60} = +0.0641 \times 10^{-6} \text{ mm}^{-1}$$

$$\Delta\varepsilon_{i,r(2),60} = +15.7 \times 10^{-6} \quad \text{and}$$

$$\Delta\kappa_{i(2),60} = +0.0641 \times 10^{-6} \text{ mm}^{-1}$$

The actions required to restrain creep due to these initial stresses during the period after $t = 60$ days are obtained from Equation 9.42:

$$\mathbf{f}_{cr,\infty} = -0.2360 \times 25,000 \times \begin{bmatrix} 358,500 \times (-216.0 \times 10^{-6}) - 243,750 \times 0.00902 \times 10^{-6} \\ -243,750 \times (-216.0 \times 10^{-6}) + 18.35 \times 10^{9} \times 0.00902 \times 10^{-6} \end{bmatrix}$$

$$-0.2069 \times 25,000 \times \begin{bmatrix} 358,500 \times (+12.6 \times 10^{-6}) - 243,750 \times 0.0402 \times 10^{-6} \\ -243,750 \times (+12.6 \times 10^{-6}) + 18.35 \times 10^{9} \times 0.0402 \times 10^{-6} \end{bmatrix}$$

$$-0.2245 \times 31,500 \times \begin{bmatrix} 358,500 \times (-1.8 \times 10^{-6}) - 243,750 \times 0.512 \times 10^{-6} \\ -243,750 \times (-1.8 \times 10^{-6}) + 18.35 \times 10^{9} \times 0.512 \times 10^{-6} \end{bmatrix}$$

$$-0.1958 \times 31,500 \times \begin{bmatrix} 358,500 \times (+4.9 \times 10^{-6}) - 243,750 \times 0.0145 \times 10^{-6} \\ -243,750 \times (+4.9 \times 10^{-6}) + 18.35 \times 10^{9} \times 0.0145 \times 10^{-6} \end{bmatrix}$$

$$-0.2873 \times 18,000 \times \begin{bmatrix} 357,000 \times (+36.9 \times 10^{-6}) - 200.8 \times 10^{6} \times 0.0459 \times 10^{-6} \\ -200.8 \times 10^{6} \times (+36.9 \times 10^{-6}) + 113.6 \times 10^{9} \times 0.0459 \times 10^{-6} \end{bmatrix}$$

$$-0.2360 \times 33,000 \times \begin{bmatrix} 358,500 \times (+15.7 \times 10^{-6}) - 243,750 \times 0.0641 \times 10^{-6} \\ -243,750 \times (+15.7 \times 10^{-6}) + 18.35 \times 10^{9} \times 0.0641 \times 10^{-6} \end{bmatrix}$$

$$-0.3043 \times 25,000 \times \begin{bmatrix} 357,000 \times (+15.7 \times 10^{-6}) - 200.8 \times 10^{6} \times 0.0641 \times 10^{-6} \\ -200.8 \times 10^{6} \times (+15.7 \times 10^{-6}) + 113.6 \times 10^{9} \times 0.0641 \times 10^{-6} \end{bmatrix}$$

$$= \begin{bmatrix} 419.0 \times 10^{3} \text{ N} \\ -102.4 \times 10^{6} \text{ Nmm} \end{bmatrix}$$

The actions required to restrain shrinkage in each concrete element and relaxation of the tendons during this time period are obtained from Equations 9.43 and 9.45, respectively:

$$\mathbf{f}_{cs,\infty} = \begin{bmatrix} 358,500 \times 18,539 \times (-300 \times 10^{-6}) + 357,000 \times 10,870 \times (-480 \times 10^{-6}) \\ -243,750 \times 18,539 \times (-300 \times 10^{-6}) - 200.8 \times 10^{6} \times 10,870 \times (-480 \times 10^{-6}) \end{bmatrix}$$

$$= \begin{bmatrix} -3,856.5 \times 10^{3} \text{ N} \\ 1,049.1 \times 10^{6} \text{ Nmm} \end{bmatrix}$$

$$\mathbf{f}_{p.rel,\infty} = \begin{bmatrix} 19.4 \times 10^{3} \text{ N} \\ 3.15 \times 10^{6} \text{ Nmm} \end{bmatrix}$$

The cross-sectional rigidities of the composite cross-section for the period $t_k = 60$ days to time infinity are obtained from Equations 9.50 through 9.52:

$$R_{A,\infty} = 358,500 \times 18,539 + 357,000 \times 10,870 + 3,000 \times 200,000$$

$$+ 1,500 \times 200,000$$

$$= 11,427 \times 10^6 \text{ N}$$

$$R_{B,\infty} = 243,750 \times 18,539 + 200.8 \times 10^6 \times 10,870 + 562.5 \times 3,000 \times 200,000$$

$$+ (-162.5) \times 1,500 \times 200,000$$

$$= 2.476 \times 10^{12} \text{ Nmm}$$

$$R_{I,\infty} = 18.35 \times 10^9 \times 18,539 + 113.6 \times 10^9 \times 10,870 + 562.5^2 \times 3,000 \times 200,000$$

$$+ (-162.5)^2 \times 1,500 \times 200,000$$

$$= 1,773.2 \times 10^{12} \text{ Nmm}^2$$

and from Equation 9.49:

$$\mathbf{F}_\infty = \frac{1}{R_{A,\infty}R_{I,\infty} - R_{B,\infty}^2}\begin{bmatrix} R_{I,\infty} & R_{B,\infty} \\ R_{B,\infty} & R_{A,\infty} \end{bmatrix}$$

$$= \begin{bmatrix} 125.5 \times 10^{-12} & 175.2 \times 10^{-15} \\ 175.2 \times 10^{-15} & 0.8086 \times 10^{-15} \end{bmatrix}$$

The vector of the change in strain that occurs after 60 days on the composite section due to creep, shrinkage and relaxation is obtained from Equation 9.48:

$$\Delta\varepsilon_{(\infty-60)} = \mathbf{F}_\infty \left(-\mathbf{f}_{cr,\infty} + \mathbf{f}_{cs,\infty} + \mathbf{f}_{p.rel,\infty} \right)$$

$$= \begin{bmatrix} 125.5 \times 10^{-12} & 175.2 \times 10^{-15} \\ 175.2 \times 10^{-15} & 0.8086 \times 10^{-15} \end{bmatrix}\begin{bmatrix} (-419.0 - 3856.5 + 219.4) \times 10^3 \\ (+102.4 + 1049.1 + 3.15) \times 10^6 \end{bmatrix}$$

$$= \begin{bmatrix} -331.8 \times 10^{-6} \\ 0.1879 \times 10^{-6} \end{bmatrix}$$

The increment of strain at the reference axis and the increment of curvature that occur between $t_k = 60$ days and time infinity are $\Delta\varepsilon_{r,\infty-60} = -331.8 \times 10^{-6}$

and $\Delta\kappa_{\infty-60} = +0.1879 \times 10^{-6}$ mm^{-1}, respectively, and the increment of strains at the top fibre of the precast section (at $y = +487.5$ mm) and at the bottom fibre (at $y = -262.5$ mm) are:

$$\Delta\varepsilon_{(1),\infty(top)} = [-331.8 - (487.5 \times 0.1879)] \times 10^{-6} = -423.4 \times 10^{-6}$$

$$\Delta\varepsilon_{(1),\infty(btm)} = [-331.8 + 262.5 \times 0.1879] \times 10^{-6} = -282.4 \times 10^{-6}$$

The increments of strains at the top fibre of the in-situ slab (at $y = +637.5$ mm) and at the bottom fibre (at $y = +487.5$ mm) are:

$$\Delta\varepsilon_{(2),\infty(top)} = [-331.8 - (637.5 \times 0.1879)] \times 10^{-6} = -451.5 \times 10^{-6}$$

$$\Delta\varepsilon_{(2),\infty(btm)} = [-331.8 - (487.5 \times 0.1879)] \times 10^{-6} = -423.4 \times 10^{-6}$$

The increments of concrete stress that develop between $t_k = 60$ days and time infinity are calculated using Equation 9.53. In the precast element:

$$\Delta\sigma_{c(1),\infty(top)} = \bar{E}_{c,eff(1),\infty}\left(\Delta\varepsilon_{(1),\infty(top)} - \Delta\varepsilon_{cs(1),\infty-60}\right) + \bar{F}_{e(1),\infty}\sigma_{c(1),60+(top)}$$

$$= 18,539 \times [-423.4 - (-300)] \times 10^{-6} + (-0.236) \times (-14.18)$$

$$= +1.06\ \text{MPa}$$

$$\Delta\sigma_{c(1),\infty(btm)} = \bar{E}_{c,eff(1),\infty}\left(\Delta\varepsilon_{(1),\infty(btm)} - \Delta\varepsilon_{cs(1),\infty-60}\right) + \bar{F}_{e(1),\infty}\sigma_{c(1),60+(btm)}$$

$$= 18,539 \times [-282.4 - (-300)] \times 10^{-6} + (-0.236) \times (+0.77)$$

$$= +0.15\ \text{MPa}$$

and in the in-situ slab:

$$\Delta\sigma_{c(2),\infty(top)} = \bar{E}_{c,eff(2),\infty}\left(\Delta\varepsilon_{(2),\infty(top)} - \Delta\varepsilon_{cs(2),\infty-60}\right) + \bar{F}_{e(2),\infty}\sigma_{c(2),60+(top)}$$

$$= 10,870 \times [-451.5 - (-480)] \times 10^{-6} - 0.3043 \times (-0.49)$$

$$= +0.46\ \text{MPa}$$

$$\Delta\sigma_{c(2),\infty(btm)} = \bar{E}_{c,eff(2),\infty}\left(\Delta\varepsilon_{(2),\infty(btm)} - \Delta\varepsilon_{cs(2),\infty-60}\right) + \bar{F}_{e(2),\infty}\sigma_{c(2),60+(btm)}$$

$$= 10870 \times [-423.4 - (-480)] \times 10^{-6} - 0.3043 \times (-0.13)$$

$$= +0.65\ \text{MPa}$$

The increment of stress in the bonded prestressing steel that develops between 60 days and time infinity is:

$$\Delta\sigma_{p,\infty} = E_p(\Delta\varepsilon_{r,(\infty-60)} - y_p\Delta\kappa_{(\infty-60)} - \Delta\varepsilon_{p.rel,(\infty-60)}) = -73.2 \text{ MPa}$$

and in the reinforcing steel in the in-situ slab:

$$\Delta\sigma_{s,\infty} = E_s(\Delta\varepsilon_{r,(\infty-60)} - y_s\Delta\kappa_{(\infty-60)}) = -87.5 \text{ MPa}$$

The total stresses and strains at time infinity are summarised as follows.
  In the precast girder:

$$\varepsilon_{(1),\infty(top)} = \varepsilon_{(1),60+(top)} + \Delta\varepsilon_{(1),\infty(top)} = -1367.1 \times 10^{-6}$$

$$\varepsilon_{(1),\infty(btm)} = \varepsilon_{(1),60+(btm)} + \Delta\varepsilon_{(1),\infty(btm)} = -672.9 \times 10^{-6}$$

$$\sigma_{c(1),\infty(top)} = \sigma_{c(1),60+(top)} + \Delta\sigma_{c(1),\infty(top)} = -13.12 \text{ MPa}$$

$$\sigma_{c(1),\infty(btm)} = \sigma_{c(1),60+(btm)} + \Delta\sigma_{c(1),\infty(btm)} = +0.91 \text{ MPa}$$

$$\sigma_{p,\infty} = \sigma_{p,60+} + \Delta\sigma_{p,\infty} = 1147 \text{ MPa}$$

In the in-situ slab:

$$\varepsilon_{(2),\infty(top)} = \varepsilon_{(2),60+(top)} + \Delta\varepsilon_{(2),\infty(top)} = -584.2 \times 10^{-6}$$

$$\varepsilon_{(2),\infty(btm)} = \varepsilon_{(2),60+(btm)} + \Delta\varepsilon_{(2),\,\infty(btm)} = -535.1 \times 10^{-6}$$

$$\sigma_{c(2),\infty(top)} = \sigma_{c(2),60+(top)} + \Delta\sigma_{c(2),\infty(top)} = +0.03 \text{ MPa}$$

$$\sigma_{c(2),\infty(btm)} = \sigma_{c(2),60+(btm)} + \Delta\sigma_{c(2),\infty(btm)} = +0.53 \text{ MPa}$$

$$\sigma_{s,\infty} = \sigma_{s,60+} + \Delta\sigma_{s,\infty} = 111.9 \text{ MPa}$$

The stress and strain distributions at time infinity are shown in Figure 9.5g.
  Note that the compressive stresses at the top of the precast member at age 60 days after the wearing surface is placed are reduced with time, much of the compression finding its way into the non-prestressed reinforcement in the in-situ slab.

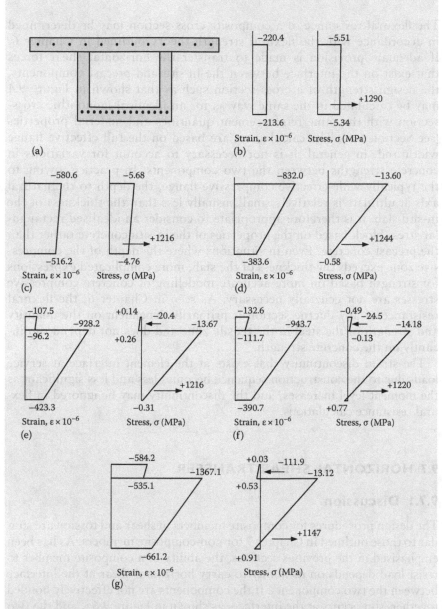

Figure 9.5 Stresses and strains on the composite cross-section of Example 9.1.
(a) Cross-section. (b) At $t = 4$ days. (c) At $t = 40$ days (before casting in-situ slab). (d) At $t = 40$ days (after casting in-situ slab). (e) At $t = 60$ days (before wearing surface). (f) At $t = 60$ days (after wearing surface). (g) At $t = \infty$.

9.6 FLEXURAL RESISTANCE

The flexural resistance of a composite cross-section may be determined in accordance with the flexural strength theory outlined in Chapter 6. If adequate provision is made to transfer the horizontal shear forces that exist on the interface between the in-situ and precast components, the design strength of a cross-section such as that shown in Figure 9.4 may be calculated in the same way as for an identical monolithic cross-section with the same reinforcement quantities and material properties (see Section 6.5). The calculations are based on the full effective flange width and, in general, it is not necessary to account for variations in concrete strengths between the two components. In practice, owing to the typically wide effective compressive flange, the depth to the natural axis at ultimate is relatively small, usually less than the thickness of the in-situ slab. It is therefore appropriate to consider an idealised rectangular stress block based on the properties of the in-situ concrete rather than the precast concrete. Even in situations where the depth of the compressive zone exceeds the thickness of the slab, more complicated expressions for strength based on more accurate modelling of concrete compressive stresses are not generally necessary. As seen in Chapter 6, the flexural resistance of any ductile section is primarily dependent on the quantity and strength of the steel in the tensile zone and does not depend significantly on the concrete strength.

The strain discontinuity that exists at the element interface at service loads due to the construction sequence becomes less and less significant as the moment level increases, and the discontinuity may be ignored in flexural resistance calculations.

9.7 HORIZONTAL SHEAR TRANSFER

9.7.1 Discussion

The design procedures for composite members in shear and torsion are similar to those outlined in Chapter 7 for non-composite members. As has been emphasised in the previous sections, the ability of a composite member to resist load depends on its ability to carry horizontal shear at the interface between the two components. If the components are not effectively bonded together, slip occurs at the interface, as shown in Figure 9.6a, and the two components act as separate beams, each carrying its share of the external loads by bending about its own centroidal axis. To ensure full composite action, slip at the interface must be prevented, and for this, there must be an effective means for transferring horizontal shear across the interface. If slip is prevented, full composite action is assured (as shown in Figure 9.6b) and the advantages of composite construction can be realised.

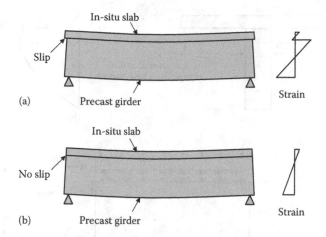

Figure 9.6 Composite and non-composite action. (a) Non-composite action. (b) Composite action.

In Section 9.2, various mechanisms for shear transfer were discussed. Natural adhesion and friction are usually sufficient to prevent slip in composite members with a wide interface between the components (such as the cross-sections shown in Figure 9.1b through d). The contact surface of the precast member is often roughened during manufacture to improve bond. Where the contact area is smaller (as on the cross-section of Figures 9.1a, e and 9.4), web reinforcement in the precast girder is often carried through the interface and anchored in the in-situ slab, thus providing increased frictional resistance (by clamping the contact surfaces together) and additional shear resistance through dowel action.

The theorem of complementary shear stress indicates that on the cross-section of an uncracked elastic composite member, the horizontal shear stress v_h at the interface between the two components is equal to the vertical shear stress at that point and is given by the well-known expression:

$$v_h = \frac{VS}{Ib_i} \tag{9.58}$$

where V is that part of the shear force caused by loads applied after the establishment of composite action; S is the first moment of the area of the in-situ element about the centroidal axis of the composite cross-section; I is the moment of inertia of the gross composite cross-section and b_i is the width of the contact surface (usually equal to the width of the top surface of the precast member)

The distribution of shear stress and the direction of the horizontal shear at the interface are shown in Figure 9.7.

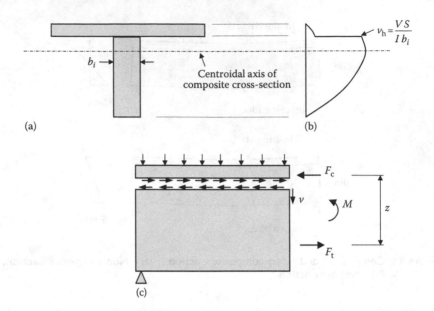

Figure 9.7 Shear stresses and actions in an elastic, uncracked, composite beam. (a) Composite cross-section. (b) Shear stresses on an uncracked, elastic section. (c) Shear on element interface.

9.7.2 Design provisions for horizontal shear

At overloads, concrete members crack and material behaviour becomes non-linear and inelastic. In design, a simple average or nominal shear stress is usually used for design strength calculations. According to EN 1992-1-1 [1], the design value of the shear stress at the interface in a composite section is given by:

$$v_{Edi} = \frac{\beta V_{Ed}}{z b_i} \tag{9.59}$$

where z is the lever arm between the compressive and tensile force resultants (as shown in Figure 9.7c) and V_{Ed} is the total transverse (vertical) shear force obtained using the appropriate factored load combination for the strength limit state (see Section 2.3.2). V_{Ed} is calculated from the total loads and not just the loads applied after the in-situ slab has hardened, because at ultimate loads, flexural cracking can cross the interface and horizontal shear resulting from all the applied load must be carried. The term β in Equation 9.59 is the ratio of the longitudinal force in the in-situ slab at the strength limit state and the total longitudinal force in either the compressive or tensile zone of the section under consideration. When the interface is located in the compression zone, β is the ratio of the compressive force in the in-situ slab, $F_{c.slab}$ (i.e. the compressive force between the extreme compressive fibre and the interface) and the total compressive force on the cross-section F_c, i.e. $\beta = F_{c.slab}/F_c$. When the interface is located in the tensile zone, β is the ratio of the sum of the tensile forces in the longitudinal reinforcement and tendons in the precast member $(F_{t.s} + F_{t.p})$ and the total tensile force on the cross-section F_t.

The design requirement in EN 1992-1-1 [1] is:

$$v_{Edi} \leq v_{Rdi} \tag{9.60}$$

The design shear resistance at the interface v_{Rdi} depends on the clamping effects produced by the shear reinforcement crossing the interface, dowel action, aggregate interlock and the effects of any transverse pressure across the interface and is given by:

$$v_{Rdi} = cf_{ctd} + \mu\sigma_n + \rho f_{yd}(\mu\sin\alpha + \cos\alpha) \leq 0.5 v f_{cd} \tag{9.61}$$

where:

c and μ are factors that depend on the roughness of the interface (classified in EN 1992-1-1 [1] as either very smooth, smooth, rough or indented). For *very smooth* interfaces (such as concrete cast against steel, plastic or specially prepared wooden moulds), $c = 0.25$ and $\mu = 0.5$. For *smooth* interfaces (such as slip formed or extruded surfaces or a free surface left without further treatment after vibration), $c = 0.35$ and $\mu = 0.6$. For *rough* interfaces (such as a concrete surface roughened by exposing the aggregate, raking or other methods giving an equivalent behaviour), $c = 0.45$ and $\mu = 0.7$. For an *indented* surface (complying with Figure 9.8), $c = 0.5$ and $\mu = 0.9$. The values for c specified earlier apply to static load conditions. Under dynamic or fatigue loading, the values of c should be halved.

f_{ctd} is the design tensile strength of the in-situ concrete (see Equation 4.12).

σ_n is the minimum normal stress on the interface caused by the minimum normal force that can act simultaneously with the shear force (positive for compression such that $\sigma_n < 0.6 f_{cd}$ and negative for tension). Note that when σ_n is tensile, the term $c f_{ctd}$ in Equation 9.61 should be taken as zero.

ρ is the reinforcement ratio A_s/A_i.

A_s is the area of reinforcement crossing the interface that is adequately anchored on each side of the interface.

A_i is the area of the interface.

α is the angle between the plane of the interface and the reinforcement crossing the interface, as shown in Figure 9.8.

v is the strength reduction factor equal to v_1 given in Equation 7.9.

If shear reinforcement (A_s in Equation 9.61) is required (i.e. the concrete and friction components in Equation 9.61 ($cf_{ctd} + \mu\sigma_n$) are insufficient on their own

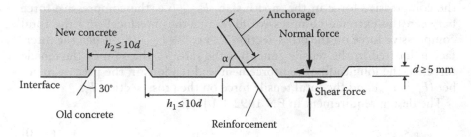

Figure 9.8 Indented interface between new and old concrete [1].

to satisfy Equation 9.60), the shear and torsional reinforcement that is already provided, and which crosses the shear plane, may be taken into account for this purpose, provided it is anchored so that it can develop its full strength at the interface. The centre-to-centre spacing of this shear reinforcement should not exceed about 3.5 times the thickness of the in-situ slab anchored by the shear reinforcement. In addition, the average thickness of the structural components on either side of the interface should not be less than about 50 mm.

EXAMPLE 9.2

The horizontal shear transfer requirements for a beam with cross-section shown in Figure 9.4 are to be determined. The beam is simply-supported over a span of 17.2 m and is subjected to the following loads:

Self-weight of precast trough girder:	8.64 kN/m
Self-weight of in-situ slab:	8.10 kN/m
Superimposed dead load:	4.05 kN/m
Transient live load:	9.60 kN/m

The behaviour of the cross-section at mid-span at service loads is calculated in Example 9.1. The effective prestressing force calculated in Example 9.1 is $P_{m,t} = A_p\sigma_{p,\infty} = 1720.5$ kN and is assumed here to be constant along the beam. Take $f_{pk} = 1860$ MPa.

The factored load combination for the strength limit state (Equation 2.1) is:

$$1.35 \times (8.64 + 8.10 + 4.05) + 1.5 \times 9.6 = 42.5 \text{ kN/m.}$$

The maximum shear force adjacent to each support is:

$$V_{Ed} = (42.5 \times 17.2)/2 = 365 \text{ kN.}$$

At the ultimate limit state in bending, using the idealised rectangular compressive stress block (see Section 6.3.2), we determine that the depth to the neutral axis at the ultimate limit state is 81.2 mm, the tensile force in the tendons is $F_{ptd} = A_p\sigma_{pud} = 2760$ kN, the compressive force in the in-situ concrete $F_{cd} = 2599$ kN and, with the non-prestressed reinforcement ($A_{s(1)}$) in the in-situ slab just above the neutral axis, the compressive force in $A_{s(1)}$ is $F_{sd(1)} = 161$ kN. The internal lever arm between the resultant compressive and tensile forces on the cross-section is $z = 765$ mm.

The neutral axis lies in the in-situ slab, just below the non-prestressed reinforcement, and so the interface between the in-situ slab and the precast girder is in the tensile zone below the neutral axis and $\beta = 1.0$.

From Equation 9.59, the design shear stress acting on the interface is:

$$v_{Edi} = \frac{1.0 \times 365 \times 10^3}{765 \times 300} = 1.59 \text{ MPa}$$

where the width of the interface $b_i = 2 \times 150 = 300$ mm.

The design shear strength at the interface v_{Rdi} is obtained from Equation 9.61. If the top surface of the precast trough has been deliberately roughened to facilitate shear transfer, EN 1992-1-1 [1] specifies $c = 0.45$ and $\mu = 0.7$. The permanent normal stress across the interface σ_n is caused by the self-weight of the in-situ slab and the superimposed dead load and is equal to $(8.10 + 4.05)/b_i = 0.0405$ MPa. Any required shear reinforcement will be at right angles to the interface, and so $\alpha = 90°$. With $f_{ck} = 25$ MPa and $f_{ctd} = 1.2$ MPa for the in-situ slab, and taking $f_{yd} = 435$ MPa, Equation 9.61 gives the design resistance of the interface in terms of the area of shear reinforcement per metre:

$$v_{Rdi} = 0.45 \times 1.2 + 0.7 \times 0.0405 + A_s \times 435 \times (0.7 \times 1.0 + 0) / (300 \times 1000)$$

$$= 0.57 + 1.015 \times 10^{-3} \times A_s$$

and from Equation 9.60:

$$A_s \geq \frac{1.59 - 0.57}{1.015 \times 10^{-3}} = 1005 \text{ mm}^2/\text{m}.$$

If 2–12 mm bars (226 mm²) cross the shear interface, one in each web, the required spacing s near each support is:

$$s \leq \frac{1000 \times 226}{1005} = 225 \text{ mm}$$

The spacing can be increased further into the span, as the shear force V_{Ed} decreases. It is important to ensure that these bars are fully anchored on each side of the shear plane.

If the contact surface were not deliberately roughened, but screeded and trowelled, $c = 0.25$ and $\mu = 0.5$, and Equation 9.61 gives:

$$v_{Rdi} = 0.25 \times 1.2 + 0.5 \times 0.0405 + A_s \times 435 \times (0.5 \times 1.0 + 0) / (300 \times 1000)$$

$$= 0.32 + 0.725 \times 10^{-3} \times A_s$$

From Equation 9.60:

$$A_s \geq \frac{1.59 - 0.32}{0.725 \times 10^{-3}} = 1752 \text{ mm}^2/\text{m}$$

and with 2–12 mm bars (226 mm²) cross the shear interface, one in each web, the required spacing s near each support is:

$$s \leq \frac{1000 \times 226}{1752} = 129 \text{ mm}$$

REFERENCES

1. EN 1992-1-1. 2004. Eurocode 2: Design of concrete structures – Part 1-1: General rules and rules for buildings. British Standards Institution, London, UK.
2. Gilbert, R.I. and Ranzi, G. 2011. *Time-Dependent Behaviour of Concrete Structures*. London, UK: Spon Press.

Chapter 10

Design procedures
for determinate beams

10.1 INTRODUCTION

The variables that must be established in the design of a prestressed concrete beam are material properties and specifications, the shape and size of the section, the amount and location of both the prestressed steel and the non-prestressed reinforcement, and the magnitude of the prestressing force. The designer is constrained by the various design requirements for the strength, serviceability, stability and durability limit states.

The optimal design is a particular combination of design variables that satisfy all the design constraints at a minimum cost. The cost of a particular design depends on local conditions at the time of construction, and variations in the costs of materials, formwork, construction expertise, labour, plant hire, transportation, etc., can change the optimal design from one site to another and also from one time to another.

It is difficult, therefore, to fix hard and fast rules to achieve the optimal design. It is difficult even to determine confidently when prestressed concrete becomes more economic than reinforced concrete or when prestressed concrete that is cracked at service loads is a better solution than uncracked prestressed concrete. However, it is possible to give some broad guidelines to achieve feasible design solutions for both fully and partially-prestressed members, i.e. for members that are either uncracked or cracked at service loads. In this chapter, such guidelines are presented and illustrated by examples.

10.2 TYPES OF SECTIONS

Many types of cross-sections are commonly used for prestressed girders. The choice depends on the nature of the applied loads, the function or usage of the member, the availability and cost of formwork, aesthetic considerations and ease of construction. Some commonly used cross-sections are shown in Figure 10.1.

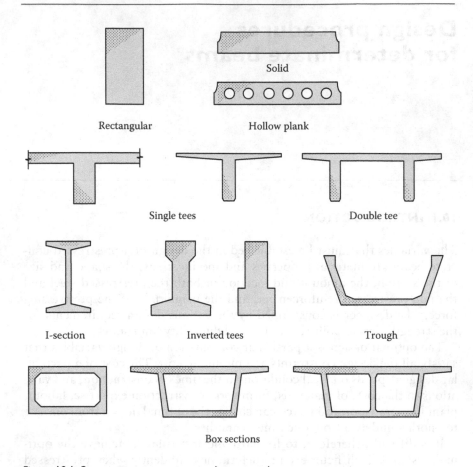

Figure 10.1 Some common prestressed concrete beam cross-sections.

Most in-situ prestressed concrete beam sections are rectangular (or tee-sections with solid slab flanges and rectangular webs). Rectangular sections are not particularly efficient in bending. The self-weight of a rectangular section is larger than for an I- or T-section of equivalent stiffness, and the prestress required to resist an external moment also tends to be larger. The formwork costs for a rectangular section, however, are generally lower and steel fixing is usually easier.

For precast prestressed concrete, where reusable formwork is available, the more efficient flanged sections are commonly used. T-sections and double T-sections are ideal for simply-supported members in situations where the self-weight of the beam is a significant part of the total load. If the moment at transfer due to self-weight (plus any other external load) is not significant, care must be taken to avoid excessive compressive stresses in the bottom fibres at transfer in T-shaped sections.

Inverted T-sections can accommodate large initial compressive forces in the lower fibres at transfer and, while being inefficient by themselves for resisting positive moment, they are usually used with a cast in-situ composite concrete deck. The resulting composite section is very efficient in positive bending.

For continuous members, where both positive and negative moments exist in different regions of the beam, I-sections and closed-box sections are efficient. Box-shaped sections are laterally stable and have found wide applications as medium- and long-span bridge girders. In addition, box sections can carry efficiently the torsional moments caused by eccentric traffic loading.

10.3 INITIAL TRIAL SECTION

10.3.1 Based on serviceability requirements

A reliable initial trial cross-section is required at the beginning of a design in order to estimate self-weight accurately and to avoid too many design iterations.

For a fully-prestressed member, i.e. a member in which tensile stress limits are set in order to eliminate cracking at service loads, Equations 5.7 and 5.8 provide estimates of the minimum section moduli required to satisfy the selected stress limits at the critical section both at transfer and under the full service loads. If the time-dependent loss of prestress is assumed (usually conservatively) to be 25%, Equation 5.7 simplifies to:

$$Z_{btm} \geq \frac{M_T - 0.75 M_o}{f_{ct,t} - 0.75 f_{cc,0}} \tag{10.1}$$

Recall that the compressive stress limit at transfer $f_{cc,0}$ in this expression is a negative number.

For a member containing a parabolic cable profile, a further guide to the selection of an initial trial section may be obtained by considering deflection requirements for the member. The mid-span deflection of an uncracked prestressed beam under a uniformly distributed unbalanced load w_{unbal} may be approximated by:

$$v = \beta \frac{w_{unbal} l^4}{E_{cm} I} + \lambda \beta \frac{w_{unbal.sus} l^4}{E_{cm} I} \tag{10.2}$$

where $w_{unbal.sus}$ is the sustained part of the unbalanced load, β is a deflection coefficient, l is the effective span of the beam, E_{cm} is the elastic modulus of concrete, I is the moment of inertia of the gross cross-section about its centroidal axis and λ is a long-term deflection multiplication factor, which should not be taken to be less than 3.0 for an uncracked prestressed member. The deflection coefficient β is equal to 5/384 for a uniformly loaded simply-supported member. For a continuous member, β depends on the support conditions, the relative lengths of the adjacent spans and the load pattern. When the variable part of the unbalanced load is not greater than the sustained part, the deflection coefficients for a continuous beam with equal adjacent spans may be taken as β = 2.75/384 for an end span and β = 1.8/384 for an interior span.

Equation 10.2 can be re-expressed as:

$$v = \beta \frac{w_{tot}l^4}{E_{cm}I} \tag{10.3}$$

where:

$$w_{tot} = w_{unbal} + \lambda w_{unbal.sus} \tag{10.4}$$

If v_{max} is the maximum permissible total deflection, then from Equation 10.3, the initial gross moment of inertia must satisfy the following:

$$I \geq \beta \frac{w_{tot}l^4}{E_{cm}v_{max}} \tag{10.5}$$

All the terms in Equation 10.5 are generally known at the start of a design, except for an estimate of λ (in Equation 10.4), which may be taken initially to equal 3 for an uncracked member. Since self-weight is usually part of the load being balanced by prestress, it does not form part of w_{tot}.

For a cracked partially-prestressed member, λ should be taken as not more than 2, for the reasons discussed in Section 5.11.4. After cracking, the effective moment of inertia I_{ef} depends on the quantity of tensile steel and the level of maximum moment. If I_{ef} is taken to be $0.5I$, which is usually conservative, an initial estimate of the gross moment of inertia of the partially-prestressed section can be obtained from:

$$I \geq 2\beta \frac{w_{tot}l^4}{E_{cm}v_{max}} \tag{10.6}$$

10.3.2 Based on strength requirements

An estimate of the section size for a prestressed member can be obtained from the flexural strength requirements of the critical section. The ultimate moment of a ductile rectangular section containing both non-prestressed and prestressed tensile steel may be found using Equation 6.21. By taking moments of the internal tensile forces in the steel about the level of the resultant compressive force in the concrete, the moment may be expressed as:

$$M_{Rd} = \sigma_{pud} A_p \left(d_p - \frac{\lambda x}{2} \right) + f_{yd} A_s \left(d_s - \frac{\lambda x}{2} \right) \tag{10.7}$$

For preliminary design purposes, this expression can be simplified if the design stress in the prestressing steel σ_{pud} is assumed (say $\sigma_{pud} = 0.9 f_{pk}/\gamma_s$) and the internal lever arm between the resultant tension and compression forces is estimated (say $0.9d$, where d is the effective depth to the resultant tensile force at the ultimate limit state). With these simplifications, Equation 10.7 becomes:

$$M_{Rd} = 0.9d \left(0.9 f_{pk} A_p/\gamma_s + f_{yd} A_s \right)$$

Dividing both sides by $f_{cd} b d^2$ gives:

$$\frac{M_{Rd}}{f_{cd} b d^2} = 0.9 \left(\frac{0.9 f_{pk}}{\gamma_s f_{cd}} \frac{A_p}{bd} + \frac{f_{yd}}{f_{cd}} \frac{A_s}{bd} \right)$$

and therefore:

$$bd^2 = \frac{M_{Rd}}{0.9 f_{cd}(q_p + q_s)} \tag{10.8}$$

where:

$$q_p = \frac{0.9 f_{pk}}{f_{cd}} \frac{A_p}{bd} \tag{10.9}$$

$$q_s = \frac{f_{yd}}{f_{cd}} \frac{A_{st}}{bd} \tag{10.10}$$

Knowing that the design resistance M_{Rd} must exceed the factored design moment M_{Ed}, Equation 10.8 becomes:

$$bd^2 \geq \frac{M_{Ed}}{0.9 f_{cd}(q_p + q_s)} \tag{10.11}$$

The quantity $q_p + q_s$ is the combined steel index, and a value of $q_p + q_s$ of about 0.3 will usually provide a ductile section. With this assumption, Equation 10.11 may be simplified to:

$$bd^2 \geq \frac{M_{Ed}}{0.27 f_{cd}} \tag{10.12}$$

Equation 10.12 can be used to obtain preliminary dimensions for an initial trial section. The design moment M_{Ed} in Equation 10.12 must include an initial estimate of self-weight.

With the cross-sectional dimensions so determined, the initial prestress and the area of prestressing steel can then be selected based on serviceability requirements. Various criteria can be adopted. For example, the prestress required to cause decompression (i.e. zero bottom fibre stress) at the section of maximum moment under full dead load could be selected. Alternatively, load balancing could be used to calculate the prestress required to produce zero deflection under a selected portion of the external load. With the level of prestress determined and the serviceability requirements for the member satisfied, the amount of non-prestressed steel required for strength is calculated.

The size of the web of a beam is frequently determined from shear strength calculations. In arriving at a preliminary cross-section for a thin-webbed member, preliminary checks in accordance with the procedures outlined in Chapter 7 should be carried out to ensure that adequate shear strength can be provided. In addition, the arrangement of the tendon anchorages at the ends of the beam often determines the shape of the section in these regions. Consideration must be given therefore to the anchorage zone requirements (in accordance with the principles discussed in Chapter 8) even in the initial stages of design.

10.4 DESIGN PROCEDURES: FULLY-PRESTRESSED BEAMS

For the design of a fully-prestressed member, stress limits both at transfer and under full loads must be selected to ensure that cracking under in-service conditions does not occur at any stage. There are relatively few situations that specifically require *no cracking* as a design requirement. Depending on the span and load combinations, however, a fully-prestressed design may well prove to be the most economical solution.

For long-span members, where self-weight is a major part of the design load, relatively large prestressing forces are required to produce an economic design and fully-prestressed members frequently result. Fully-prestressed construction is also desirable if a crack-free or water-tight structure is required or if the structure needs to possess high fatigue strength. In building structures, however, where the spans are generally small to medium, full prestressing may lead to excessive camber, and partial prestressing, where cracking may occur at service loads, is often a better solution.

When the critical sections have been proportioned so that the selected stress limits are satisfied at all stages of loading, checks must be made on the magnitude of the losses of prestress, the deflection and the flexural, shear and torsional strengths. In addition, the anchorage zone must be designed.

10.4.1 Beams with varying eccentricity

The following steps will usually lead to the satisfactory design of a statically determinate, fully-prestressed beam with a draped tendon profile.

1. Determine the loads on the beam both at transfer and under the most severe load combination for the serviceability limit states. Next, determine the moments at the critical section(s) both at transfer and under the full service loads (M_o and M_T, respectively). An initial estimate of self-weight is required here.
2. Make an initial selection of concrete strength and establish material properties. Using Equation 10.1, choose an initial trial cross-section.
3. Select the maximum permissible total deflection v_{max} caused by the estimated unbalanced loads w_{unbal}. This is a second serviceability requirement in addition to the no cracking requirement that prompted the fully-prestressed design. Next use Equation 10.5 to check that the gross moment of inertia of the section selected in Step 2 is adequate.
4. Estimate the time-dependent losses of prestress (see Section 5.10.3), and, using the procedure outlined in Section 5.4.1, determine the prestressing force and eccentricity at the critical section(s). With due consideration of the anchorage zone and other construction requirements, select the size and number of prestressing tendons.
5. Establish suitable cable profile(s) by assuming the friction losses and by obtaining bounds to the cable eccentricity using Equations 5.14 through 5.17.
6. Calculate both the immediate and time-dependent losses of prestress. Ensure that the calculated losses are less than those assumed in steps 4 and 5. Repeat steps 4 and 5, if necessary.
7. Check the deflection at transfer and the final long-term deflection under maximum and minimum loads. If necessary, consider the inclusion of non-prestressed steel to reduce time-dependent deformations (top steel in the span regions to reduce downward deflection, bottom steel to reduce time-dependent camber). Adjust the section size or the prestress level (or both), if the calculated deflection is excessive. Where an accurate estimate of time-dependent deflection is required, the time analysis described in Section 5.7 is recommended.
8. Check the ultimate strength in bending at each critical section (in accordance with the discussion in Chapter 6). If necessary, additional non-prestressed tensile reinforcement may be used to increase strength. Add compressive reinforcement to improve ductility, as required.
9. Check the shear strength of the beam (and torsional strength if applicable) in accordance with the provisions outlined in Chapter 7. Design suitable shear reinforcement where required.
10. Design the anchorage zone using the procedures presented in Chapter 8.

Note: Durability and fire protection requirements are usually satisfied by an appropriate choice of concrete strength and cover to the tendons made very early in the design procedure (usually at Step 2).

EXAMPLE 10.1 FULLY-PRESTRESSED POST-TENSIONED DESIGN (DRAPED TENDON)

A slab and beam floor system consists of post-tensioned, simply-supported T-beams spanning 18.5 m and spaced 4 m apart. A 140 mm thick, continuous reinforced concrete, one-way slab spans from beam to beam. An elevation and a cross-section of a typical T-beam are shown in Figure 10.2. The beam is to be designed as a fully-prestressed member. The floor supports a superimposed permanent dead load of 2 kPa and a live load of 3 kPa (of which 1 kPa is considered permanent). Material properties are f_{ck} = 35 MPa at 28 days and 25 MPa at transfer, f_{pk} = 1,860 MPa, E_{cm} = 34,000 MPa at 28 days and E_{cm} = 31,000 MPa at transfer and E_p = 195,000 MPa. The maximum deflection of the beam at mid-span is to be limited to span/500.

For this fully-prestressed design, the following stress limits have been selected:

At transfer: $f_{ct,0}$ = 1.25 MPa   and $f_{cc,0}$ = −12.5 MPa
After all losses: $f_{ct,t}$ = 1.5 MPa   and $f_{cc,t}$ = −16.0 MPa

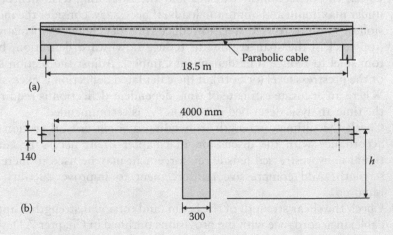

Figure 10.2 Beam details for Example 10.1. (a) Elevation. (b) Cross-section.

(1) Mid-span moments:

Due to self-weight: To estimate the self-weight of the floor w_{sw}, an initial trial depth $h = 1100$ mm is assumed (about span/17). If the concrete floor weighs 24 kN/m³:

$$w_{sw} = 24 \times [4 \times 0.14 + 0.3 \times (1.1 - 0.14)] = 20.4 \text{ kN/m}$$

and the mid-span moment due to self-weight is:

$$M_{sw} = \frac{20.4 \times 18.5^2}{8} = 871 \text{kNm}$$

Due to 2.0 kPa superimposed dead load:

$$w_G = 2 \times 4 = 8 \text{ kN/m} \quad \text{and} \quad M_G = \frac{8 \times 18.5^2}{8} = 342 \text{ kNm}$$

Due to 3.0 kPa live load:

$$w_Q = 3 \times 4 = 12 \text{ kN/m} \quad \text{and} \quad M_Q = \frac{12 \times 18.5^2}{8} = 513 \text{ kNm}$$

At transfer:

$$M_0 = M_{sw} = 871 \text{ kNm}$$

Under full loads:

$$M_T = M_{sw} + M_G + M_Q = 1726 \text{ kNm}$$

(2) Trial section size:

From Equation 10.1:

$$Z_{btm} \geq \frac{(1726 - 0.75 \times 871) \times 10^6}{1.5 - 0.75 \times (-12.5)} = 98.6 \times 10^6 \text{ mm}^3$$

Choose the trial cross-section shown in Figure 10.3.

The revised self-weight is 20.7 kN/m and therefore the revised design moments are $M_0 = 886$ kN m and $M_T = 1741$ kN m.

Note: This section just satisfies the requirement for the effective width of T-beam flanges in EN 1992-1-1 [1], namely that the overhanging part of the effective flange width does not exceed the value given by Equation 6.33 or half the clear distance to the next parallel beam.

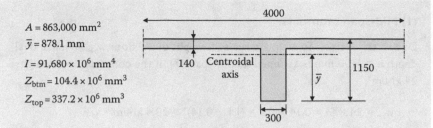

$A = 863,000 \text{ mm}^2$

$\bar{y} = 878.1 \text{ mm}$

$I = 91,680 \times 10^6 \text{ mm}^4$

$Z_{btm} = 104.4 \times 10^6 \text{ mm}^3$

$Z_{top} = 337.2 \times 10^6 \text{ mm}^3$

Figure 10.3 Trial cross-section (Example 10.1).

(3) Check deflection requirements:

For this particular floor, the maximum deflection v_{max} is limited to span/500 = 37 mm. If it is assumed that only the self-weight of the floor is balanced by prestress, the unbalanced load is:

$$w_{unbal} = w_G + w_Q = 20 \text{ kN/m}$$

With one-third of the live load specified as permanent, the sustained unbalanced load is taken to be (refer Section 2.3.4):

$$w_{ub.sus} = w_G + \psi_2 \, w_Q = 8 + 0.333 \times 12 = 12 \text{ kN/m}$$

With the long-term deflection multiplier taken as $\lambda = 3$, Equation 10.4 gives:

$$w_{tot} = 20 + 3 \times 12 = 56 \text{ kN/m}$$

and from Equation 10.5:

$$I \geq \frac{5}{384} \frac{56 \times 18,500^4}{34,000 \times 37} = 67,890 \times 10^6 \text{ mm}^4$$

The trial cross-section satisfies this requirement and excessive deflection is unlikely.

(4) Determine the prestressing force and eccentricity required at mid-span:

The procedure outlined in Section 5.4.1 is used for the satisfaction of the selected stress limits. The section properties α_{top} and α_{btm} are given by:

$$\alpha_{top} = A/Z_{top} = 0.00256 \quad \text{and} \quad \alpha_{btm} = A/Z_{btm} = 0.00827$$

and Equations 5.3 and 5.4 provide upper limits on the magnitude of prestress at transfer:

$$P_{m0} \leq \frac{863,000 \times 1.25 + 0.00256 \times 886 \times 10^6}{0.00256e - 1} = \frac{3.347 \times 10^6}{0.00256e - 1} \tag{10.1.1}$$

$$P_{m0} \leq \frac{-863,000 \times (-12.5) + 0.00827 \times 886 \times 10^6}{0.00827e + 1} = \frac{18.11 \times 10^6}{0.00827e + 1} \tag{10.1.2}$$

Equations 5.5 and 5.6 provide lower limits on the prestress under full service loads. If the time-dependent loss of prestress is assumed to be 20% (i.e. $\Omega = 0.80$), then:

$$P_{m0} \geq \frac{-863,000 \times 1.5 + 0.00827 \times 1741 \times 10^6}{0.8 \times (0.00827e + 1)} = \frac{16.38 \times 10^6}{0.00827e + 1} \tag{10.1.3}$$

$$P_{m0} \geq \frac{863,000 \times (-16.0) + 0.00256 \times 1741 \times 10^6}{0.8 \times (0.00256e - 1)} = \frac{-11.69 \times 10^6}{0.00256e - 1} \tag{10.1.4}$$

Two cables are assumed with duct diameters of 80 mm and with a 40 mm minimum cover to the ducts. The position of the ducts at mid-span and the location of the resultant prestressing force P are illustrated in Figure 10.4. The resultant force in each tendon is assumed to be located at one quarter of the duct diameter below the top of the duct.

The maximum eccentricity to the resultant prestressing force is therefore:

$$e_{max} = 878 - 155 = 723 \text{ mm}$$

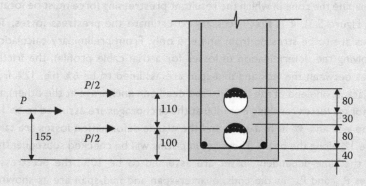

Figure 10.4 Cable locations and relevant dimensions at mid-span.

With this eccentricity at mid-span, Equations 10.1.1 through 10.1.4 give, respectively:

$$P_{m0} \leq 3,935 \text{ kN} \quad P_{m0} \leq 2,596 \text{ kN} \quad P_{m0} \geq 2,347 \text{ kN} \quad P_{m0} \geq -13,743 \text{ kN}$$

and the minimum required prestressing force at mid-span that satisfies all four equations is obtained from Equation 10.1.3:

$$P_{m0} = 2347 \text{ kN}$$

If the immediate losses at mid-span are assumed to be 10%, then the required jacking force is

$$P_j = P_{m0}/0.9 = 2608 \text{ kN}$$

From Table 4.8, the cross-sectional area of a 12.5 mm diameter seven-wire strand is 93.0 mm^2, f_{pk} = 1860 MPa and $f_{p0.1k}$ = 1600 MPa. The maximum stress in the tendon at jacking is the smaller of 0.8 f_{pk} = 1488 MPa or 0.9 $f_{p0.1k}$ = 1440 MPa (see Section 5.3). Therefore, the maximum jacking force per strand is 1440 × 93 × 10^{-3} = 133.9 kN. The minimum number of seven-wire strands is therefore 2608/133.9 = 19.5.

Try two cables each containing 10 strands, i.e. A_p = 1860 mm^2 (or A_p = 930 mm^2/cable) with a jacking force of 2608/20 = 130.4 kN/strand.

(5) Establish cable profiles:

Since the member is simply-supported and uniformly loaded, and because the friction losses are only small, parabolic cable profiles with a sufficiently small resultant eccentricity at each end and an eccentricity of 723 mm at mid-span will satisfy the stress limits at every section along the beam. In order to determine the zone in which the resultant prestressing force must be located (see Figure 5.3), it is first necessary to estimate the prestress losses. The cables are to be stressed from one end only. From preliminary calculations (involving the determination of losses for a trial cable profile), the friction losses between the jack and mid-span are assumed to be 6% (i.e. 12% from the jack at one end of the beam to the dead end anchorage at the other), the anchorage losses resulting from slip at the anchorages are assumed to be 14% at the jack and 3% at mid-span, and the elastic deformation losses are taken to be 1% along the beam. These assumptions will be checked subsequently.

If the time-dependent losses are assumed to be 20%, the prestressing forces P_{m0} and $P_{m,t}$ at the ends, quarter-span and mid-span are as shown in Table 10.1. Also tabulated are the moments at each section at transfer and

Table 10.1 Bounds on the eccentricity of prestress (Example 10.1)

Distance from jack (mm)	0	4625	9250	13,875	18,500
Estimated short-term losses (%)	15	13	10	10	13
P_{m0} (kN)	2217	2269	2347	2,347	2,269
$P_{m,t}$ (kN)	1774	1815	1878	1,878	1,815
M_o (kNm)	0	664	886	664	0
M_T (kNm)	0	1306	1741	1,306	0
e_{max} (mm)	468	747	813	718	454
e_{min} (mm)	−209	512	723	491	−207

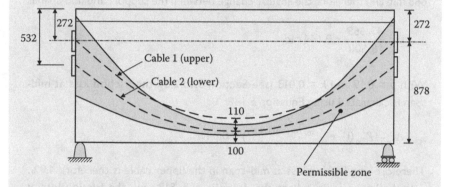

Figure 10.5 Parabolic cable profiles (Example 10.1).

under full loads, the maximum eccentricity (determined in this case from Equation 5.15) and the minimum eccentricity (determined from Equation 5.16 in this example).

The permissible zone, in which the resultant force in the prestressing steel must be located, is shown in Figure 10.5. The individual cable profiles are also shown. The cables are separated sufficiently at the ends of the beam to easily accommodate the anchorages for the two cables.

(6) Check losses of prestress:

Immediate losses:

Elastic deformation: At mid-span, the initial prestress in each cable is $P_{m0} = 2347/2 = 1174$ kN. The upper cable is the first to be stressed and therefore suffers elastic deformation losses when the second (lower) cable is subsequently stressed. The prestressing force in the lower cable causes an axial compressive strain at the centroidal axis of cross-section at transfer of $\varepsilon_a = P_{m0}/(AE_{cm})$, and with the average value of P_{m0} in the bottom cable taken to be

1174 kN, ε_a = 43.9 × 10⁻⁶. The overall shortening of the member due to this elastic strain causes a shortening of the unbonded top cable and an elastic deformation loss of about:

$$\Delta\sigma_p A_p = \varepsilon_a E_p A_p = 43.9 \times 10^{-6} \times 195,000 \times 930 \times 10^{-3} = 8.0 \text{ kN}$$

This is about 0.31% of the total jacking force. The loss of force in the lower cable due to elastic shortening is zero.

Friction losses: The change in the slope of the tendon between the support and mid-span is obtained using Equation 1.6. For the upper cable, the drape is 668 mm and therefore the angular change between the support and mid-span is

$$\theta = \frac{4 \times 668}{18,500} = 0.144 \text{ rad}$$

With μ = 0.19 and k = 0.013 (see Section 5.10.2.3), the friction loss at mid-span is calculated using Equation 5.148:

$$\Delta P_\mu = P_{max}(1 - e^{-0.19(0.144 + 0.013 \times 9.25)}) = 0.049 P_{max}$$

Therefore, the friction loss at mid-span in the upper cable is therefore 4.9%.

In the lower cable, where the drape is only 518 mm, the friction loss at mid-span is 4.3%. The average loss of the prestressing force at mid-span due to friction is therefore 0.046 x 2608 = 120 kN. This loss is less than that assumed in step 5.

Anchorage losses: The loss of prestress caused by Δ_{slip} = 6 mm at the wedges at the jacking end is calculated in accordance with the discussion in Section 5.10.2.4. With the average friction loss at mid-span of 4.6%, the slope of the prestressing line (see Figure 5.25) is:

$$\beta = \frac{0.046 P_j}{l/2} = \frac{0.046 \times 2608 \times 10^3}{9250} = 13.0 \text{ N/mm}$$

The length of beam L_{di} over which the anchorage slip affects the prestress is found using Equation 5.150:

$$L_{di} = \sqrt{\frac{195,000 \times 2 \times 930 \times 6}{13.0}} = 12,940 \text{ mm}$$

Equation 5.151 gives the immediate loss of prestress at the anchorage caused by Δ_{slip}:

$$(\Delta P_{di})_{L_{di}} = 2 \times 13.0 \times 12,940 \times 10^{-3} = 336.4 \text{ kN} (=12.9\% \text{ loss})$$

and the anchorage loss at mid-span is:

$$(\Delta P_{di})_{mid} = 2\beta(L_{di} - l/2) = 2 \times 13.0 \times (12,940 - 9,250) \times 10^{-3}$$
$$= 95.9 \text{ kN} (= 3.7\%)$$

The anchorage losses are compatible with the assumptions made in step 5.

Jacking force: From step 4 and as listed in Table 10.1, the required prestress at mid-span immediately after transfer is $P_{m0} = 2347$ kN. Adding the elastic shortening, friction and anchorage losses, the minimum force required at the jack is:

$$P_j = 2347 + 8.0 + 120 + 95.9 = 2571 \text{ kN} = 1286 \text{ kN/cable}$$

and this is close to the value assumed in steps 4 and 5. The minimum tendon stress at the jack is:

$$\sigma_{pj} = \frac{P_j}{A_p} = \frac{1286 \times 10^3}{930} = 1383 \text{ MPa} = 0.743 f_{pk} = 0.864 f_{p0.1k}$$

which satisfies the maximum jacking stress requirements of EN 1992-1-1 [1].

Time-dependent losses:

An accurate time analysis of the cross-section at mid-span can be carried out using the procedure outlined in Section 5.7.3 (and illustrated in Example 5.5). In this example, the more approximate procedures discussed in Section 5.10.3 are used to check time-dependent losses. First, we need to estimate the creep and shrinkage characteristics of the concrete, and to do this, we will use the procedures specified in Sections 4.2.5.3 and 4.2.5.4.

The hypothetical thickness of the web of this beam is defined in Section 4.2.5.3 and is taken as:

$$h_0 = \frac{2A_c}{u} = \frac{2 \times 300 \times 1150}{2 \times (300 + 1150 - 140)} = 263 \text{ mm}$$

Assuming cement class N, strength class C35/45, inside conditions (RH = 50%) and the age at first loading (transfer) t_0 = 10 days, Figure 4.7 gives the final creep coefficient $\varphi(\infty, t_0)$ = 2.4. The autogenous shrinkage and drying shrinkage strains at time infinity are $\varepsilon_{ca}(\infty)$ = −63 × 10^{-6} and $\varepsilon_{cd}(\infty)$ = −355 × 10^{-6} (given by Equations 4.31 and 4.32, respectively), and, from Equation 4.29, the final design shrinkage strain $\varepsilon_{cs}(\infty)$ = −418 × 10^{-6}.

With $\sigma_{pi} = \sigma_{pm0} = P_{m0}/A_p$ = 1262 MPa for these post-tensioned low-relaxation strands (Class 2), $\mu = \sigma_{pi}/f_{pk}$ = 0.68 and ρ_{1000} = 0.025, from Equation 4.35, after 30 years:

$$\Delta\sigma_{p,r} = \left[-0.66 \times 2.5 \times e^{9.1 \times 0.68} \left(\frac{263,000}{1,000} \right)^{0.75(1-0.68)} \times 10^{-5} \right]$$

$$\times 1,262 = -38.6\ \text{MPa}$$

The concrete stress at the level of the prestressing steel at mid-span (i.e. at e = 723 mm) immediately after the application of the full sustained load (32.7 kN/m) is:

$$\sigma_{c.QP} = -\frac{P_{m0}}{A} - \frac{P_{m0}e^2}{I} + \frac{M_{sus}e}{I}$$

$$= -\frac{2,347 \times 10^3}{863,000} - \frac{2,347 \times 10^3 \times 723^2}{91,680 \times 10^6} + \frac{1,399 \times 10^6 \times 723}{91,680 \times 10^6} = -5.07\ \text{MPa}$$

The total time-dependent loss of steel stress at mid-span is approximated using Equation 5.152:

$$\Delta\sigma_{p,c+s+r} = \frac{-418 \times 10^{-6} \times 195,000 + 0.8 \times -38.6 + \dfrac{195,000}{34,000} \times 2.4 \times -5.07}{1 + \dfrac{195,000 \times 1,860}{34,000 \times 863,000} \left(1 + \dfrac{863,000}{91,680 \times 10^6} \times 723^2 \right) \left[1 + 0.8 \times 2.4 \right]}$$

$$= -150\ \text{MPa}$$

which is 11.9% of the prestress immediately after transfer. This is smaller than the time-dependent losses of 20% assumed in Steps 3, 4 and 5, and a slightly smaller value of P_{m0}, and hence P_i, could be selected. However, given the approximate nature of these serviceability calculations, the original estimate of P_i is considered satisfactory. Should a more accurate estimate of the time-dependent losses of prestress be required, the analysis procedure outlined in Section 5.7.3 is recommended.

(7) Deflection check:

At transfer: The average drape for the two cables is 593 mm, and the transverse force exerted on the beam by the draped tendons at transfer is obtained using Equation 1.7:

$$\frac{8P_{m0}e}{l^2} = \frac{8 \times 2,347 \times 10^3 \times 593}{18,500^2} = 32.5 \text{ N/mm} (\text{kN/m}) \uparrow$$

Note that the prestressing force at mid-span is taken as an average for the span.

The self-weight of the floor was calculated in Step 2 as $w_{sw} = 20.7$ kN/m ↓. Immediately after transfer, $f_{ck} = 25$ MPa, and the elastic modulus of concrete at this time is $E_{cm} = 31,000$ MPa. The mid-span deflection at transfer is therefore:

$$v_0 = \frac{5}{384} \frac{(32.5 - 20.7) \times 18,500^4}{31,000 \times 91,680 \times 10^6} = 6.3 \text{ mm} (\uparrow)$$

For most structures, an upward deflection of this magnitude at transfer will be satisfactory.

Under full loads: The effective prestress at mid-span after all losses is $P_{m,t} = 2040$ kN. The transverse load exerted on the beam by the tendons is therefore:

$$\frac{8 \times 2,040 \times 10^3 \times 593}{18,500^2} = 28.3 \text{ N/mm} (\text{kN/m}) \uparrow$$

The sustained gravity loads are $w_{sus} = w_{sw} + w_G + \psi_\ell w_Q = 32.7$ kN/m (↓), and the short-term curvature and deflection at mid-span caused by all the sustained loads are:

$$\kappa_{sus,0} = \frac{(w_{sus} - w_{p,t})l^2}{8E_{cm}I} = \frac{(32.7 - 28.3) \times 18,500^2}{8 \times 34,000 \times 91,680 \times 10^6} = 0.0604 \times 10^{-6} \text{ mm}^{-1}$$

$$v_{sus,0} = \frac{5}{384} \frac{(w_{sus} - w_{p,t})l^4}{E_{cm}I} = \frac{5}{384} \frac{(32.7 - 28.3) \times 18,500^4}{34,000 \times 91,680 \times 10^6} = 2.2 \text{ mm} (\downarrow)$$

Under the sustained loads, the initial curvature is small on all sections, and the short-term and long-term deflections will also be small.

The creep-induced curvature may be approximated using Equation 5.183. The member is uncracked, and with $\rho = A_p/(b_w d_p) = 0.0062$, the factor α $(= \alpha_2)$ in Equation 5.185 is taken as 1.24:

$$\kappa_{cc} = \frac{2.4}{1.24} \times 0.0604 \times 10^{-6} = 0.117 \times 10^{-6} \text{ mm}^{-1}$$

With the creep-induced curvature at each end of the member equal to zero, the creep deflection is obtained from Equation 5.165:

$$v_{cc} = \frac{18,500^2}{96} \times (0 + 10 \times 0.117 \times 10^{-6} + 0) = 4.2 \text{ mm } (\downarrow)$$

From Equation 5.189, the shrinkage coefficient $k_r = 0.38$ and an estimate of the average shrinkage-induced curvature at mid-span is obtained from Equation 5.187:

$$\kappa_{cs} = -\frac{0.38 \times (-418 \times 10^{-6})}{1150} = 0.138 \times 10^{-6} \text{ mm}^{-1}$$

At each support, the prestressing steel is located at or near the centroidal axis and the shrinkage-induced curvature is zero. This positive load-independent curvature at mid-span causes a downward deflection of (Equation 5.165):

$$v_{cs} = \frac{18,500^2}{96} \times (0 + 10 \times 0.138 \times 10^{-6} + 0) = 4.9 \text{ mm } (\downarrow)$$

The final deflection due to the sustained load and shrinkage is therefore:

$$v_{sus,0} + v_{cc} + v_{cs} = 11.3 \text{ mm } (\downarrow)$$

The deflection that occurs on application of the 2.0 kPa variable live load (= 8 kN/m) is:

$$v_{var,0} = \frac{5}{384} \frac{8 \times 18,500^4}{34,000 \times 91,680 \times 10^6} = 3.9 \text{ mm } (\downarrow)$$

It is evident that the beam performs satisfactorily at service loads with a maximum final deflection of $v_{max} = 11.3 + 3.9 = 15.2$ mm ($\downarrow$) = span/1217. This conclusion was foreshadowed in the preliminary deflection check in step 3.

(8) Check resistance in bending at mid-span:

Using the load factors specified in Section 2.3.2 (Equation 2.1), the design load is:

$$w_{Ed} = 1.35(w_{sw} + w_G) + 1.5w_Q = 56.75 \text{ kN/m}$$

and the design moment at mid-span is:

$$M_{Ed} = \frac{56.75 \times 18.5^2}{8} = 2428 \text{ kNm}$$

The cross-section at mid-span contains a total area of prestressing steel A_p = 1860 mm^2 at an effective depth d_p = 995 mm. The design moment resistance is calculated using the procedure outlined in Section 6.3. With f_{ck} = 35 MPa, Equations 6.2 and 6.4 give, respectively, λ = 0.8 and η = 1.0. Adopting the conservative (Line 2) for the idealised design stress–strain diagram (see Figure 4.12), we will initially assume $\sigma_{pud} = f_{pd} = f_{p0.1k}/\gamma_s$ = 1391 MPa. If the neutral axis lies within the slab flange, the depth to the neutral axis is:

$$x = \frac{f_{pd}A_p}{\eta f_{cd}b\lambda} = \frac{1391 \times 1860}{1.0 \times 23.33 \times 4000 \times 0.8} = 34.7 \text{ mm}$$

which is in fact within the flange. For this section, the quantity of tensile steel is only small and the member is very ductile. The strain in the prestressing steel ε_{pud} is much greater than f_{pd}/E_p and therefore the assumption that $\sigma_{pud} = f_{pd}$ is reasonable.

By taking moments of the internal forces about any point on the cross-section (for example, the level of the resultant compressive force located $\lambda x/2$ below the top surface), the design resistance is:

$$M_{Rd} = 1860 \times 1391 \times \left(995 - \frac{0.8 \times 34.7}{2} \right) = 2538 \text{ kNm} > M_{Ed}$$

Therefore, the cross-section at mid-span has adequate flexural strength and no non-prestressed longitudinal steel is required. At least two non-prestressed longitudinal reinforcement bars will be located at the top and bottom of the web of the beam in the corners of the transverse stirrups that are required for shear.

(9) Check shear strength:

As in step 8, w_{Ed} = 56.75 kN/m. The design shear resistance is here checked at the cross-sections 1 m and 2 m from the support (at the end with the live-end anchorage).

At x = 1.0 m,

V_{Ed} = 468.2 kN   and M_{Ed} = 496.6 kNm

The average depth of the prestressing steel below the centroidal axis of the cross-section is $y = e$ = 251 mm, the effective depth to the resultant tension in the tendon is d_p = 523 mm and the average slope of the tendons is y' = 0.114 rad (refer to Figure 10.5). The prestress immediately after transfer at this cross-section is P_{m0} = 2228 kN (see Table 10.1), and the total time-dependent loss at this section (calculated using Equation 5.182) is 299.8 kN. The effective prestress is therefore $P_{m,t}$ = 1928 kN, and the vertical component of prestress is $P_v = P_{m,t}\,y'$ = 220 kN.

Check whether the section has cracked in bending under the full design bending moment:

$$\sigma_{btm} = -\frac{P_{m,t}}{A} - \frac{P_{m,t}ey_b}{I} + \frac{M_{Ed}y_b}{I} = -2.23 - 4.64 + 4.75 = -2.12 \text{ MPa}$$

and therefore the cross-section has not cracked. From Equation 7.4:

$$V_{Rd,c} = \frac{91,680 \times 10^6 \times (300 - 0.5 \times 80)}{115.7 \times 10^6}\sqrt{(1.47)^2 + 1.0 \times 2.23 \times 1.47}$$

$$= 480.5 \text{ kN} > V_{Ed}$$

Therefore, no calculated shear reinforcement is required at this cross-section.

At x = 2.0 m:

V_{Ed} = 411.5 kN   and M_{Ed} = 936.4 kNm

$y = e$ = 359 mm, d_p = 631 mm, y' = 0.101 rad, P_{m0} = 2239 kN

$P_{m,t}$ = 1938 kN   and $P_v = P_{m,t}\,y'$ = 220 kN

Now σ_{btm} = −0.01 MPa and therefore the cross-section has not cracked. With $V_{Rd,c}$ = 481.8 kN > V_{Ed}, no calculated shear reinforcement is required at this cross-section.

Equation 7.17 gives the minimum shear reinforcement requirements. If 10 mm closed stirrups are used (two vertical legs with A_{sv} = 157 mm² and f_{yd} = 435 MPa):

$$\frac{A_{sw.min}}{s} = \frac{0.08 \times 300 \times \sin 90 \times \sqrt{35}}{435} = 0.326$$

The stirrup spacing must satisfy the following: $s \leq 157/0.326 = 482$ mm.

Use 10 mm closed stirrups at 480 mm maximum centres throughout.

(10) Design anchorage zone:

The bearing plates at the end of each cable are 220 mm square as shown in Figure 10.6. The centroid of each plate lies on the vertical axis of symmetry, the upper plate being located on the centroidal axis of the cross-section and the lower plate centred 260 mm below the centroidal axis, as shown.

The distribution of forces on the anchorage zone after the upper cable is stressed is shown in Figure 10.7a, together with the bursting moments induced within the anchorage zone. The depth of the symmetrical prism behind the upper anchorage plate is 544 mm as shown. The transverse tension within the symmetrical prism caused by the bursting moment behind the anchorage plate (M_b = 142.6 kNm) is:

$$F_{bt} = \frac{M_b}{h_e/2} = \frac{142.8 \times 10^6}{544/2} \times 10^{-3} = 525.0 \text{ kN}$$

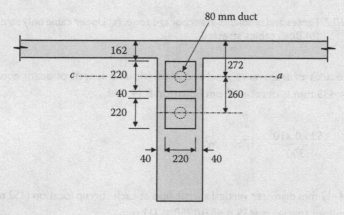

Figure 10.6 End elevation showing size and location of bearing plates.

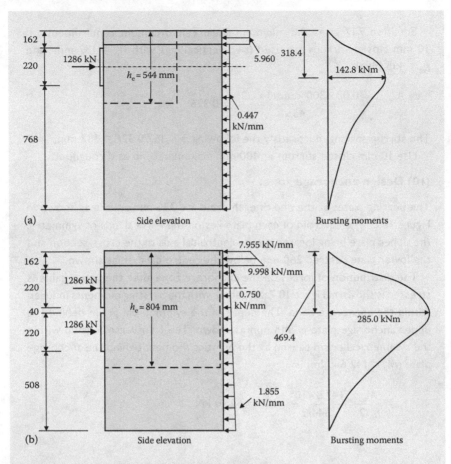

Figure 10.7 Forces and moments in anchorage zone. (a) Upper cable only stressed. (b) Both cables stressed.

and the area of transverse steel required within a length of beam equal to $0.8h_e = 435$ mm is obtained from Equation 8.14:

$$A_{sb} = \frac{525.0 \times 10^3}{300} = 1750 \text{ mm}^2$$

Using 4–12 mm diameter vertical stirrup legs at each stirrup location (452 mm²), the required spacing is $(435 \times 452)/1750 = 112$ mm.

The distribution of forces on the anchorage zone when both cables are stressed is shown in Figure 10.7b. The maximum bursting moment is 285.0 kNm, and the depth of the symmetrical prism behind the combined anchorage plates is 804 mm. The vertical tension and the required area of transverse steel (needed within a length of beam equal to $0.8h_e = 643$ mm) are:

$$T_b = \frac{285.0 \times 10^6}{804/2} \times 10^{-3} = 709 \text{ kN} \quad \text{and} \quad A_{sb} = \frac{709 \times 10^3}{300} = 2363 \text{ mm}^2$$

The maximum spacing of the vertical stirrups (452 mm²/stirrup location) is $(643 \times 452)/2363 = 123$ mm.

Use two 12 mm diameter stirrups every 100 mm from the end face of the beam to 800 mm therefrom.

The horizontal dispersion of prestress into the slab flange creates transverse tension in the slab, as indicated in the plan in Figure 10.8. From Figure 10.7b, the total force in the flange is 1257 kN when both cables are stressed. From the truss analogy shown, the transverse tension is 314.3 kN and the required area of steel is:

$$A_s = (314.3 \times 10^3)/300 = 1048 \text{ mm}^2$$

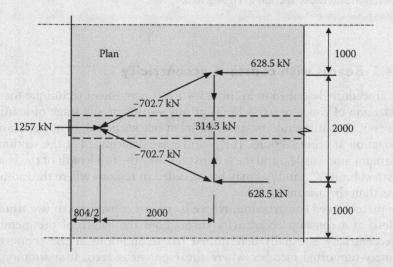

Figure 10.8 Idealised horizontal truss within slab flange.

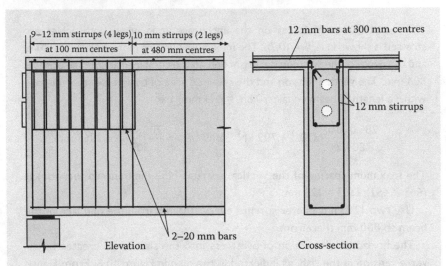

Figure 10.9 Reinforcement details in the anchorage zone.

This steel must be placed horizontally in the slab within a length of 0.8 × 4000 = 3200 mm. Use 12 mm diameter bars at (3200 × 113/1048) = 345 mm centres within 4 m of the free edge of the slab.

The reinforcement details within the anchorage zone are shown in the elevation and cross-section in Figure 10.9.

10.4.2 Beams with constant eccentricity

The procedure described in Section 5.4.1 is a convenient technique for the satisfaction of concrete stress limits at any section at any stage of loading. However, the satisfaction of stress limits at one section does not guarantee satisfaction at other sections. If P_{m0} and e are determined at the section of maximum moment M_0, and if e is constant over the full length of the beam, the stress limits $f_{cc,0}$ and $f_{ct,0}$ may be exceeded in regions where the moment is less than the maximum value.

In pretensioned construction, where it is most convenient to use straight tendons at a constant eccentricity throughout the length of the member, the eccentricity is usually determined from conditions at the support of a simply-supported member where the moment is zero. In a simple pre-tensioned beam of constant cross-section, the stress distributions at the support and the section of maximum moment (M_0 at transfer and M_T under the maximum in-service loads) are shown in Figure 10.10. At transfer, the maximum concrete tensile and compressive stresses both occur at

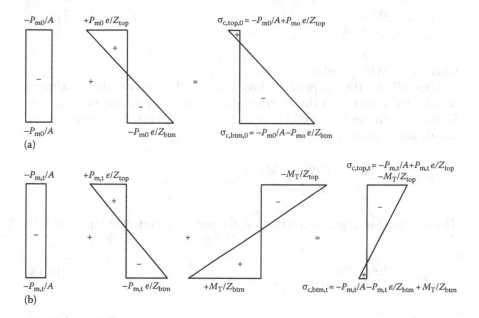

Figure 10.10 Concrete stresses in members with constant eccentricity of prestress. (a) At the support, immediately after transfer. (b) At mid-span after all losses and under full loads ($P_{m,t} = \Omega P_{m0}$).

the support. To guard against unwanted cracking at the support, the top fibre tensile stress must be less than the tensile stress limit $f_{ct,0}$ and the compressive bottom fibre stress should also be limited by the compressive stress limit $f_{cc,0}$. Remembering that $f_{cc,0}$ is negative:

$$\sigma_{c,top,0} = -\frac{P_{m0}}{A} + \frac{P_{m0}e}{Z_{top}} \le f_{ct,0} \tag{10.13}$$

$$\sigma_{c,btm,0} = -\frac{P_{m0}}{A} - \frac{P_{m0}e}{Z_{btm}} \ge f_{cc,0} \tag{10.14}$$

By rearranging Equations 10.13 and 10.14 to express P_{m0} as a linear function of e, the following design equations similar to Equations 5.3 and 5.4 (but with $M_0 = 0$) are obtained:

$$P_{m0} \le \frac{Af_{ct,0}}{\alpha_{top}e - 1} \tag{10.15}$$

$$P_{m0} \leq \frac{-Af_{cc,0}}{\alpha_{btm}e + 1} \tag{10.16}$$

where $\alpha_{top} = A/Z_{top}$ and $\alpha_{btm} = A/Z_{btm}$.

After all the time-dependent losses have taken place, the maximum tensile stress occurs in the bottom concrete fibre at mid-span ($\sigma_{c,btm,t}$ in Figure 10.10b) and, if cracking is to be avoided, must be limited to the tensile stress limit $f_{ct,t}$:

$$\sigma_{c,btm,t} = -\frac{\Omega P_{m0}}{A} - \frac{\Omega P_{m0}e}{Z_{btm}} + \frac{M_T}{Z_{btm}} \leq f_{ct,t} \tag{10.17}$$

This may be rearranged to give the design equation (identical to Equation 5.5):

$$P_{m0} \geq \frac{-Af_{ct,t} + \alpha_{btm}M_T}{\Omega(\alpha_{btm}e + 1)} \tag{10.18}$$

By selecting a value of P_{m0} that satisfies Equations 10.15, 10.16 and 10.18, the selected stress limits will be satisfied both at the support at transfer and at the critical section of maximum moment under the full service loads. The compressive stress limit $f_{cc,t}$ at the critical section is rarely of concern in a pretensioned member of constant cross-section.

To find the minimum sized cross-section required to satisfy the selected stress limits both at the support and at mid-span at all stages of loading, Equation 10.14 may be substituted into Equation 10.17 to give:

$$\Omega f_{cc,0} + \frac{M_T}{Z_{btm}} \leq f_{ct,t}$$

and therefore:

$$Z_{btm} \geq \frac{M_T}{f_{ct,t} - \Omega f_{cc,0}} \tag{10.19}$$

Equation 10.19 can be used to select an initial trial cross-section, and then the required prestressing force and the maximum permissible eccentricity can be determined using Equations 10.15, 10.16 and 10.18.

Note the difference between Equation 10.1 (where $\Omega = 0.75$) and Equation 10.19. The minimum section modulus obtained from Equation 10.1 is controlled by the *incremental moment* $(M_T - \Omega M_0)$ since the satisfaction of stress limits is considered only at the critical section. The stress limits on all other sections are automatically satisfied by suitably varying the eccentricity along the span. If the eccentricity varies such that $P_{m0}e$ is numerically equal to the moment at transfer M_0 at all sections, then only the change in moment $(M_T - \Omega M_0)$ places demands on the flexural rigidity of the member. The moment ΩM_0 is balanced by the eccentricity of prestress. However, for a beam with constant eccentricity, e is controlled by the stress limits at the support (where M_0 is zero). It is therefore the total moment at the critical section M_T that controls the minimum section modulus, as indicated in Equation 10.19.

In order to avoid excessive concrete stresses at the supports at transfer, tendons are often *debonded* near the ends of pretensioned members. In this way, a constant eccentricity greater than the limit obtained from either Equation 10.13 or 10.14 is possible.

For a simply-supported member containing straight tendons at a constant eccentricity, the following design steps are appropriate:

1. Determine the loads on the beam both at transfer and under the most severe load combination for the serviceability limit states. Hence, determine the moments M_0 and M_T at the critical section (an initial estimate of self-weight is required here).
2. Make an initial selection of concrete strength and establish material properties. Using Equation 10.19, choose an initial trial cross-section.
3. Estimate the time-dependent losses and use Equations 10.15, 10.16 and 10.18 to determine the prestressing force and eccentricity at the critical section.
4. Calculate both the immediate and time-dependent losses. Ensure that the calculated losses are less than those assumed in step 3. Repeat step 3, if necessary.
5. Check the deflection at transfer and the final long-term deflection under maximum and minimum loads. Consider the inclusion of non-prestressed steel to reduce the long-term deformation, if necessary. Adjust section size and/or prestress level, if necessary.
6. Check the design flexural resistance at the critical sections. Calculate the quantities of non-prestressed reinforcement required for strength and ductility.
7. Check the design shear resistance of the beam (and the design torsional resistance, if applicable) in accordance with the provisions outlined in Chapter 7. Design suitable stirrups where required.
8. Design the anchorage zone using the procedures presented in Chapter 8.

EXAMPLE 10.2   FULLY-PRESTRESSED PRETENSIONED DESIGN (STRAIGHT TENDONS)

Simply-supported, fully-prestressed planks, with a typical cross-section shown in Figure 10.11, are to be designed to span 6.5 m. The planks are to be placed side by side to form a precast floor and are to be pretensioned with straight tendons at a constant eccentricity. The planks are assumed to be long enough for the full prestress to develop at each support (although this is frequently not the case in practice). The floor is to be subjected to a superimposed dead load of 1.2 kPa and a live load of 3.0 kPa (of which 0.7 kPa may be considered to be permanent and the remainder transitory). As in Example 10.1, material properties are f_{ck} = 35 MPa at 28 days and 25 MPa at transfer, f_{pk} = 1,860 MPa, E_{cm} = 34,000 MPa at 28 days and E_{cm} = 31,000 MPa at transfer and E_p = 195,000 MPa, and the selected stress limits are $f_{ct,0}$ = 1.25 MPa, $f_{cc,0}$ = −12.5 MPa, $f_{ct,t}$ = 1.5 MPa and $f_{cc,t}$ = −16.0 MPa.

(1) Mid-span moments

Due to self-weight: If the initial depth of the plank is assumed to be h = span/40 = 160 mm, and the plank is assumed to weigh 24 kN/m³, then w_{sw} = 24 × 0.16 × 1.05 = 4.03 kN/m and at mid-span:

$$M_{sw} = \frac{4.03 \times 6.5^2}{8} = 21.28 \text{ kNm}$$

Due to superimposed dead and live load:

$$w_G = 1.2 \times 1.05 = 1.26 \text{ kN/m} \quad \text{and} \quad M_G = \frac{1.26 \times 6.5^2}{8} = 6.65 \text{ kNm}$$

$$w_Q = 3 \times 1.05 = 3.15 \text{ kN/m} \quad \text{and} \quad M_Q = \frac{3.15 \times 6.5^2}{8} = 16.64 \text{ kNm}$$

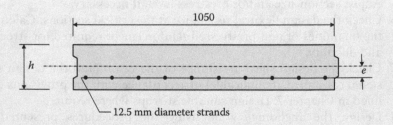

Figure 10.11 Cross-section of pretension plank (Example 10.2).

At transfer:

$$M_0 = M_{sw} = 21.28 \text{ kNm}$$

Under full loads:

$$M_T = M_{sw} + M_G + M_Q = 44.57 \text{ kNm}$$

(2) Trial section size

From Equation 10.19:

$$Z_{btm} \geq \frac{44.57 \times 10^6}{1.5 - 0.75 \times (-12.5)} = 4.098 \times 10^6 \text{ mm}^3$$

and therefore:

$$h \geq \sqrt{\frac{Z_{btm} \times 6}{b}} = \sqrt{\frac{4.098 \times 10^6 \times 6}{1050}} = 153 \text{ mm}$$

Try $h = 160$ mm as originally assumed.

(3) Determine the prestressing force and the eccentricity

With $h = 160$ mm, the section properties are $A = 168 \times 10^3$ mm^2, $I = 358.4 \times 10^6$ mm^4, $Z_{top} = Z_{btm} = 4.48 \times 10^6$ mm^3 and $\alpha_{top} = \alpha_{btm} = 0.0375$. Substituting into Equations 10.15 and 10.16 gives the upper limits to P_{m0}:

$$P_{m0} \leq \frac{168,000 \times 1.25}{0.0375e - 1} = \frac{210 \times 10^3}{0.0375e - 1}$$

$$P_{m0} \leq \frac{-168,000 \times (-12.5)}{0.0375e + 1} = \frac{2,100 \times 10^3}{0.0375e + 1}$$

and substituting into Equation 10.18 gives the lower limit to P_{m0}:

$$P_{m0} \geq \frac{-168,000 \times 1.5 + 0.0375 \times 44.57 \times 10^6}{0.75 \times (0.0375e + 1)} = \frac{1,893 \times 10^3}{(0.0375e + 1)}$$

By equating the maximum value of P_{im0} from Equation 10.15 with the minimum value from Equation 10.18, we obtain the maximum eccentricity:

$$\frac{210 \times 10^3}{0.0375e_{max} - 1} = \frac{1893 \times 10^3}{(0.0375e_{max} + 1)}$$

and solving gives $e_{max} = 33.4$ mm.

Taking e = 33 mm, the corresponding minimum prestress P_{m0} is obtained from Equation 5.18:

$$P_{m0} \geq \frac{1893 \times 10^3}{(0.0375 \times 33 + 1)} = 846 \text{ kN}$$

Assuming 5% immediate losses at mid-span, the minimum jacking force is $P_j = P_{m0}/0.95 = 891$ kN. Using 12.5 mm diameter seven-wire strands each with an area of 93.0 mm² (see Table 4.8), f_{pk} = 1860 MPa and $f_{p0.1k}$ = 1600 MPa. The maximum stress in the tendon at jacking is the smaller of $0.8f_{pk}$ = 1488 MPa or $0.9f_{p0.1k}$ = 1440 MPa (see Section 5.3). Therefore, the maximum jacking force per strand is 1440 × 93 × 10^{-3} = 133.9 kN. The minimum number of seven-wire strands is therefore 891/133.9 = 6.7.

Use seven 12.5 mm diameter strands at e = 33 mm (i.e. A_p = 7 × 93.0 = 651 mm², d_p = 113 mm), with an initial jacking stress of $\sigma_{pi} = P_j/A_p$ = 891 × $10^3/651$ = 1369 MPa.

(4) Calculate losses of prestress

Immediate losses: For this pretensioned member with straight tendons, the immediate loss of prestress is due to elastic shortening. The concrete stress at the steel level at mid-span immediately after transfer is:

$$\sigma_{cp.0} = -\frac{846 \times 10^3}{168 \times 10^3} - \frac{846 \times 10^3 \times 33^2}{358.4 \times 10^6} + \frac{21.28 \times 10^6 \times 33}{358.4 \times 10^6} = -5.65 \text{ MPa}$$

and from Equation 5.146, the loss of prestress at mid-span due to elastic shortening is:

$$\Delta P_{el} = \frac{195,000}{31,000} \times (-5.65) \times 651 \times 10^{-3} = -23.1 \text{ kN}$$

Note that at the supports, where the moment caused by external loads is zero, $\sigma_{cp.0}$ = −7.61 MPa and ΔP_{el} = −31.2 kN.

Jacking force: From step 3, the required minimum prestress at mid-span immediately after transfer is P_{m0} = 846 kN and adding the elastic shortening losses, the minimum force required at the jack is:

$$P_j = 846 + 23.1 = 869.1 \text{ kN} = 124.2 \text{ kN/strand}$$

With this jacking force, the stress in the strand at the jack is:

$$\sigma_{pj} = \frac{P_j}{A_p} = \frac{869.1 \times 10^3}{651} = 1335 \text{ MPa} = 0.718 f_{pk}$$

which is less than 0.9 $f_{p0.1k}$ and is therefore acceptable.

In summary:

$$P_i = 869.1 \text{ kN}$$

$$P_{m0} = 846 \text{ kN at mid-span (and } P_{m0} = 838 \text{ kN at the supports)}$$

Time-dependent losses: The hypothetical thickness of an isolated plank is $h_0 = 138.8$ mm. Assuming cement class N, strength class C35/45, inside conditions (RH = 50%) and that the age at first loading (transfer) is $t_0 = 7$ days, Figure 4.7 gives the final creep coefficient $\varphi(\infty, t_0) = 3.0$. The autogenous shrinkage and drying shrinkage strains at time infinity are $\varepsilon_{ca}(\infty) = -63 \times 10^{-6}$ and $\varepsilon_{cd}(\infty) = -424 \times 10^{-6}$ (given by Equations 4.31 and 4.32, respectively), and, from Equation 4.29, the final design shrinkage strain $\varepsilon_{cs}(\infty) = -487 \times 10^{-6}$.

With $\sigma_{pi} = P_i/A_p = 1369$ MPa for the pretensioned low-relaxation strand (Class 2), $\mu = \sigma_{pi}/f_{pk} = 0.736$ and $\rho_{1000} = 0.025$, from Equation 4.35, after 30 years:

$$\Delta\sigma_{p,r} = \left[-0.66 \times 2.5 \times e^{9.1 \times 0.736} \left(\frac{263,000}{1,000} \right)^{0.75(1-0.736)} \times 10^{-5} \right]$$

$$\times 1369 = -55.2 \text{ MPa}$$

The sustained load is $w_{sus} = 6.03$ kN/m, and the sustained moment at mid-span is $M_{sus} = 31.8$ kNm. The concrete stress at the centroid of the prestressing steel at mid-span (at e = 33 mm) immediately after the application of the full sustained load is:

$$\sigma_{c.QP} = -\frac{P_{m0}}{A} - \frac{P_{m0}e^2}{I} + \frac{M_{sus}e}{I} = -4.68 \text{ MPa}$$

At the supports, where the moment caused by external loads is zero, $\sigma_{c.QP} = -7.53$ MPa.

The total time-dependent loss of steel stress at mid-span is approximated using Equation 5.152:

$$\Delta\sigma_{p,c+s+r} = \frac{-487 \times 10^{-6} \times 195,000 + 0.8 \times -55.2 + \dfrac{195,000}{34,000} \times 3.0 \times -4.68}{1 + \dfrac{195,000 \times 651}{34,000 \times 168,000} \left(1 + \dfrac{168,000}{358.4 \times 10^6} \times 33^2 \right) \left[1 + 0.8 \times 3.0 \right]}$$

$$= -197 \text{ MPa}$$

which is 15% of the prestress immediately after transfer. This is smaller than the time-dependent losses of 25% assumed in step 3 and is therefore acceptable. The total time-dependent loss at the supports is −241 MPa.

The prestressing force after all losses $P_{m,t}$ is at mid-span is:

$$P_{m,t} = P_{m0} + A_p \Delta \sigma_{p,c+s+r} = 846 - 651 \times 197 \times 10^{-3} = 718 \text{ kN}$$

and at the supports:

$$P_{m,t} = P_{m0} + A_p \Delta \sigma_{p,c+s+r} = 838 - 651 \times 241 \times 10^{-3} = 681 \text{ kN}$$

(5) Deflection check

At transfer: The curvature immediately after transfer at each support is:

$$\kappa_{0.s} = \frac{-P_{m0}e}{E_{cm,0}I} = \frac{-838 \times 10^3 \times 33}{31,000 \times 358.4 \times 10^6} = -2.49 \times 10^{-6} \text{ mm}^{-1}$$

and at mid-span is:

$$\kappa_{0.m} = \frac{M_o - P_{m0}e}{E_{cm,0}I} = \frac{21.28 \times 10^6 - 846 \times 10^3 \times 33}{31,000 \times 358.4 \times 10^6} = -0.598 \times 10^{-6} \text{ mm}^{-1}$$

The corresponding deflection at mid-span may be calculated using Equation 5.165:

$$v_0 = \frac{6500^2}{96} \left[-2.49 + 10 \times (-0.598) - 2.49 \right] \times 10^{-6} = -4.8 \text{ mm } (\uparrow)$$

which is likely to be satisfactory in most practical situations.

Under full loads: The instantaneous curvature caused by the prestress at the supports is:

$$\kappa_{sus,0.s} = \frac{-P_{m,t}e}{E_{cm,t}I} = \frac{-682 \times 10^3 \times 33}{34,000 \times 358.4 \times 10^6} = -1.85 \times 10^{-6} \text{ mm}^{-1}$$

With $\rho = 0.0055$, Equation 5.185 gives $\alpha = 1.22$ and the final creep-induced curvature at the supports is estimated using Equation 5.183:

$$\kappa_{cc.s} = \kappa_{sus,0.s} \frac{\varphi(\infty, t_0)}{\alpha} = -1.85 \times 10^{-6} \times \frac{3.0}{1.22} = -4.54 \times 10^{-6} \text{ mm}^{-1}$$

The final load-dependent curvature at the supports is:

$$\kappa_{sus.s} = \kappa_{sus,0.s} + \kappa_{cc.s} = -6.39 \times 10^{-6} \text{ mm}^{-1}$$

The moment at mid-span caused by the sustained loads is $M_{sus} = M_{sw} + M_G + (0.7/3.0) M_Q = 31.8$ kNm, and the instantaneous curvature caused by the prestress and the sustained moment is:

$$\kappa_{sus,0.m} = \frac{M_{sus} - P_{m,t}e}{E_{cm}I} = \frac{31.8 \times 10^6 - 719 \times 10^3 \times 33}{34,000 \times 358.4 \times 10^6} = 0.663 \times 10^{-6} \text{ mm}^{-1}$$

With $\alpha = 1.22$, the final creep-induced curvature at mid-span is:

$$\kappa_{cc.m} = 0.663 \times 10^{-6} \times \frac{3.0}{1.22} = 1.64 \times 10^{-6} \text{ mm}^{-1}$$

The final load-dependent curvature at mid-span is:

$$\kappa_{sus.m} = \kappa_{sus,0.m} + \kappa_{cc.m} = 2.30 \times 10^{-6} \text{ mm}^{-1}$$

The moment at mid-span due to the variable part of the live load is $(2.3/3.0)$ $M_Q = 12.76$ kNm and the corresponding curvature at mid-span is:

$$\kappa_{var.m} = \frac{M_{var}}{E_{cm}I} = \frac{12.76 \times 10^6}{34,000 \times 358.4 \times 10^6} = 1.05 \times 10^{-6} \text{ mm}^{-1}$$

The shrinkage-induced curvature is constant along the span (since the bonded steel is at a constant eccentricity). From Equation 5.189, the shrinkage coefficient $k_r = 0.19$, and, from Equation 5.187, the final shrinkage-induced curvature at mid-span is:

$$\kappa_{cs} = \frac{0.19 \times 487 \times 10^{-6}}{160} = 0.58 \times 10^{-6} \text{ mm}^{-1}$$

The final curvatures at each end and at mid-span are the sum of the load-dependent and shrinkage curvatures:

$$\kappa_s = \kappa_{sus.s} + \kappa_{cs} = -5.84 \times 10^{-6} \text{ mm}^{-1}$$

$$\kappa_m = \kappa_{sus.m} + \kappa_{var.m} + \kappa_{cs} = +3.93 \times 10^{-6} \text{ mm}^{-1}$$

From Equation 5.165, the final maximum mid-span deflection is:

$$v_{max} = \frac{6500^2}{96}\left(-5.84 + 10 \times 3.93 - 5.84\right) \times 10^{-6}$$

$$= +12.2 \text{ mm } (\downarrow) = \text{span}/535$$

A deflection of this magnitude is probably satisfactory, provided that the floor does not support any brittle partitions or finishes.

(6) Check resistance in bending at mid-span

Using the same procedure as outlined in step 8 of Example 8.1, the design resistance in bending of the cross-section containing $A_p = 651$ mm^2 at $d_p = 113$ mm is $M_{Rd} = 103.7$ kNm, and this is greater than $M_{Ed} = 62.7$ kNm. The flexural resistance of the plank is adequate.

(7) Check shear resistance

For this wide shallow plank, the design shear force V_{Ed} is much less than the design resistance $V_{Rd,c}$ on each cross-section and no transverse steel is required.

10.5 DESIGN PROCEDURES: PARTIALLY-PRESTRESSED BEAMS

In the design of a partially-prestressed member, concrete stresses at transfer should be checked to ensure undesirable cracking or excessive compressive stresses do not occur during and immediately after the stressing operation. However, under full service loads, cracking is permitted and so a smaller level of prestress than that required for a fully-prestressed structure may be adopted. It is often convenient to approach the design from a design strength point of view in much the same way as for a conventionally reinforced member. Equations 10.6 and 10.12 can both be used to select an initial section size in which tensile reinforcement (both prestressed and non-prestressed) may be added to provide adequate strength and ductility. The various serviceability requirements can then be used to determine the level of prestress. The designer may choose to limit tension under the sustained load or some portion of it. Alternatively, the designer may select a part of the total load to be balanced by the prestress. Under this balanced load, the curvature induced on a cross-section by the eccentric prestress is equal and opposite to the curvature caused by the load. Losses are calculated and the area of prestressing steel is determined.

It should be remembered that the cross-section obtained using Equation 10.12 is a trial section only. Serviceability requirements may indicate that a larger section is needed or a smaller section would be satisfactory. If the latter is the case, the strength and ductility requirements can usually still be met by the inclusion of non-prestressed reinforcement, either compressive or tensile or both.

After establishing the dimensions of the cross-section, and after the magnitude of the prestressing force and the size and location of the prestressed steel have been determined, the non-prestressed steel required to provide the necessary additional strength and ductility is calculated. Checks are made with regard to deflection and crack control, and finally, the shear reinforcement and anchorage zones are designed.

The following steps usually lead to a satisfactory design.

1. Determine the loads on the beam including an initial estimate of self-weight. Hence, determine the in-service moments at the critical section, both at transfer M_o and under the full loads M_T. Also calculate the design moment M_{Ed} at the critical section.
2. Make an initial selection of concrete strength and establish material properties. Using Equation 10.12, determine suitable section dimensions. Care should be taken when using Equation 10.12. If the neutral axis at design resistance is outside the flange in a T-beam or I-beam, the approximation of a rectangular compression zone may not be acceptable. For long-span, lightly loaded members, deflection and not strength will usually control the size of the section.
3. By selecting a suitable load to be balanced, the unbalanced load can be calculated and Equation 10.6 can be used to check the initial trial section selected in step 2. Adjust section dimensions, if necessary.
4. Determine the prestressing force, the area of prestressing steel and the cable profile to suit the serviceability requirements. For example, no tension may be required under a portion of the service load, such as the dead load. Alternatively, the load at which deflection is zero may be the design criterion.
5. Calculate the immediate and time-dependent losses of prestress and ensure that the serviceability requirements adopted in step 4 and the stress limits at transfer are satisfied.
6. To supplement the prestressing steel determined in step 4, calculate the non-prestressed reinforcement (if any) required to provide adequate flexural strength and ductility.
7. Check crack control and deflections both at transfer and under full loads. A cracked section analysis is usually required to determine I_{ef} and to check the increment of steel stress after cracking.
8. Design for shear (and torsion) at the critical sections in accordance with the design provisions in Chapter 7.
9. Design the anchorage zone using the procedures outlined in Chapter 8.

Example 10.3   Partially-prestressed beam (draped tendon)

The fully-prestressed T-beam designed in Example 10.1 is redesigned here as a partially-prestressed beam. A section and an elevation of the beam are shown in Figure 10.2, and the material properties and floor loadings are as described in Example 10.1. At transfer, the stress limits are $f_{ct,0} = 1.25$ MPa and $f_{cc,0} = -12.5$ MPa.

(1) Mid-span moments

As in step 1 of Example 10.1, $w_G = 8$ kN/m, $w_Q = 12$ kN/m, $M_G = 342$ kNm and $M_Q = 513$ kNm.

Since the deflection of the fully-prestressed beam designed in Example 10.1 is only small, a section of similar size may be acceptable even after cracking. The same section will be assumed here in the estimate of self-weight. Therefore, $w_{sw} = 20.7$ kN/m, $M_{sw} = 886$ kNm and the moments at mid-span at transfer and under full loads are as calculated previously:

$$M_0 = 886 \text{ kNm} \quad \text{and} \quad M_T = 1741 \text{ kNm}$$

The design moment resistance at mid-span is calculated as in Step 8 of Example 10.1, i.e. $M_{Ed} = 2428$ kNm.

(2) Trial section size based on strength considerations

From Equation 10.12:

$$bd^2 \geq \frac{2428 \times 10^6}{0.27 \times 23.33} = 385.5 \times 10^6 \text{ mm}^3$$

For $b = 4000$ mm, the required effective depth is $d > 310.5$ mm.

Clearly, strength and ductility are easily satisfied (as is evident in Step 8 of Example 10.1). Deflection requirements will control the beam depth.

(3) Trial section size based on acceptable deflection

In Example 10.1, the balanced load was $w_{bal} = 28.3$ kN/m (see Step 7). If cracking is permitted, then less prestress will be provided than in Example 10.1. For this cracked, partially-prestressed member, we will adopt enough prestress to balance a load of $w_{bal} = 20$ kN/m. Therefore, the maximum unbalanced load is $w_{unbal} = w_{sw} + w_G + w_Q - w_{bal} = 20.7$ kN/m and the sustained unbalanced load is $w_{unbal.sus} = w_{sw} + w_G + \psi_2 w_Q - 20 = 12.7$ kN/m. The choice of balanced load is somewhat arbitrary, with $w_{bal} = 28.6$ kN/m representing the balanced load for a fully-prestressed design (Example 10.1) and $w_{bal} = 0$ representing a reinforced concrete design (with zero prestress).

Anticipating the inclusion of some compression steel to limit long-term deflection and hence taking $\lambda = 1.5$, Equation 10.4 gives:

$$w_{tot} = 20.7 + 1.5 \times 12.7 = 39.8 \text{ kN/m}$$

If the maximum total deflection v_{max} is to be limited to 50 mm (= span/370), then from Equation 10.6, the initial gross moment of inertia must satisfy the following:

$$I \geq 2 \times \frac{5}{384} \times \frac{39.8 \times 18,500^4}{34,000 \times 50} = 71,420 \times 10^6 \text{ mm}^4$$

Choose the same trial section as was used for the fully-prestressed design (as shown in Figure 10.3).

(4) Determine prestressing force and cable profile

A single prestressing cable is to be used, with sufficient prestress to balance a load of 20 kN/m. The cable is to have a parabolic profile with zero eccentricity at each support and $e = 778$ mm at mid-span (i.e. at mid-span, $d_p = 1050$ mm and the duct has the same cover at mid-span as the lower cable shown in Figure 10.5). The duct diameter is 80 mm with 40 mm concrete cover to the duct.

The effective prestress required at mid-span to balance $w_{bal} = 20$ kN/m is calculated using Equation 1.7:

$$P_{m,t} = \frac{w_{bal} l^2}{8e} = \frac{20.0 \times 18.5^2}{8 \times 0.778} = 1,100 \text{ kN}$$

Since the initial stress in the concrete at the steel level is lower than that in Example 10.1, due to the reduced prestressing force, the creep losses will be lower. The time-dependent losses are here assumed to be 8%. If the immediate losses at mid-span (friction plus anchorage draw-in) are assumed to be 10%, the prestressing force at mid-span immediately after transfer P_{m0} and the required jacking force P_j are:

$$P_{m0} = \frac{P_{m,t}}{0.92} = 1196 \text{ kN} \quad \text{and} \quad P_j = \frac{P_{m0}}{0.9} = 1329 \text{ kN}$$

From Table 4.8, the cross-sectional area of a 12.5 mm diameter seven-wire strand is 93.0 mm^2 and the maximum stress in the tendon at jacking is $0.9 f_{p0.1k} = 1440$ MPa. Therefore, the maximum jacking force per strand is $1440 \times 93 \times 10^{-3} = 133.9$ kN.

The minimum number of seven-wire strands is therefore 1329/133.9 = 9.9. Try ten 12.7 mm diameter ordinary seven-wire strands (A_p = 930 mm^2).

(5) Calculate losses of prestress

Immediate losses: With only one prestressing cable, elastic deformation losses are zero. Using the same procedures as demonstrated in Example 10.1, the friction loss between the jack and mid-span is 5.3% and the anchorage (draw-in) loss at mid-span is 3.1%. The total immediate loss is therefore 8.4% and this is less than the 10% assumed earlier.

Time-dependent losses: The concrete has the same shrinkage and creep characteristics as in Example 10.1, i.e. the final creep coefficient $\varphi(\infty, t_0) = 2.4$ and the final design shrinkage strain $\varepsilon_{cs}(\infty) = -418 \times 10^{-6}$. The hypothetical thickness of the web is h_0 = 263 mm.

With $\sigma_{pi} = \sigma_{pm0} = P_{m0}/A_p$ = 1286 MPa for these post-tensioned low-relaxation strands (Class 2), $\mu = \sigma_{pi}/f_{pk} = 0.69$ and $\rho_{1000} = 0.025$, from Equation 4.35, after 30 years:

$$\Delta\sigma_{p,r} = \left[-0.66 \times 2.5 \times e^{9.1 \times 0.69} \left(\frac{263,000}{1,000} \right)^{0.75(1-0.69)} \times 10^{-5} \right] \times 1,286$$

$$= -41.3 \, \text{MPa}$$

Assuming that the concrete stress at the level of the prestressing steel at mid-span (i.e. at e = 778 mm) immediately after the application of the full sustained load (32.7 kN/m) is compressive:

$$\sigma_{c,QP} = -\frac{P_{m0}}{A} - \frac{P_{m0}e^2}{I} + \frac{M_{sus}e}{I}$$

$$= -\frac{1,196 \times 10^3}{863,000} - \frac{1,196 \times 10^3 \times 778^2}{91,680 \times 10^6} + \frac{1,399 \times 10^6 \times 778}{91,680 \times 10^6}$$

$$= +2.59 \, \text{MPa (tensile)}$$

This means the concrete crack will open at the steel level under the full service load and the stress in the concrete at the tendon level is zero, i.e. $\sigma_{c,QP}$ = 0. The total time-dependent loss of steel stress at mid-span is approximated using Equation 5.152:

$$\Delta\sigma_{p,c+s+r} = \frac{-418 \times 10^{-6} \times 195,000 + 0.8 \times -41.3 + \dfrac{195,000}{34,000} \times 2.4 \times 0}{1 + \dfrac{195,000 \times 930}{34,000 \times 863,000} \left(1 + \dfrac{863,000}{91,680 \times 10^6} \times 778^2 \right) \left[1 + 0.8 \times 2.4 \right]}$$

$$= -102.2 \, \text{MPa}$$

This is 7.9% of the prestress immediately after transfer and is less than the 8% assumed in Step 4.

By comparison with the beam in Example 10.1 (that has almost double the jacking force), the concrete stress limits at transfer are clearly satisfied.

(6) Design for flexural strength

As stated in Step 1, the design moment at mid-span is M_{Ed} = 2410 kNm. Using the procedures outlined in Chapter 6 (and used in Example 10.1), the design resistance of the cross-section containing A_p = 930 mm² at d_p = 1050 mm is M_{Rd1} = 1349 kN (with x = 17.3 mm). Clearly, additional non-prestressed tensile steel is required to ensure adequate strength. If the depth of the non-prestressed tensile reinforcement is d_o = 1080 mm, then the required steel area may be obtained from:

$$A_s = \frac{M_{Ed} - M_{Rd1}}{f_{yd}(d_o - \lambda x/2)} = \frac{(2410 - 1349) \times 10^6}{435 \times (1080 - 0.8 \times 17.3/2)} = 2273 \text{ mm}^2$$

Try four 28 mm diameter bottom reinforcing bars (A_{st} = 2464 mm², f_{yk} = 500 MPa) in two layers, as shown in Figure 10.12.

Checking the strength of this cross-section gives F_{ptd} = 1293 kN, F_{sd} = 1072 kN, x = 31.7 mm and M_{Rd} = 2485 kNm > M_{Ed}. The proposed section at mid-span shown in Figure 10.12 has adequate strength and ductility.

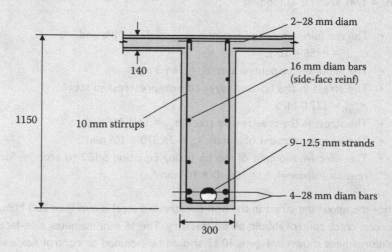

Figure 10.12 Proposed steel layout at mid-span (Example 10.3).

(7) Check deflection and crack control

The maximum moment at mid-span due to the full service load is M_T = 1741 kNm, and the moment at mid-span caused by the sustained load is M_{sus} = 1399 kNm. After all losses, the prestressing force at mid-span is $P_{m,t}$ = 1100 kN. The design tensile strength of concrete is taken to be f_{ctm} = 3.2 MPa, and, with the web reinforcement ratios, $\rho_w = (A_s + A_p)/(b_w d) = (2464 + 930)/$ $(300 \times 1063) = 0.0106$ and $\rho_{cw} = A_{sc}/(b_w d) = 1232/(300 \times 1063) = 0.0039$, the tensile stress that develops in the bottom fibre due to shrinkage may be approximated by Equation 5.179:

$$\sigma_{cs} = \left(\frac{2.5 \times 0.0106 - 0.8 \times 0.0039}{1 + 50 \times 0.0106} \times 200,000 \times 418 \times 10^{-6} \right) = 1.28 \text{ MPa}$$

The cracking moment may be approximated using Equation 5.178:

$$M_{cr} = \left[104.4 \times 10^6 \times \left(3.2 - 1.28 + \frac{1,100 \times 10^3}{863,000} \right) + 1,100 \times 10^3 \times 778 \right] \times 10^{-6}$$

$$= 1,189 \text{ kNm}$$

Cracking occurs at mid-span since the cracking moment is less than the sustained moment. Using the cracked section analysis described in Section 5.8.2, the response of the cracked section at mid-span to the full service moment (M_T = 1741 kNm) is as follows:

- The top fibre stress and strain: $\sigma_{c,0(top)}$ = −5.31 MPa and $\varepsilon_{0(top)}$ = −156 × 10⁻⁶.
- The depth to the neutral axis: d_n = 164.3 mm.
- The stress in the bottom layer of non-prestressed steel: $\sigma_{s(1),0}$ = 177.0 MPa.
- The stress in the prestressed steel: $\sigma_{p,0}$ = 1411 MPa.
- The average moment of inertia: I_{av} = 36,370 × 10⁶ mm⁴.
- The effective moment of inertia (using Equation 5.182 to account for tension stiffening): I_{ef} = 50,610 × 10⁶ mm⁴.

Since the maximum stress in the non-prestressed steel is less than 200 MPa, flexural crack control should be satisfactory. The 16 mm diameter side-face reinforcement shown in Figure 10.12 should be included to control flexural cracking in the web of the beam above the bottom steel.

The upward transverse force exerted by the prestress on the member is $w_{bal} = 20$ kN/m, and the maximum gravity load is 40.7 kN/m. An estimate of the maximum short-term deflection v_0 caused by the full service load is:

$$v_0 = \frac{5}{384} \times \frac{(40.7 - 20.0) \times 18,500^4}{34,000 \times 50,610 \times 10^6} = 18.3 \text{ mm } (\downarrow)$$

Under the sustained loads, the loss of stiffness due to cracking will not be as great. The cracks will partially close, and the depth of the compression zone will increase as the variable live load is removed. For the calculation of the short-term deflection due to the sustained loads (32.7 kN/m), the magnitude of I_{ef} is higher than that used previously. However, using $I_{ef} = 50,610 \times 10^6$ mm^4 will result in a conservative overestimate of deflection:

$$v_{0.sus} = \frac{5}{384} \times \frac{(32.7 - 20.0) \times 18,500^4}{34,000 \times 50,610 \times 10^6} = 11.3 \text{ mm } (\downarrow)$$

With the tensile web reinforcement ratios $\rho = 0.0106$ and the area of compression reinforcement $A_{sc} = 1232$ mm^2, Equations 5.184 and 5.185 give $\alpha_1 = 5.98$ and $\alpha_2 = 1.51$. With the depth to the neutral axis on the cracked section (ignoring prestress), d_{n1} is determined as 98.8 mm, and Equation 5.186 gives $\alpha = 2.83$. The creep-induced deflection may be approximated by:

$$v_{cc} = \frac{\varphi(\infty, t_0)}{\alpha} v_{0.sus} = \frac{2.4}{2.83} \times 11.3 = 9.6 \text{ mm } (\downarrow)$$

At the supports, where the cross-section is uncracked and the prestressing steel is located at the centroidal axis and $\rho = A_{st}/(b_w d_o) = 0.0075$, Equation 5.189 gives $k_r = 0.22$, and an estimate of the average shrinkage-induced curvature is obtained from Equation 5.187:

$$\kappa_{cs.s} = -\frac{0.22 \times (-418 \times 10^{-6})}{1150} = 0.080 \times 10^{-6} \text{ mm}^{-1}$$

At mid-span, where the cross-section has cracked and the prestressing cable is near the bottom of the section, and $\rho = 0.0106$, Equations 5.188 and 5.190 give $k_{r1} = 0.826$ and $k_{r2} = 0.390$, and from Equation 5.191, $k_r = 0.65$. An estimate of the average shrinkage-induced curvature at mid-span is given by Equation 5.187:

$$\kappa_{cs,m} = -\frac{0.65 \times (-418 \times 10^{-6})}{1150} = 0.236 \times 10^{-6} \text{ mm}^{-1}$$

The shrinkage-induced deflection is obtained from Equation 5.165:

$$v_{cs} = \frac{18,500^2}{96}\left(0.080 + 10 \times 0.236 + 0.080\right) \times 10^{-6} = 9.0 \text{ mm } (\downarrow)$$

The maximum final deflection is therefore:

$$v_0 + v_{cc} + v_{cs} = 36.9 \text{ mm } (\downarrow) = \text{span}/501$$

Deflections of this order may be acceptable for most floor types and occupancies.

The design for shear strength and the design of the anchorage zone for this beam are similar to the procedures illustrated in steps 9 and 10 of Example 10.1.

It should be noted that the same cross-sectional dimensions are required for both the fully-prestressed solution in Example 10.1 and the partially-prestressed solution in Example 10.3, provided the deflection calculated earlier is acceptable. Both satisfy strength and serviceability requirements. With significantly less prestress, the partially-prestressed beam is most likely to be the more economical solution.

REFERENCE

1. EN 1992-1-1. 2004. Eurocode 2: Design of concrete structures – Part 1-1: General rules and rules for buildings. British Standards Institution, London, UK.

Chapter 11

Statically indeterminate members

11.1 INTRODUCTION

The previous chapters have been concerned with the behaviour of individual cross-sections and the analysis and design of statically determinate members. In such members, the deformation of individual cross-sections can take place without restraint being introduced at the supports, and reactions and internal actions can be determined using only the principles of statics. For any set of loads on a statically determinate structure, there is one set of reactions and internal actions that satisfies equilibrium, i.e. there is a single load path.

In this chapter, attention is turned towards the analysis and design of statically indeterminate or continuous members, where the number of unknown reactions is greater than the number of equilibrium equations available from statics. The internal actions in a continuous member depend on the relative stiffnesses of the individual regions and, in structural analysis, consideration must be given to the material properties, the geometry of the structure and geometric compatibility, as well as equilibrium. For any set of loads applied to a statically indeterminate structure, there are an infinite number of sets of reactions and internal actions that satisfy equilibrium, but only one set that also satisfies geometric compatibility and the stress–strain relationships for the constituent materials at each point in the structure. Imposed deformations cause internal actions in statically indeterminate members and methods for determining the internal actions caused by both imposed loads and imposed deformations are required for structural design.

By comparison with simply-supported members, continuous members enjoy certain structural and aesthetic advantages. Maximum bending moments are significantly smaller and deflections are substantially reduced for a given span and load. The reduced demand on strength and the increase in overall stiffness permit a shallower cross-section for an indeterminate member for any given serviceability requirement, and this leads to greater flexibility in sizing members for aesthetic considerations.

441

In reinforced concrete structures, these advantages are often achieved without an additional cost, since continuity is an easily achieved consequence of in-situ construction. Prestressed concrete, on the other hand, is very often not cast in-situ, but is precast, and continuity is not a naturally achieved consequence. In precast construction, continuity is obtained with extra expense and care in construction. When prestressed concrete is cast in-situ, or when continuity can be achieved by stressing precast units together over several supports, continuity can result in significant cost savings. By using single cables for several spans, the number of anchorages can be reduced significantly, as can the labour costs associated with the stressing operation.

Continuity provides increased resistance to transient loads and also to progressive collapse resulting from wind, explosion or earthquake. In continuous structures, failure of one member or cross-section does not necessarily jeopardise the entire structure, and a redistribution of internal actions may occur. When overload of the structure or member in one area occurs, a redistribution of forces may take place, provided that the structure is sufficiently ductile and an alternative load path is available.

In addition to the obvious advantages of continuous construction, there are several notable disadvantages. Some of the disadvantages are common to all continuous structures, and others are specific to the characteristics of prestressed concrete. Among the disadvantages common to all continuous beams and frames are: (1) the occurrence of a region of both high shear and high moment adjacent to each internal support; (2) high localised moment peaks over the internal supports; and (3) the possibility of high moments and shears resulting from imposed deformations caused by foundation or support settlement, temperature changes and restrained shrinkage.

In continuous beams of prestressed concrete, the quantity of prestressed reinforcement can often be determined from conditions at mid-span, with additional non-prestressed reinforcement included at each interior support to provide the additional strength required in these regions. The length of beam associated with the high local moment at each interior support is relatively small, so that only short lengths of non-prestressed reinforcement are usually required. In this way, economical partially-prestressed concrete continuous structures can be proportioned.

When cables are stressed over several spans in a continuous member, the loss of prestress caused by friction along the duct may be large. The tendon profile usually follows the moment diagram and the tendon suffers relatively large angular changes as the sign of the moment changes along the member from span to span and the distance from the jacking end of the tendon increases. In the design of long continuous members, the loss of prestress that occurs during the stressing operation must therefore be carefully checked. Attention must also be given to the accommodation of the axial deformation that takes place as the member is stressed. Prestressed concrete members shorten as a result of the longitudinal prestress, and this can require special structural details at the supports of continuous members to allow for this movement.

There are other disadvantages or potential problems that may arise as a result of continuous construction. Often beams are built into columns or walls in order to obtain continuity, thereby introducing large additional lateral forces and moments in these supporting elements.

Perhaps the most significant difference between the behaviour of statically indeterminate and statically determinate prestressed concrete structures is the restraining actions that may develop in continuous structures as a result of imposed deformations. As a statically indeterminate structure is prestressed, the supports provide restraint to the deformations caused by prestress (both axial shortening and curvature) and reactions may be introduced at the supports. The supports also provide restraint to volume changes of the concrete caused by temperature variations and shrinkage. The reactions induced at the supports during the prestressing operation are self-equilibrating and they introduce additional moments and shears in a continuous member, called *secondary moments and shears*. These secondary actions may or may not be significant in design. Methods for determining the magnitudes of the secondary effects and their implications in the design for both strength and serviceability are discussed in this chapter.

11.2 TENDON PROFILES

The tendon profile used in a continuous structure is selected primarily to maximise the beneficial effects of prestress and to minimise the disadvantages discussed in Section 11.1. The shape of the profile may be influenced by the techniques adopted for construction. Construction techniques for prestressed concrete structures have changed considerably over the past half century with many outstanding and innovative developments. Continuity can be achieved in many ways. Some of the more common construction techniques and the associated tendon profiles are briefly discussed here. The methods presented later in the chapter for the analysis of continuous structures are not dependent, however, on the method of construction.

Figure 11.1a represents the most basic tendon configuration for continuous members and is used extensively in slabs and relatively short, lightly loaded beams. Because of the straight soffit, simplicity of formwork is the main advantage of this type of construction. The main disadvantage is the high immediate loss of prestress caused by friction between the tendon and the duct. With the tendon profile following the shape of the moment diagram, the tendon undergoes relatively large angular changes over the length of the member. Tensioning from both ends can be used to reduce the maximum friction loss in long continuous members.

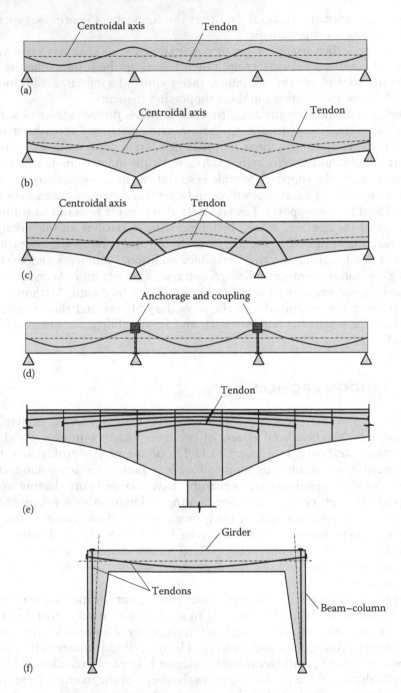

Figure 11.1 Representative tendon profiles. (a) Prismatic beam. (b) Haunched beam
with continuous tendon. (c) Haunched beam with overlapping tendons.
(d) Segmental beam construction. (e) Cantilever construction. (f) Portal frame.

Figure 11.1b indicates an arrangement that has considerable use in longer span structures subjected to heavy applied loads. By haunching the beam as shown, large eccentricities of prestress can be obtained in the regions of high negative moment. This arrangement permits the use of shallower cross-sections in the mid-span region, and the reduced cable drape can lead to smaller friction losses.

Techniques for overlapping tendons or providing cap cables are numerous. Figure 11.1c shows a tendon layout where the regions of high negative moment are provided with extra prestressing. Continuity of the structure is maintained even though there may be considerable variation of prestress along the member. This general technique can eliminate some of the disadvantages associated with the profiles shown in Figures 11.1a and b where the prestressing force is gradually decreasing along the member. However, any structural benefits that arise by tendon layouts of the type shown in Figure 11.1c are gained at the expense of extra prestressing and additional anchorages.

Many types of segmental construction are available and a typical case is represented in Figure 11.1d. Precast or cast in-situ segments are stressed together using prestress couplers to achieve continuity. The couplers and hydraulic jacks are accommodated during the stressing operation within cavities located in the end surface of the individual segments. The cavities can later be filled with concrete, cement grout or other suitable compounds, as necessary.

In large-span structures, such as bridges spanning highways, rivers and valleys, construction techniques are required where falsework is restricted to a minimum. The *cantilever construction method* permits the erection of prestressed concrete segments without the need for major falsework systems. Figure 11.1e illustrates diagrammatically the tendon profiles for a method of construction where precast elements are positioned alternatively on either side of the pier and stressed against the previously placed elements, as shown. The structure is designed initially to sustain the erection forces and construction loads as simple balanced cantilevers on each side of the pier. When the structure is completed and the cantilevers from adjacent piers are joined, the design service loads are resisted by the resulting continuous haunched girders. Construction and erection techniques, such as *balanced cantilevered construction*, are continually evolving and considerable ingenuity is evident in the development of these applications.

Figure 11.1f shows a typical tendon profile for a prestressed concrete portal frame. Prestressed concrete portal frames have generally not had widespread use. With the sudden change of direction of the member axis at each corner of the frame, it is difficult to prestress the columns and beams in a continuous fashion. The horizontal beam and vertical columns are therefore usually stressed separately, with the beam and column tendons crossing at the frame corners and the anchorages positioned on the end and top outside faces of the frame, as shown.

11.3 CONTINUOUS BEAMS

11.3.1 Effects of prestress

As mentioned in Section 11.1, the deformation caused by prestress in a statically determinate member is free to take place without any restraint from the supports. In statically indeterminate members, however, this is not necessarily the case. The redundant supports impose additional geometric constraints, such as zero deflection at intermediate supports (or some prescribed non-zero settlement) or zero slope at a built-in end. During the stressing operation, the geometric constraints may cause additional reactions to develop at the supports, which in turn change the distribution and magnitude of the moments and shears in the member. The magnitudes of these additional reactions (usually called *hyperstatic reactions*) depend on the magnitude of the prestressing force, the support configuration and the tendon profile. For a particular structure, a prestressing tendon with a profile that does not cause hyperstatic reactions is called a *concordant tendon*. Concordant tendons are discussed further in Section 11.3.2.

The moment induced by prestress on a particular cross-section in a statically indeterminate structure may be considered to be made up of two components:

1. The first component is the product of the prestressing force P and its eccentricity from the centroidal axis e. This is the moment that acts on the concrete part of the cross-section when the geometric constraints imposed by the redundant supports are removed. The moment Pe is known as the *primary moment*; and
2. The second component is the moment caused by the hyperstatic reactions, i.e. the additional moment required to achieve deformations that are compatible with the support conditions of the indeterminate structure. The moments caused by the hyperstatic reactions are the *secondary moments*.

In a similar way, the shear force caused by prestress on a cross-section in a statically indeterminate member can be divided into primary and secondary components. The primary shear force in the concrete is equal to the prestressing force P times the slope θ of the tendon at the cross-section under consideration. For a member containing only horizontal tendons ($\theta = 0$), the primary shear force on each cross-section is zero. The secondary shear force at a cross-section is caused by the hyperstatic reactions.

The resultant internal actions caused by prestress at any cross-section are the algebraic sums of the primary and secondary effects.

Since the secondary effects are caused by hyperstatic reactions at the supports, it follows that the secondary moments always vary linearly between the supports in a continuous prestressed concrete member and the secondary shear forces are constant in each span.

11.3.2 Determination of secondary effects using virtual work

In the design and analysis of continuous prestressed concrete members, it is usual to make the following simplifying assumptions (none of which introduce significant errors for normal applications):

1. concrete behaves in a linear-elastic manner within the range of stresses considered;
2. plane sections remain plane throughout the full range of loading;
3. the effects of external loading and prestress on the member can be calculated separately and added to obtain the final conditions, i.e. the principle of superposition is valid; and
4. the magnitude of the eccentricity of prestress is small in comparison with the member length, and hence the horizontal component of the prestressing force is assumed to be equal to the prestressing force at every cross-section.

Consider the two-span beam shown in Figure 11.2a with straight prestressing tendons at a constant eccentricity e below the centroidal axis.

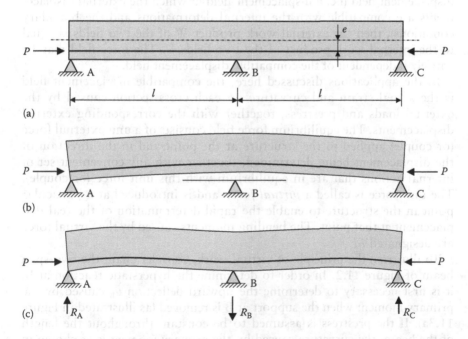

Figure 11.2 Two-span prestressed beam with straight tendon (constant e). (a) Beam elevation. (b) Unrestrained deflection due to primary moment (support B removed). (c) Restrained deformation.

Prior to prestressing, the beam rests on the three supports at A, B and C. On each cross-section, prestress causes an axial force P on the concrete and a negative primary moment Pe. If the support at B was removed, the hogging curvature associated with the primary moment would cause the beam to deflect upwards at B, as shown in Figure 11.2b. In the real beam, the deflection at B is zero, as indicated in Figure 11.2c. To satisfy this geometric constraint, a downward reaction is induced at support B, together with equilibrating upward reactions at supports A and C.

To determine the magnitude of these hyperstatic reactions, one of a number of different methods of structural analysis can be used. For one- or two-fold indeterminate structures, the force method (or flexibility method) is a convenient approach. For multiply redundant structures, a displacement method (such as moment distribution) is more appropriate.

Moment-area methods can be used for estimating the deflection of beams from known curvatures. The principle of virtual work can also be used and is often more convenient. The principle is briefly outlined in the following text. For a more comprehensive discussion of virtual work, the reader is referred to text books on structural analysis, such as Reference [1].

The principle of virtual work states that if a structure is subjected to an equilibrium force field (i.e. a force field in which the external forces are in equilibrium with the internal actions) and a geometrically consistent displacement field (i.e. a displacement field in which the external displacements are compatible with the internal deformations and the boundary conditions), then the external work product W of the two fields is equal to the internal work product of the two fields U. The force field may be entirely independent of the compatible displacement field.

In the applications discussed here, the compatible displacement field is the actual strain and curvature on each cross-section caused by the external loads and prestress, together with the corresponding external displacements. The equilibrium force field consists of a unit external force (or couple) applied to the structure at the point and in the direction of the displacement being determined, together with any convenient set of internal actions that are in equilibrium with this unit force (or couple). The unit force is called a *virtual force* and is introduced at a particular point in the structure to enable the rapid determination of the real displacement at that point. The bending moments caused by the virtual force are designated $\bar{M}$.

To illustrate the principle of virtual work, consider again the two-span beam of Figure 11.2. In order to determine the hyperstatic reaction at B, it is first necessary to determine the upward deflection v_B caused by the primary moment when the support at B is removed (as illustrated in Figure 11.3a). If the prestress is assumed to be constant throughout the length of the beam, the curvature caused by the primary moment is as shown in

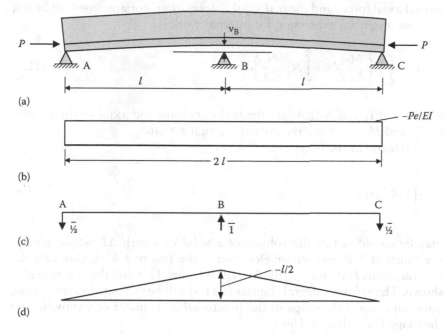

Figure 11.3 Virtual forces on a two-span beam. (a) Unrestrained deflection due to primary moment (support B removed). (b) Curvature caused by primary moment. (c) Virtual force at position of redundant support and virtual reactions. (d) Virtual moment diagram $\bar{M}$.

Figure 11.3b. A unit virtual force is introduced at B in the direction of v_B, as indicated in Figure 11.3c, and the corresponding virtual moments are illustrated in Figure 11.3d.

The external work is the product of the virtual forces and their corresponding displacements:

$$W = 1 \times v_\mathrm{B} = v_\mathrm{B} \tag{11.1}$$

In this example, the internal work is the integral over the length of the beam of the product of the virtual moments $\bar{M}$ and the real deformations $-Pe/EI$:

$$U = \int_0^{2l} \bar{M}\left(\frac{-Pe}{EI}\right)\mathrm{d}x \tag{11.2}$$

If the virtual force applied to a structure produces virtual axial forces $\bar{N}$, in addition to virtual bending moments, then internal work is also done by the virtual axial forces and the real axial deformation. For any length of beam, Δl, a more general expression for internal work is:

$$U = \int_0^{\Delta l} \bar{M}\left(\frac{M}{EI}\right) dx + \int_0^{\Delta l} \bar{N}\left(\frac{N}{EA}\right) dx \tag{11.3}$$

where $M/(EI)$ and $N/(EA)$ are the real curvature and axial strain, respectively, and $\bar{M}$ and $\bar{N}$ are the virtual internal actions.

An integral of the form:

$$\int_0^{\Delta l} \bar{F}(x)F(x)\, dx \tag{11.4}$$

may be considered as the volume of a solid of length ΔL whose plan is the function $F(x)$ and whose elevation is the function $\bar{F}(x)$. Consider the two functions $F(x)$ and $\bar{F}(x)$ illustrated in Figure 11.4 and the notation also shown. The volume integral (Equation 11.4) can be evaluated exactly using Simpson's rule if the shape of the function $F(x)$ is linear or parabolic and the shape $\bar{F}(x)$ is linear. Thus:

$$\int_0^{\Delta l} \bar{F}(x)F(x)\, dx = \frac{\Delta l}{6}\left(\bar{F}_L F_L + 4\bar{F}_M F_M + \bar{F}_R F_R\right) \tag{11.5}$$

In the example considered here, the function $F(x)$ is constant and equal to $-Pe/EI$ (i.e. $F_L = F_M = F_R = -Pe/EI$), and the function $\bar{F}(x)$ is the virtual moment diagram $\bar{M}$, which is also negative and varies linearly from A to B and linearly from B to C, as shown in Figure 11.3d. Evaluating the internal work in the spans AB and BC, Equation 11.2 gives:

$$U = U_{AB} + U_{BC} = 2 \times \int_0^l \bar{M}\left(\frac{-Pe}{EI}\right) dx \tag{11.6}$$

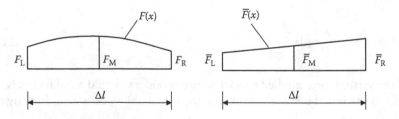

Figure 11.4 Notation for volume integration.

With $\bar{F}_L = 0$, $\bar{F}_M = -l/4$ and $\bar{F}_R = -l/2$, Equation 11.5 gives:

$$U = 2 \times \frac{l}{6}\left[\left(0 \times \frac{-Pe}{EI}\right) + \left(4 \times \frac{-l}{4} \times \frac{-Pe}{EI}\right) + \left(\frac{-l}{2} \times \frac{-Pe}{EI}\right)\right] = \frac{Pel^2}{2EI} \quad (11.7)$$

The principle of virtual work states that:

$$W = U \quad (11.8)$$

and substituting Equations 11.1 and 11.7 in Equation 11.8 gives:

$$v_B = \frac{Pel^2}{2EI} \quad (11.9)$$

It is next necessary to calculate the magnitude of the redundant reaction, R_B, required to restore compatibility at B, i.e. the value of R_B required to produce a downward deflection at B equal in magnitude to the upward deflection given in Equation 11.9. It is convenient to calculate the *flexibility coefficient* f_B associated with the released structure. The flexibility coefficient f_B is the deflection at B caused by a unit value of the redundant reaction at B. The curvature diagram caused by a unit vertical force at B has the same shape as the moment diagram shown in Figure 11.3d. That is, the curvature diagram caused by a unit force at B (M/EI) and the virtual moment diagram $\bar{M}$ have the same shape and the same sign. Using the principle of virtual work and Equation 11.5 to evaluate the volume integral, we get:

$$f_B = \int_0^{2l} \bar{M}\left(\frac{M}{EI}\right)dx = 2 \times \frac{l}{6EI}\left(\left(4 \times \frac{-l}{4} \times \frac{-l}{4}\right) + \left(\frac{-l}{2} \times \frac{-l}{2}\right)\right) = \frac{l^3}{6EI} \quad (11.10)$$

Compatibility requires that the deflection of the real beam at B is zero:

$$v_B + f_B R_B = 0 \quad (11.11)$$

and therefore:

$$R_B = -\frac{v_B}{f_B} = -\frac{Pel^2}{2EI} \times \frac{6EI}{l^3} = -\frac{3Pe}{l} \quad (11.12)$$

The negative sign indicates that the hyperstatic reaction is downwards (or opposite in direction to the unit virtual force at B). With the hyperstatic reactions thus calculated, the secondary moments and shears are determined readily. The effects of prestress on the two-span beam under

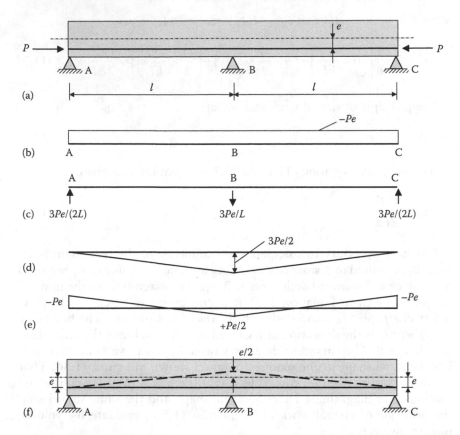

Figure 11.5 Effects of prestress. (a) Elevation of two-span beam with constant eccentricity. (b) Primary moment diagram. (c) Hyperstatic reactions. (d) Secondary moment diagram. (e) Total moment diagram caused by prestress (primary + secondary). (f) The pressure line.

consideration are shown in Figure 11.5. In a statically determinate beam under the action of prestress only, the resultant force on the concrete at a particular cross-section is a compressive force F_C equal in magnitude to the prestressing force and located at the position of the tendon. The distance of the force F_C from the centroidal axis is therefore equal to the primary moment divided by the prestressing force $Pe/P = e$. In a statically indeterminate member, if secondary moments exist at a section, the location of F_C does not coincide with the position of the tendon. The distance of F_C from the centroidal axis is the total moment due to prestress (primary plus secondary) divided by the prestressing force.

For the beam shown in Figure 11.5a, the total moment due to prestress is illustrated in Figure 11.5e. The position of the stress resultant F_C varies as the total moment varies along the beam. At the two exterior supports (ends A and C), F_C is located at the tendon level (i.e. a distance e below the centroidal axis), since the secondary moment at each end is zero. At the interior support B, the secondary moment is $3Pe/2$, and F_C is located at $e/2$ above the centroidal axis (or $3e/2$ above the tendon level). In general, at any section of a continuous beam, the distance of F_C from the level of the tendon is equal to the secondary moment divided by the prestressing force. If the position of F_C at each section is plotted along the beam, a line known as the *pressure line* is obtained. The pressure line for the beam of Figure 11.5a is shown in Figure 11.5f.

If the prestressing force produces hyperstatic reactions, and hence secondary moments, the pressure line does not coincide with the tendon profile. If, however, the pressure line and the tendon profile do coincide at every section along a beam, there are no secondary moments and the tendon profile is said to be *concordant*. In a statically determinate member, of course, the pressure line and the tendon profile always coincide.

11.3.3 Linear transformation of a tendon profile

The two-span beam shown in Figure 11.6 is similar to the beam in Figure 11.2a (and Figure 11.5a), except that the eccentricity of the tendon is not constant but varies linearly in each span. At the exterior supports, the eccentricity is e (as in the previous examples) and at the interior support the eccentricity is ke, where k is arbitrary. If the tendon is above the centroidal axis at B, as shown, k is negative.

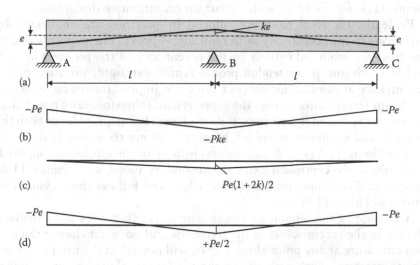

Figure 11.6 Moments induced by prestress in a two-span beam with linearly varying tendon profile. (a) Linear variation of tendon eccentricity. (b) Primary moment diagram. (c) Secondary moment diagram. (d) Total moment diagram caused by prestress.

The primary moment at a section is the product of the prestressing force and the tendon eccentricity and is shown in Figure 11.6b. If the support at B is removed, the deflection at B (v_B) caused by the primary moment may be calculated using the principle of virtual work. The virtual moment diagram $\overline{M}$ is shown in Figure 11.3d.

Using Equation 11.5 to perform the required volume integration, we get:

$$v_B = 2 \times \frac{l}{6EI}\left[\left(4 \times \frac{-l}{4} \times \left(\frac{-Pe - Pke}{2}\right)\right) + \left(\frac{-l}{2} \times (-Pke)\right)\right] = \frac{Pel^2}{EI}\left(\frac{1 + 2k}{6}\right)$$

(11.13)

The flexibility coefficient associated with a release at support B is given by Equation 11.10, and the compatibility condition of zero deflection at the interior support is expressed by Equation 11.11. Substituting Equations 11.10 and 11.13 into Equation 11.11 gives the hyperstatic reaction at B:

$$R_B = -\frac{Pe}{l}(1 + 2k)$$

(11.14)

The secondary moments produced by this downward reaction at B are shown in Figure 11.6c. The secondary moment at the interior support is $(R_B \times 2l)/4 = Pe(1 + 2k)/2$. Adding the primary and secondary moment diagrams gives the total moment diagram produced by prestress and is shown in Figure 11.6d. This is identical with the total moment diagram shown in Figure 11.5e for the beam with a constant eccentricity e throughout.

Evidently, the total moments induced by prestress are unaffected by variations in the eccentricity at the interior support. The moments due to prestress are produced entirely by the eccentricity of the prestress at each end of the beam. If the tendon profile remains straight, variation of the eccentricity at the interior support does not impose transverse loads on the beam (except directly over the supports) and therefore does not change the moments caused by prestress. It does change the magnitudes of both the primary and secondary moments, however, but not their sum. If the value of k in Figure 11.6 is -0.5 (i.e. the eccentricity of the tendon at support B is $e/2$ above the centroidal axis), the secondary moments in Figure 11.6c disappear. The tendon profile is concordant and follows the pressure line shown in Figure 11.5f.

A change in the tendon profile in any beam that does not involve a change in the eccentricities at the free ends and does not change the tendon curvature at any point along a span will not affect the total moments due to prestress. Such a change in the tendon profile is known as *linear transformation*, since it involves a change in the tendon eccentricity at each cross-section by an amount that is linearly proportional to the distance of the cross-section from the end of each span.

Linear transformation can be used in any beam to reduce or eliminate secondary moments. For any statically indeterminate beam, the tendon profile in each span can be made concordant by linearly transforming the profile so that the total moment diagram and the primary moment diagram are the same. The tendon profile and the pressure line for the beam will then coincide.

Stresses at any section in an uncracked structure due to the prestressing force can be calculated as follows:

$$\sigma = -\frac{P}{A} \pm \frac{Pe^*y}{I} \qquad\qquad (11.15)$$

The term e^* is the eccentricity of the pressure line from the centroidal axis of the member, and not the actual eccentricity of the tendon (unless the tendon is concordant and the pressure line and tendon profile coincide). The significance of the pressure line is now apparent. It is the location of the concrete stress resultant caused by the axial prestress, the moment caused by the tendon eccentricity and the moment caused by the hyperstatic support reactions.

11.3.4 Analysis using equivalent loads

In the previous section, the force method was used to determine the hyperstatic reaction in a onefold indeterminate structure. This method is useful for simple structures, but is not practical for manual solution when the number of redundants is greater than two or three.

A procedure more suited to determining the effects of prestress in highly indeterminate structures is the *equivalent load method*. In this method, the forces imposed on the concrete by the prestressing tendons are considered as externally applied loads. The structure is then analysed under the action of these *equivalent loads* using moment distribution or an alternative method of structural analysis. The equivalent loads include the loads imposed on the concrete at the tendon anchorage (which may include the axial prestress, the shear force resulting from a sloping tendon and the moment due to an eccentrically placed anchorage) and the transverse forces exerted on the member wherever the tendon changes direction. Commonly occurring tendon profiles and their equivalent loads are illustrated in Figure 11.7.

The total moment caused by prestress at any cross-section is obtained by analysing the structure under the action of the equivalent loads in each span. The moment due to prestress is caused only by moments applied at each end of a member (due to an eccentrically located tendon anchorage) and by transverse loads resulting from changes in the direction of the tendon anywhere between the supports. Changes in tendon direction at a support (such as at support B in Figure 11.6a) do not affect the moment caused by prestress, since the transverse load passes directly into the support. This is why the total moments caused by prestress in the beams of Figures 11.5a and 11.6a are identical.

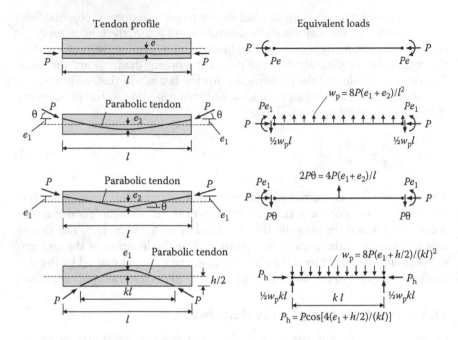

Figure 11.7 Tendon profiles and equivalent loads.

The primary moment at any section is the product of the prestress and its eccentricity Pe. The secondary moment may therefore be calculated by subtracting the primary moment from the total moment caused by the equivalent loads.

11.3.4.1 Moment distribution

Moment distribution is a relaxation procedure developed by Hardy Cross [2] for the analysis of statically indeterminate beams and frames. It is a displacement method of analysis that is ideally suited to manual calculation. Although the method has been replaced in many applications by other analysis procedures implemented in computer programs, it remains a valuable tool for practising engineers because it is simple, easy to use and provides an insight into the physical behaviour of the structure.

Initially, the rotational stiffness of each member framing into each joint in the structure is calculated. Joints in the structure are then locked against rotation by the introduction of imaginary restraints. With the joints locked, fixed-end moments (FEMs) develop at the ends of each loaded member. At a locked joint, the imaginary restraint exerts a moment on the structure equal to the unbalanced moment, which is the resultant of all the FEMs at the joint. The joints are then released, one at a time, by applying a moment to the joint equal and opposite to the unbalanced moment. This balancing moment is distributed to the members framing into the joint in proportion to their rotational stiffnesses. After the unbalanced moment at a joint has been balanced, the joint is relocked. The moment distributed to each member at a released joint induces a carry-over moment at the far end of the member. These carry-over moments are the source of new unbalanced moments at adjacent locked joints. Each joint is unlocked, balanced and then relocked, in turn, and the process is repeated until the unbalanced moments at every joint are negligible. The final moment in a particular member at a joint is obtained by summing the initial FEM and all the increments of distributed and carry-over moments. With the moment at each end of a member thus calculated, the moments and shears at any point along the member can be obtained from statics.

Consider the member AB shown in Figure 11.8a. When the couple M_{AB} is applied to the rotationally released end at A, the member deforms as shown and a moment M_{BA} is induced at the fixed support B at the far end of the member. The relationships between the applied couple M_{AB} and the rotation at A (θ_A) and between the couples at A and B may be expressed as:

$$M_{AB} = k_{AB}\theta_A \tag{11.16}$$

and

$$M_{BA} = CM_{AB} \tag{11.17}$$

where k_{AB} is the stiffness coefficient for the member AB and the term C is the carry-over factor. For a prismatic member, it is a simple matter (using virtual work) to show that for the beam in Figure 11.8a:

$$k_{AB} = \frac{4EI}{l} \tag{11.18}$$

$$C = 0.5 \tag{11.19}$$

Expressions for the stiffness coefficient and carry-over factor for members with other support conditions are shown in Figure 11.8b through d. FEMs for members carrying distributed and concentrated loads are shown in Figure 11.8e through i.

The stiffness coefficient for each member framing into a joint in a continuous beam or frame is calculated and summed to obtain the total rotational stiffness of the joint Σk. The *distribution factor* for a member at the joint is the fraction of the total balancing moment distributed to that particular member each time the joint is released. Since each member meeting at a joint rotates by the same amount, the distribution factor for member AB is $k_{AB}/\Sigma k$. The sum of the distribution factors for each member at a joint is therefore unity.

An example of moment distribution applied to a continuous beam is given in the following example. For a more detailed description of moment distribution, the reader is referred to textbooks on structural analysis such as Reference [1].

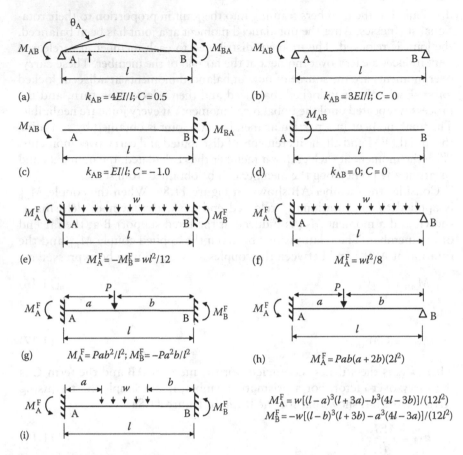

Figure 11.8 Stiffness coefficients, carry-over factors and fixed-end moments for prismatic members. (a) Propped cantilever with moment applied at simple support. (b) Simply-supported beam with moment at one end. (c) Cantilever with moment at free end. (d) Cantilever with moment at fixed end. (e) Uniformly loaded fixed-ended beam. (f) Uniformly loaded propped cantilever. (g) Fixed-ended beam with concentrated load. (h) Propped cantilever with concentrated load. (i) Fixed-ended beam with part uniformly distributed load.

EXAMPLE 11.1 CONTINUOUS BEAM

The continuous beam shown in Figure 11.9a has a rectangular cross-section 400 mm wide and 900 mm deep. If the prestressing force is assumed to be constant along the length of the beam and equal to 1800 kN, calculate the bending moment and shear force diagrams induced by prestress. The tendon profile shown in Figure 11.9a is adopted for illustrative purposes only. In practice, a post-tensioned tendon profile with sharp kinks or sudden changes in direction would not be used.

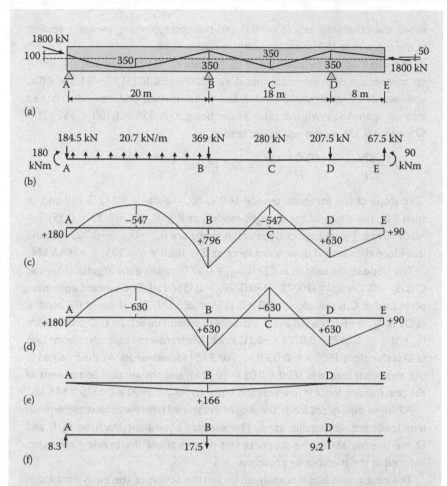

Figure 11.9 Equivalent loads and actions induced by prestress (Example 11.1).
(a) Beam elevation and idealised tendon profile. (b) Equivalent loads
exerted on concrete beam by tendon. (c) Total moment caused by prestress, M_p (kNm). (d) Primary moment, Pe (kNm). (e) Secondary moment,
$M_p - Pe$ (kNm). (f) Hyperstatic reactions caused by prestress (kN).

Relatively short lengths of more gradually curved tendons would be used instead of
the kinks shown at B, C and D. The results of an analysis using the idealised tendon
profile do, however, provide a reasonable approximation of the behaviour of a
more practical beam with continuous curved profiles at B, C and D.

In span AB, the shape of the parabolic tendon is $y = 0.00575x^2 - 0.1025\,x +$
0.1 and its slope is $y' = dy/dx = 0.0115x - 0.1025$, where x is the distance (in
metres) along the beam from support A and y is the depth (in metres) of the
tendon above the centroidal axis. At support A ($x = 0$), the tendon is 100 mm

above the centroidal axis (y = +0.1) and the corresponding moment applied at the support is 180 kNm, as shown in Figure 11.9b.

The slope of the tendon at A is θ_A = dy/dx = −0.1025 rad and the vertical component of prestress is therefore 1800 × (−0.1025) = −184.5 kN (i.e. downwards). The parabolic tendon exerts an upward uniformly distributed load on span AB. With the cable drape being z_d = 350 + [(100 + 350)/2] = 575 mm = 0.575 m, the equivalent load w_p is:

$$w_p = \frac{8Pz_d}{l^2} = \frac{8 \times 1800 \times 0.575}{20^2} = 20.7 \text{ kN/m}(\uparrow)$$

The slope of the parabolic tendon at B is θ_{BA} = dy/dx = +0.1275 rad and, in span BD, the slope of the straight tendon at B is θ_{BC} = −(0.35 + 0.35)/9 = −0.0778 rad. The change of slope at B is therefore $\theta_{BC} - \theta_{BA}$ = −0.205 rad, and therefore the vertical downward force at B is 1800 × (−0.205) = −369.5 kN.

The slope of the tendon in CD is θ_{CD} = 0.0778 rad and the angular change at C is $\theta_C = \theta_{CD} - \theta_{BC}$ = 0.0778 − (−0.0778) = 0.1556 rad. The upward equivalent point load at C is therefore 1800 × 0.1556 = 280 kN. The slope of the tendon in DE is θ_{DE} = −0.0375 rad and the change in tendon direction at D is therefore $\theta_{DE} - \theta_{CD}$ = −0.0375 − 0.0778 = −0.1153 rad. The transverse equivalent point load at D is therefore 1800 × (−0.1153) = −207.5 kN (downwards). At the free end E, the equivalent couple is 1800 × 0.05 = 90 kNm and the vertical component of the prestressing force is upwards and equal to $P\theta_{DE}$ = 1800 × 0.0375 = 67.5 kN.

All these equivalent loads are shown in Figure 11.9b. Note that the equivalent loads are self-equilibrating. The vertical equivalent loads at A, B and D are directly above the supports and do not affect the bending moments induced in the member by prestress.

The continuous beam is analysed under the action of the equivalent loads using moment distribution as outlined in Table 11.1.

The total moment diagram caused by prestress (as calculated in Table 11.1) and the primary moments are illustrated in Figures 11.9c and d, respectively. The secondary moment diagram in Figure 11.9e is obtained by subtracting the primary moments from the total moments and the hyperstatic reactions shown in Figure 11.9f are deduced from the secondary moment diagram.

The secondary shear force diagram corresponding to the hyperstatic reactions is illustrated in Figure 11.10a. The total shear force diagram is obtained from statics using the total moments calculated by moment distribution and is given in Figure 11.10b. By subtracting the secondary shear force from the total shear force at each section, the primary shear force diagram shown in Figure 11.10c is obtained. Note that the primary shear force at any section is the vertical component of prestress $P\theta$.

Table 11.1 Moment distribution table (Example 11.1)

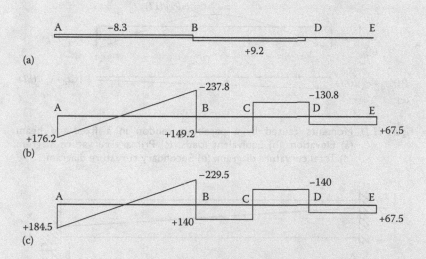

	AB		BA	BD		DB	DE	ED
Stiffness coefficients			$\dfrac{3EI}{20}$	$\dfrac{4EI}{18}$		$\dfrac{3EI}{18}$	0	
Carry-over factor	0.5		0	0.5		0.5	0	
Distribution factor	0		0.403	0.597		1.0	0	0
FEM (kNm)	−180		+1035	−630		+630	−630	+90
			−90					
			−127	−188				
						−94		
						+94		
				+47				
			−19	−28				
						−14		
						+14		
				+7				
			−3	−4				
Final moments (kNm)	−180		+796	−796		+630	−630	+90

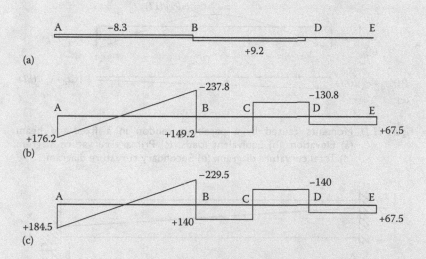

Figure 11.10 Shear force components caused by prestress (Example 11.1). (a) Secondary shear force diagram (kN). (b) Total shear force diagram (kN). (c) Primary shear force diagram $P\theta$ (kN).

EXAMPLE 11.2   FIXED-END BEAMS

The beams shown in Figures 11.11a, 11.12 and 11.13a are rotationally, restrained at each end but are not restrained axially. Determine the moments induced by prestress in each member. Assume that the prestressing force is constant throughout and the member has a constant EI.

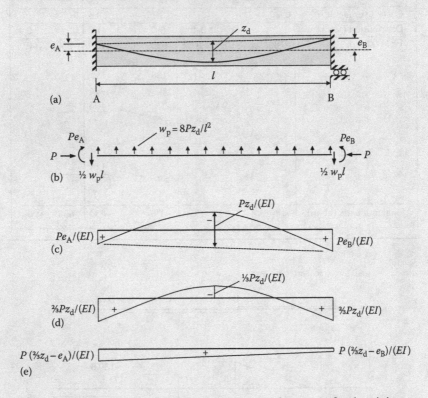

(a) A   B

(b) Pe_A $w_p = 8Pz_d/l^2$ Pe_B
P $\frac{1}{2}w_p l$ $\frac{1}{2}w_p l$ P

(c) $Pe_A/(EI)$ $Pz_d/(EI)$ $Pe_B/(EI)$

(d) $\frac{2}{3}Pz_d/(EI)$ $\frac{1}{3}Pz_d/(EI)$ $\frac{2}{3}Pz_d/(EI)$

(e) $P(\frac{2}{3}z_d - e_A)/(EI)$ $P(\frac{2}{3}z_d - e_B)/(EI)$

Figure 11.11 Moments caused by a parabolic tendon in a fixed-end beam. (a) Elevation. (b) Equivalent loads. (c) Primary curvature diagram. (d) Total curvature diagram. (e) Secondary curvature diagram.

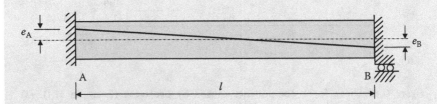

(a) A   B e_B

Figure 11.12 Fixed-end beam with straight tendon.

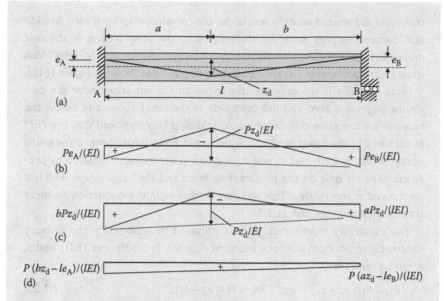

Figure 11.13 Curvature induced by a harped tendon in a fixed-end beam. (a) Elevation. (b) Primary curvature diagram. (c) Total curvature diagram. (d) Secondary curvature diagram.

Case (a): The beam shown in Figure 11.11a is prestressed with a parabolic tendon profile with unequal end eccentricities. The equivalent loads on the structure are illustrated in Figure 11.11b, with end moments of Pe_A and Pe_B, as shown, and an equivalent uniformly distributed upward load of $w_p = 8Pz_d/l^2$.

If the rotational restraints at each end of the beam are released, the curvature is due entirely to the primary moment and is directly proportional to the tendon eccentricity, as shown in Figure 11.11c. The total curvature diagram is obtained by adding the curvature caused by the primary moments to the curvature caused by the restraining secondary moments at each end of the beam M_A^s and M_B^s, respectively. The secondary curvature caused by these secondary moments varies linearly over the length of the beam (provided EI is uniform) so that the total curvature involves a linear shift in the baseline of the primary curvature diagram (see Figure 11.11d).

From structural analysis, for example, using virtual work, it is possible to calculate the restraining moments M_A^s and M_B^s required to produce zero slope at each end of the beam, i.e. $\theta_A = \theta_B = 0$. However, because the beam is fixed-ended, the moment-area theorems reduce the problem to one that can be solved by inspection. Since the slopes at each end are identical, the net area under the total curvature diagram must be zero, i.e. the baseline in Figure 11.11c must be translated and rotated until the area under the curvature diagram is zero. In addition, because support A lies on the tangent to the beam axis at B, the first moment of the final curvature diagram about support A must also be zero. With these two requirements, the total curvature diagram is as shown in Figure 11.11d.

Note that this is the only solution in which the net area under the curvature diagram is zero and the centroids of the areas above and below the baseline are the same distance from A. It should also be noted that the FEM at each end of the beam is $\frac{2}{3}Pz_d = w_p l^2/12$, which is independent of the initial eccentricities at each end e_A and e_A. Evidently, the moment induced by prestress depends only on the prestressing force and the cable drape, and not on the end eccentricities. This was foreshadowed in the discussion of linear transformation in Section 11.3.3.

The secondary moment diagram is obtained by subtracting the primary moment diagram from the total moment diagram. From Figures 11.11c and d, it can be seen that:

$$M_A^s = P(\tfrac{2}{3} z_d - e_A) \quad \text{and} \quad M_B^s = P(\tfrac{2}{3} z_d - e_B) \tag{11.20}$$

The secondary curvature diagram caused by the linearly varying secondary moments is shown in Figure 11.11e.

Case (b): The beam in Figure 11.12 is prestressed with a single straight tendon with arbitrary end eccentricities. This beam is essentially the same as that in the previous example, except that the tendon drape is zero. To satisfy the moment-area theorems in this case, the baseline of the total curvature diagram coincides with the primary curvature diagram, i.e. the total moment induced by prestress is everywhere zero, and the primary and secondary moments at each cross-section are equal in magnitude and opposite in sign. By substituting $z_d = 0$ in Equation 11.20, the secondary moments at each end of the beam of Figure 11.12 are:

$$M_A^s = -M_A^p = -Pe_A \quad \text{and} \quad M_B^s = -M_B^p = -Pe_B \tag{11.21}$$

Case (c): The beam in Figure 11.13a is prestressed with the harped tendon shown. The primary curvature diagram is shown in Figure 11.13b, and the total curvature diagram, established, for example, by satisfaction of the moment-area theorems, is illustrated in Figure 11.13c. As for the previous case, the total curvature (moment) induced by prestress is independent of the end eccentricities e_A and e_B. The curvature induced by the secondary moments is given in Figure 11.13d, and the secondary moments at each support are:

$$M_A^s = P\left(\frac{b}{l} z_d - e_A\right) \quad \text{and} \quad M_B^s = P\left(\frac{a}{l} z_d - e_B\right)$$

11.3.5 Practical tendon profiles

In a span of a continuous beam, it is rarely possible to use a tendon profile that consists of a single parabola, as shown in Figure 11.11a. A more realistic tendon profile consists of a series of segments each with a different shape. Frequently, the tendon profile is a series of parabolic segments, concave in the spans and convex over the interior supports, as illustrated in Figure 11.1a. The convex segments are required to avoid sharp kinks in the tendon at the supports.

Consider the span shown in Figure 11.14, with a tendon profile consisting of three parabolic segments. Adjacent segments are said to be compatible at their point of intersection if the slope of each segment is the same. Compatible segments are desirable to avoid kinks in the tendon profile. In Figure 11.14, B is the point of maximum eccentricity (e_1) and is located a distance of $\alpha_1 l$ from the interior support. Parabolas 1 and 2 each have zero slope at B. The point of inflection at C between the concave parabola 2 and the convex parabola 3 is located a distance $\alpha_2 l$ from the interior support. Parabolas 2 and 3 have the same slope at C. Over the internal support at D, the eccentricity is e_2 and the slope of parabola 3 is zero. By equating the slopes of parabolas 2 and 3 at C, it can be shown that:

$$h_i = \frac{\alpha_2}{\alpha_1}(e_1 + e_2) \tag{11.22}$$

and that the point C lies on the straight line joining the points of maximum eccentricity, B and D. The slope of parabolas 2 and 3 at C is:

$$\theta_C = \frac{2(e_1 + e_2)}{\alpha_1 l} \tag{11.23}$$

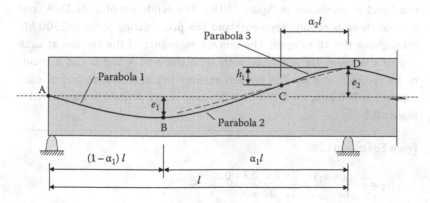

Figure 11.14 Tendon profile with parabolic segments.

The curvatures of each of the three parabolic segments (κ_{p1}, κ_{p2} and κ_{p3}) are given by:

$$\kappa_{p1} = \frac{1}{r_1} = \frac{2e_1}{l^2(1-\alpha_1)^2} \quad \text{(concave)} \tag{11.24}$$

$$\kappa_{p2} = \frac{1}{r_2} = \frac{2(e_1 + e_2 - h_i)}{l^2(\alpha_1 - \alpha_2)^2} \quad \text{(concave)} \tag{11.25}$$

$$\kappa_{p3} = \frac{1}{r_3} = \frac{2(e_1 + e_2)}{\alpha_1 \alpha_2 l^2} \quad \text{(convex)} \tag{11.26}$$

where r_1, r_2 and r_3 are the radii of curvature of parabolas 1, 2 and 3, respectively.

The length of the convex parabola $\alpha_2 l$ should be selected so that the radius of curvature of the tendon is such that the duct containing the tendon can be bent to the desired profile without damage. For a multi-strand system, r_3 should be greater than about 75ϕ, where ϕ is the inside diameter of the duct.

Equations 11.22 through 11.26 are useful for the calculation of the equivalent loads imposed by a realistic draped tendon profile and the determination of the effects of these loads on the behaviour of a continuous structure.

EXAMPLE 11.3

Determine the total and secondary moments caused by prestress in the fixed-end beam shown in Figure 11.15a. The tendon profile ACDEB consists of three parabolic segments and the prestressing force is 2500 kN throughout the 16 m span. The convex segments of the tendon at each end of the beam are identical, with zero slope at A and B and a radius of curvature $r_3 = 8$ m. The tendon eccentricity at mid-span and at each support is 300 mm, i.e. $e_1 = e_2 = 0.3$ m and α_1 (as defined in Figure 11.14) equals 0.5.

From Equation 11.26:

$$\alpha_2 = \frac{2r_3(e_1 + e_2)}{\alpha_1 l^2} = \frac{2 \times 8 \times (0.3 + 0.3)}{0.5 \times 16^2} = 0.075$$

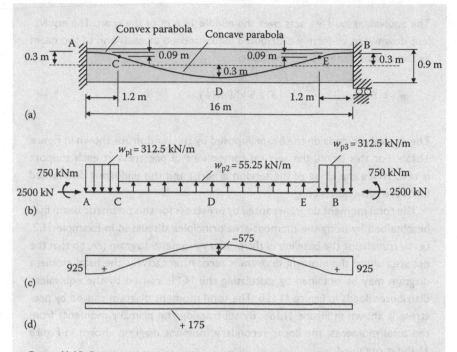

(a)

(b)

(c)

(d)

Figure 11.15 Fixed-end beam with curvilinear tendon profile (Example 11.3). (a) Elevation. (b) Equivalent loads. (c) Total moment diagram caused by prestress (kNm). (d) Secondary moment diagram (kNm).

The convex parabolic segments therefore extend for a distance $\alpha_2 l = 1.2$ m at each end of the span, as shown. The depth of the points of inflection (points C and E) below the tendon level at each support is obtained using Equation 11.22:

$$h_i = \frac{0.075}{0.5}(0.3+0.3) = 0.09 \text{ m}$$

The curvature of the concave parabolic segment CDE extending over the middle $16 - (2 \times 1.2) = 13.6$ m of the span is given by Equation 11.25:

$$\kappa_{p2} = \frac{2 \times (0.3 + 0.3 - 0.09)}{16^2 \times (0.5 - 0.075)^2} = 0.0221 \text{ m}^{-1}$$

and the equivalent uniformly distributed upward load exerted by the concrete tendon is:

$$w_{p2} = P\kappa_{p2} = 2500 \times 0.0221 = 55.25 \text{ kN/m}(\uparrow)$$

The equivalent load w_{p2} acts over the middle 13.6 m of the span. The equivalent downward uniformly distributed load imposed at each end of the beam by the convex tendons AC and EB is:

$$w_{p3} = P\kappa_{p3} = \frac{P}{r_3} = \frac{2500}{8} = 312.5 \text{ kN/m}(\downarrow)$$

The equivalent loads on the beam imposed by the tendon are shown in Figure 11.15b. For this beam, the vertical component of prestress at each support is zero (since the slope of the tendon is zero) and the uniformly distributed loads are self-equilibrating.

The total moment diagram caused by prestress for this prismatic beam may be obtained by using the moment-area principles discussed in Example 11.2, i.e. by translating the baseline of the primary moment diagram (Pe) so that the net area under the moment diagram is zero. Alternatively, the total moment diagram may be obtained by calculating the FEMs caused by the equivalent distributed loads in Figure 11.15b. The total moment diagram caused by prestress is shown in Figure 11.15c. By subtracting the primary moments from the total moments, the linear secondary moment diagram shown in Figure 11.15d is obtained.

If an idealised tendon such as that shown in Figure 11.11a was used to model this more realistic profile (with $e_A = e_B = 0.3$ m and $z_d = 0.6$ m), the total moment at each end (see Figure 11.11d) is:

$$\frac{2Pz_d}{3} = \frac{2 \times 2500 \times 0.6}{3} = 1000 \text{ kN}$$

which is about 8% higher than the value shown in Figure 11.15c.

11.3.6 Members with varying cross-sectional properties

The techniques presented for the analysis of continuous structures hold equally well for members with non-uniform section properties. Section properties may vary owing to haunching or changes in member depth (as illustrated in Figures 11.1b, c and e), or from varying web and flange thicknesses, or simply from cracking in regions of high moment.

Increasing the member depth by haunching is frequently used to increase the tendon eccentricity in the peak moment regions at the interior supports. In such members, the position of the centroidal axis

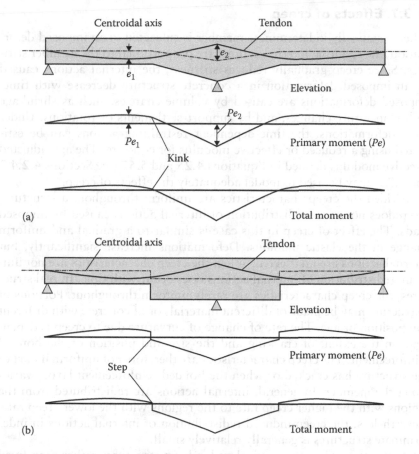

Figure 11.16 Moments induced by prestress in haunched members. (a) Member with tapering soffit. (b) Member with stepped soffit.

varies along the member. If the tendon profile is a smooth curve and the centroidal axis suffers sharp changes in direction or abrupt steps (where the member depth changes suddenly), the total moment diagram caused by prestress also exhibits corresponding kinks or steps, as shown in Figure 11.16.

To determine the FEMs and carry-over factors for members with varying section properties and to calculate the member displacements, the principle of virtual work may be used. The internal work is readily calculated using Equation 11.3 by expressing the section properties (EI and EA) as functions of position x. By dividing the structure into small segments, Equation 11.5 can be used in many practical problems to provide a close approximation of the volume integral for internal work in a non-prismatic member.

11.3.7 Effects of creep

When a statically indeterminate member is subjected to an imposed deformation, the resulting internal actions are proportional to the member stiffness. Since creep gradually reduces stiffness, the internal actions caused by an imposed deformation in a concrete structure decrease with time. Imposed deformations are caused by volume changes, such as shrinkage and temperature changes, and by support settlements or rotations. Under these deformations, the time-dependent restraining actions can be estimated using a reduced or effective modulus for concrete. The age-adjusted effective modulus defined in Equations 4.25 and 5.57 (see Sections 4.2.4.3 and 5.7.2) may be used to model adequately the effects of creep.

Provided the creep characteristics are uniform throughout a structure, creep does not cause redistribution of internal actions caused by imposed loads. The effect of creep in this case is similar to a gradual and uniform change in the elastic modulus. Deformations increase significantly, but internal actions are unaffected. When the creep characteristics are not uniform, redistribution of internal actions does occur with time. In real structures, the creep characteristics are rarely uniform throughout. Portions of a structure may be made of different materials or of concrete with different composition or age. The rate of change of curvature due to creep is dependent on the extent of cracking and the size and position of the bonded reinforcement. The creep characteristics are therefore not uniform if part of the structure has cracked or when the bonded reinforcement layout varies along the member. In general, internal actions are redistributed from the regions with the higher creep rate to the regions with the lower creep rate. Nevertheless, the creep-induced redistribution of internal actions in indeterminate structures is generally relatively small.

Since prestress imposes equivalent loads on structures rather than fixed deformations, the internal actions caused by prestress are not significantly affected by creep. The internal actions are affected in so far as creep causes a reduction of the prestressing force by anything between zero and about 12%. Hyperstatic reactions induced by prestress in indeterminate structures are not therefore significantly relieved by creep.

If the structural system changes after the application of some of the prestress, creep may cause a change in the hyperstatic reactions. For example, the two-span beam shown in Figure 11.17 is fabricated as two precast units of length l and joined together at the interior support by a cast in-situ joint. Creep causes a gradual development of hyperstatic reactions with time and the resulting secondary moments and shears. After the in-situ joint is constructed, the structure is essentially the same as that shown in Figure 11.5a.

Before the joint in Figure 11.17 is cast, the two precast units are simply-supported, with zero deflection and some non-zero slope at the interior support. Immediately after the joint is made and continuity is established, the primary moment in the structure is the same as that shown in Figure 11.5b,

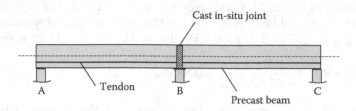

Figure 11.17 Providing continuity at an interior support.

but the secondary moment at B (and elsewhere) is zero. With time, creep causes a gradual change in the curvature on each cross-section. If the support at B was released, the member would gradually deflect upwards due to the creep-induced hogging curvature associated with the primary moment Pe. If it is assumed that the creep characteristics are uniform and that the prestressing force is constant throughout, the time-dependent upward deflection caused by prestress is obtained by multiplying the deflection given in Equation 11.9 by the creep coefficient (adjusted to include the restraint offered to creep by the bonded reinforcement):

$$v_{\mathrm{B}}(t) = \frac{Pel^2}{2E_{\mathrm{cm}}I}\big[\varphi(t_k,t_0)/\alpha\big] \tag{11.27}$$

where the parameter α was introduced in Equation 5.183, and defined in Section 5.11.4.1, and is typically in the range 1.1–1.5 for an uncracked member containing bonded reinforcement.

The short-term deflection at B caused by a unit value of the redundant force applied at the release B is given in Equation 11.10. Owing to creep, however, the redundant at B is gradually applied to the structure. It is therefore appropriate to use the age-adjusted effective modulus ($\bar{E}_{\mathrm{c.eff}}$ given in Equation 4.25) to determine the corresponding time-dependent deformations (elastic plus creep). Substituting $\bar{E}_{\mathrm{c.eff}}$ for E in Equation 11.10 gives:

$$f_{\mathrm{B}}(t) = \frac{l^3[1 + \chi\varphi(t_k,t_0)/\alpha]}{6E_{\mathrm{cm}}I} \tag{11.28}$$

To enforce the compatibility condition that the deflection at B is zero, Equation 11.11 gives:

$$R_{\mathrm{B}}(t) = -\frac{v_{\mathrm{B}}(t)}{f_{\mathrm{B}}(t)} = -\frac{3Pe}{l}\frac{\varphi(t_k,t_0)/\alpha}{1 + \chi\varphi(t_k,t_0)/\alpha} = R_{\mathrm{B}}\frac{\varphi(t_k,t_0)/\alpha}{1 + \chi\varphi(t_k,t_0)/\alpha} \tag{11.29}$$

where $R_{\mathrm{B}}(t)$ is the creep-induced hyperstatic reaction at B and R_{B} is the hyperstatic reaction that would have developed at B if the structure was initially continuous and then prestressed with a straight tendon. The reaction

R_B is shown in Figure 11.5 and given in Equation 11.12. For a prestressed element with typical long-term values of the creep, aging and restraint coefficients (say $\varphi\ (t_k, 7) = 2.5$, $\chi = 0.65$ and $\alpha = 1.2$), Equation 11.29 gives:

$$R_B(t) = R_B \frac{2.5/1.2}{1 + 0.65 \times 2.5/1.2} = 0.885 R_B$$

In general, if R is any hyperstatic reaction or the restrained internal action that would occur at a point due to prestress in a continuous member and $R(t)$ is the corresponding creep-induced value if the member is made continuous only after the application of the prestress, then:

$$R(t) = \frac{\varphi(t_k, t_0)/\alpha}{1 + \chi\varphi(t_k, t_0)/\alpha} R \tag{11.30}$$

If the creep characteristics are uniform throughout the structure, then Equation 11.30 may be applied to systems with any number of redundants.

When continuity is provided at the interior supports of a series of simple precast beams not only is the time-dependent deformation caused by prestress restrained, but the deformation due to the external loads is also restrained. For all external loads applied after continuity has been established, the effects can be calculated by moment distribution or an equivalent method of analysis. Due to the loads applied prior to casting the joints, when the precast units are simply-supported (such as self-weight), the moment at each interior support is initially zero. However, after the joint has been cast, the creep-induced deformation resulting from the self-weight moments in the spans is restrained and moments develop at the supports. For the beam shown in Figure 11.5a, the moment at B due to self-weight is $M_B = w_{sw}l^2/8$. For the segmental beam shown in Figure 11.17, it can be shown that the restraining moment that develops at support B due to creep and self-weight is:

$$M_B(t) = \frac{\varphi(t_k, t_0)/\alpha}{1 + \chi\varphi(t_k, t_0)/\alpha} M_B$$

EXAMPLE 11.4

Consider two simply-supported pretensioned concrete planks, erected over two adjacent spans as shown in Figure 11.17. An in-situ reinforced concrete joint is cast at the interior support to provide continuity. Each plank is 1000 mm wide by 150 mm thick and pretensioned with straight strands at a constant depth of 110 mm below the top fibre, with $A_p = 400$ mm² and $P_{init} = 500$ kN. A typical cross-section is shown in Figure 11.18. The span of each plank is $l = 6$ m. If it is assumed that the continuity is provided immediately after the transfer of prestress, the reactions that develop with time

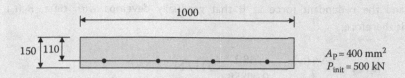

Figure 11.18 Cross-section of planks (Example 11.4).

due to creep are to be determined. For convenience and to better illustrate the effects of creep in this situation, shrinkage is not considered and the self-weight of the planks is also ignored. The material properties are $E_{cm,0}$ = 30,000 MPa, f_{ctm} = 3.0 MPa, φ (t_k, t_0) = 2.0, $\chi(t_k, t_0)$ = 0.65, ε_{cs} = 0 and $\bar{E}_{c.eff}$ = 13,040 MPa. From Equation 5.185, α =1.15.

Immediately after transfer, and before any creep has taken place, the curvature on each cross-section caused by the eccentric prestress may be calculated using the procedure outlined in Section 5.6.2 giving κ_0 = −2.02 × 10^{-6} mm^{-1}. The top and bottom fibre concrete stresses are $\sigma_{top,0}$ = 1.30 MPa and $\sigma_{btm,0}$ = −7.81 MPa, so that cracking has not occurred at transfer.

In the absence of any restraint at the supports, the curvature on each cross-section would change with time due to creep from κ_0 to κ_k = −5.81 × 10^{-6} mm^{-1} (calculated using the procedure outlined in Section 5.7.3). If the support at B in Figure 11.17 was removed, the deflection at B ($\Delta v_B(t_k)$) that would occur with time due to the change in curvature due to creep (($\kappa_k)_{cr}$ = κ_k − κ_0 = −3.79 × 10^{-6} mm^{-1}) can be calculated from Equation 5.165 for the planks that are now spanning 12 m:

$$\Delta v_B(t_k) = \frac{12,000^2}{96}\left[-3.79 + 10 \times (-3.79) - 3.79\right] \times 10^{-6} = -68.2 \text{ mm (i.e. upward)}$$

The deflection at B caused by a unit value of the vertical redundant reaction force gradually applied at the release is calculated using the procedures of Section 11.3.7 and for this example is given by Equation 11.28. With the second moment of area of the age-adjusted transformed cross-section $\bar{I}$ = 288.3 × 10^6 mm^4, Equation 11.28 gives:

$$\bar{f}_{11} = \frac{l^3}{6\bar{E}_{c.eff}\bar{I}} = \frac{6,000^3}{6 \times 13,040 \times 288.3 \times 10^6} = 9.58 \times 10^{-3} \text{ mm/N}$$

and the redundant force at B that gradually develops with time, $R_B(t_k)$, is therefore:

$$R_B(t_k) = -\frac{\Delta v_B(t_k)}{\bar{f}_{11}} = -\frac{-68.2}{0.00958} = 7117\,\text{N}$$

The reaction that would have developed at B due to prestress immediately after transfer, if the member had initially been continuous, is $R_B = 8621$ N (Equation 11.12), and the approximation of Equation 11.29 gives:

$$R_B(t_k) \approx \frac{2.0/1.15}{1 + 0.65 \times 2.0/1.15} \times 8621 = 7038\,\text{N}$$

In this example, Equation 11.29 gives a value for $R_B(t_k)$ that is within 2% of the value determined earlier and provides a quick and reasonable estimate of the effects of creep. The secondary moment at B that develops with time due to prestress is 21.4 kNm, and this is 83% of the secondary moment that would have developed if the two planks had been continuous at transfer.

11.4 STATICALLY INDETERMINATE FRAMES

The equivalent load method is a convenient approach for the determination of primary and secondary moments in framed structures. In the treatment of continuous beams in the previous section, it was assumed that all members were free to undergo axial shortening. This is often not the case in real structures. When the horizontal member of a portal frame, for example, is prestressed, significant restraint to axial shortening may be provided by the flexural stiffness of the vertical columns. Moment distribution can be used to determine the internal actions that develop in the structure as a result of the axial restraint.

Consider the single-bay portal frame shown in Figure 11.19a. Owing to the axial shortening of the girder BC during prestressing, the top of each column moves laterally by an amount Δ. The FEMs induced in the structure are shown in Figure 11.19b. If the girder BC was free to shorten (i.e. was unrestrained by the columns), the displacement Δ that would occur immediately after the application of a prestressing force P to the girder is:

$$\Delta = \frac{P}{E_{cm}A_b}\frac{l_b}{2} \tag{11.31}$$

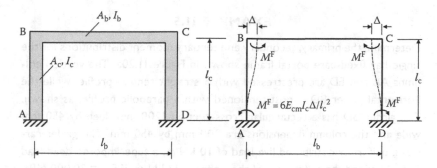

Figure 11.19 Fixed-end moments in a fixed-base frame due to axial shortening of the girder.

This value of Δ is usually used as a starting point in the analysis. The FEMs in the supporting columns due to a relative lateral end displacement of Δ are given by:

$$M^F = \frac{6E_{cm}I_c}{l_c^2}\Delta \tag{11.32}$$

and a moment distribution is performed to calculate the restraining actions produced by the FEMs. If the base of the frame at A was pinned rather than fixed, the FEM at B due to the rotation at A would be $(3E_{cm}I_c/l_c^2)\Delta$. In addition to bending in the beam and in the columns, an outward horizontal reaction is induced at the base of each column and the girder BC is therefore subjected to tension. The tension in BC will reduce the assumed axial shortening, usually by a small amount when the columns are relatively slender. If the reduction in Δ is significant, a second iteration could be performed using the reduced value for Δ to obtain a revised estimate of the FEMs and, hence, a more accurate estimate of the axial restraint.

The magnitude of the axial restraining actions depends on the relative stiffness of the columns and girder. The stiffer the columns, the greater is the restraint to axial shortening of the girder, and hence the larger is the reduction of prestress in the girder. On the other hand, slender columns offer less resistance to deformation and less restraint to the girder.

Axial shortening of the girder BC can also occur due to creep and shrinkage. A time analysis to include these effects can be made by using the age-adjusted effective modulus for concrete, instead of the elastic modulus, to model the gradually applied restraining actions caused by creep and shrinkage.

The internal actions that arise in a prestressed structure as a result of the restraint to axial deformation are sometimes called *tertiary effects*. These effects are added to the primary and secondary effects (calculated using the equivalent load method) to obtain the total effect of prestress in a framed structure.

EXAMPLE 11.5

Determine the primary, secondary and tertiary moment distributions for the single-bay fixed-base portal frame shown in Figure 11.20a. The vertical columns AB and ED are prestressed with a straight tendon profile, while the horizontal girder BD is post-tensioned with a parabolic profile, as shown. The girder BD has a rectangular cross-section 1200 mm deep by 450 mm wide and the column dimensions are 900 mm by 450 mm. The girder carries a uniformly distributed live load of 10 kN/m, a superimposed dead load of 5 kN/m and the self-weight of the girder is 13 kN/m. If E_{cm} = 30,000 MPa,

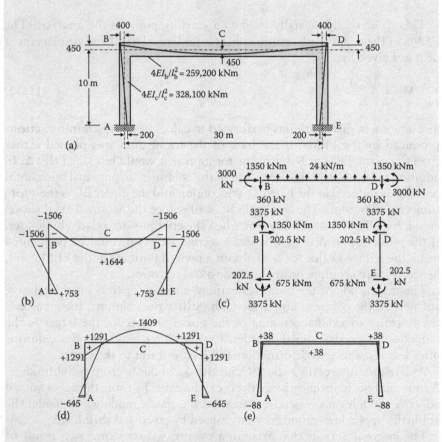

Figure 11.20 Actions in fixed-base portal frame (Example 11.5). (a) Elevation. (b) Moments due to gravity loads (kNm). (c) Equivalent loads due to prestress. (d) Primary + secondary moments due to prestress (kNm). (e) Moments caused by axial restraint (kNm).

the moments caused by the total uniformly distributed load on the girder (live load + dead load + self-weight = 28 kN/m) are calculated using moment distribution and are shown in Figure 11.20b.

By satisfying the serviceability requirements (as discussed in Chapter 5), an estimate of the prestressing force and the tendon profile can be made for both the girder BD and the columns. For the girder, the tendon profile shown in Figure 11.20a is selected and the effective prestress $P_{m,t.BD}$ required to balance the self-weight plus dead load is determined:

$$P_{m,t.BD} = \frac{18 \times 30^2}{8 \times 0.9} = 2250 \text{ kN}$$

If the time-dependent losses are taken as 25%, the average prestressing force in the girder immediately after transfer is $P_{m0.BD} = 3000$ kN.

To determine the effective prestress in the columns, the primary moments in the girder and in the columns at the corner connections B and D are taken to be the same. If the eccentricity in the column at B (to the centroidal axis of the column) is 400 mm, as shown in Figure 11.20a, then $0.4P_{m,t.AB} = 0.45P_{m,t.BD}$. Therefore:

$$P_{m,t.AB} = \frac{2250 \times 0.45}{0.4} = 2531 \text{ kN}$$

The time-dependent losses in the columns are also taken as 25% and the prestressing force immediately after transfer is therefore $P_{m0.AB} = 3375$ kN.

The equivalent load method and moment distribution are used here to calculate the primary and secondary moments caused by prestress. The equivalent loads imposed by the tendon on the concrete members immediately after prestressing are shown in Figure 11.20c. The FEM caused by prestress at each end of span BD is obtained using the results of the fixed-end beam analysed in Example 11.2 – case (a) (and illustrated in Figure 11.11) and is given by:

$$M_{BD}^F = \frac{2P_{m0.BD}z_{d.BD}}{3} = \frac{2 \times 3000 \times 0.9}{3} = 1800 \text{ kNm}$$

The FEMs in the vertical columns due to the straight tendon profile are zero, as was determined for the fixed-end beam analysed in Example 11.2 – case (b). From a moment distribution, the primary and secondary moments caused by prestress are calculated and are illustrated in Figure 11.20d.

To calculate the tertiary effect of axial restraint, the axial shortening of BD immediately after prestressing is estimated using Equation 11.31:

$$\Delta = \frac{3,000 \times 10^3}{30,000 \times 1,200 \times 450} \frac{30,000}{2} = 2.78 \text{ mm}$$

The FEM in the columns is obtained from Equation 11.32 and is given by:

$$M^F = \frac{6 \times 30,000 \times 900^3 \times 450}{10,000^2 \times 12} \times 2.78 = 136.8 \text{ kNm}$$

Moment distribution produces the tertiary moments shown in Figure 11.20e. The restraining tensile axial force induced in the girder BD is only 12.6 kN and, compared with the initial prestress, is insignificant in this case.

11.5 DESIGN OF CONTINUOUS BEAMS

11.5.1 General

The design procedures outlined in Chapter 10 for statically determinate beams can be extended readily to cover the design of indeterminate beams. The selection of tendon profile and magnitude of prestress in a continuous beam is based on serviceability considerations, as is the case for determinate beams. Load balancing is a commonly used technique for making an initial estimate of the level of prestress required to control deflections. The design of individual cross-sections for bending and shear strength, the estimation of losses of prestress and the design of the anchorage zones are the same for all types of beams, irrespective of the number of redundants.

In continuous beams, the satisfaction of concrete stress limits for crack control must involve consideration of both the primary and secondary moments caused by prestress. Concrete stresses resulting from prestress should be calculated using the pressure line, rather than the tendon profile, as the position of the resultant prestress in the concrete.

Because of the relatively large number of dependent and related variables, the design of continuous beams tends to be more iterative than the design of simple beams and more dependent on the experience and engineering judgement of the designer. A thorough understanding of the behaviour of continuous prestressed beams and knowledge of the implications of each design decision are of great benefit.

11.5.2 Service load range: Before cracking

Prior to cracking, the behaviour of a continuous beam is essentially linear and the principle of superposition can be used in the analysis. This means that the internal actions and deformations caused by prestress and those caused by the external loads can be calculated separately using linear analyses and the combined effects obtained by simple summation.

Just as for simple beams, a designer must ensure that a continuous beam is serviceable at the two critical loading stages: immediately after transfer (when the prestress is at its maximum and the applied service loads are usually small) and under the full loads after all losses have taken place (when the prestress is at a minimum and the applied loads are at a maximum).

In order to obtain a good estimate of the in-service behaviour, the prestressing force must be accurately known at each cross-section. This involves a reliable estimate of losses, both short-term and long-term. It is also important to know the load at which flexural cracking is likely to occur. In Sections 5.7.3 and 5.7.4, it was observed that creep and shrinkage gradually relieve the concrete of some of its initial prestress and transfer the resultant compression from the concrete to the bonded reinforcement. Therefore, a reliable estimate of the cracking moment at a particular cross-section must involve consideration of the time-dependent effects of creep and shrinkage.

Prior to cracking, load balancing can be used in design to establish a suitable effective prestressing force and tendon profile. The concept of load balancing was introduced in Section 1.4.3 and involves balancing a pre-selected portion of the applied load (and self-weight) with the transverse equivalent load imposed on the beam by the draped tendons. Under the balanced load w_{bal}, the curvature on each cross-section is zero, the beam does not therefore deflect, and each cross-section is subjected only to the longitudinal axial prestress applied at the anchorages.

By selecting a parabolic tendon profile with the sag z_d as large as cover requirements permit, the minimum prestressing force required to balance w_{bal} is calculated by rearranging Equation 1.7 to give:

$$P = \frac{w_{bal} l^2}{8 z_d} \tag{11.33}$$

In order to control the final deflection of a continuous beam, the balanced load w_{bal} is often taken to be the sustained or permanent load (or some significant percentage of it).

Because of its simplicity, load balancing is probably the most popular approach for determining the prestressing force in a continuous member. Control of deflection is an obvious attraction. However, load balancing does not guard against cracking caused by the unbalanced loads, and it does not ensure that individual cross-sections possess adequate strength.

If the balanced load is small, and hence the prestressing force and prestressing steel quantities are also small, significant quantities of non-prestressed steel may be required to increase the strength of the critical cross-sections and to limit crack widths under full service loads.

At service loads prior to cracking, the concrete stresses on any cross-section of a continuous beam can be calculated easily by considering only the unbalanced load and the longitudinal prestress. The transverse loads imposed on the beam by the draped tendons have been effectively cancelled by w_{bal}. The total moment diagram due to prestress (primary + secondary moments) is equal and opposite to the moment diagram caused by w_{bal}. The primary and secondary moments induced by prestress need not, therefore, enter into the calculations, and there is no need to calculate the hyperstatic reactions at this stage (at least for the determination of concrete stresses). In Example 11.6, the load balancing approach is applied to a two-span continuous member.

In the discussion to this point, the prestressing force has been assumed to be constant throughout the member. In long members, friction losses may be significant and the assumption of constant prestress may lead to serious errors. To account for variations in the prestressing force with distance from the anchorage, a continuous member may be divided into segments. Within each segment, the prestressing force may be assumed constant and equal to its value at the midpoint of the segment. In many cases, it may be acceptable to adopt each individual span as a segment of constant prestress. In other cases, it may be necessary to choose smaller segments to model the effects of variations in prestress more accurately.

It is possible, although rarely necessary, to calculate the equivalent loads due to a continuously varying prestressing force. With the shape of the tendon profile throughout the member and the variation of prestress due to friction and draw-in determined previously, the transverse equivalent load at any point is equal to the curvature of the tendon (obtained by differentiating the equation for the tendon shape twice) times the prestressing force at that point. The effect of prestress due to these non-uniform equivalent transverse loads can then be determined using the same procedures as for uniform loads.

EXAMPLE 11.6 LOAD BALANCING

The idealised parabolic tendons in the two-span beam shown in Figure 11.21 are required to balance a uniformly distributed gravity load of 20 kN/m. The beam cross-section is rectangular: 800 mm deep and 300 mm wide. Determine the concrete stress distribution on the cross-section at B over the interior support when the total uniformly distributed gravity load is 25 kN/m. Assume that the prestressing force is constant throughout.

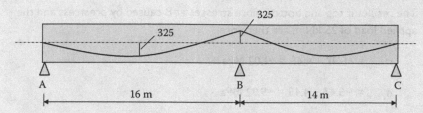

Figure 11.21 Two-span beam (Example 11.6).

In span AB, the tendon sag is $z_{d.AB}$ = 325 + (0.5 × 325) = 487.5 mm and the required prestressing force is obtained from Equation 11.33:

$$P = \frac{20 \times 16^2}{8 \times 0.4875} = 1313 \text{ kN}$$

If P is constant throughout, the required sag in BC may also be obtained from Equation 11.33:

$$z_{d.BC} = \frac{20 \times 14^2}{8 \times 1313} = 0.373 \text{ m} = 373 \text{ mm}$$

and the eccentricity of the tendon at the midpoint of the span BC is equal to:

373 − (0.5 × 325) = 210.5 mm (below the centroidal axis).

Under the balanced load of 20 kN/m, the beam is subjected only to the axial prestress applied at each anchorage. The concrete stress on every cross-section is uniform and equal to:

$$\sigma = -\frac{P}{A} = -\frac{1313 \times 10^3}{800 \times 300} = -5.47 \text{ MPa}$$

The bending moment at B due to the uniformly distributed unbalanced load of 5 kN/m is −142.5 kNm (obtained by moment distribution or an equivalent method of analysis), and the extreme fibre concrete stresses at B caused by the unbalanced moment are:

$$\sigma = \pm \frac{M}{Z} = \pm \frac{142.5 \times 10^6 \times 6}{800^2 \times 300} = \pm 4.45 \text{ MPa}$$

The resultant top and bottom fibre stresses at B caused by prestress and the applied load of 25 kN/m are therefore:

$$\sigma_{c,top} = -5.47 + 4.45 = -1.02 \text{ MPa}$$

$$\sigma_{c,btm} = -5.47 - 4.45 = -9.92 \text{ MPa}$$

The same result could have been obtained by adding the total stresses caused by the equivalent loads (longitudinal plus transverse forces imposed by prestress) to the stresses caused by a uniformly distributed gravity load of 25 kN/m.

11.5.3 Service load range: After cracking

When the balanced load is relatively small, the unbalanced load may cause cracking in the peak moment regions over the interior supports and at mid-span. When cracking occurs, the stiffness of the member is reduced in the vicinity of the cracks. The change in relative stiffness between the positive and negative moment regions causes a redistribution of bending moments. In prestressed members, the reduction of stiffness caused by cracking in a particular region is not as great as in an equivalent reinforced concrete member and the redistribution of bending moments under short-term service loads can usually be ignored. It is therefore usual to calculate beam moments using a linear analysis both before and after cracking.

The effect of cracking should not be ignored, however, when calculating the deflection of the member. A cracked section analysis (see Section 5.8.3) can be used to determine the effective moment of inertia of the cracked section (see Section 5.11.3) and the corresponding initial curvature. After calculating the initial curvature at each end and at the midpoint of a span, the short-term deflection may be determined using Equation 5.165.

Under the sustained loads, the extent of cracking is usually not great. In many partially-prestressed members, the cracks over the interior supports (caused by the peak loads) are completely closed for most of the life of the member. The time-dependent change in curvature caused by creep, shrinkage and relaxation at each support and at mid-span can be calculated using the time analysis of Section 5.7.3 (or Section 5.9.2 if the cracks remain open under the permanent loads). With the final curvature determined at the critical sections, the long-term deflection can also be calculated using Equation 5.165.

Alternatively, long-term deflections may be estimated from the short-term deflections using the approximate expressions outlined in Section 5.11.4.

The control of flexural cracking in a cracked prestressed beam is easily achieved by suitably detailing the bonded reinforcement in the cracked region, as discussed in Sections 5.12.1 and 5.12.2.

11.5.4 Overload range and design resistance in bending

11.5.4.1 Behaviour

The behaviour of a continuous beam in the overload range depends on the ductility of the cross-sections in the regions of maximum moment. If the cross-sections are ductile, their moment–curvature relationships are similar to that shown in Figure 11.22.

Consider the propped cantilever shown in Figure 11.23a. Each cross-section is assumed to possess a ductile moment–curvature relationship. At service loads, bending moments in the beam, even in the post-cracking range, may

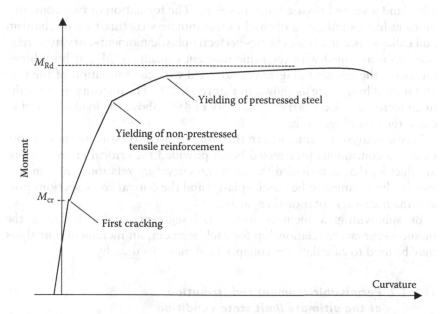

Figure 11.22 Moment–curvature relationship for a ductile prestressed cross-section.

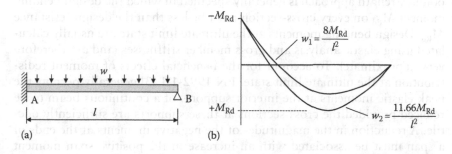

Figure 11.23 Moment redistribution in a propped cantilever. (a) Beam elevation. (b) Bending moment diagrams.

be approximated reasonably using elastic analysis. The magnitude of the negative elastic moment at A caused by the uniformly distributed load w is $wl^2/8$. When the load w causes yielding of the reinforcement on the cross-section at A, a sudden loss of stiffness occurs (as illustrated by the kinks in the moment–curvature relationship in Figure 11.22). Any further increase in load will cause large increases in curvature at A, but only small increases in moment. A constant-moment hinge (or *plastic hinge*) develops at A as the moment capacity is all but exhausted and the curvature becomes large. In reality, the moment at the hinge is not constant, but the rate of increase in moment with curvature in the post-yield range is very small. As loading increases and the moment at the support A remains constant or nearly so, the moment at mid-span increases until it too reaches the design resistance $+M_{Rd}$ and a second plastic hinge develops. The formation of two constant-moment hinges reduces a onefold indeterminate structure to a mechanism and collapse occurs. If an elastic–perfectly plastic moment–curvature relationship is assumed with the same moment capacity $\pm M_{Rd}$ at both hinge locations, the moment diagrams associated with the formation of the first and second hinges are as shown in Figure 11.23b. The ductility at A results in an increase in load carrying capacity of 46% above the load required to cause the first hinge to form.

Plastic analysis techniques can therefore be used to estimate the collapse load of a continuous prestressed beam provided the critical cross-sections are ductile, that is provided the moment–curvature relationships can reasonably be assumed to be elastic–plastic and the critical cross-sections possess the necessary rotational capacity.

By subdividing a member into small segments and calculating the moment–curvature relationship for each segment, an incremental analysis may be used to calculate the collapse load more accurately.

11.5.4.2 Permissible moment redistribution at the ultimate limit state condition

For the design of prestressed concrete continuous structures, a lower bound strength approach is generally specified in which the design bending moment M_{Ed} on every cross-section must be less than the design resistance M_{Rd}. Design bending moments at the ultimate limit state are usually calculated using elastic analysis and gross member stiffnesses (and are therefore very approximate). To account for the beneficial effects of moment redistribution at the ultimate limit state, EN 1992-1-1 [3] generally permits the peak elastic moments at the interior supports of a continuous beam to be reduced provided the cross-sections at these supports are sufficiently ductile. A reduction in the magnitudes of the negative moments at the ends of a span must be associated with an increase in the positive span moment and the resulting distribution of moments must remain in equilibrium with the applied loads. According to EN 1992-1-1 [3], the elastically determined bending moments at any interior support may be reduced by redistribution, provided the designer can show that there is adequate rotational capacity in the peak moment regions. This requirement is deemed to be satisfied for continuous beams and slabs provided:

1. the member is predominantly subject to flexure;
2. the ratios of the lengths of adjacent spans are in the range 0.5–2.0;
3. the ratio of the redistributed moment to the elastic bending moment δ satisfies:

$$\delta \geq 0.44 + 1.25\left(0.6 + \frac{0.0014}{\varepsilon_{cu2}}\right)\left(\frac{x_u}{d}\right) \quad \text{for } f_{ck} \leq 50 \text{ MPa} \qquad (11.34)$$

$$\delta \geq 0.54 + 1.25\left(0.6 + \frac{0.0014}{\varepsilon_{cu2}}\right)\left(\frac{x_u}{d}\right) \quad \text{for } f_{ck} > 50 \text{ MPa} \qquad (11.35)$$

4. in addition, δ must be greater than 0.7 where Class B and Class C reinforcement and tendons are used and δ must be greater than 0.8 where Class A reinforcement is used; and
5. the static equilibrium of the structure after redistribution is used to evaluate all action effects for strength design, including shear checks.

In Equations 11.34 and 11.35, x_u is the depth of the neutral axis at the ultimate limit state after redistribution, d is the effective depth of the section and ε_{cu2} is the ultimate strain specified in Table 4.2.

EN 1992-1-1 [3] cautions that redistribution should not be carried out in situations where the rotational capacity cannot be assessed confidently, such as in the corners of prestressed frames. In addition, for the design of columns in framed structures, the elastic moments obtained from the frame analysis should be used without any redistribution.

11.5.4.3 Secondary effects at the ultimate limit state condition

The design moment M_{Ed} may be calculated as the sum of the moments caused by the factored design load (dead, live, etc., as outlined in Section 2.3.2, Equation 2.2) and the moments resulting from the hyperstatic reactions caused by prestress (with a load factor of 1.0).

Earlier in this chapter, the hyperstatic reactions and the resulting secondary moments were calculated using linear-elastic analysis. Primary moments, secondary moments and the moments caused by the applied loads were calculated separately and summed to obtain the combined effect. Superposition is only applicable, however, when the member behaviour is linear. At overloads, behaviour is highly non-linear and it is not possible to distinguish between the moments caused by the applied loads and those caused by the hyperstatic reactions. Consider the ductile propped cantilever in Figure 11.23. After the formation of the first plastic hinge at A, the beam becomes determinate for all subsequent load increments. With no rotational restraint at the hinge at A, the magnitude of the secondary moment is not at all clear. The total moment and shears can only be determined using a refined analysis that accurately takes into account the various sources of material non-linearity. It is meaningless to try to subdivide the total moments into individual components. The treatment of secondary moments at the ultimate limit state has been studied extensively, see References [4]–[7].

Provided that the structure is ductile and moment redistribution occurs as the collapse load is approached, secondary moments can be ignored in design strength calculations. After all, the inclusion of an uncertain estimate of the secondary moment generally amounts to nothing more than an increase in the support moments and a decrease in the span moment or vice versa, i.e. a redistribution of moments. Since the moments due to the factored design loads at the strength limit state are calculated using elastic analysis, there is no guarantee that the inclusion of the secondary moments (also calculated using gross stiffnesses) will provide better agreement with the actual moments in the structure after moment redistribution.

On the other hand, if the critical section at an interior support is non-ductile, its design needs to be carefully considered. It is usually possible to avoid non-ductile sections by the inclusion of sufficient quantities of compressive reinforcement. If non-ductile sections cannot be avoided, it is recommended that secondary moments (calculated using linear-elastic analysis and gross stiffnesses) are considered at the ultimate limit state. Where the secondary moment at an interior support has the same sign as the moment caused by the applied loads, it is usually conservative to include the secondary moment (with a load factor of 1.0) in the calculation of the design moment M_{Ed}. Where the secondary moment is of opposite sign to the moment caused by the applied loads, it is usually conservative to ignore its effect.

11.5.5 Steps in design

A suitable design sequence for a continuous prestressed concrete member is as follows:

1. Determine the loads on the beam both at transfer and under the most severe load combination for the serviceability limit states. Make an initial selection of concrete strength and establish material properties.

   Using approximate analysis techniques, estimate the maximum design moments at the critical sections in order to make an initial estimate of the cross-section size and self-weight. The moment and deflection coefficients given in Figure 11.24 may prove useful.

   Determine appropriate cross-section sizes at the critical sections. The discussion in Section 10.3 is relevant here. Equation 10.11 may

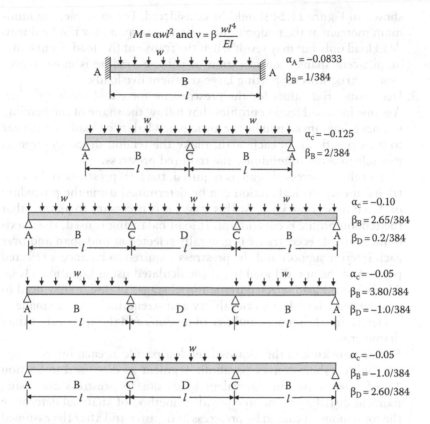

$$M = \alpha wl^2 \text{ and } v = \beta \frac{wl^4}{EI}$$

$\alpha_A = -0.0833$
$\beta_B = 1/384$

$\alpha_c = -0.125$
$\beta_B = 2/384$

$\alpha_c = -0.10$
$\beta_B = 2.65/384$
$\beta_D = 0.2/384$

$\alpha_c = -0.05$
$\beta_B = 3.80/384$
$\beta_D = -1.0/384$

$\alpha_c = -0.05$
$\beta_B = -1.0/384$
$\beta_D = 2.60/384$

Figure 11.24 Moment and deflection coefficients for equal span elastic beams.

be used to obtain cross-sectional dimensions that are suitable from the point of view of flexural strength and ductility. By estimating the maximum unbalanced load, the sustained part of the unbalanced load and by specifying a maximum deflection limit for the structure, a minimum moment of inertia may be selected from Equation 10.5 (if the member is to be crack free) or Equation 10.6 (if cracking does occur). If a fully-prestressed (uncracked) beam is required, Equation 10.1 can be used to determine the minimum section modulus at each critical section.

For continuous beams in building structures, the span-to-depth ratio is usually in the range 24–30, but this depends on the load level and the type of cross-section.

2. Determine the bending moment and shear force envelopes both at transfer and under the full service loads. These envelopes should include the effects of self-weight, superimposed permanent dead and live loads and the maximum and minimum values caused by transient loads. Where they are significant, pattern loadings such as those

shown in Figure 11.24 should be considered. For example, the minimum moment at the midpoint of a particular span may not be due to dead load only but may result when the transient live load occurs only on adjacent spans. Consideration of pattern loading is most important in structures supporting large transient live loads.

3. Determine trial values for the prestressing force and tendon profile. Assume idealised tendon profiles that follow the shape of the bending moment diagram caused by the anticipated balanced loads (or as near to it as practical). In each span, make the tendon drape as large as possible in order to minimise the required prestress.

   If a fully-prestressed beam is required, the trial prestress and eccentricity at each critical section can be determined using the procedure outlined in Section 5.4.1. At this stage, it is necessary to assume that the tendon profile is concordant. If load balancing is used, the maximum available eccentricity is generally selected at mid-span and over each interior support and the prestress required to balance a selected portion of the applied load (w_{bal}) is calculated using Equation 11.33. The balanced load selected in the initial stages of design may need to be adjusted later when serviceability and strength checks are made.

   Determine the number and size of tendons and the appropriate duct diameter(s).

4. Replace the kink in the idealised tendon profile at each interior support with a short convex parabolic segment as discussed in Section 11.3.5. Determine the equivalent loads due to prestress and using moment distribution (or an equivalent method of analysis) determine the total moment caused by prestress at transfer and after the assumed time-dependent losses. By subtracting the primary moments from the total moments, calculate the secondary moment diagram and, from statics, determine the hyperstatic reactions at each support.

5. Concrete stresses at any cross-section caused by prestress (including both primary and secondary effects) and the applied loads may now be checked at transfer and after all losses. If the beam is fully-prestressed, the trial estimate of prestress made in Step 3 was based on the assumption of a concordant tendon profile and secondary moments were ignored. If secondary moments are significant, stresses calculated here may not be within acceptable limits and a variation of either the prestressing force or the eccentricity may be required.

6. Calculate the losses of prestress and check the assumptions made earlier.

7. Check the design bending resistance at each critical section. If necessary, additional non-prestressed tensile reinforcement may be used to increase strength. Add compressive reinforcement to improve ductility, if required. Some moment redistribution at the ultimate limit state may be permissible to reduce peak negative moments at interior supports, provided that cross-sections at the supports have adequate ductility.

8. Check the deflection at transfer and the final long-term deflection. For partially-prestressed designs, check crack control in regions of peak moment. Consider the inclusion of non-prestressed steel to reduce time-dependent deformations, if necessary. Adjust the section size or the prestress level (or both), if the calculated deflection is excessive.

9. Check the design shear resistance of beam (and the torsional resistance, if applicable) in accordance with the provisions of Chapter 7. Design suitable shear reinforcement where required.

10. Design the anchorage zone using the procedures presented in Chapter 8.

Note: Durability and fire protection requirements are usually satisfied by an appropriate choice of concrete strength and cover to the tendons in Steps 1 and 3.

EXAMPLE 11.7

Design the four-span beam shown in Figure 11.25. The beam has a uniform I-shaped cross-section and carries a uniformly distributed dead load of 25 kN/m (not including self-weight) and a transient live load of 20 kN/m. Controlled cracking is to be permitted at peak loads. The beam is prestressed by jacking simultaneously from each end, thereby maintaining symmetry of the prestressing force about the central support C and avoiding excessive friction losses. At 28 days, f_{ck} = 50 MPa (f_{cd} = 33.3 MPa); at transfer, $f_{ck}(t)$ = 40 MPa ($f_{cd}(t)$ = 26.67 MPa) and f_{pk} = 1860 MPa.

---

(1) and (2): The bending moments caused by the applied loads must first be determined. Because the beam is symmetrical about the central support at C, the bending moment envelopes can be constructed from the moment diagrams shown in Figure 11.26 caused by the distributed load patterns shown. These moment diagrams were calculated for a unit distributed load (1 kN/m) using moment distribution.

If the self-weight is estimated at 15 kN/m, the total dead load is 40 kN/m, and the factored design loads are (Equation 2.2):

$$\gamma_G w_G = 1.35 \times 40 = 54 \text{ kN/m} \quad \text{and} \quad \gamma_Q w_Q = 1.5 \times 20 = 30 \text{ kN/m}$$

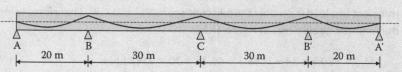

A		B		C		B'		A'
	20 m		30 m		30 m		20 m	

Figure 11.25 Elevation of beam (Example 11.7).

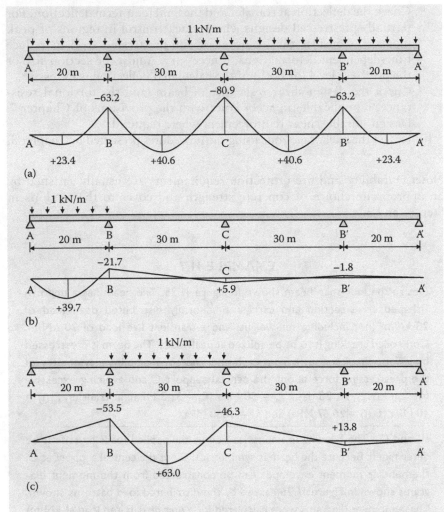

Figure 11.26 Bending moment diagrams (kNm) due to unit distributed loads (Example 11.7). (a) Load Case 1: (1 kN/m throughout). (b) Load Case 2: (1 kN/m on span AB only). (c) Load Case 3: (1 kN/m on span BC only).

The maximum design moment M_{Ed} occurs over the support C, when the transient live load is on only the adjacent spans BC and CB'. Therefore, using the moment coefficients in Figure 11.26:

$$M_{Ed} = -80.9 \times 54 + (-46.3 - 46.3) \times 30 = 7147 \text{ kNm}$$

The overall dimensions of the cross-section are estimated using Equation 10.12 (which is valid provided the compressive stress block at the ultimate limit state is within the flange of the I-section):

$$bd^2 \geq \frac{7147 \times 10^6}{0.27 \times 33.3} = 794 \times 10^6 \text{ mm}^3$$

Try $b = 750$ mm, $d = 1100$ mm and $h = 1250$ mm.

The span-to-depth ratio (l/h) for the interior span is 24 which should prove acceptable from a serviceability point of view.

To obtain a trial flange thickness, find the depth of the compressive stress block. The volume of the stress block is $F_{cd} = \eta f_{cd} \lambda x b$ (see Equation 6.6), with $\lambda = 0.8$ and $\eta = 1.0$ for 50 MPa concrete (Equations 6.2 and 6.4, respectively). With the lever arm between F_{cd} and the resultant tensile force taken to be $0.85d$, then:

$$\lambda x = \frac{M_{Ed}}{0.85d\eta f_{cd}b} = \frac{7147 \times 10^6}{0.85 \times 1100 \times 1.0 \times 33.33 \times 750} = 306 \text{ mm}$$

Adopt a tapering flange 250 mm thick at the tip and 350 mm thick at the web.

To ensure that the web width is adequate for shear, it is necessary to ensure that web-crushing does not occur. If the vertical component of the prestressing force P_v is ignored, then based on Equation 7.8:

$$V_{Ed} \leq V_{Rd,max} = \frac{\alpha_{cw} b_w z v_1 f_{cd}}{(\cot \theta_v + \tan \theta_v)} \tag{11.7.1}$$

The maximum shear force V_{Ed} also occurs adjacent to support C when live load is applied to spans BC and CB′ and is equal to 1251 kN. Assuming the average axial prestress is approximated by $\sigma_{cp} = 5$ MPa, Equation 7.10 gives $\alpha_{cw} = 1.15$, the distance between the tension and compression chords is $z = 0.8h = 1000$ mm, $v_1 = 0.48$ (from Equation 7.9) and $\cot \theta_v = 2.5$, Equation 11.7.1 can be rearranged to give:

$$b_w \geq \frac{(2.5 + 0.4) \times 1251 \times 10^3}{1.15 \times 1000 \times 0.48 \times 33.33} = 197 \text{ mm}$$

It is advisable to select a web width significantly greater than this minimum value in order to avoid unnecessarily large quantities of transverse steel and the resulting steel congestion. Duct diameters of about 100 mm are anticipated, with only one duct in any horizontal plane through the web. With these considerations, the web width is taken to be $b_w = 300$ mm.

The trial cross-section and section properties are shown in Figure 11.27. The self-weight is $24 \times 0.645 = 15.5$ kN/m, which is a little more than 3% higher than originally assumed. The revised value of M_{Ed} at C is 7201 kNm.

(3) If 100 mm ducts are assumed (side by side in the flanges), the cover to the reinforcement is 40 mm, and 12 mm stirrups are used, the maximum eccentricity over an interior support and at mid-span is:

$$e_{max} = 625 - 40 - 12 - (\tfrac{3}{4} \times 100) = 498 \text{ mm}$$

The maximum sag (or drape) in the spans BC and CB′ is therefore:

$$(z_{d.BC})_{max} = 2 \times 498 = 996 \text{ mm}$$

The balanced load is taken to be 32 kN/m (which is equal to self-weight plus about two-thirds of the additional dead load). From Equation 11.33, the required average effective prestress in span BC is:

$$(P_{m,t})_{BC} = \frac{32 \times 30^2}{8 \times 0.996} = 3614 \text{ kN}$$

If the friction loss between the midpoint of span AB and the midpoint of BC is assumed to be 15% (to be subsequently checked), then:

$$(P_{m,t})_{AB} = \frac{3614}{0.85} = 4252 \text{ kN}$$

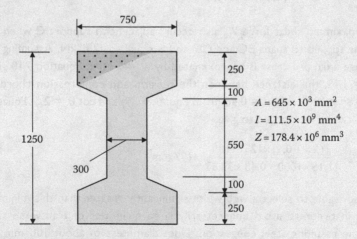

$A = 645 \times 10^3 \text{ mm}^2$

$I = 111.5 \times 10^9 \text{ mm}^4$

$Z = 178.4 \times 10^6 \text{ mm}^3$

Figure 11.27 Trial section dimensions and properties.

and the required drape in span AB is:

$$z_{d.AB} = \frac{32 \times 20^2}{8 \times 4252} = 0.376 \text{ m} = 376 \text{ mm}$$

The idealised tendon profiles for spans AB and BC are shown in Figure 11.28, together with the corresponding tendon slopes and friction losses (calculated from Equation 5.148 with $\mu = 0.19$ and $k = 0.01$). The friction losses at the mid-span of BC are 13.8%, and if the time-dependent losses in BC are assumed to be 15%, then the required jacking force is:

$$P_j = \frac{3614}{0.862 \times 0.85} = 4932 \text{ kN}$$

From Table 4.8, the cross-sectional area of a 12.5 mm diameter seven-wire strand is 93.0 mm², $f_{pk} = 1860$ MPa and $f_{p0.1k} = 1600$ MPa. The maximum stress in the tendon at jacking is the smaller of 0.8 $f_{pk} = 1488$ MPa or 0.9 $f_{p0.1k} = 1440$ MPa (see Section 5.3). Therefore, the maximum jacking force per strand is 1440 × 93 × 10^{-3} = 133.9 kN. The minimum number of strands is therefore 4932/133.9 = 37.

Try two cables each containing 19 strands ($A_p = 1767$ mm²/cable).

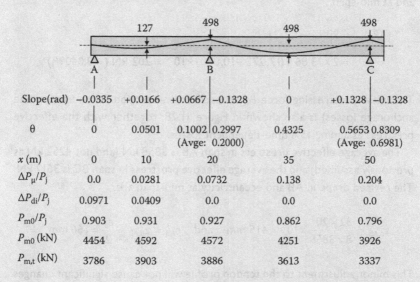

	x =	127	498		498		498
			A		B		C
Slope(rad)	−0.0335	+0.0166	+0.0667	−0.1328	0	+0.1328	−0.1328
θ	0	0.0501	0.1002	0.2997	0.4325	0.5653	0.8309
			(Avge: 0.2000)			(Avge: 0.6981)	
x (m)	0	10	20		35	50	
$\Delta P_\mu/P_j$	0.0	0.0281	0.0732		0.138	0.204	
$\Delta P_{di}/P_j$	0.0971	0.0409	0.0		0.0	0.0	
P_{m0}/P_j	0.903	0.931	0.927		0.862	0.796	
P_{m0} (kN)	4454	4592	4572		4251	3926	
$P_{m,t}$ (kN)	3786	3903	3886		3613	3337	

Figure 11.28 Friction losses and tendon forces.

The two cables are to be positioned so that they are located side by side in the top flange over the interior supports and in the bottom flange at mid-span of BC but are located one above the other in the web. The position of the resultant tension in the tendons should follow the desired tendon profile.

The loss of prestress due to a 6 mm draw-in at the anchorage is calculated as outlined in Section 5.10.2.4. The slope of the prestress line adjacent to the anchorage at A is:

$$\beta = \frac{0.0281 \times P_i}{l_{AB}/2} = 13.86 \text{ N/mm}$$

and, from Equation 5.150, the length of beam associated with the draw-in losses is:

$$L_{di} = \sqrt{\frac{195,000 \times 2 \times 1,767 \times 6}{13.86}} = 17,272 \text{ mm}$$

The loss of force at the jack due to slip at the anchorage is given by Equation 5.151:

$$(\Delta P_{di})_{L_{di}} = 2\beta L_{di} = 2 \times 13.86 \times 17,272 \times 10^{-3} = 479 \text{ kN} \ (= 0.0971 P_j)$$

and at mid-span:

$$(\Delta P_{di})_{AB} = 2\beta(L_{di} - l_{AB}/2)$$
$$= 2 \times 13.86 \times (17,272 - 10,000) \times 10^{-3} = 202 \text{ kN} \ (= 0.0409 P_i)$$

The initial prestressing force P_{m0} at each critical section (after friction and anchorage losses) is also shown in Figure 11.28, together with the effective prestress assuming 15% time-dependent losses.

The average effective prestress in span AB is 3858 kN (and not 4252 kN as previously assumed) and the average effective prestress in span BC is 3612 kN. The revised drape in AB and eccentricity at mid-span are:

$$z_{d.AB} = \frac{32 \times 20^2}{8 \times 3858} \times 10^3 = 415 \text{ mm} \quad \text{and} \quad e_{AB} = z_{d.AB} - \frac{e_B}{2} = 166 \text{ mm}$$

This minor adjustment to the tendon profile will not cause significant changes in the friction losses.

(4) The beam is next analysed under the equivalent transverse loads caused by the effective prestress. The sharp kinks in the tendons over the supports B and C are replaced by short lengths with a convex parabolic shape, as illustrated and analysed in Example 11.3. In this example, it is assumed that the idealised tendons provide a close enough estimate of moments due to prestress.

The equivalent uniformly distributed transverse load due to the effective prestress is approximately 32 kN/m (upwards). Using the moment diagram in Figure 11.26a, the total moments due to prestress at B and C are:

$$(M_{pt})_B = +63.2 \times 32 = 2022 \text{ kNm}$$

$$(M_{pt})_C = +80.9 \times 32 = 2589 \text{ kNm}$$

The secondary moments at B and C are obtained by subtracting the primary moments corresponding to the average prestress in each span (as was used for the calculation of total moments):

$$(M_{ps})_B = (M_{pt})_B - (P_{m,t}e)_B$$

$$= 2022 - 0.5 \times (3858 + 3612) \times 0.498 = 162 \text{ kNm}$$

$$(M_{ps})_C = (M_{pt})_C - (P_{m,t}e)_C = 2589 - 3612 \times 0.498 = 790 \text{ kNm}$$

The total and secondary moment diagrams are shown in Figure 11.29, together with the corresponding hyperstatic reactions. It should be noted that, in the real beam, the equivalent transverse load varies along the beam as the prestressing force varies and the moment diagrams shown in Figure 11.29 are only approximate. A more accurate estimate of the moments due to prestress and the hyperstatic reactions can be made by dividing each span into smaller segments (say four per span) and assuming constant prestress in each of these segments.

(5) It is prudent to check the concrete stresses at transfer. The equivalent transverse load at transfer is $32/0.85 = 37.6$ kN/m ($\uparrow$) and the self-weight is 15.5 kN/m ($\downarrow$). Therefore, the unbalanced load is 22.1 kN/m ($\uparrow$). At support C, the moment caused by the uniformly distributed unbalanced load is (see Figure 11.26a):

$$(M_{unbal})_C = +80.9 \times 22.1 = 1788 \text{ kNm}$$

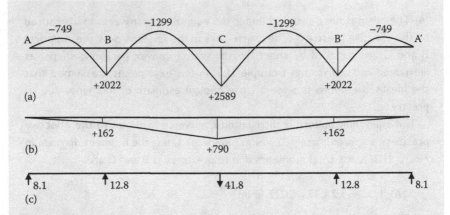

Figure 11.29 Moments and reactions caused by the average effective prestress. (a) Total moment caused by prestress (kNm). (b) Secondary moments (kNm). (c) Hyperstatic reactions (kN).

and the initial prestressing force at C is 3926 kN. The extreme fibre concrete stresses immediately after transfer at C are:

$$\sigma_{top} = -\frac{3926 \times 10^3}{645 \times 10^3} - \frac{1788 \times 10^6}{178.4 \times 10^6} = -6.09 - 10.02 = -16.11 \text{ MPa}$$

$$\sigma_{btm} = -\frac{3926 \times 10^3}{645 \times 10^3} + \frac{1788 \times 10^6}{178.4 \times 10^6} = -6.09 + 10.02 = +3.93 \text{ MPa}$$

The mean tensile strength at transfer is f_{ctm} = 3.5 MPa and, therefore, cracking at support C is likely to occur at transfer. Bonded reinforcement should therefore be provided in the bottom of the member over support C to control cracking at transfer. For this level of tension, it is reasonable to calculate the resultant tensile force on the concrete (assuming no cracking) and supply enough non-prestressed steel to carry this tension with a steel stress of 200 MPa. In this case, the tensile zone on the uncracked section lies entirely within the 250 thick bottom flange and the resultant tension (determined from the calculated stress distribution) is 361 kN and therefore:

$$(A_s)_{btm} = \frac{361 \times 10^3}{200} = 1806 \text{ mm}^2$$

Use four 25 mm diameter reinforcing bars (1964 mm²) or equivalent.

As an alternative to the inclusion of this non-prestressed reinforcement, the member might be *stage stressed*, where only part of the prestress is initially transferred to the concrete when just the self-weight is acting and the remaining prestress is applied when the sustained dead load (or part of it) is in place. This would avoid the situation considered earlier where the maximum prestress is applied when the minimum external load is acting.

Similar calculations are required to check for cracking at other sections at transfer. At support B, $(M_{unbal})_B = +63.2 \times 22.1 = 1397$ kNm, $P_{m0} = 4572$ kN, and $\sigma_{btm} = 0.74$ MPa. Cracking will not occur at B at transfer. Evidently, support C is the only location where cracking is likely to occur at transfer.

For this partially-prestressed beam, before conditions under full loads can be checked (using cracked section analyses in the cracked regions), it is necessary to determine the amount of non-prestressed steel required for strength.

(6) In this example, the time-dependent losses estimated earlier are assumed to be satisfactory. In practice, of course, losses should be calculated using the procedures of Section 5.10.3 and illustrated in Example 10.1.

(7) The strength of each cross-section is now checked. For the purpose of this example, calculations are provided only for the critical section at support C. From Step 3, the design moment due to the factored dead plus live loads (assuming elastic analysis) is -7201 kNm. The secondary moment can be included with a load factor of 1.0. Therefore:

$$M_{Ed} = -7201 + 790 = -6411 \text{ kNm}$$

The inclusion of the secondary moment here is equivalent to a redistribution of moment at C of 11%. The secondary moment will cause a corresponding increase in the positive moments in the adjacent spans. If the cross-section at C is ductile, a further redistribution of moment may be permissible (as outlined in Section 11.5.4.2). No additional redistribution of moment is considered here.

The minimum required design moment resistance at C is $M_{Rd} = 6411$ kNm. Using the procedure outlined in Section 6.3.4, the strength of a cross-section with flange width $b = 750$ mm and containing $A_p = 3534$ mm^2 at $d_p = 1123$ mm is:

$$M_{Rd1} = 5037 \text{ kNm (with } x = 245.8 \text{ mm)}$$

Additional non-prestressed tensile reinforcement $(A_s)_{top}$ is required in the top of the cross-section at C. If the distance from the reinforcement to the compressive face is $d_o = 1185$ mm, then A_s can be calculated using Equation 6.25:

$$(A_s)_{top} \geq \frac{M_{Ed} - M_{Rd1}}{f_{yd}z} = \frac{(6411 - 5037) \times 10^6}{435 \times 0.9 \times (1185 - 0.8 \times 245.8)} = 3551 \text{ mm}^2$$

Use eight 25 mm diameter bars in the top over support C (3928 mm²). This is in addition to the four 25 mm bars required in the bottom of the section for crack control at transfer. These bottom bars in the compressive zone at the ultimate limit state will improve ductility. From a strength analysis of the proposed cross-section, with reinforcement details shown in Figure 11.30, the section at C satisfies the strength and moment redistribution requirements of EN 1992-1-1 [3].

Similar calculations show that four 25 mm diameter bars are required in the negative moment region over the first interior support at B and B', but at the mid-span region in all spans, the prestressing steel alone provides adequate moment capacity.

(8) It is necessary to check crack control under full service loads. Results are provided for the cross-section at support C. With the effective prestress balancing 32 kN/m, the unbalanced sustained load is $w_{unbal.sus}$ = 25 + 15.5 − 32 = 8.5 kN/m and the unbalanced transient load is 20 kN/m. Using the moment coefficients in Figure 11.26 and the transient live load only on spans BC and CB', the maximum unbalanced moment at support C is:

$$(M_{unbal})_C = 8.5 \times (-80.9) + 20 \times (-92.6) = -2540 \text{ kNm}$$

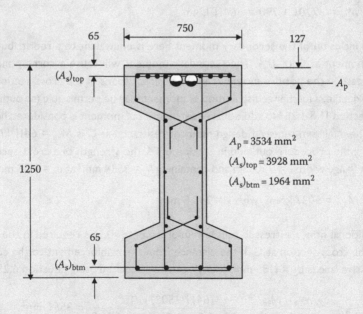

$A_p = 3534 \text{ mm}^2$

$(A_s)_{top} = 3928 \text{ mm}^2$

$(A_s)_{btm} = 1964 \text{ mm}^2$

Figure 11.30 Reinforcement details for cross-section at support C (Example 11.7).

With the effective prestress at C, $P_{m,t}$ = 3337 kN, the extreme top fibre stress under the unbalanced loads is:

$$\sigma_{c,top} = -\frac{3337 \times 10^3}{645 \times 10^3} + \frac{2540 \times 10^6}{178.4 \times 10^6} = -5.17 + 14.24 = +9.07 \text{ MPa}$$

With the tensile strength taken as f_{ctm} = 3.5 MPa, cracking will occur under the full unbalanced moment. The error associated with estimates of the cracking moment based on elastic stress calculation may be significantly large. As was discussed in Section 5.7.3 and illustrated in Example 5.5, creep and shrinkage may cause a large redistribution of stress on the cross-section with time, particularly when the cross-section contains significant quantities of non-prestressed reinforcement (as is the case here). If a more accurate estimate of stresses is required, a time analysis is recommended (see Section 5.7.3).

A cracked section analysis, similar to that outlined in Section 5.8.3, is required to calculate the loss of stiffness due to cracking and the increment of tensile steel stress, in order to check crack control. The maximum in-service moment at C is equal to the sum of the moment caused by the full external service loads and the secondary moment:

$$M_C = -80.9 \times (25 + 15.5) + 20 \times (-92.6) + 790 = -4338 \text{ kNm}$$

A cracked section analysis reveals that the tensile stress in the non-prestressed top steel at this moment is only 106 MPa, which is much less than the increment of 200 MPa specified in EN 1992-1-1 [3] and given in Table 5.5 for 25 mm diameter bars if the maximum crack width is to be limited to 0.3 mm. Crack widths should therefore be acceptably small.

This design example is taken no further here. Deflections are unlikely to be excessive, but should be checked using the procedures outlined in Section 5.11. The design for shear and the design of the anchorage zones are in accordance with the discussions in Chapter 10.

REFERENCES

1. Ranzi, G. and Gilbert, R I. 2015. *Structural Analysis: Principles, Methods and Modelling.* Boca Raton, FL: CRC Press, Taylor & Francis Group.
2. Cross, H. 1930. Analysis of continuous frames by distributing fixed-end moments. *Transactions of the American Society of Civil Engineers*, 96(1793), 1–10.
3. EN 1992-1-1. 2004. Eurocode 2: Design of concrete structures – Part 1-1: General rules and rules for buildings. British Standards Institution, London, UK.

4. Lin, T.Y. and Thornton, K. 1972. Secondary moments and moment redistribution in continuous prestressed concrete beams. *Journal of the Prestressed Concrete Institute*, 17(1), 1–20.
5. Mattock, A.H. 1972. Secondary moments and moment redistribution in continuous prestressed concrete beams. Discussion of Lin and Thornton 1972. *Journal of the Prestressed Concrete Institute*, 17(4), 86–88.
6. Nilson, A.H. 1978. *Design of Prestressed Concrete*. New York: Wiley.
7. Warner, R.F. and Faulkes, K.A. 1983. Overload behaviour and design of continuous prestressed concrete beams. Presented at the *International Symposium on Non-Linearity and Continuity in Prestressed Concrete*, University of Waterloo, Waterloo, Ontario, Canada.

Chapter 12

Two-way slabs

Behaviour and design

12.1 INTRODUCTION

Post-tensioned concrete floors form a large proportion of all prestressed concrete construction and are economically competitive with reinforced concrete slabs in most practical medium- to long-span situations.

Prestressing overcomes many of the disadvantages associated with reinforced concrete slabs. Deflection, which is almost always the governing design consideration, is better controlled in post-tensioned slabs. A designer is better able to reduce or even eliminate deflection by a careful choice of prestress. More slender slab systems are therefore possible, and this may result in increased head room or reduced floor-to-floor heights. Prestress also inhibits cracking and may be used to produce crack-free and watertight floors. Prestressed slabs generally have simple uncluttered steel layouts. Steel fixing and concrete placing are therefore quicker and easier. In addition, prestress improves punching shear (see Chapter 7) and reduces formwork stripping times and formwork costs. On the other hand, prestressing often produces significant axial shortening of slabs and careful attention to the detailing of movement joints is frequently necessary.

In this chapter, the analysis and design of the following common types of prestressed concrete slab systems are discussed (each type is illustrated in Figure 12.1):

1. *one-way slabs*;
2. *edge-supported two-way slabs* are rectangular slab panels supported on all four edges by either walls or beams. Each panel edge may be either continuous or discontinuous;
3. *flat plate slabs* are continuous slabs of constant thickness supported by a rectangular grid of columns;

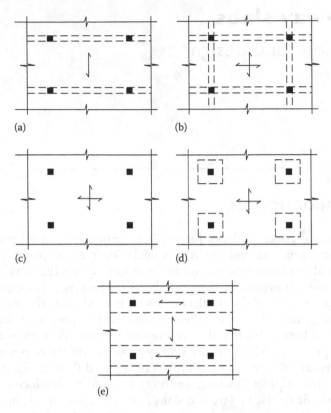

Figure 12.1 Plan views of different slab systems. (a) One-way slab. (b) Edge-supported two-way slab. (c) Flat plate. (d) Flat slab with drop panels. (e) Band beam and slab.

4. *flat slabs with drop panels* are similar to flat plate slabs but with local increases in the slab thickness (drop panels) over each supporting column; and

5. *band-beam and slab systems* comprise wide shallow continuous prestressed beams in one direction (generally the longer span) with one-way prestressed or reinforced slabs in the transverse direction (generally the shorter span).

Almost all prestressed slabs are post-tensioned using draped tendons. Use is often made of flat-ducted tendons consisting five or less strands in a flat sheath with fan-shaped anchorages, as shown in Figure 12.2. Individual strands are usually stressed one at a time using light handheld hydraulic jacks. The flat ducts are structurally efficient and allow maximum tendon eccentricity and drape. These ducts are most often grouted after stressing to provide bond between the steel and the concrete.

In North America and elsewhere, unbonded construction is often used for slabs. Single plastic-coated greased tendons are generally used, resulting in

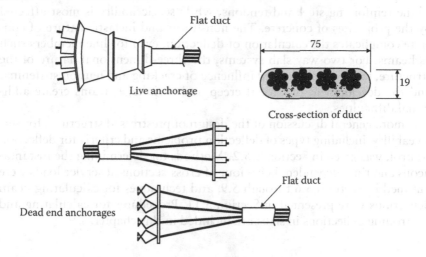

Flat duct

Live anchorage

75

19

Cross-section of duct

Flat duct

Dead end anchorages

Figure 12.2 Details of typical flat-ducted tendons and anchorages.

slightly lower costs, small increases in available tendon drape, the elimination of the grouting operation (therefore reducing cycle times) and reduced friction losses. However, unbonded construction also leads to reduced flexural strength, reduced crack control (additional bonded reinforcement is often required), possible safety problems if a tendon is lost or damaged (by corrosion, fire, accident) and increased demolition problems. Single strands are also more difficult to fix to profile.

Prestressed concrete slabs are typically thin in relation to their spans. If a slab is too thin, it may suffer excessively large deflections when fully loaded or exhibit excessive camber after transfer. The initial selection of the thickness of a slab is usually governed by the serviceability requirements for the member. The selection is often based on personal experience or on recommended maximum span-to-depth ratios. While providing a useful starting point in design, such a selection of slab thickness does not necessarily ensure that serviceability requirements are satisfied. Deflections at all critical stages in the slab's history must be calculated and limited to acceptable design values. Failure to predict deflections adequately has frequently resulted in serviceability problems.

In slab design, *excessive deflection* is a relatively common type of failure. This is particularly true for slabs supporting relatively large transitory live loads or for slabs not subjected to their full service loads until some considerable time after transfer. EN 1992-1-1 [1] requires that the camber, deflection and vibration frequency and amplitude of slabs must be within acceptable limits at service loads. In general, however, little guidance is given as to how this is to be done.

The service load behaviour of a concrete structure can be predicted far less reliably than its strength. Strength depends primarily on the properties of the reinforcing steel and tendons, while serviceability is most affected by the properties of concrete. The non-linear and inelastic nature of concrete complicates the calculation of deflection, even for line members such as beams. For two-way slab systems, the three-dimensional nature of the structure, the less well-defined influence of cracking and tension stiffening, and the development of biaxial creep and shrinkage strains create additional difficulties.

A more general discussion of the design of prestressed structures for serviceability, including types of deflection problems and criteria for deflection control, was given in Section 2.5.2. Methods for determining the instantaneous and time-dependent behaviour of cross-sections at service loads were outlined in Sections 5.6 through 5.9, and techniques for calculating beam deflections were presented in Section 5.11. Procedures for calculating and controlling deflections in slabs are included in this chapter.

12.2 EFFECTS OF PRESTRESS

As discussed previously, the prestressing operation results in the imposition of both longitudinal and transverse forces on post-tensioned members. The concentrated longitudinal prestress P produces a complex stress distribution immediately behind the anchorage and the design of the anchorage zone requires careful attention (see Chapter 8). At sections further away from the anchorage, the longitudinal prestress applied at the anchorage causes a linearly varying compressive stress over the depth of the slab. If the longitudinal prestress is applied at the centroidal axis (which is generally at the mid-depth of the slab), this compressive stress is uniform over the slab thickness and equal to P/A.

We have already seen that wherever a change in direction of the tendon occurs, a transverse force is imposed on the member. For a parabolic tendon profile, such as that shown in Figure 12.3a, the curvature is constant along the tendon and hence the transverse force imposed on the member is uniform along its length (if P is assumed to be constant). From Equation 1.7, the uniformly distributed transverse force is:

$$w_p = \frac{8Pz_d}{l^2} \tag{12.1}$$

where z_d is the sag of the parabolic tendon and l is the span. If the cable spacing is uniform across the width of a slab and P is the prestressing force per unit width of slab, then w_p is the uniform upward load per unit area.

The cable profile shown in Figure 12.3a, with the sharp kink located over the internal support, is an approximation of the more realistic and practical profile shown in Figure 12.3b. The difference between the effects of the idealised and practical profiles is discussed in Section 11.3.5 for

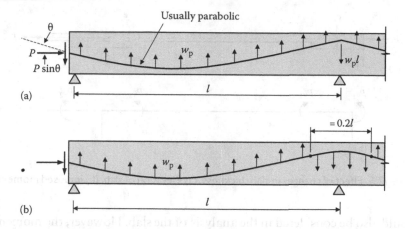

(a)

(b)

Figure 12.3 Tendon profiles in a continuous slab. (a) Idealised tendon profile. (b) Practical tendon profile.

continuous beams. The idealised profile is more convenient for the analysis and design of continuous slabs, and the error introduced by the idealisation is usually not significant.

The transverse load w_p causes moments and shears that usually tend to be opposite in sign to those produced by the external loads. In Figure 12.4, the elevation of a prestressing tendon in a continuous slab is shown, together with the transverse loads imposed on the slab by the tendon in each span. If the slab is a two-way slab, with prestressing tendons placed in two orthogonal directions, the total transverse load caused by the prestress is the sum of w_p for the tendons in each direction.

The longitudinal prestress applied at the anchorage may also induce moments and shears in a slab. At changes in slab thickness, such as those that occur in a flat slab with drop panels, the anchorage force P is eccentric with respect to the centroidal axis of the section, as shown in Figure 12.5a. The moments caused by this eccentricity are indicated in Figure 12.5b and

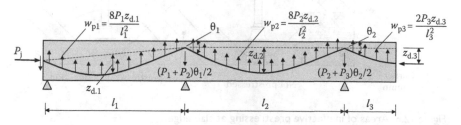

Figure 12.4 Transverse loads imposed by tendons in one direction.

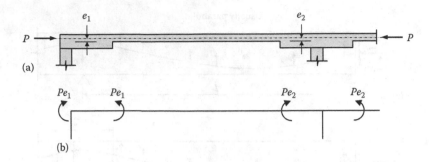

(a)

(b)

Figure 12.5 Effect of changes in slab thickness. (a) Elevation of slab. (b) Imposed moments.

should also be considered in the analysis of the slab. However, the moments produced by relatively small changes in slab thickness tend to be small compared with those caused by cable curvature and, if the thickening is below the slab, it is conservative to ignore them.

At some distance from the slab edge, the concentrated anchorage forces have dispersed and the slab is uniformly stressed. The so-called angle of dispersion 2β (shown in Figure 12.6) determines the extent of the slab where the prestress is not effective. Specifications of the angle of dispersion vary considerably. It is claimed in some trade literature [2] that tests have shown 2β to be 120°. In EN 1992-1-1 [1], β is specified conservatively as 33.7° (so that the angle of dispersion is 67.4°).

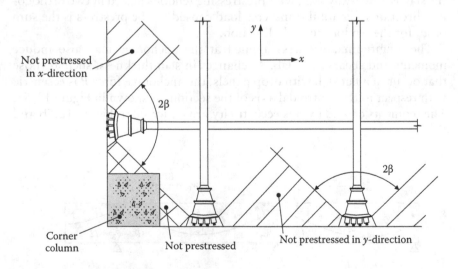

Figure 12.6 Areas of ineffective prestressing at slab edge.

Care must be taken in the design of the hatched areas of the slab shown in Figure 12.6, where the prestress in one or both directions is not effective. It is good practice to include a small quantity of bonded non-prestressed reinforcement in the bottom of the slab perpendicular to the free edge in all exterior spans. An area of non-prestressed steel of about $0.0015bh$ is usually sufficient for crack control, where b and h are the width and depth of the slab. In addition, when checking the punching shear strength at the corner column in Figure 12.6, the beneficial effect of prestress is not available over the full critical shear perimeter. At sections remote from the slab edge, the average P/A stresses are uniform across the entire slab width and do not depend on either the value of β or variations of cable spacing from one region of the slab to another.

12.3 BALANCED LOAD STAGE

Under transverse loads, two-way panels deform into dish-shaped surfaces, as shown in Figure 12.7. The slab is curved in both principal directions and therefore bending moments exist in both directions. In addition, part of the applied load is resisted by twisting moments which develop in the slab at all locations except the lines of symmetry.

The prestressing tendons are usually placed in two directions parallel to the panel edges, each tendon providing resistance to a share of the applied load. The transverse load on the slab produced by the tendons in one direction adds to (or subtracts from) the transverse load imparted by the tendons in the perpendicular direction. For edge-supported slabs, the portion of the load to be carried by tendons in each direction is more or less arbitrary, the only strict requirement is the satisfaction of statics. For flat slabs, the total load must be carried by tendons in each direction from column line to column line.

The concept of utilising the transverse forces resulting from the curvature of the draped tendons to balance a selected portion of the applied load is useful from the point of view of controlling deflections. In addition to providing the basis for establishing a suitable tendon profile, load balancing allows the determination of the prestressing force required to produce zero deflection in a slab panel under the selected balanced load.

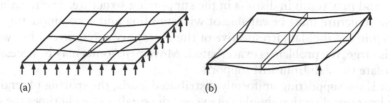

(a) (b)

Figure 12.7 Deformation of interior two-way slab panels. (a) Edge-supported slab. (b) Flat slab.

At the balanced load, the slab is essentially flat (no curvature) and is subjected only to the effects of the prestressing forces applied at the anchorages. A slab of uniform thickness is subjected only to uniform compression (P/A) in the directions of the orthogonal tendons. With the state of the slab under the balanced load confidently known, the deflection due to the unbalanced portion of the load may be calculated by one of the techniques discussed later in this chapter. The calculation of the deflection of a prestressed slab is usually more reliable than for a conventionally reinforced slab, because only a portion of the total service load needs to be considered (the unbalanced portion) and, unlike reinforced concrete slabs, prestressed slabs are often uncracked at service loads.

To minimise deflection problems, the external load to be balanced is usually a significant portion of the sustained or permanent service load. If all the permanent load is balanced, the sustained concrete stress (P/A) is uniform over the slab depth. A uniform compressive stress distribution produces uniform creep strain and, hence, little long-term load-dependent curvature or deflection. Bonded reinforcement does, of course, provide restraint to both creep and shrinkage and causes a change of curvature with time if the steel is eccentric to the slab centroid. However, the quantity of bonded steel in prestressed slabs is generally relatively small and the time-dependent curvature caused by this restraint does not usually cause significant deflection.

Problems can arise if a relatively heavy dead load is to be applied at some time after stressing. Excessive camber after transfer, which continues to increase with time owing to creep, may cause problems prior to the application of the fully balanced load. In such a case, the designer may consider *stage stressing* as a viable solution.

The magnitude of the average concrete compressive stress after all losses can indicate potential serviceability problems. If P/A is too low, the prestress may not be sufficient to prevent or control cracking due to shrinkage, temperature changes and the unbalanced loads. Some codes of practice specify minimum limits on the average concrete compressive stress after all losses. Using flat-ducted tendons containing four or more strands, prestressing levels are typically in the range $P/A = 1.2–2.6$ MPa in each direction of a two-way slab.

If the average prestress is high, axial deformation of the slab may be large and may result in distress in the supporting structure. The remainder of the structure must be capable of withstanding and accommodating the shortening of the slab, irrespective of the average concrete stress, but when P/A is large, the problem is exacerbated. Movement joints may be necessary to isolate the slab from stiff supports.

For slabs supporting uniformly distributed loads, the spacing of tendons in at least one direction should not exceed the smaller of eight times the slab thickness and 1.5 m [3], particularly for slabs containing less than minimum quantities of conventional tensile reinforcement (i.e. less than about 0.15% of the cross-sectional area of the slab).

12.4 INITIAL SIZING OF SLABS

12.4.1 Existing guidelines

At the beginning of the design of a suspended post-tensioned floor, the designer must select an appropriate floor thickness. The floor must be stiff enough to avoid excessive deflection or camber, and it must have adequate fire resistance and durability.

In its recommendations for the design of post-tensioned slabs, the Post-Tensioning Institute [4] suggested that the span-to-depth ratios given in Table 12.1 have proved acceptable, in terms of both performance and economy. Note that for flat plates and flat slabs with drop panels, the longer of the two orthogonal spans is used in the determination of the span-to-depth ratio, while for edge-supported slabs, the shorter span is used.

The minimum slab thickness that will prove acceptable in any situation depends on a variety of factors, not the least of which is the level of the superimposed load and the occupancy. The effect of load level on the limiting span-to-depth ratio of flat slabs in building structures is illustrated in Figure 12.8. The minimum thickness obtained from Table 12.1 may not therefore be appropriate in some situations. On the other hand, thinner slabs may be acceptable, if the calculated deflections, camber and vibration frequency and amplitude are acceptable. In addition to the satisfaction of serviceability requirements, strength requirements, such as punching shear at supporting columns, must also be satisfied. Fire resistance and durability requirements must also be considered.

A slab exposed to fire must retain its structural adequacy and integrity for a particular *fire resistance period* (FRP). It must also be sufficiently thick to limit the temperature on one side, when exposed to fire on the other side, i.e. it must provide a suitable FRP for insulation. As discussed in Section 2.7, the FRP required for a particular structure is generally specified by the local building authority and depends on the type of structure and its occupancy. The minimum dimensions of a slab exposed to fire are specified in EN 1992-1-2 [5] and were given in Table 2.10.

Table 12.1 Limiting span-to-depth ratios [4]

Floor system	Span-to-depth ratio l/h
Flat plate	45
Flat slab with drop panels	50
One-way slab	48
Edge-supported slab	55
Waffle slab	35
Band beams ($b \approx 3h$)	30

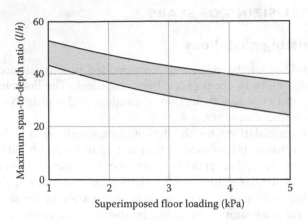

Figure 12.8 Effect of superimposed load on maximum *l/h* for flat slabs.

12.4.2 Serviceability approach for the calculation of slab thickness

For uniformly loaded slabs, a better initial estimate of slab thickness which should ensure adequate stiffness and satisfactory service load behaviour can be made using a procedure originally developed for reinforced concrete slabs [6] and extended to cover post-tensioned floor systems [7]. By rearranging the expression for the deflection of a span, a simple equation is developed for the span-to-depth ratio that is required to satisfy any specified deflection limit.

If it is assumed that a prestressed concrete slab is essentially uncracked at service loads, which is most often the case, the procedure for estimating the overall depth of the slab is relatively simple. Figure 12.9 shows typical interior panels of a one-way slab, a two-way edge-supported slab and two-way flat slabs. *Equivalent one-way slab strips* are also defined and illustrated for each slab type. For a one-way slab, the mid-span deflection is found by analysing a strip of unit width as shown in Figure 12.9a. For an edge-supported slab, the deflection at the centre of the panel may be calculated from an equivalent slab strip through the centre of the panel in the short direction, as shown in Figure 12.9b. For the flat plate and flat slab panels shown in Figures 12.9c and d, the deflection at the midpoint of the long span on the column line is found by analysing a unit-wide strip located on the column line.

The stiffness of these equivalent slab strips must be adjusted for each slab type so that the deflection of the one-way strip at mid-span is the same as the deflection of the two-way slab at that point. For an edge-supported slab, for example, the stiffness of the equivalent slab strip is increased significantly to realistically model the actual slab deflection at the mid-panel. For a flat slab, the stiffness of the slab strip must be reduced, if the maximum deflection at the centre of the panel is to be controlled rather than

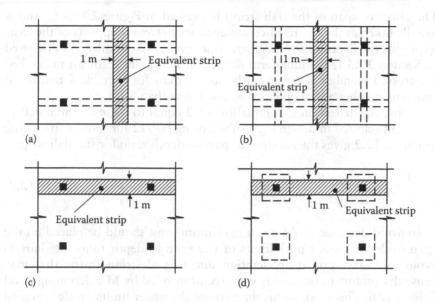

Figure 12.9 Slab types and equivalent slab strips [7]. (a) One-way slab. (b) Edge-supported two-way slab. (c) Flat plate. (d) Flat slab with drop panels.

the deflection on the column line. The stiffness adjustment is made using a *slab system factor K* that was originally calibrated using a non-linear finite element model [8,9].

By rearranging the equation for the mid-span deflection of a one-way slab, an expression can be obtained for the minimum slab thickness required to satisfy any specified deflection limit. The maximum deflection caused by the unbalanced uniformly distributed service loads on an uncracked one-way prestressed slab strip of width *b*, depth *h* and effective span *l* may be estimated using Equation 10.2, which is reproduced and renumbered here for ease of reference:

$$v = \beta \frac{w_{unbal} l^4}{E_{cm} I} + \lambda \beta \frac{w_{unbal.sus} l^4}{E_{cm} I} \tag{12.2}$$

where E_{cm} is the elastic modulus of concrete, I is the gross moment of inertia of the cross-section, w_{unbal} is the unbalanced service load per unit length and $w_{unbal.sus}$ is the sustained portion of the unbalanced load per unit length. In the design of a slab, $w_{unbal.sus}$ should not be taken less than 25% of the self-weight of the member. This is to ensure that at least a small long-term deflection is predicted by Equation 12.2. A small long-term deflection is inevitable, even for the case when an attempt is made to balance the entire sustained load by prestress. The term β in Equation 12.2 is a deflection coefficient that depends on the support conditions and the type of load.

The effective span of the slab strip l is defined in Figure 2.1 as l_{eff} and is usually taken as the centre-to-centre distance between supports or the clear span plus the depth of the member, whichever is the smaller. As discussed in Section 10.3.1, the long-term deflection multiplier λ for an uncracked prestressed member is significantly higher than for a cracked reinforced concrete member and should not be taken less than 3.

Setting the deflection v in Equation 12.2 equal to the maximum deflection limit selected in design v_{max}, substituting $bh^3/12$ for I and rearranging Equation 12.2 gives the maximum span-to-depth ratio for the slab strip:

$$\frac{l}{h} \leq \left[\frac{(v_{max}/l)bE_{cm}}{12\beta(w_{unbal} + \lambda w_{unbal.sus})} \right]^{1/3} \tag{12.3}$$

To avoid dynamic problems, a maximum limit should be placed on the span-to-depth ratio. Upper limits of the span-to-depth ratio for slabs to avoid excessive vertical acceleration due to pedestrian traffic that may cause discomfort to occupants were recommended by Mickleborough and Gilbert [10]. This work forms the basis of the upper limits on l/h specified in Equation 12.4.

For prestressed concrete slabs, an estimate of the minimum slab thickness may be obtained by applying Equation 12.3 to the slab strips in Figure 12.9. Equation 12.3 can be re-expressed as follows:

$$\frac{l}{h} \leq K \left[\frac{(v_{max}/l)1000E_{cm}}{w_{unbal} + \lambda w_{unbal.sus}} \right]^{1/3} \tag{12.4}$$

<50 for one-way slabs and flat slabs
<55 for two-way edge-supported slabs

The width of the equivalent slab strip b has been taken as 1000 mm and, in the absence of any other information, the long-term deflection multiplier λ should be taken not less than 3. The loads w_{unbal} and $w_{unbal.sus}$ are in kPa (i.e. kN/m^2 or N/mm^2) and E_{cm} is in MPa. The term K is the *slab system factor* that accounts for the support conditions of the slab panel, the aspect ratio of the panel, the load dispersion and the torsional stiffness of the slab. For each slab type, values of K are presented and discussed in the following.

12.4.2.1 Slab system factor, K

One-way slabs: For a one-way slab, K depends only on the support conditions and the most critical pattern of unbalanced load. From Equation 12.3:

$$K = \left(\frac{1}{12\beta} \right)^{1/3} \tag{12.5}$$

For a continuous slab, β should be determined for the distribution of unbalanced load which causes the largest deflection in each span. For most slabs, a large percentage of the sustained load (including self-weight) is balanced by the prestress and much of the unbalanced load is transitory. Pattern loading must therefore be considered in the determination of β. For a simply-supported span β = 5/384, and from Equation 12.5, K = 1.86. For a fully loaded *end span* of a one-way slab that is continuous over three or more equal spans and with the adjacent interior span unloaded, β = 3.5/384 (determined from an elastic analysis) and, therefore, K = 2.09. For a fully loaded *interior span* of a continuous one-way slab with adjacent spans unloaded, β = 2.6/384 and K = 2.31.

Flat slabs: For flat slabs, the values of K given earlier must be modified to account for the variation of curvature across the panel width. The moments, and hence curvatures, in the uncracked slab are greater close to the column line than those near the mid-panel of the slab in the middle strip region. For this reason, the deflection of the slab on the column line will be greater than the deflection of a one-way slab of similar span and continuity. If the deflections of the equivalent slab strips in Figure 12.9c and d are to represent accurately the deflection of the real slab on the column line, a greater than average share of the total load on the slab must be assigned to the column strip (of which the equivalent strip forms a part). A reasonable assumption is that 65% of the total load on the slab is carried by column strips, and so the value for K for a flat slab becomes:

$$K = \left(\frac{1}{15.6\beta}\right)^{1/3} \tag{12.6}$$

For an end span, with β = 3.5/384, Equation 12.6 gives K = 1.92. For an interior span, with β = 2.6/384, the slab system factor is K = 2.12.

For a slab containing drop panels that extend at least $l/6$ in each direction on each side of the support centre line and that have an overall depth not less than 1.3 times the slab thickness beyond the drops, the values of K given earlier may be increased by 10%. If the maximum deflection at the centre of the panel is to be limited (rather than the deflection on the long-span column line), the values of K for an end span and for an interior span should be reduced to 1.75 and 1.90, respectively.

Edge-supported two-way slabs: For an edge-supported slab, values for K must be modified to account for the fact that only a portion of the total load is carried by the slab in the short-span direction. In addition, torsional stiffness, as well as compressive membrane action, increases the overall slab stiffness. In an earlier study of span-to-depth limits for reinforced concrete slabs, a non-linear finite element model was used to quantify these effects [6]. Values of K depend on the aspect ratio of the rectangular edge-supported panel and the support conditions of all edges and are given in Table 12.2.

Table 12.2 Values of K for an uncracked two-way edge-supported slab [7]

| Support conditions for slab panel | Slab system factor, K | | | |
| | Ratio of long span to short span | | | |
	1.0	1.25	1.5	2.0
4 edges continuous	3.0	2.6	2.4	2.3
1 short edge discontinuous	2.8	2.5	2.4	2.3
1 long edge discontinuous	2.8	2.4	2.3	2.2
2 short edges discontinuous	2.6	2.4	2.3	2.3
2 long edges discontinuous	2.6	2.2	2.0	1.9
2 adjacent edges discontinuous	2.5	2.3	2.2	2.1
2 short +1 long edge discontinuous	2.4	2.3	2.2	2.1
2 long +1 short edge discontinuous	2.4	2.2	2.1	1.9
4 edges discontinuous	2.3	2.1	2.0	1.9

12.4.3 Discussion

Equation 12.4 forms the basis of a useful approach for the preliminary design of prestressed slabs. When the load to be balanced and the deflection limit have been selected, an estimate of slab depth can readily be made. All parameters required for input into Equation 12.4 are usually known at the beginning of the design. An iterative approach may be required if an estimate of self-weight is needed, i.e. if the load to be balanced is less than self-weight.

Deflections at various stages in the slab history may still have to be calculated, particularly if the unbalanced load causes significant cracking or if an unusual load history is expected. Serviceability problems, however, can be minimised by a careful choice of slab depth using Equation 12.4. This involves an understanding of the derivation of the equation and its limitations. If, for example, a designer decides to minimise deflection by balancing the entire sustained load, it would be unwise to set the sustained part of the unbalanced load $w_{unbal.sus}$ to zero. In the real slab, of course, the magnitude of the sustained unbalanced load varies as the prestressing force varies with time and does not remain zero. Restraint to creep and shrinkage caused by the eccentric bonded steel will inevitably cause some time-dependent deflection (or camber). In such cases, selection of a slab depth greater than that indicated by Equation 12.4 would be prudent. It is suggested that $w_{unbal.sus}$ should not be taken to be less than 0.25 times the self-weight of the slab. As with the rest of the design process, sound engineering judgement is required.

EXAMPLE 12.1

Determine preliminary estimates of the thickness of a post-tensioned flat slab floor for an office building. The supporting columns are 400 mm by 400 mm in section and are regularly spaced at 9.8 m centres in one direction and 7.8 m centres in the orthogonal direction. Drop panels extending span/6 in each direction are located over each interior column. The slab supports a dead load of 1 kPa (in addition to self-weight) and a service live load of 3.0 kPa (of which 1.0 kPa is sustained or permanent). The self-weight of the slab only is to be balanced by prestress. Therefore, the unbalanced loads are:

$$w_{unbal} = 4.0 \text{ kPa} \quad \text{and} \quad w_{unbal.sus} = 2.0 \text{ kPa}$$

In this example, the longer effective span is calculated as clear span + h. If h is initially assumed to be about 200 mm, then $l = 9800 - 400 + 200 = 9600$ mm. The elastic modulus for concrete is taken as $E_{cm} = 33,000$ MPa.

In Case (a), the maximum deflection on the column line in the long-span direction is first limited to span/250, and then in Case (b), it is limited to span/500.

Case (a): The deflection in an exterior or edge panel of the slab will control the thickness. From Equation 12.6, $K = 1.92$ for an end span and this may be increased by 10% to account for the stiffening effect of the drop panels, i.e. $K = 2.11$. With $\lambda = 3$, Equation 12.4 gives:

$$\frac{9,600}{h} \le 2.11 \times \left[\frac{(1/250) \times 1,000 \times 33,000}{4.0 + 3 \times 2.0} \right]^{1/3} = 49.8$$

$\therefore h \ge 193$ mm plus drop panels.

Case (b): If the slab supports brittle partitions and the deflection limit is taken to be span/500, a thicker slab than that required for (a) will be needed. Assuming $h = 250$ mm, the revised effective span is $l = 9650$ mm and, with $\lambda = 3$, Equation 12.4 gives:

$$\frac{9,650}{h} \le 2.11 \times \left[\frac{(1/500) \times 1,000 \times 33,000}{4.0 + 3 \times 2.0} \right]^{1/3} = 39.6$$

$\therefore h \ge 244$ mm plus drop panels.

EXAMPLE 12.2

Evaluate the slab thickness required for an edge panel of a two-way slab with short and long effective spans of 8.5 and 11 m, respectively. The slab is continuously supported on all four edges by stiff beams and is discontinuous on one long edge only. The slab must carry a dead load of 1.25 kPa (plus self-weight) and a service live load of 3 kPa (of which 1 kPa is sustained). As in the previous example, only the self-weight is to be balanced by prestress, and therefore:

$$w_{unbal} = 4.25 \text{ kPa} \quad \text{and} \quad w_{unbal.sus} = 2.25 \text{ kPa}$$

The maximum mid-panel deflection is limited to $v_{max} = 20$ mm and the elastic modulus for concrete is $E_{cm} = 33,000$ MPa.

With an aspect ratio of 11.0/8.5 = 1.29, the slab system factor is obtained by interpolation from Table 12.2, i.e. $K = 2.4$. With $\lambda = 3$, Equation 12.4 gives:

$$\frac{8,500}{h} \leq 2.4 \times \left[\frac{(20/8,500) \times 1,000 \times 33,000}{4.25 + 3 \times 2.25} \right]^{1/3} = 46.0$$

$$\therefore h \geq 185 \text{ mm}.$$

12.5 OTHER SERVICEABILITY CONSIDERATIONS

12.5.1 Cracking and crack control in prestressed slabs

The effect of cracking in slabs is to reduce the flexural stiffness of the highly stressed regions and thus to increase the deflection. Prior to cracking, deflection calculations are usually based on the moment of inertia of the gross concrete section I_g neglecting the contributions of the reinforcement. After cracking, the effective moment of inertia I_{ef}, which is less than I_g, is used. In Section 5.8, the analysis of a cracked prestressed section was presented and procedures for calculating the cracked moment of inertia and for including the tension stiffening effect were discussed in Section 5.11.3. Using these procedures, the effective moment of inertia of a cracked region of the slab can be calculated.

For prestressed concrete flat slabs, flexural cracking at service loads, if it occurs, is usually confined to the negative moment column strip region above the supports. A mat of non-prestressed reinforcement is often placed in the top of the slab over the column supports for crack control and to increase both the stiffness and the strength of this highly stressed region. This mat of crack control reinforcement should consist of bars in each direction (or welded wire mesh) with each bar or wire being continuous across the column line and extending at least 0.25 times the clear span in each direction from the face of the column. An area of crack control reinforcement of about 0.001 times the gross area of the cross-section is usually sufficient.

The mechanism of flexural cracking in a statically indeterminate two-way slab is complex. The direction of flexural cracking is affected to some extent by the spacing and the type of bonded reinforcement, the level of prestress in each direction, the support conditions and the level and distribution of the applied loads. However, for slabs containing conventionally tied bonded reinforcement at practical spacings in both directions, flexural cracks occur in the direction perpendicular to the direction of principal tension.

If the level of prestress in a slab is sufficiently high to ensure that the tensile stresses in a slab in bending are always less than the tensile strength of concrete, flexural cracking will not occur. If the level of prestress is not sufficient, cracking occurs and bonded reinforcement at reasonable centres is necessary to control the cracks adequately. Because slabs tend to be very lightly reinforced, the maximum moments at service loads are rarely very much larger than the cracking moment. However, when cracking occurs, the stress in the bonded reinforcement increases and crack widths may become excessive if too little bonded steel is present or the steel spacing is too wide.

The requirements for flexural crack control in EN 1992-1-1 [1] were discussed in Sections 5.12.1 and 5.12.2, and the maximum increment of stress in the steel near the tensile face (as the load is increased from its value when the extreme concrete tensile fibre is at zero stress to the short-term service load value) was given in Tables 5.5 and 5.6. In addition, the requirements for the control of *direct tension cracking* in slabs due to restrained shrinkage and temperature changes were outlined in Section 5.12.4.

12.5.2 Long-term deflections

As discussed in Chapter 5, long-term deflections due to creep and shrinkage are influenced by many variables, including load intensity, mix proportions, slab thickness, age of slab at first loading, curing conditions, quantity of compressive steel, relative humidity and temperature.

In most prestressed slabs, the majority of the sustained load is most often balanced by the transverse force exerted by the tendons on the slab. Under this balanced load, the time-dependent deflection will not be zero because of the restraint to creep and shrinkage offered by eccentrically located bonded reinforcement. Therefore, the use of a simple deflection multiplier to calculate long-term deflection is not satisfactory.

In Sections 4.2.5.3 and 4.2.5.4, the procedures specified in EN 1992-1-1 [1] for calculating the final creep coefficient of concrete $\varphi(\infty,t_0)$ and the final shrinkage strain $\varepsilon_{cs}(\infty)$ are presented, and the procedures for the determination of the long-term behaviour of uncracked and cracked prestressed cross-sections are presented in Sections 5.7 and 5.9. Alternative and more approximate expressions for estimating the creep and shrinkage components of the long-term deflection of beams are given in Section 5.11.4. Similar equations for determining deflections of slab strips are presented in the following.

For uncracked prestressed concrete slab cross-sections, which usually have low quantities of steel, the increase in curvature due to creep is nearly proportional to the increase in strain due to creep. This is in contrast with the behaviour of a cracked reinforced concrete cross-section. If we set α in Equation 5.183 to unity on every cross-section, the final creep-induced deflection v_{cc} may be approximated by:

$$v_{cc} = \varphi(\infty, t_0)v_{sus.0} \tag{12.7}$$

where $v_{sus.0}$ is the short-term deflection produced by the sustained portion of the unbalanced load. Typical values for the final creep coefficient for concrete in post-tensioned slabs $\varphi(\infty,t_0)$ are in the range 2.5–3.0.

The average deflection due to shrinkage of an equivalent slab strip (in the case of edge-supported slabs) or the wide beam (as discussed subsequently in Section 12.9.6 for the case of flat slabs) may be obtained from:

$$v_{cs} = \beta \kappa_{cs} l^2 \tag{12.8}$$

where κ_{cs} is the average shrinkage-induced curvature, l is the effective span of the slab strip under consideration and β depends on the support conditions and equals 0.125 for a simply-supported span, 0.090 for an end span of a continuous member and 0.065 for an interior span of a continuous member.

The shrinkage curvature κ_{cs} is non-zero wherever the eccentricity of the bonded steel area is non-zero and varies along the span as the eccentricity of the draped tendons varies. A simple and very approximate estimate of the average shrinkage curvature for a fully-prestressed slab, which will usually produce reasonable results, is:

$$\kappa_{cs} = \frac{0.3\varepsilon_{cs}}{h} \tag{12.9}$$

For a cracked partially-prestressed slab, with significant quantities of non-prestressed conventional reinforcement, the value of κ_{cs} is usually at least 100% higher than that indicated earlier.

12.6 DESIGN APPROACH: GENERAL

After making an initial selection of the slab thickness, the second step in slab design is to determine the amount and distribution of prestress. Load balancing is generally used to this end. A portion of the load on a slab is balanced by the transverse forces imposed by the draped tendons in each direction. Under the balanced load, the slab remains plane (without curvature) and is subjected only to the resultant longitudinal compressive P/A stresses. It is the remaining unbalanced load that enters into the calculation of service load behaviour, particularly for the estimation of load-dependent deflections and for checking the extent of cracking and crack control.

At ultimate limit state conditions, when the slab behaviour is non-linear and superposition is no longer valid, the full factored design load must be considered. The factored design moments and shears at each critical section must be calculated and compared with the design strength of the section, as discussed in Chapters 6 (for flexure) and 7 (for shear). Slabs are usually very ductile and redistribution of moments occurs as the collapse load of the slab is approached. Under these conditions, secondary moments can usually be ignored.

In the following sections, procedures for the calculation of design moments and shears at the critical sections in the various slab types are presented. In addition, techniques and recommendations are also presented for the determination of the magnitude of the prestressing force required in each direction to balance the desired load.

12.7 ONE-WAY SLABS

A one-way slab is generally designed as a beam with cables running in the direction of the span at uniform centres. A slab strip of unit width is analysed using simple beam theory. In any span, the maximum cable sag z_d depends on the concrete cover requirements and the tendon dimensions. When z_d is determined, the prestressing force required to balance an external load w_{bal} is calculated from Equation 11.33, which is restated and renumbered here for ease of reference:

$$P = \frac{w_{bal} l^2}{8z_d} \tag{12.10}$$

In the transverse direction, conventional reinforcement may be used to control shrinkage and temperature cracking (see Section 5.12.4) and to distribute local load concentrations. Not infrequently, the slab is prestressed in the transverse direction to eliminate the possibility of shrinkage cracking parallel to the span and to ensure a watertight and crack-free slab.

12.8 TWO-WAY EDGE-SUPPORTED SLABS

12.8.1 Load balancing

Consider the interior panel of the two-way edge-supported slab shown in Figure 12.10. The panel is supported on all sides by walls or beams and contains parabolic tendons in both the *x*- and *y*-directions. If the cables in each direction are uniformly spaced, then from Equation 12.1, the upward forces per unit area exerted by the tendons in each direction are:

$$w_{px} = \frac{8P_x z_{d.x}}{l_x^2} \tag{12.11}$$

and

$$w_{py} = \frac{8P_y z_{d.y}}{l_y^2} \tag{12.12}$$

where P_x and P_y are the prestressing forces in each direction per unit width, and $z_{d.x}$ and $z_{d.y}$ are the cable drapes in each direction.

The uniformly distributed downward load to be balanced per unit area w_{bal} is calculated as:

$$w_{bal} = w_{px} + w_{py} \tag{12.13}$$

In practice, perfect load balancing is not possible, since external loads are rarely perfectly uniformly distributed. However, for practical purposes, adequate

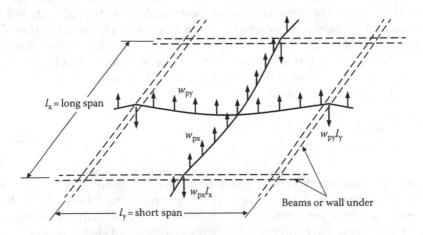

Figure 12.10 Interior edge-supported slab panel.

load balancing can be achieved. Any combination of w_{px} and w_{py} that satisfies Equation 12.13 can be used to make up the balanced load. The smallest quantity of prestressing steel will result if all the loads are balanced by cables in the short-span direction, i.e. $w_{bal} = w_{py}$. However, under unbalanced loads, serviceability problems in the form of unsightly cracking may result. It is often preferable to distribute the prestress in much the same way as the load is distributed to the supports in an elastic slab, i.e. more prestress in the short-span direction than in the long-span direction. The balanced load resisted by tendons in the short direction may be estimated by:

$$w_{py} = \frac{l_x^4}{\delta l_y^4 + l_x^4} w_{bal} \tag{12.14}$$

where δ depends on the support conditions and is given by:

$\quad \delta = 1.0 \quad$ for 4 edges continuous or discontinuous
$\qquad = 1.0 \quad$ for 2 adjacent edges discontinuous
$\qquad = 2.0 \quad$ for 1 long edge discontinuous
$\qquad = 0.5 \quad$ for 1 short edge discontinuous
$\qquad = 2.5 \quad$ for 2 long +1 short edge discontinuous
$\qquad = 0.4 \quad$ for 2 short +1 long edge discontinuous
$\qquad = 5.0 \quad$ for 2 long edges discontinuous
$\qquad = 0.2 \quad$ for 2 short edges discontinuous

Equation 12.14 is the expression obtained for that portion of any external load which is carried in the short-span direction if twisting moments are ignored and if the mid-span deflections of the two orthogonal unit-wide strips through the slab centre are equated.

With w_{px} and w_{py} selected, the prestressing force per unit width in each direction is calculated using Equations 12.11 and 12.12 as:

$$P_x = \frac{w_{px} l_x^2}{8z_{d.x}} \tag{12.15}$$

and

$$P_y = \frac{w_{py} l_y^2}{8z_{d.y}} \tag{12.16}$$

Equilibrium dictates that the downward forces per unit length exerted over each edge support by the reversal of cable curvature (as shown in Figure 12.10) are:

$w_{px} l_x$ (kN/m)   carried by the short-span supporting beams or walls

$w_{py} l_y$ (kN/m)   carried by the long-span supporting beams or walls

The total force imposed by the slab tendons that must be carried by the edge beams is therefore:

$$w_{px}l_xl_y + w_{py}l_yl_x = w_{bal}l_xl_y$$

and this is equal to the total upward force exerted by the slab cables. Therefore, for this two-way slab system, to carry the balanced load to the supporting columns, resistance must be provided for twice the total load to be balanced (i.e. the slab tendons must resist $w_{bal}l_xl_y$ and the supporting beams must resist $w_{bal}l_xl_y$). This requirement is true for all two-way floor systems, irrespective of construction type or material.

At the balanced load condition, when the transverse forces imposed by the cables exactly balance the applied external loads, the slab is subjected only to the compressive stresses imposed by the longitudinal prestress in each direction, i.e. $\sigma_x = P_x/h$ and $\sigma_y = P_y/h$, where h is the slab thickness.

12.8.2 Methods of analysis

For any service load above (or below) the balanced load, moments are induced in the slab and, if large enough, these moments may lead to cracking or excessive deflection. A reliable technique for estimating slab moments is therefore required to check in-service behaviour under the unbalanced loads. In addition, reliable estimates of the maximum moments and shears caused by the full factored dead and live loads must be made in order to check the flexural and shear strength of a slab.

A simplified method for the analysis of reinforced two-way edge-supported rectangular slabs subjected to uniformly distributed design ultimate loads [11,12] is described here. In the absence of more accurate methods of analysis, the moment coefficients specified in Table 12.3 may be used to determine the design moments in prestressed concrete edge-supported slabs.

If w_{Ed} is the factored design load per unit area at the strength limit state, the positive design moments per unit width at the mid-span of the slab in each direction are:

$$M_{Ed.x} = \beta_x w_{Ed} l_y^2 \tag{12.17}$$

and

$$M_{Ed.y} = \beta_y w_{Ed} l_y^2 \tag{12.18}$$

where l_y is the short effective span and β_x and β_y are the moment coefficients that depend on the support conditions and the aspect ratio of the panel (i.e. l_x/l_y). The values for β_x and β_y are given in Table 12.3 or may be obtained from:

$$\beta_x = \frac{2[\sqrt{3 + (\gamma_y/\gamma_x)^2} - \gamma_y/\gamma_x]^2}{9\gamma_x^2} \tag{12.19}$$

Table 12.3 Design moment coefficients for rectangular edge-supported slabs [12]

Edge conditions of the rectangular slab panel	Short-span coefficient β_y								Long-span coefficient β_x for all values of l_x/l_y
	Aspect ratio l_x/l_y								
	1.0	1.1	1.2	1.3	1.4	1.5	1.75	≥2.0	
1. Four edges continuous	0.024	0.028	0.032	0.035	0.037	0.040	0.044	0.048	0.024
2. One short edge discontinuous	0.028	0.032	0.036	0.038	0.041	0.043	0.047	0.050	0.028
3. One long edge discontinuous	0.028	0.035	0.041	0.046	0.050	0.054	0.061	0.066	0.028
4. Two short edges discontinuous	0.034	0.038	0.040	0.043	0.045	0.047	0.050	0.053	0.034
5. Two long edges discontinuous	0.034	0.046	0.056	0.065	0.072	0.078	0.091	0.100	0.034
6. Two adjacent edges discontinuous	0.035	0.041	0.046	0.051	0.055	0.058	0.065	0.070	0.035
7. Three edges discontinuous (one long edge continuous)	0.043	0.049	0.053	0.057	0.061	0.064	0.069	0.074	0.043
8. Three edges discontinuous (one short edge continuous)	0.043	0.054	0.064	0.072	0.078	0.084	0.096	0.105	0.043
9. Four edges discontinuous	0.056	0.066	0.074	0.081	0.087	0.093	0.103	0.111	0.056

and

$$\beta_y = \frac{\beta_x l_y}{l_x} + \frac{2[1 - (l_y/l_x)]}{3\gamma_x^2} \qquad (12.20)$$

where,

$\gamma_y = 2.0$ if both short edges are discontinuous
$\gamma_y = 2.5$ if one short edge is discontinuous
$\gamma_y = 3.0$ if both short edges are continuous
$\gamma_x = 2.0$ if both long edges are discontinuous
$\gamma_x = 2.5$ if one long edge is discontinuous
$\gamma_x = 3.0$ if both long edges are continuous

The magnitudes of the negative design moments at a continuous edge and at a discontinuous edge are taken to be 1.33 times the positive mid-span value and 0.5 times the positive mid-span value, respectively.

The moment coefficients of Table 12.3 may also be used for serviceability calculations, with the moments caused by the unbalanced load taken as:

$$M_{\text{unbal.x}} = \beta_x w_{\text{unbal}} l_y^2 \tag{12.21}$$

and

$$M_{\text{unbal.y}} = \beta_y w_{\text{unbal}} l_y^2 \tag{12.22}$$

These values of unbalanced moments may be used to check for cracking at service loads. However, for edge-supported prestressed concrete slabs, cracking is unlikely at service loads. Even reinforced concrete slabs that are continuously supported on all edges are often uncracked at service loads. If cracking is detected, then an average effective moment of inertia I_{ef} that reflects the loss of stiffness due to cracking should be used in deflection calculations (see Section 5.11.3). For the determination of deflection of the slab strip in the shorter-span direction (shown in Figure 12.9b), a weighted average value of I_{ef} should be used and may be taken as 0.7 times the value of I_{ef} at mid-span plus 0.3 times the average of the values of I_{ef} at each end of the span. For an exterior span, a reasonable weighted average is 0.85 times the mid-span value plus 0.15 times the value at the continuous end. If a particular region is uncracked, I_{ef} for this region should be taken as I_g.

For the purposes of calculating the shear forces in a slab or the forces applied to the supporting walls or beams, the uniformly distributed load on the slab is allocated to the supports as shown in Figure 12.11.

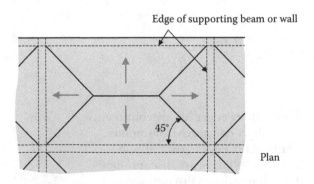

Figure 12.11 Distribution of shear forces in an edge-supported slab.

EXAMPLE 12.3

Design an exterior panel of a 180 mm thick two-way floor slab for a retail store. The rectangular panel is supported on four edges by stiff beams and is discontinuous on one long edge as shown in Figure 12.12a. The slab is post-tensioned in both directions using the draped parabolic cable profiles shown

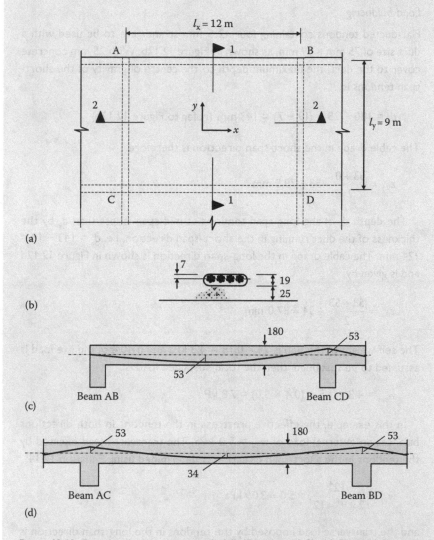

(a)

(b)

(c)

(d)

Figure 12.12 Details of edge-supported slab (Example 12.3). (a) Plan. (b) Section through a typical duct in y-direction at mid-span (dimensions in mm). (c) Tendon profile in y-direction (Section 1–1). (d) Tendon profile in x-direction (Section 2-2).

in Figures 12.12c and d. The slab supports a dead load of 1.5 kPa in addition to its own self-weight and the live load is 5.0 kPa. The level of prestress required to balance a uniformly distributed load of 5.0 kPa is required. Relevant material properties are f_{ck} = 40 MPa, f_{ctm} = 3.5 MPa, E_{cm} = 35,000 MPa, f_{pk} = 1,860 MPa and E_p = 195,000 MPa.

Load balancing:

Flat-ducted tendons containing four 12.5 mm strands are to be used with a duct size of 75 mm × 19 mm, as shown in Figure 12.12b. With 25 mm concrete cover to the duct, the maximum depth to the centre of gravity of the short-span tendons is:

$$d_y = 180 - 25 - (19 - 7) = 143 \text{ mm (refer to Figure 12.12b)}$$

The cable drape in the short-span direction is therefore:

$$z_{d.y} = \frac{53 + 0}{2} + 53 = 79.5 \text{ mm}$$

The depth d_x of the long-span tendons at mid-span is less than d_y by the thickness of the duct running in the short-span direction, i.e. d_x = 143 − 19 = 124 mm. The cable drape in the long-span direction is shown in Figure 12.12d and is given by:

$$z_{d.x} = \frac{53 + 53}{2} + 34 = 87.0 \text{ mm}$$

The self-weight of the slab is 24 × 0.18 = 4.3 kPa and if 40% of the live load is assumed to be sustained, then the total sustained load is:

$$w_{sus} = 4.3 + 1.5 + (0.4 \times 5.0) = 7.8 \text{ kPa}$$

In this example, the effective prestress in the tendons in both directions balances an external load of w_{bal} = 5.0 kPa. The transverse load exerted by the tendons in the short-span direction is determined using Equation 12.14:

$$w_{py} = \frac{12^4}{2 \times 9^4 + 12^4} \times 5.0 = 3.06 \text{ kPa}$$

and the transverse load imposed by the tendons in the long-span direction is calculated using Equation 12.13:

$$w_{px} = w_{bal} - w_{py} = 5.0 - 3.06 = 1.94 \text{ kPa}$$

The effective prestress in each direction is obtained from Equations 12.15 and 12.16:

$$P_y = \frac{3.06 \times 9^2}{8 \times 0.0795} = 390 \text{ kN/m} \quad \text{and} \quad P_x = \frac{1.94 \times 12^2}{8 \times 0.087} = 401 \text{ kN/m}$$

To determine the jacking forces and cable spacing in each direction, both the time-dependent losses and friction losses must be calculated. For the purpose of this example, it is assumed that the time-dependent losses in each direction are 15% and the immediate losses (friction, anchorage, etc.) in the y-direction are 8% and in the x-direction are 12%. Immediately after transfer, before the time-dependent losses have taken place, the prestressing forces at mid-span in each direction are:

$$P_{m0.y} = \frac{390}{0.85} = 459 \text{ kN/m} \quad \text{and} \quad P_{m0.x} = \frac{401}{0.85} = 472 \text{ kN/m}$$

and at the jack:

$$P_{j.y} = \frac{459}{0.92} = 499 \text{ kN/m} \quad \text{and} \quad P_{j.x} = \frac{472}{0.88} = 536 \text{ kN/m}$$

Using four 12.5 mm strands/tendon, $A_p = 372$ mm²/tendon and the maximum jacking force/tendon is 0.9 $f_{p0.1k} \times A_p = 1440 \times 372 = 536$ kN, and the required tendon spacing in each direction (rounded down to the nearest 10 mm) is therefore:

$$s_y = \frac{1000 \times 536}{499} = 1070 \text{ mm} \quad \text{and} \quad s_x = \frac{1000 \times 536}{536} = 1000 \text{ mm}$$

We will select a tendon spacing of 1000 mm in each direction.

This simply means that the tendons in the y-direction will balance slightly more load than previously assumed. With one tendon in each direction per metre width, the revised prestressing forces at the jack per metre width are $P_y = P_x = 536$ kN/m and at mid-span, after all losses, are:

$$P_{m.t.y} = 0.85 \times 0.92 \times 536 = 419 \text{ kN/m}$$

and

$$P_{m.t.x} = 0.85 \times 0.88 \times 536 = 401 \text{ kN/m}$$

The load to be balanced is revised using Equations 12.11 and 12.12:

$$w_{py} = \frac{8 \times 419 \times 0.0795}{9^2} = 3.29 \text{ kPa}$$

and

$$w_{px} = \frac{8 \times 401 \times 0.087}{12^2} = 1.94 \text{ kPa}$$

and therefore $w_{bal} = 3.29 + 1.94 = 5.23$ kPa.

Estimate maximum moment due to unbalanced load:

The maximum unbalanced transverse load to be considered for short-term serviceability calculations is:

$$w_{unbal} = w_{sw} + w_G + \psi_1 w_Q - w_{bal} = 4.30 + 1.5 + (0.7 \times 5.0) - 5.23 = 4.07 \text{ kPa}$$

Under this unbalanced load, the maximum moment occurs over the beam support CD. Using the moment coefficients for edge-supported slabs in Table 12.3, the maximum moment is approximated by:

$$M_{CD} = -1.33 \times 0.047 \times 4.07 \times 9^2 = -20.6 \text{ kNm/m}$$

Check for cracking:

In the y-direction over support CD, the concrete stresses in the top and bottom fibres caused by the maximum moment M_{CD} after all losses are:

$$\sigma_{c.top} = -\frac{P_{m,t,y}}{A} + \frac{M_{CD}}{Z} = -2.33 + 3.81 = +1.48 \text{ MPa (tension)}$$

$$\sigma_{c.btm} = -\frac{P_{m,t,y}}{A} - \frac{M_{CD}}{Z} = -2.33 - 3.81 = -6.14 \text{ MPa (compression)}$$

where A is the area of the gross cross-section per metre width (= 180×10^3 mm^2/m) and Z is the section modulus per metre width (= 5.4×10^6 mm^3/m).

Both tensile and compressive stresses are relatively low. Even though the moment used in these calculations is an average and not a peak moment, if cracking does occur, it will be localised and the resulting loss of stiffness will be small. Deflection calculations may be based on the properties of the uncracked cross-section.

Estimate maximum total deflection:

The deflection at the mid-panel of the slab can be estimated using the so-called crossing beam analogy, in which the deflections of a pair of orthogonal beams (slab strips) through the centre of the panel are equated. The fraction of the unbalanced load carried by the strip in the short-span direction is given by an equation similar to Equation 12.14. With:

$$w_{unbal.y} = \frac{12^4}{2.0 \times 9^4 + 12^4} \times 4.07 = 0.61 \times 4.07 = 2.48 \text{ kN/m}$$

and with the deflection coefficient β taken as 2.6/384 (in accordance with the discussion in Section 12.4.2), the corresponding short-term deflection at mid-span of this 1 m wide slab strip in the short-span direction through the mid-panel (assuming the variable live load is removed from the adjacent slab panel) is approximated by:

$$v_0 = \frac{2.6}{384} \frac{w_{unbal.y} l_y^4}{E_{cm} I} = \frac{2.6}{384} \frac{2.48 \times 9,000^4}{35,000 \times 486 \times 10^6} = 6.48 \text{ mm}$$

The sustained portion of the unbalanced load on the slab strip is:

$$\frac{l_x^4}{\delta l_y^4 + l_x^4} \times (w_{sw} + w_G + \gamma_2 w_Q - w_{bal}) = 0.61 \times [4.30 + 1.5 + (0.4 \times 5.0) - 5.23]$$

$$= 1.57 \text{ kPa}$$

and the corresponding short-term deflection is:

$$v_{sus.0} = \frac{1.57}{2.48} \times v_0 = 4.10 \text{ mm}$$

Assuming a final creep coefficient $\varphi(\infty, t_0) = 2.5$ and conservatively ignoring the restraint provided by any bonded reinforcement, the creep-induced deflection given by Equation 12.7 may be estimated using:

$$v_{cc} = 2.5 \times 4.10 = 10.25 \text{ mm}$$

The final shrinkage strain is assumed to be $\varepsilon_{cs} = 0.0005$. The shrinkage curvature κ_{cs} is non-zero wherever the eccentricity of the steel area is non-zero and varies along the span as the eccentricity of the draped tendons varies. A simple and very approximate estimate of the average final shrinkage curvature is made using Equation 12.9:

$$\kappa_{cs} = \frac{0.3\varepsilon_{cs}}{h} = \frac{0.3 \times 0.0005}{180} = 0.83 \times 10^{-6} \text{ mm}^{-1}$$

The average deflection of the slab strip due to shrinkage is given by Equation 12.8:

$$v_{cs} = 0.090 \times 0.83 \times 10^{-6} \times 9000^2 = 6.08 \text{ mm}$$

The maximum total deflection of the slab strip is therefore:

$$v_{tot} = v_0 + v_{cc} + v_{cs} = 6.48 + 10.25 + 6.08 = 22.8 \text{ mm} = \text{span}/395$$

and the long-term to short-term deflection ratio is $\lambda = 2.52$. This deflection is likely to be satisfactory for a retail floor.

It is of value to examine the slab thickness predicted by Equation 12.4, if the limiting deflection is taken to be 22.8 mm. For this edge-supported slab panel, the slab system factor is obtained from Table 12.2 as $K = 2.37$, the unbalanced load $w_{unbal} = 4.07$ kPa, and the sustained part of the unbalanced load is $w_{unbal.sus} = 2.48$ kPa. With $\lambda = 2.52$, the minimum slab thickness required to limit the total deflection to 22.8 mm is obtained from Equation 12.4 as:

$$\frac{9,000}{h} \le 2.37 \times \left[\frac{(22.8/9,000) \times 1,000 \times 35,000}{4.07 + 2.52 \times 2.48} \right]^{1/3} = 48.5$$

$\therefore h \ge 185$ mm.

In this example, Equation 12.4 is consistent with the deflection calculation procedure and just conservative.

Check flexural strength:

It is necessary to check the design strength of the slab. As previously calculated, the dead load is $1.5 + 4.3 = 5.8$ kPa and the live load is 5.0 kPa. The factored design load (using the load factors specified in Equation 2.2) is:

$$w_{Ed} = 1.35 \times 5.8 + 1.5 \times 5.0 = 15.33 \text{ kPa}$$

The design moments at mid-span in each direction are obtained from Equations 12.17 and 12.18, with $\beta_y = 0.047$ and $\beta_x = 0.028$ taken from Table 12.3:

$$M_{Ed.y} = 0.047 \times 15.33 \times 9^2 = 58.4 \text{ kNm/m}$$

$$M_{Ed.x} = 0.028 \times 15.33 \times 9^2 = 34.8 \text{ kNm/m}$$

The maximum design moment occurs over the beam support CD (the long continuous edge) and is:

$$(M_{Ed.x})_{CD} = -1.33 \times 58.4 = -77.7 \text{ kNm/m}$$

A safe lower bound solution to the problem of adequate strength is obtained if the design strength of the slab at this section exceeds the design moment.

The resistance per metre width of the 180 mm thick slab containing tendons at 1000 mm centres (i.e. $A_p = 372 \text{ mm}^2/\text{m}$) at an effective depth of 143 mm is obtained using the procedures discussed in Chapter 6. Such an analysis indicates that the cross-section is ductile, with the depth to the neutral axis of $x_1 = 24.3$ mm (or $0.17d$). The tensile force in the steel is 517.5 kN/m ($\sigma_{pud1} = f_{pd} = f_{p0.1k}/\gamma_s = 1391$ MPa) and the magnitude of the design resistance is given by Equation 6.24:

$$M_{Rd1} = 1391 \times 372 \times \left(143 - \frac{0.8 \times 24.3}{2}\right) \times 10^{-6} = 69.0 \text{ kNm/m}$$

With $M_{Rd1} < (M_{Ed.x})_{CD}$, conventional reinforcement is required to supplement the prestressing steel over the beam support CD. From Equation 6.25, with the internal lever arm $z_2 = 111.2$ mm (Equation 6.26), the required area of additional non-prestressed steel is approximated by:

$$A_s = \frac{(M_{Ed.x})_{CD} - M_{Rd1}}{f_{yd} z_2} = \frac{(77.7 - 69.0) \times 10^6}{435 \times 111.2} = 180 \text{ mm}^2/\text{m}$$

Use 12 mm diameter bars ($f_{yk} = 500$ MPa) at 450 mm centres ($A_{st} = 251 \text{ mm}^2/\text{m}$) as additional steel in the top of the slab over beam support CD. If this steel had been necessary for crack control, a maximum spacing of around 300 mm would have been recommended, but it is included specifically to increase the design resistance and calculations have indicated that crack control will not be a problem.

Checking flexural strength at other critical sections indicates that:

(a) At mid-span in the y-direction:

$$M_{Ed.y} = 58.4 \text{ kNm/m} \quad \text{and} \quad M_{Rd1} = 69.0 \text{ kNm/m} \ (d = 143 \text{ mm})$$

∴ No additional bottom reinforcement is required at mid-span in the y-direction.

(b) At mid-span in the x-direction:

$$M_{Ed.x} = 34.8 \text{ kNm/m} \quad \text{and} \quad M_{Rd1} = 59.1 \text{ kNm/m} \ (d = 124 \text{ mm})$$

∴ No additional bottom reinforcement is required at mid-span in the x-direction.

(c) At the short continuous support in the x-direction:

$$(M_{Ed.x})_{AC} = 1.33 \times 32.8 = 46.3 \text{ kNm/m} \quad \text{and}$$

$$M_{Rd1} = 69.0 \text{ kNm/m} \ (d = 143 \text{ mm})$$

∴ No additional top reinforcement is required in the x-direction over AC and BD.

Summary of reinforcement requirements:

Tendons consisting of four 12.5 mm strands at 1000 mm centres in each direction are used with the profiles shown in Figures 12.12c and d. In addition, 12 mm diameter non-prestressed reinforcing bars in the y-direction at 450 mm centres are also placed in the top of the slab over the long support CD (extending on each side of the beam to the point 0.3 times the clear span in the x-direction from the face of the support).

Check shear strength:

In accordance with Figure 12.11, the maximum shear in the slab occurs at the face of the long support near its mid-length, where:

$$V_{Ed} = w_{Ed} l_y/2 = 15.33 \times 9/2 = 69.0 \text{ kN/m}$$

The contribution of the concrete to the shear resistance $V_{Rd,c}$ in the region of low moment at the face of the discontinuous support is given by Equation 7.4.

With f_{ctd} = 1.67 MPa, σ_{cp} = $P_{m,t,y}/A$ = 2.33 MPa, I = 486 × 10⁶ mm⁴/m and S = 4.05 × 10⁶ mm³/m:

$$V_{Rd,c} = \left(\frac{486 \times 10^6 \times 1000}{4.05 \times 10^6} \sqrt{(1.67)^2 + 1.0 \times 2.33 \times 1.67} \right) \times 10^{-3} = 310 \text{ kN}$$

Clearly, V_{Ed} is much less than $V_{Rd,c}$ and the shear resistance is ample here. Shear resistances at all other sections are also satisfactory. Shear is rarely a problem in edge-supported slabs.

12.9 FLAT PLATE SLABS

12.9.1 Load balancing

Flat plates behave in a similar manner to edge-supported slabs except that the *edge beams* are strips of slab located on the column lines, as shown in Figure 12.13. The edge beams have the same depth as the remainder of the slab panel, and therefore the system tends to be less stiff and more prone to serviceability problems. The load paths for both the flat plate and the edge-supported slabs are, however, essentially the same (compare Figures 12.10 and 12.13).

In the flat plate panel of Figure 12.13, the total load to be balanced is $w_{bal}l_x l_y$. The upward forces per unit area exerted by the slab tendons in each direction are given by Equations 12.11 and 12.12, and the slab tendons impose a total upward force of:

$$w_{px}l_x l_y + w_{py}l_y l_x = w_{bal}l_x l_y$$

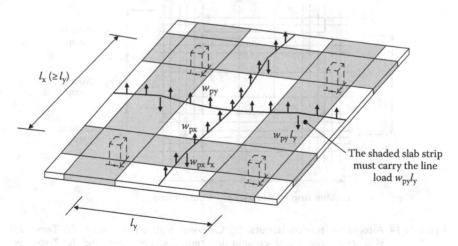

Figure 12.13 Interior flat plate panel.

Just as for edge-supported slabs, the slab tendons may be distributed arbitrarily between the x- and y-directions provided that adequate additional tendons are placed in the slab strips to balance the line loads $w_{py}l_y$ and $w_{px}l_x$ shown on the column lines in Figure 12.13. These additional *column line* tendons correspond to the *beam* tendons in an edge-supported slab system. For perfect load balancing, the column line tendons would have to be placed within the width of slab in which the slab tendons exert downward load due to reverse curvature. However, this is not a strict requirement and considerable variation in tendon spacing can occur without noticeably affecting slab behaviour. Column line tendons are frequently spread out over a width of slab as large as one half the shorter span, as indicated in Figure 12.14c.

The total upward force that must be provided in the slab along the column lines is:

$$w_{px}l_xl_y + w_{py}l_yl_x = w_{bal}l_xl_y$$

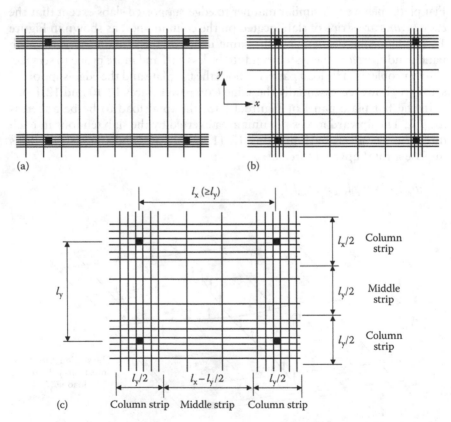

Figure 12.14 Alternative tendon layouts. (a) One-way slab arrangement. (b) Two-way slab arrangement with column line tendons in narrow band. (c) Two-way slab arrangement with column line tendons distributed over column strip.

Therefore, prestressing tendons (slab tendons plus column line tendons) must be provided in each panel to give a total upward force of $2w_{bal}l_xl_y$. The slab tendons and column line tendons *in each direction* must, between them, provide an upward force equal to the total load to be balanced $w_{bal}l_xl_y$. For example, in the slab system shown in Figure 12.14a, the entire load to be balanced is carried by slab tendons in the y-direction, i.e. $w_{py} = w_{bal}$ and $w_{px} = 0$. This entire load is deposited as a line load on the column lines in the x-direction and must be balanced by column line tendons in this vicinity. This slab has in effect been designed as a one-way slab spanning in the y-direction and supported by shallow heavily stressed slab strips on the x-direction column lines.

The two-way system shown in Figure 12.14b is more likely to perform better under unbalanced loads, particularly when the orthogonal spans l_x and l_y are similar and the panel is roughly square. In practice, however, steel congestion over the supporting columns and minimum spacing requirements (often determined by the size of the anchorages) make the concentration of tendons on the column lines impossible. Figure 12.14c shows a more practical and generally acceptable layout. Approximately 75% of the tendons in each direction are located in the column strips, as shown, the remainder being uniformly spread across the middle strip regions.

If the tendon layout is such that the upward force on the slab is approximately uniform, then at the balanced load the slab has zero deflection and is subjected only to uniform compression caused by the longitudinal prestress in each direction applied at the anchorages. Under unbalanced loads, moments and shears are induced in the slab. To calculate the moments and stresses due to unbalanced service loads and to calculate the factored design moments and shears in the slab (in order to check for strength), one of the methods described in the following sections may be adopted.

12.9.2 Behaviour under unbalanced load

Figure 12.15 illustrates the distribution of moments caused by an unbalanced uniformly distributed load w_{unbal} on an internal panel of a flat plate. The moment diagram in the direction of span l_x is shown in Figure 12.15b. The slab in this direction is considered as a wide, shallow beam of width l_y and span l_x and carrying a load $w_{unbal}l_y$ per unit length. The relative magnitudes of the negative moments M_{1-2} and M_{3-4} and positive moment M_{5-6} are found by elastic frame analysis (see Section 12.9.3) or more approximate recommendations (see Section 12.9.4). Whichever method is used, the *total static moment* M_o is fixed by statics and is given by:

$$M_o = \frac{w_{unbal}l_yl_x^2}{8} \tag{12.23}$$

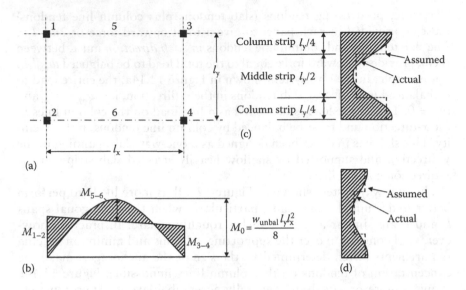

Figure 12.15 Distribution of moments in a flat plate. (a) Plan. (b) Moments in long
direction. (c) Distribution of M_{1-2} across panel. (d) Distribution of
M_{5-6} across panel.

In Figure 12.15c and d, variations in elastic moments across the panel at
the column lines and at mid-span, respectively, are shown. At the column
lines, where curvature is a maximum, the moment is also a maximum. On
panel centre line, where curvature is a minimum, so too is the moment. In
design, it is convenient to divide the panel into column and middle strips
and to assume that the moment is constant in each strip as shown. The
column strips in the l_x direction are defined as strips of width $0.25l_y$, but
not greater than $0.25l_x$, on each side of the column centre line. The middle
strips are the slab strips between the column strips.

It may appear from the moment diagrams that at the design loads, the
best distribution of tendons (and hence strength) is one in which tendons
are widely spaced in the middle strips and closer together in the column
strips, as shown in Figure 12.14c. However, provided that the slab is ductile, redistribution of moments takes place as the ultimate condition is
approached and the final distribution of moments depends very much on
the layout of the bonded steel.

After the slab cracks and throughout the overload range, superposition
is no longer applicable and the concepts of balanced and unbalanced loads
are not meaningful. As discussed in Section 11.5.4, for ultimate limit state
design, when the load factors are applied to the dead and live load moments,
codes of practice often insist that secondary moments are considered with
a load factor of 1.0. However, provided that the slab is ductile, and slabs
are usually very ductile, secondary moments may be ignored in the strength

calculations. The difficulty in accurately estimating slab moments, particularly in the overload range, is rendered relatively unimportant by the ductile nature of slabs.

12.9.3 Frame analysis

A commonly used technique for the analysis of flat plates in building structures is the *equivalent frame method* (or *idealised frame method*). The structure is idealised into a set of parallel two-dimensional frames running in two orthogonal directions through the building. Each frame consists of a series of vertical columns spanned by horizontal *beams*. These idealised beams consist of the strip of slab of width on each side of the column line equal to half the distance to the adjacent parallel row of columns and include any floor beams forming part of the floor system. The member stiffnesses are determined and the frames are analysed under any desired gravity loading using a linear-elastic frame analysis. EN 1992-1-1 [1] permits the stiffness of the members to be based on gross cross-sections and, for vertical gravity loads, the stiffness may be based on the full width of the panels. For flat slab structures subjected to lateral (horizontal) loads, the stiffness of the horizontal (floor) members should be based on 40% of the full width of the panels. EN 1992-1-1 [1] states that this reduction of stiffness reflects the increased flexibility of the slab–column connections compared to column–beam connections in beam and slab floor systems. For a flat plate building in which shear walls, or some other bracing system, are provided to resist all lateral loads, it is usually permissible to analyse each floor of the building separately, with the columns above and below the slab assumed to be fixed at their remote ends.

The equivalent frame method provides a relatively crude model of structural behaviour, with inaccuracies being associated with each of the following assumptions: (1) a two-way plate is idealised by orthogonal one-way strips; (2) the stiffness of a cracked slab may be significantly less than that based on gross sections; and (3) a linear-elastic analysis is applied to a structure that is non-linear and inelastic both at service loads and at overloads. A simple estimate of member stiffness, for example, based on gross section properties, will lead to an estimate of frame moments that satisfies equilibrium and usually provides an acceptable solution. When such a frame analysis is used to check bending strength, an equilibrium load path is established that will prove to be a satisfactory basis for design, provided that the slab is ductile and the moment distribution in the real slab can redistribute towards that established in the analysis. For strength design, it is usually sufficient to analyse the slab assuming that the full factored load is applied to all spans.

For the determination of the design moments at each critical section of the frame at the serviceability limit states, variations of the load intensities on individual spans should be considered, including pattern loading whereby the transient load is applied to some spans and not to others.

Loading patterns to be considered to assess deflection and cracking should include at least the following:

1. where the loading pattern is known, the frame should be analysed under that known loading. This includes the factored permanent dead load G; and
2. with regard to the live loads Q, where the pattern of loaded and unloaded spans is variable, the factored live load should be applied:
   a. on alternate spans (this will permit the determination of the maximum factored positive moment near the middle of the loaded spans)
   b. on two adjacent spans (this will permit the determination of the maximum factored negative moment at the interior support between the loaded spans)
   c. on all spans

The frame moments calculated at the critical sections of the idealised horizontal members are distributed across the floor slab into the column and middle strips (as defined in the previous section). EN 1992-1-1 [1] apportions the total frame moments to column and middle strips in accordance with Table 12.4. Studies have shown that the performance of reinforced concrete flat slabs both at service loads and at overloads is little affected by variations in the fraction of the total frame moment that is assigned to the column strip [13], provided that the slab is ductile and capable of the necessary moment redistribution.

With the in-service moments caused by the unbalanced loads determined at all critical regions in the slab, checks for cracking and crack control and calculations of deflection may be undertaken.

When the design resistances of the column and middle strips are being checked, it is advisable to ensure that the depth to the neutral axis at any section does not exceed about $0.25d$. This will ensure sufficient ductility for the slab to redistribute bending moments towards the bending moment diagram predicted by the idealised frame analysis and will also allow the designer to safely ignore the secondary moments. There are obvious advantages in

Table 12.4 Fraction of frame moments distributed to the column and middle strips [1]

	Negative moments (%)	Positive moments (%)
Column strip	60–80	50–70
Middle strip	40–20	50–30

Note: The sum of moments resisted by the column and middle strips at any location must always equal the frame moment at that location.

allocating a large fraction of the negative moment at the supports to the column strip. The resulting increased steel quantities stiffen and strengthen this critical region of the slab, thereby improving punching shear and crack control. In prestressed flat slabs, only the column strip regions over the interior columns are likely to experience significant cracking.

12.9.4 Direct design method

A simple semi-empirical approach for the analysis of prestressed flat plates is the 'direct design method'. The following limitations are often imposed on the use of the direct design method [12]:

1. there are at least two continuous spans in each direction;
2. the support grid is rectangular, or nearly so (individual supports may be offset up to a maximum of 10% of the span in the direction of the offset);
3. the ratio of the longer span to the shorter span measured centre-to-centre of supports within any panel is not greater than 2.0;
4. in each direction, successive span lengths do not differ by more than one-third of the longer span and in no case is an end span longer than the adjacent interior span;
5. gravity loads are essentially uniformly distributed. Lateral loads are resisted by shear walls or braced vertical elements and do not enter into the analyses;
6. the live load does not exceed twice the dead load; and
7. low-ductility reinforcement is not used as the flexural reinforcement.

The slab is divided into design strips in each direction and each strip is designed one span at a time. The total static moment M_o in each span of the design strip is calculated from:

$$M_o = \frac{w_{Ed} l_t l_{eff}^2}{8} \tag{12.24}$$

where w_{Ed} is the design load per unit area (factored for strength), l_{eff} is the effective span (i.e. the lesser of the centre-to-centre distance between supports and $l_n + h$), l_n is the clear span between the faces of the supports, h is the overall slab thickness and l_t is the width of the design strip measured transverse to the direction of bending. For an interior design strip, l_t is equal to the average of the centre-to-centre distance between the supports of the adjacent transverse spans. For an edge design strip, l_t is measured from the slab edge to the point halfway to the centre line of the next interior and parallel row of supports.

The static moment M_o is shared between the supports (negative moments) and the mid-span (positive moment). At any critical section, the design

Table 12.5 Design moment factors for an end span [1]

	Negative moment factor at		Positive moment factor
	Exterior support	Interior support	
Flat slabs with exterior edge unrestrained	0.0	0.80	0.60
Flat slabs with exterior edge restrained by columns only	0.25	0.75	0.50
Flat slabs with exterior edge restrained by spandrel beams and columns	0.30	0.70	0.50
Flat slabs with exterior edge fully restrained	0.65	0.65	0.35
Beam and slab construction	0.15	0.75	0.55

Table 12.6 Design moment factors for an interior span [1]

Type of slab system	Negative moment factor	Positive moment factor
All types	0.65	0.35

moment may be determined by multiplying M_o by the relevant factor given in Table 12.5 or 12.6, as appropriate. At any interior support, the floor slab should be designed to resist the larger of the two negative design moments determined for the two adjacent spans unless the unbalanced moment is distributed to the adjoining members in accordance with their relative stiffnesses.

The positive and negative design moments are next distributed to the column and middle strips using the column strip moment factor from Table 12.4.

12.9.5 Shear resistance

Punching shear strength requirements often control the thickness of a flat slab at the supporting columns and must always be checked. The shear resistance of the slabs was discussed in Section 7.4 and methods for designing the slab–column intersection were presented.

If frame analyses are performed to check the flexural resistance of a slab, the design moment M_{Ed} transferred from the slab to a column and the design shear V_{Ed} are obtained from the relevant analyses. M_{Ed} is that part of the unbalanced slab bending moments that is transferred into the column at the support. If the direct design method is used for the slab design, M_{Ed} and V_{Ed} must be calculated separately. The shear force crossing the critical shear perimeter around a column support may be taken as the product of the factored design load w_{Ed} and the plan area of slab supported by the column and located outside the critical shear perimeter. At an interior support, M_{Ed} may be taken as [1]:

$$M_{Ed} = 0.06[(1.35w_G + 0.75w_Q)l_t(l_{eff})^2 - 1.35w_Gl_t(l'_{eff})^2] \qquad (12.25)$$

where w_G and w_Q are the uniformly distributed dead and live loads on the slab (per unit area), l_t is the transverse width of the slab as defined in the text below Equation 12.24, l_{eff} and l'_{eff} are the longer and shorter effective spans on either side of the column. For an edge column, M_{Ed} is equal to the design moment at the exterior edge of the slab and may be taken as $0.25M_o$ (where M_o is the static moment for the end span of the slab calculated using Equation 12.24).

When detailing the slab–column connection, it is advisable to have at least two prestressing tendons crossing the critical shear perimeter in each direction. Additional well anchored non-prestressed reinforcement crossing the critical perimeter will also prove beneficial (both in terms of crack control and ductility) in the event of unexpected overloads.

12.9.6 Deflection calculations

The deflection of a uniformly loaded flat slab may be estimated using the *wide beam method* which was formalised by Nilson and Walters [14]. Originally developed for reinforced concrete slabs, the method is particularly appropriate for prestressed flat slabs which are usually uncracked at service loads [15]. The basis of the method is illustrated in Figure 12.16. Deflections of the two-way slab are calculated by considering separately the slab deformations in each direction. The contributions in each direction are then added to obtain the total deflection.

In Figure 12.16a, the slab is considered to act as a wide shallow beam of width equal to the smaller panel dimension l_y and span equal to the longer panel effective span l_x. This wide beam is assumed to rest on unyielding supports. Because of variations in the moments caused by the unbalanced loads and the flexural rigidity across the width of the slab, all unit strips in the x-direction will not deform identically. Unbalanced moments and hence curvatures in the regions near the column lines (the column strip) are greater than in the middle strips. This is particularly so for uncracked prestressed concrete slabs or prestressed slabs that are cracked only in the column strips. The deflection on the column line is therefore greater than that at the panel centre. The slab is next considered to act as a wide shallow beam spanning in the y-direction, as shown in Figure 12.16b. Once again, the effect of variation of moment across the wide beam is shown.

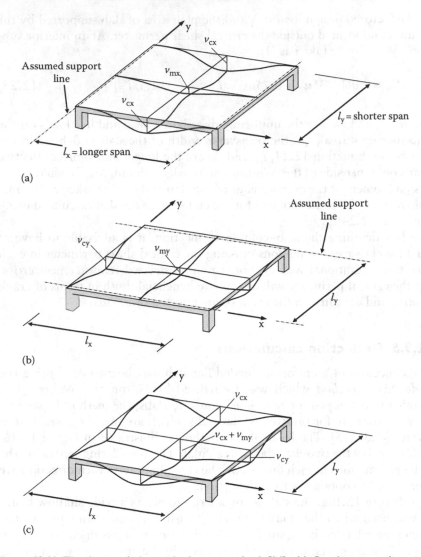

Figure 12.16 The basis of the wide beam method [14]. (a) Bending in *x*-direction. (b) Bending in *y*-direction. (c) Combined bending.

The mid-panel deflection is the sum of the mid-span deflection of the column strip in the long direction and that of the middle strip in the short direction, as shown in Figure 12.16c:

$$v_{mid} = v_{cx} + v_{my} \tag{12.26}$$

The method can be used irrespective of whether the moments in each direction are determined by the equivalent frame method, the frame analysis based on gross stiffnesses or the direct design method (see Sections 12.9.3 and 12.9.4). The definition of column and middle strips, the longitudinal moments in the slab, the lateral moment distribution coefficients and other details are the same as for the moment analysis so that most of the information required for the calculation of deflection is already available.

The actual deflection calculations are more easily performed for strips of floor in either direction bounded by the panel centre lines, as is used for the moment analysis. In each direction, an average deflection v_{avge} at mid-span of the wide beam is calculated from the previously determined moment diagram and the moment of inertia of the entire wide beam I_{beam} using the deflection calculation procedures outlined in Section 5.11. The effect of the moment variation across the wide beam, as well as possible differences in column and middle strip sizes and rigidities, is accounted for by multiplying the average deflection by the ratio of the curvature of the relevant strip to the curvature of the wide beam. For example, for the wide beam in the x-direction, the column and middle strip deflections are, respectively:

$$v_{cx} = v_{avge.x} \frac{M_{col}}{M_{beam}} \frac{E_{cm}I_{beam}}{E_{cm}I_{col}} \tag{12.27}$$

and

$$v_{mx} = v_{avge.x} \frac{M_{mid}}{M_{beam}} \frac{E_{cm}I_{beam}}{E_{cm}I_{mid}} \tag{12.28}$$

It is usual to assume that M_{col}/M_{beam} is about 0.7 and therefore M_{mid}/M_{beam} is about 0.3. If cracking is detected in the column strip, the effective moment of inertia of the cracked cross-section can be calculated using the analysis described in Section 5.11.3. The effective moment of inertia of the column strip I_{col} is calculated as the average of I_{ef} at the negative moment region at each end of the strip, which may include the loss of stiffness due to cracking and/or the stiffening effect of a drop panel, and the positive moment region, which is usually uncracked. I_{col} is then added to the moment of inertia of the middle strip I_{mid} (which is also usually uncracked and therefore based on gross section properties) to form the effective moment of inertia of the wide beam I_{beam}. These quantities are then used in the calculation of the short-term column and middle strip deflections in each direction using Equations 12.27 and 12.28. A reasonable estimate of the weighted average effective moment of inertia of an interior span of the wide beam is obtained by taking 0.7 times the value at mid-span plus 0.3 times the average of the values at each end of the span. For an exterior span, a reasonable weighted average is 0.85 times the mid-span value plus 0.15 times the value at the continuous end. This recommendation may also be used for the calculation of I_{col} for a cracked column strip.

The moment of inertia of the wide beam is, of course, always the sum of I_{col} and I_{mid}. Long-term deflections due to sustained unbalanced loads can also be calculated in each direction using the procedures outlined in Section 5.11.4.

Nilson and Walters [14] originally proposed to analyse a fixed-ended beam and then calculate the deflection produced by rotation at the supports. This does not significantly improve the accuracy of the model, and the additional complication is not warranted.

EXAMPLE 12.4

Determine the tendons required in the 220 mm thick flat slab shown in Figure 12.17. The live load on the slab is 3.0 kPa and the dead load is 1.0 kPa plus the slab self-weight. All columns are 600 mm by 600 mm and are 4 m long above and below the slab. At the top of each column, a 300 mm column capital is used to increase the supported area, as shown. In this example, the

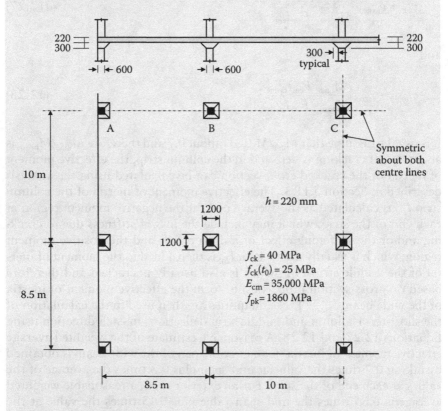

Figure 12.17 Plan and section of flat plate (Example 12.4).

dead load w_G is to be effectively balanced by prestress and is given by $w_G =$ 1 kPa + self-weight = 1 + (24 × 0.22) = 6.3 kPa.

1. Checking punching shear:

Before proceeding too far into the design, it is prudent to make a preliminary check of punching shear at typical interior and exterior columns. Consider the interior column B in Figure 12.17. The area of slab supported by the column is 10 × (8.5 + 10)/2 = 92.5 m². Using the strength load factors specified in EN 1992-1-1 [1] (Equation 2.2), the factored design load is:

$$w_{Ed} = 1.35w_G + 1.5w_Q = (1.35 \times 6.3) + (1.5 \times 3.0) = 13.0 \text{ kN/m}^2$$

and therefore the shear force crossing the critical section may be approximated by:

$$V_{Ed} \approx 13.0 \times 92.5 = 1203 \text{ kN}$$

From Equation 12.25, the design moment transferred to the column may be taken as:

$$M_{Ed} = 0.06[(1.35 \times 6.3 + 0.75 \times 3.0) \times 10 \times 9.02^2 - 1.35 \times 6.3 \times 10 \times 7.52^2]$$

$$= 236 \text{ kNm}$$

The design value of the maximum punching shear resistance on any control section is obtained from Equation 7.45:

$$v_{Rd,max} = 0.3 \times \left[1 - \frac{40}{250}\right] \times 26.67 = 6.72 \text{ MPa}$$

Referring to Figure 7.18, $l_H = 300$ mm, $c = 600$ mm and $l_1 = l_2 = 1200$ mm. The average effective depth around the shear perimeter is taken to be $d_{eff} =$ 220 − 50 = 170 mm, and from Equation 7.38, $r_{cont} = 2 \times 170 + 0.56 \times 1200 =$ 1012 mm. The critical shear perimeter is located at a distance r_{cont} from the centroid of the column (constructed so that its length is minimised). In each span direction, the critical shear perimeter is located at 412 mm from the edge of the loaded area (i.e. the edge of the column capital). The critical shear perimeter is therefore:

$$u_1 = 4 \times 600 + 2\pi \times 412 = 4989 \text{ mm}$$

The design punching shear resistance for a slab without shear reinforcement is obtained from Equation 7.42. With $k = 2.0$, $v_{min} = 0.035 \times k^{1.5} \times f_{ck}^{0.5} = 0.626$. We will assume that $\rho_1 = 0.006$ and the average prestress in the concrete is assumed to be $\sigma_{cp} = 2.4$ MPa. These assumptions will need to be checked subsequently. Equation 7.42 gives:

$$v_{Rd,c} = 0.12 \times 2.0 \times (100 \times 0.006 \times 40)^{1/3} + 0.1 \times 2.4 = 0.932 \text{ MPa}$$
$$(>v_{min} + k_1\sigma_{cp})$$

From Equation 7.49:

$$W_1 = 0.5 \times 600^2 + 600 \times 600 + 4 \times 600 \times 170 + 16 \times 170^2 + 2\pi \times 170 \times 600$$

$$= 2051 \times 10^3 \text{ mm}^2$$

and from Equation 7.47 and Table 7.2:

$$\beta = 1 + 0.6 \times \frac{236 \times 10^6}{1203 \times 10^3} \times \frac{4989}{2051 \times 10^3} = 1.286$$

From Equation 7.46, the maximum shear stress on the basic control perimeter is:

$$v_{Ed} = 1.286 \times \frac{1203 \times 10^3}{4989 \times 170} = 1.82 \text{ MPa}$$

which is much less than $v_{Rd,max}$, but greater than $v_{Rd,c}$, and therefore shear reinforcement is required.

Punching shear at edge and corner columns should similarly be checked.

2. *Establish cable profiles:*

Using four 12.5 mm strands in a flat duct, with 25 mm concrete cover to the duct (the same as in Figure 12.12b), the maximum depth to the centre of gravity of the strand is:

$$d_p = 220 - (25 + 19 - 7) = 183 \text{ mm}$$

and the corresponding eccentricity is $e = 73$ mm. The maximum cable drapes in an exterior span and in an interior span are, respectively:

$$(z_{d.max})_{ext} = \frac{73}{2} + 73 = 109.5 \text{ mm}$$

and

$$(z_{d.max})_{int} = \frac{73+73}{2} + 73 = 146 \text{ mm}$$

Consider the trial cable profile shown in Figure 12.18. For the purposes of this example, it is assumed that jacking occurs simultaneously from both ends of a tendon so that the prestressing force in a tendon is symmetrical with respect to the centre line of the structure (shown in Figure 12.17). The friction losses have been calculated using Equation 5.148 with $\mu = 0.19$ and $k = 0.016$ for flat ducts, and the losses due to a 6 mm draw-in at the anchorage are calculated as outlined in Section 5.10.2.4 using Equation 5.151. The immediate losses (friction + draw-in) are also shown in Figure 12.18.

3. *Determine tendon layout:*

It is assumed here that the average time-dependent loss of prestress in each low-relaxation tendon is 15%. Of course, this assumption should be checked.

The effective prestressing forces per metre width required to balance 6.3 kPa using the fully available drape in the exterior span (AB) and in the interior span (BC) are found using Equation 12.10:

$$(P_{m,t})_{AB} = \frac{6.3 \times 8.5^2}{8 \times 0.1095} = 520 \text{ kN/m}$$

$$(P_{m,t})_{BC} = \frac{6.3 \times 10^2}{8 \times 0.146} = 540 \text{ kN/m}$$

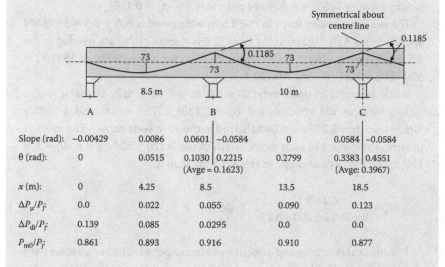

Slope (rad):	−0.00429	0.0086	0.0601	−0.0584	0	0.0584	−0.0584
θ (rad):	0	0.0515	0.1030	0.2215	0.2799	0.3383	0.4551
			(Avge = 0.1623)			(Avge: 0.3967)	
x (m):	0	4.25	8.5		13.5	18.5	
$\Delta P_{\mu}/P_j$:	0.0	0.022	0.055		0.090	0.123	
$\Delta P_{di}/P_j$:	0.139	0.085	0.0295		0.0	0.0	
P_{m0}/P_j:	0.861	0.893	0.916		0.910	0.877	

Figure 12.18 Cable profile and immediate loss details (Example 12.4).

and the corresponding forces required at the jack prior to the instantaneous and time-dependent losses are:

$$P_j = \frac{520}{0.890 \times 0.85} = 687 \text{ kN/m}$$

(required to balance 6.3 kPa in the exterior span)

$$P_j = \frac{540}{0.901 \times 0.85} = 705 \text{ kN/m}$$

(required to balance 6.3 kPa in the interior span)

The jacking force is therefore governed by the requirements for the interior span.

For the 10 m wide panel, the total jacking force required is $705 \times 10 = 7050$ kN. If the maximum stress in the tendon is $0.9f_{p0.1k} = 1440$ MPa, the total area of prestressing steel is therefore:

$$A_p = \frac{7050 \times 10^3}{1440} = 4896 \text{ mm}^2$$

At least 14 flat-ducted cables are required in the 10 m wide panel (with the area of prestressing steel in each cable $A_p = 372$ mm²/cable) with an initial jacking force of $7050/14 = 504$ kN per cable (i.e. $\sigma_{pj} = 0.728f_{pk}$).

The required jacking force in the 8.5 m wide panel is $705 \times 8.5 = 5993$ kN and therefore $A_p = 4162$ mm². At least 12 flat-ducted cables are needed in the 8.5 m wide panels ($A_p = 4462$ mm²) with an initial jacking force of $5993/12 = 500$ kN per cable (i.e. $\sigma_{pj} = 0.722 f_{pk}$).

In the interests of uniformity, all tendons will be initially stressed with a jacking force of 504 kN/cable (i.e. $\sigma_{pj} = 0.728f_{pk}$). This means that a slightly higher load than 6.3 kPa will be balanced in the 10 m wide panels. The average prestress at the jack in each metre width of slab is $(26 \times 504)/(8.5 + 10) = 708$ kN/m, and the revised drape in the exterior span is:

$$h_{AB} = \frac{6.3 \times 8.5^2}{8 \times 708 \times 0.890 \times 0.85} = 0.106 \text{ m}$$

The final cable profile and effective prestress per panel after all losses are shown in Figure 12.19.

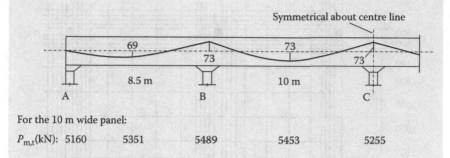

For the 10 m wide panel:

$P_{m,t}$(kN): 5160   5351   5489   5453   5255

Figure 12.19 Cable profile and effective prestress (Example 12.4).

The maximum average stress in the concrete due to the longitudinal anchorage force after the deferred losses is:

$$\frac{P}{A} = \frac{5,489 \times 10^3}{10,000 \times 220} = 2.50 \text{ MPa}$$

which is within the range mentioned in Section 12.3 as being typical for flat slabs.

The cable layout for the slab is shown on the plan in Figure 12.20. For effective load balancing, about 75% of the cables are located in the column strips. The minimum spacing of tendons is usually governed by the size of the anchorage and is taken here as 300 mm, while a maximum spacing of 1600 mm has also been adopted.

4. *Serviceability considerations*:

In practice, the time-dependent losses in the post-tensioned tendons should now be checked using the procedures outlined in Section 5.10.3 and illustrated previously in Examples 10.1 and 10.3. For the purposes of this example, we will assume that the losses have been checked at the critical sections and are as assumed in step 3. We will now analyse the slab under the unbalanced loads and check for the likelihood of cracking.

Considering the 10 m wide frame on column line ABC in Figure 12.17, the effective spans (clear spans + slab depth) are:

For end span AB: $(l_{eff})_{AB}$ = 8.5 − 0.6 − 0.6 + 0.22 = 7.52 m

For interior span BC: $(l_{eff})_{BC}$ = 10.0 − 0.6 − 0.6 + 0.22 = 9.02 m

With the full dead load balanced by the effective prestress, the maximum unbalanced load is 3 kPa (of which 1.0 kPa is assumed to be sustained).

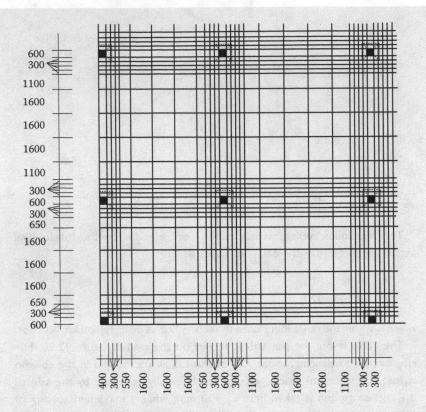

Figure 12.20 Tendon layout (Example 12.4).

The total static moments in the end span AB and in the interior span BC caused by the maximum unbalanced load are obtained using Equation 12.23:

$$(M_o)_{AB} = \frac{3 \times 10 \times 7.52^2}{8} = 212.1 \text{ kNm}$$

and

$$(M_o)_{BC} = \frac{3 \times 10 \times 9.02^2}{8} = 305.1 \text{ kNm}$$

The moment diagrams for each span caused by the unbalanced load obtained using the direct design method are shown in Figure 12.21.

Check for cracking: For the 5 m wide and 220 mm deep column strip, the properties of the gross cross-section are $A = 1100 \times 10^3$ mm^2, $I = 4437 \times 10^6$ mm^4 and $Z = 40.33 \times 10^6$ mm^3. Taking 75% of the negative frame moment

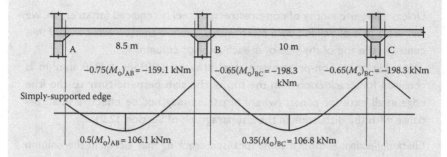

Figure 12.21 Bending moment diagram for flat slab frame ABC (Example 12.4).

at support C to be carried in the 5 m wide column strip, the column strip moment at C is $(M_C)_{col} = 0.75 \times (-198.3) = -148.7$ kNm and the effective prestress in the column strip at C is $P_{m,t} = 0.85 \times 0.877 \times P_j = 2628$ kN. The tensile stress in the top concrete fibre is:

$$\sigma_{c,top} = -\frac{P_{m,t}}{A} - \frac{(M_C)_{col}}{Z} = -2.39 + 3.69 = +1.30 \text{ MPa (tension)}$$

This is less than the lower characteristic tensile strength of the concrete $f_{ctk,0.05}$ and therefore cracking is unlikely and will not cause any significant loss of stiffness (see the discussion on flexural crack control in Section 12.5.1). The tensile stresses at the top of the slab in the column strip over support B and in the bottom of the slab in the column strip at the mid-spans of AB and BC will be less than that calculated earlier because in each case the magnitude of the effective prestress is larger than that at support C and the magnitude of the unbalance moment is smaller.

Although the tensile stresses in the column strips are less than the tensile strength of concrete, some local cracking over the interior column support is likely since peak moments are much higher than average values. A mat of conventional non-prestressed reinforcement (rectangular in plan) is here provided over the interior supporting columns to ensure crack control. The steel in each direction will be continuous across the column support line and extend to 25% of the clear span in each direction (see discussion in Section 12.5.1). The non-prestressed steel area in each direction for crack control is at least:

$$A_{s.x} = A_{s.y} = 0.001 \times 220 \times 1000 = 220 \text{ mm}^2/\text{m}$$

Unless a larger quantity of non-prestressed steel is required for strength, we will use a mat of welded wire mesh with 7.6 mm diameter wires at 200 mm centre in the top of the slab over each interior column.

In addition, a non-prestressed steel area of $0.0015bh = 330$ mm²/m is required for crack control in the top of the slab perpendicular to the free edge in all exterior panels (where prestress may not be effective in accordance with the discussion in the last paragraph of Section 12.2).

Check deflection: Although some localised cracking may occur in the column strip over the interior columns of this slab, it will not be sufficient to significantly affect the slab stiffness and deflection may be calculated using the properties of the gross cross-section.

We will check deflection in the end span and in the interior spans of the 10 m wide slab strip centred on column line ABC in Figure 12.17 and subjected to the unbalanced load. For this uncracked *wide beam*: $I_{col} = 4437 \times 10^6$ mm⁴, $I_{mid} = 4437 \times 10^6$ mm⁴ and $I_{beam} = 8873 \times 10^6$ mm⁴. The maximum average deflection v_{avge} at mid-span of the wide beam occurs when adjacent spans are unloaded. In accordance with the discussion in Section 12.4.2 (concerning effect of pattern loading on the β factor in Equation 12.6), the appropriate deflection coefficient for the end span is $\beta = 3.5/384$ and for the interior span is $\beta = 2.6/384$. Therefore, due to the maximum unbalanced load:

$$(v_{avge})_{AB} = \frac{3.5}{384} \frac{w_{unbal} l_t (l_{eff})^4_{AB}}{E_{cm} I_{beam}} = \frac{3.5}{384} \frac{3.0 \times 10 \times 7,520^4}{35,000 \times 8,873 \times 10^6} = 2.81 \text{ mm}$$

$$(v_{avge})_{BC} = \frac{3.5}{384} \frac{w_{unbal} l_t (l_{eff})^4_{BC}}{E_{cm} I_{beam}} = \frac{2.6}{384} \frac{3.0 \times 10 \times 9,020^4}{35,000 \times 8,873 \times 10^6} = 4.33 \text{ mm}$$

Taking 70% of moment in the column strip, the deflections of the column strip and the middle strip are obtained from Equations 12.27 and 12.28:

$$(v_c)_{AB} = (v_{avge})_{AB} \times 0.7 \times \frac{l_{beam}}{l_{col}} = 2.81 \times 0.7 \times 2 = 3.93 \text{ mm}$$

$$(v_m)_{AB} = (v_{avge})_{AB} \times 0.3 \times \frac{l_{beam}}{l_{mid}} = 2.81 \times 0.3 \times 2 = 1.69 \text{ mm}$$

$$(v_c)_{BC} = 4.33 \times 0.7 \times 2 = 6.06 \text{ mm}$$

$$(v_m)_{BC} = 4.33 \times 0.3 \times 2 = 2.60 \text{ mm}$$

Due to symmetry, the column and middle strip deflections in the orthogonal direction are the same for this uncracked slab.

The maximum short-term deflection at the midpoint of the panels due to the unbalanced load is obtained by adding the column strip deflection to the x-direction with the middle strip deflection in the y-direction (Equation 12.26). For the edge panel, adjacent to column line AB (with $l_x = 10$ m and $l_y = 8.5$ m), the maximum short-term deflection is:

$$(v_{i.mid})_{AB} = 6.06 + 1.69 = 7.75 \text{ mm}$$

while for the internal panel adjacent to column line BC (with $l_x = 10$ m and $l_y = 10$ m), the maximum short-term deflection is:

$$(v_{i.mid})_{BC} = 6.06 + 2.60 = 8.66 \text{ mm}$$

The sustained portion of the unbalanced load $w_{unbal.sus} = 1.0$ kPa ($= 0.33 \, w_{unbal}$) and the short-term mid-panel deflections produced by the sustained unbalanced load are therefore one-third of the values given earlier. Assuming the final creep coefficient $\varphi(\infty, t_0) = 2.5$, the creep-induced deflection at the midpoint of each panel is estimated from Equation 12.7:

$$(v_{cc.mid})_{AB} = 2.5 \times 0.333 \times 7.75 = 6.46 \text{ mm}$$

$$(v_{cc.mid})_{BC} = 2.5 \times 0.333 \times 8.66 = 7.22 \text{ mm}$$

Assuming the final shrinkage strain in the concrete is $\varepsilon_{cs}^* = 0.0005$, the average shrinkage curvature κ_{cs} in each direction is estimated using Equation 12.9:

$$\kappa_{cs} = \frac{0.3 \times 0.0005}{220} = 0.68 \times 10^{-6} \text{ mm}^{-1}$$

The average deflections due to shrinkage on the column centre lines are obtained from Equation 12.8:

$$(v_{cs})_{AB} = 0.090 \times 0.68 \times 10^{-6} \times 7520^2 = 3.46 \text{ mm}$$

$$(v_{cs})_{BC} = 0.065 \times 0.68 \times 10^{-6} \times 9020^2 = 3.60 \text{ mm}$$

and the shrinkage deflections at the midpoints of the panels adjacent to column lines AB and BC are the sum of the shrinkage deflection in each direction:

$$(v_{cs.mid})_{AB} = (v_{cs})_{AB} + (v_{cs})_{BC} = 7.06 \text{ mm}$$

$$(v_{cs.mid})_{BC} = (v_{cs})_{BC} + (v_{cs})_{BC} = 7.20 \text{ mm}$$

Therefore, the maximum total deflections at the midpoints of the panels adjacent to column lines AB and BC are:

$$(v_{mid})_{AB} = (v_{i.mid})_{AB} + (v_{cc.mid})_{AB} + (v_{cs.mid})_{AB} = 7.75 + 6.46 + 7.06 = 21.3 \text{ mm}$$

$$(v_{mid})_{BC} = (v_{i.mid})_{BC} + (v_{cc.mid})_{BC} + (v_{cs.mid})_{BC} = 8.66 + 7.22 + 7.20 = 23.1 \text{ mm}$$

As the longer effective span in each of these two panels is l_{eff} = 9.02 m, the maximum deflection is l_{eff}/390 and this should be satisfactory for most occupancies.

5. *Check shear and flexural resistance*:

With the level of prestress determined, punching shear should also be checked at both exterior and interior columns in accordance with the proce-dure outlined in Section 7.4. The dimensions of the column capitals may need to be modified and shear reinforcement may be required particularly in the spandrel strips along each free edge.

The flexural resistance of the slab must also be checked. For the pur-poses of this example, the design flexural resistance of the interior panel will be compared with the design moments determined from the direct design method. As calculated in step 1, w_{Ed} =13.0 kPa, the panel width is l_t = 10 m and the effective span of an interior panel is l_{eff} = 9.02 m. From Equation 12.24, the total static moment is:

$$M_o = \frac{13.0 \times 10 \times 9.02^2}{8} = 1322 \text{ kNm}$$

From Table 12.6, the negative support moment is:

$$0.65 M_o = 859 \text{ kNm}$$

Because both the positive and negative moment capacities are similar (each having the same quantity of prestressed steel at the same effective depth), only the negative moment region needs to be checked. From Table 12.5, the design negative moment in the column strip at the support is taken as:

$$M_{Ed} = 0.7 \times 859 = 601 \text{ kNm}$$

The 5 m wide column strip contains 10 cables (A_p = 3720 mm^2) at an effective depth of 183 mm. The following results are obtained for the column strip at the column support in accordance with the design strength procedures outlined in Chapter 6:

$$\sigma_{pud} = 1391 \text{ MPa}; \quad F_{pt} = 5175 \text{ kN}; \quad x = 48.5 \text{ mm} = 0.265d_p;$$
$$M_{Rd} = 847 \text{ kNm}$$

The design resistance of the column strip M_{Rd} is substantially greater than M_{Ed} and therefore the slab possesses adequate flexural strength at this location. The strength is also adequate at all other regions in the slab.

12.9.7 Yield line analysis of flat plates

Yield line analysis is a convenient tool for calculating the load required to cause flexural failure in reinforced concrete slabs. The procedure was described in detail by Johansen [16,17] and is a plastic method for the analysis of two-way slabs, with yield lines (or plastic hinge lines) developing in the slab and reducing the slab to a mechanism.

Typical yield line patterns for a variety of slab types subjected to uniformly distributed loads are shown in Figure 12.22. The yield lines divide the slab into rigid segments. At collapse, each segment rotates about an axis of rotation that is either a fully supported edge or a straight line through one or more point supports, as shown. All deformation is assumed to take place on the yield lines between the rigid segments or on the axes of rotation. The yield line pattern, or the collapse mechanism, for a particular slab must be compatible with the support conditions.

The principle of virtual work is used to determine the collapse load corresponding to any possible yield line pattern. For a particular layout of yield lines, a compatible virtual displacement system is postulated. Symmetry in the slab and yield line pattern should be reflected in the virtual displacement system. The external work W done by all the external forces as the slab undergoes its virtual displacement is equal to the internal work U. The internal work associated with a particular yield line is the product of the total bending moment on the yield line and the angular rotation that takes place at the line. Since all internal deformation takes place on the yield lines, the internal work U is the sum of the work done on all yield lines.

In reinforced concrete slabs with isotropic reinforcement, the moment of resistance or plastic moment m_u (per unit length) is constant along any yield line and the internal work associated with any of the collapse mechanisms shown in Figure 12.22 is easily calculated. In prestressed concrete slabs, the depth of the orthogonal prestressing tendons may vary from point to point along a particular yield line and the calculation of U is more difficult.

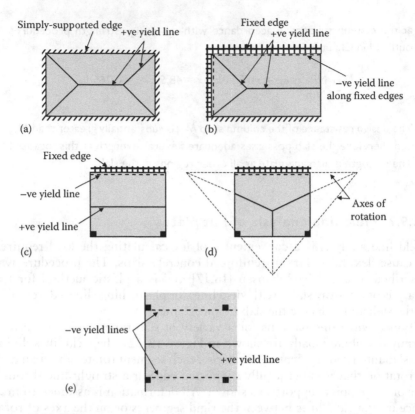

Figure 12.22 Plan views of slabs showing typical yield line patterns. (a) Four simply-supported edges. (b) Two fixed, one simple and one free edge. (c) One fixed edge and two corner supports - Pattern 1. (d) One fixed edge and two corner supports - Pattern 2. (e) Edge panel of a flat plate.

For flat plate structures, however, with the yield line patterns shown in Figures 12.22e and 12.23, the prestressing tendons crossing a particular yield line do so at the same effective depth, the plastic moment per unit length of the yield line is constant provided the tensile reinforcement and tendon areas per unit length are constant and the collapse load is readily calculated.

Consider the interior span of Figure 12.23a. If it is assumed conservatively that the columns are point supports and that the negative yield lines pass through the support centre lines and if the slab strip shown is given a unit vertical displacement at the position of the positive yield line, the external work done by the collapse loads w_u (in kN/m^2) acting on the slab strip is the total load on the strip times its average virtual displacement (which in this case is 0.5):

$$W = \frac{w_u l_t l}{2}$$

(12.29)

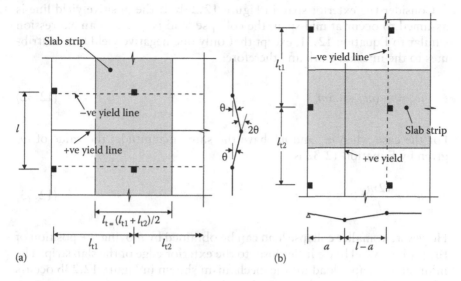

Figure 12.23 Yield line analysis of a flat plate. (a) Interior span. (b) Exterior span.

The internal work done at the negative yield line at each end of the span is the total moment $m'_u l_t$ times the angular change at the yield line $\theta\ (= 1/(l/2) = 2/l)$. At the positive yield line, the angular change is $2\theta\ (= 4/l)$ and the internal work is $m_u l_t \times 4/l$. The total internal work on all yield lines is therefore:

$$U = m_u l_t \frac{4}{l} + 2m'_u l_t \frac{2}{l} = \frac{4l_t(m_u + m'_u)}{l} \tag{12.30}$$

The principle of virtual forces states that $W = U$ and therefore:

$$w_u = \frac{8}{l^2}(m_u + m'_u) \tag{12.31}$$

where m_u and m'_u are the moment resistances per unit length along the positive and negative yield lines, respectively.

When calculating m_u and m'_u, it is reasonable to assume that the total quantity of prestressed and non-prestressed steel crossing the yield line is uniformly distributed across the slab strip, even though this is unlikely to be the case.

The amount of non-prestressed steel and the depth of the prestressed tendons may be different at each end of an interior span, and hence the value of m'_u at each negative yield line may be different. When this is the case, the positive yield line will not be located at mid-span. The correct position is the one that corresponds to the smallest collapse load w_u.

Consider the exterior span in Figure 12.23b. If the positive yield line is assumed to occur at mid-span, the collapse load is given by an expression similar to Equation 12.31, except that only one negative yield line contributes to the internal work and therefore:

$$w_u = \frac{8}{l^2}(m_u + 0.5m'_u) \tag{12.32}$$

For the case when m_u and m'_u have the same magnitude, the value of w_u given by Equation 12.32 is:

$$w_u = \frac{12m_u}{l^2} \tag{12.33}$$

However, a smaller collapse load can be obtained by moving the position of the positive yield line a little closer to the exterior edge of the slab strip. The minimum collapse load for the mechanism shown in Figure 12.23b occurs when $a = 0.414L$, and the internal work is:

$$U_i = m_u l_t \left(\frac{1}{0.414l} + \frac{1}{0.586l} \right) + m'_u l_t \frac{1}{0.586l} = \frac{5.83l_t m_u}{l}$$

The external work is still given by Equation 12.29. Equating the internal and external work gives:

$$w_u = \frac{11.66m_u}{l^2} \tag{12.34}$$

The collapse loads predicted by both Equations 12.33 and 12.34 are close enough to suggest that, for practical purposes, the positive yield line in this mechanism may be assumed to be at mid-span.

Yield line analysis is therefore an 'upper bound approach' and predicts a collapse load that is equal to or greater than the theoretically correct value. It is important to check that another yield line pattern corresponding to a lower collapse load does not exist. In flat plates, a fan-shaped yield line pattern may occur locally in the slab around a column (or in the vicinity of any concentrated load), as shown in Figure 12.24.

The concentrated load P_u at which the fan mode shown in Figure 12.24c occurs is:

$$P_u = 2\pi(m_u + m'_u) \tag{12.35}$$

The loads required to cause the fan mechanisms around the columns in Figure 12.24a and b increase as the column dimensions increase. Fan mechanisms

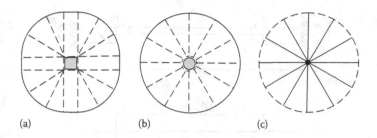

(a)   (b)   (c)

Figure 12.24   Fan mechanisms at columns or under concentrated loads. (a) Rectangular column (under slab). (b) Circular column (under slab). (c) Concentrated load (on top of the slab).

may be critical in cases where the column dimensions are both less than about 6% of the span in each direction [18].

Although yield analysis theoretically provides an upper bound to the collapse load, slabs tested to failure frequently (almost invariably) carry very much more load than that predicted. When slab deflections become large, in-plane forces develop in most slabs and the applied load is resisted by *membrane action* in addition to bending. The collapse load predicted by yield line analysis is therefore usually rendered conservative by membrane action. Although yield line analysis provides a useful measure of flexural strength, it does not provide any information regarding serviceability. Service load behaviour must be examined separately.

12.10 FLAT SLABS WITH DROP PANELS

Flat slabs with drop panels behave and are analysed similarly to flat plates. The addition of drop panels improves the structural behaviour both at service loads and at overloads. Drop panels stiffen the slab, thereby reducing deflection. Drop panels also increase the flexural and shear strength of the slab by providing additional depth at the slab–column intersection. The extent of cracking in the negative moment region over the column is also reduced. The slab thickness outside the drop panel may be significantly reduced from that required for a flat plate. Drop panels, however, interrupt ceiling lines and are often undesirable from an architectural point of view.

Drop panels increase the slab stiffness in the regions over the columns and therefore affect the distribution of slab moments caused by unbalanced loads. The negative or hogging moments over the columns tend to be larger and the span moments tend to be smaller than the corresponding moments in a flat plate.

Building codes often place minimum limits on the dimensions of drop panels. For example, to include the effect of drop panels when sizing a slab by limiting the span-to-depth ratios using Equation 12.4, drop panels

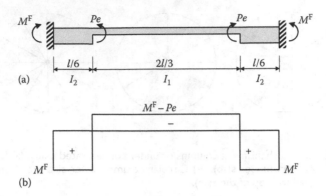

Figure 12.25 Bending moments due to eccentricity of longitudinal prestress. (a) Equivalent
loads. (b) Bending moment diagram.

should extend a distance equal to at least one-sixth of the span on each
side of the column centre line and the projection of the drop below the slab
should be at least 30% of the slab thickness beyond the drop [12].

In Figure 12.25, the moments introduced into a slab by the change in
eccentricity of the horizontal prestressing force at the drop panels are illustrated. These may be readily included in the slab analysis. The fixed end
moment at each support of the span shown in Figure 12.25a is given by:

$$M^F = \frac{2Pe}{(I_1/I_2) + 2} \qquad (12.36)$$

and the resultant bending moment diagram is shown in Figure 12.25b.
The moments of inertia of the various slab regions I_1 and I_2 are defined in
Figure 12.25a. The moments in the drop panel due to this effect are positive and those in the span are negative, as shown, and, although usually relatively small, tend to reduce the moments caused by the unbalanced loads.

12.11 BAND-BEAM AND SLAB SYSTEMS

Band-beam floors are a popular form of prestressed concrete construction. A one-way prestressed or reinforced concrete slab is supported by
wide shallow beams (slab bands or band beams) spanning in the transverse
direction. The system is particularly appropriate when the spans in one
direction are significantly larger than those in the other direction.

The slab bands, which usually span in the long direction, have a depth
commonly about two to three times the slab thickness and a width that
may be as wide as the drop panels in a flat slab. A section through a typical
band-beam floor is shown in Figure 12.26. The one-way slab is normally

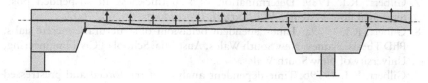

Figure 12.26 Band-beam and slab floor system.

considered to have an effective span equal to the clear span (from band edge to band edge) plus the slab depth. If the slab is prestressed, the tendons are usually designed using a load balancing approach and have a constant eccentricity over the slab bands with a parabolic drape through the effective span as shown in Figure 12.26. The depth and width of the band beams should be carefully checked to ensure that the reaction from the slab, deposited near the edge of the band, can be safely carried back to the column line.

The prestressing forces at the slab tendon anchorages will also induce moments at the change of depth from slab to slab band in the same way as was discussed for drop panels.

The slab band is normally designed to carry the full load in the transverse direction (usually the long-span direction). The prestressing tendons in this direction are concentrated in the slab bands, and these are also usually designed by load balancing. Because the prestress disperses out into the slab over the full panel width, the prestress anchorage should be located at the centroid of the T-section comprising the slab band and a slab flange equal in width to the full panel.

When checking serviceability and strength of the slab band, the effective flange width of the T-section is usually assumed to be equal to the width of the column strip as defined for a flat plate in Section 12.9.2.

REFERENCES

1. EN 1992-1-1. 2004. Eurocode 2: Design of concrete structures – Part 1–1: General rules and rules for buildings. British Standards Institution, London, UK.
2. VSL Prestressing (Aust.) Pty. Ltd. 1988. *Slab Systems*, 2nd edn. Sydney, New South Wales, Australia: V.S.L.
3. ACI 318-14M. 2014. Building code requirements for reinforced concrete. Detroit, MI: American Concrete Institute.
4. Post-Tensioning Institute. 1977. Design of post-tensioned slabs. Glenview, IL: Post-Tensioning Institute.
5. EN 1992-1-2. 2004. Eurocode 2: Design of concrete structures – Part 1-2: General rules – Structural fire design. British Standards Institution, London, UK.
6. Gilbert, R.I. 1985. Deflection control of slabs using allowable span to depth ratios. *ACI Journal, Proceedings*, 82, 67–72.

7. Gilbert, R.I. 1989. Determination of slab thickness in suspended post-tensioned floor systems. *ACI Journal, Proceedings*, 86, 602–607.
8. Gilbert, R.I. 1979a. Time-dependent behaviour of structural concrete slabs. PhD Thesis. Sydney, New South Wales, Australia: School of Civil Engineering, University of New South Wales.
9. Gilbert, R.I. 1979b. Time-dependent analysis of reinforced and prestressed concrete slabs. *Proceedings Third International Conference in Australia on Finite Elements Methods*. Sydney, New South Wales, Australia: University of New South Wales, Unisearch Ltd, pp. 215–230.
10. Mickleborough, N.C. and Gilbert, R.I. 1986. Control of concrete floor slab vibration by L/D limits. *Proceedings of the 10th Australian Conference on the Mechanics of Structures and Materials*, University of Adelaide, Adelaide, South Australia, Australia, pp. 51–56.
11. Wood, R.H. 1968. The reinforcement of slabs in accordance with a pre-determined field of moments. *Concrete*, 2(2), 69–76.
12. AS3600. 2009. Australian standard for concrete structures. Standards Australia, Sydney, New South Wales, Australia.
13. Gilbert, R.I. 1984. Effect of reinforcement distribution on the serviceability of reinforced concrete flat slabs. *Proceedings of the Ninth Australasian Conference on the Mechanics of Structures and Materials*, University of Sydney, Sydney, New South Wales, Australia, pp. 210–214.
14. Nilson, A.H. and Walters, D.B. 1975. Deflection of two-way floor systems by the equivalent frame method. *ACI Journal*, 72, 210–218.
15. Nawy, E.G. and Chakrabarti, P. 1976. Deflection of prestressed concrete flat plates. *Journal of the PCI*, 21, 86–102.
16. Johansen, K.W. 1962. *Yield-Line Theory*. London, UK: Cement and Concrete Association.
17. Johansen, K.W. 1972. *Yield-Line Formulae for Slabs*. London, UK: Cement and Concrete Association.
18. Ritz, P., Matt, P., Tellenbach, Ch., Schlub, P. and Aeberhard, H.U. 1981. *Post-Tensioned Concrete in Building Construction – Post-Tensioned Slabs*. Berne, Switzerland: Losinger.

Chapter 13

Compression and tension members

13.1 TYPES OF COMPRESSION MEMBERS

Many structural members are subjected to longitudinal compression, including columns and walls in buildings, bridge piers, foundation piles, poles, towers, shafts and web and chord members in trusses. The idea of applying prestress to a compression member may at first seem unnecessary or even unwise. In addition to axial compression, however, these members are often subjected to significant bending moments. Bending in compression members can result from a variety of load types. Moments are induced in the columns in framed structures by the gravity loads on the floor systems. Lateral loads on buildings and bridges cause bending in columns and piers and lateral earth pressures bend foundation piles. Even members that are intended to be axially loaded may be subjected to unintentional bending caused by eccentric external loading or by initial crookedness of the member itself. Most codes of practice specify a minimum eccentricity for use in design. All compression members must therefore be designed for combined bending and compression.

Prestress can be used to overcome the tension caused by bending and therefore reduce or eliminate cracking at service loads. By eliminating cracking, prestress can be used to reduce the lateral deflection of columns and piles and greatly improve the durability of these elements. Prestress also improves the handling of slender precast members and is used to overcome the tension due to rebound in driven piles. The strength of compression members is dependent on the strength of the concrete and considerable advantage can be gained by using concrete with high mechanical properties. Prestressed columns and piles are therefore commonly precast in an environment where quality control and supervision are of a high standard.

If a structural member is subjected primarily to axial compression, with little or no bending, prestress causes a small reduction in the load carrying capacity. For most prestressed concrete columns, the level of prestress is usually between 1.5 and 5 MPa, which is low enough not to cause significant reductions in strength. When the eccentricity of the applied

563

load is large and bending is significant, however, prestress results in an increase in the moment resistance, in addition to improved behaviour at service loads.

13.2 CLASSIFICATION AND BEHAVIOUR OF COMPRESSION MEMBERS

Consider the pin-ended column shown in Figure 13.1. The column is subjected to an external compressive force P applied at an initial eccentricity e_o. When P is first applied, the column shortens and deflects laterally by an amount δ_i. The bending moment at each end of the column is Pe_o, but at the column mid-length the moment is $P(e_o + \delta_i)$. The moment at any section away from the column ends depends on the lateral deflection of the column, which in turn depends on the length of the column and its flexural stiffness. The initial moment Pe_o is called the primary moment and the moment caused by the lateral displacement of the column $P\delta_i$ is the secondary moment. As the applied load P increases, so too does the lateral displacement δ_i. The rate of increase of the secondary moment $P\delta_i$ is therefore faster than the rate of increase of P. This non-linear increase in the internal actions is brought about by the change in geometry of the column and is referred to as *geometric non-linearity*.

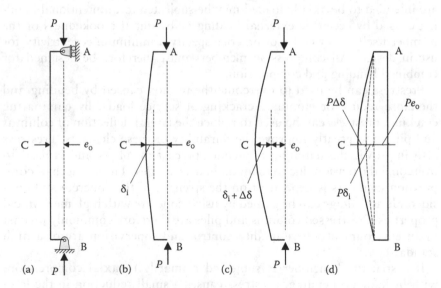

Figure 13.1 Deformation and moments in a slender pin-ended column. (a) Elevation. (b) Deformation at $t = 0$. (c) Deformation at time t. (d) Moments.

For a reinforced or prestressed concrete column under sustained loads, the member suffers additional lateral deflection due to creep. This time-dependent deformation leads to additional bending in the member, and this in turn causes the column to deflect further. During a period of sustained loading, an additional deflection $\Delta\delta$ develops and the resulting gradual increase in secondary moment with time $P(\delta_i + \Delta\delta)$ reduces the load carrying capacity.

Columns are usually classified into two categories according to their length or slenderness. Short (or stocky) columns are compression members in which the secondary moments are insignificant, i.e. columns that are geometrically linear. Long (or slender) columns are geometrically non-linear and the secondary moment is significant, i.e. the lateral deflection of the column is enough to cause a significant increase in the bending moment at the critical section and, hence, a reduction in strength. For given cross-sectional and material properties, the magnitude of the secondary moment depends on the length of the column and its support conditions. The secondary moment in a long column may be as great as or greater than the primary moment and the load carrying capacity is much less than that of a short column with the same cross-section.

The resistance of a stocky column is equal to the resistance of its cross section when a compressive load is applied at an eccentricity e_o. The resistance depends only on the cross-sectional dimensions, the quantity and distribution of the steel reinforcement (both prestressed and non-prestressed) and the compressive strengths of both concrete and steel. Many practical concrete columns in buildings are, in fact, stocky columns. The analysis of a prestressed concrete column cross-section at the ultimate limit state is presented in Section 13.3.

The resistance of a slender column is also determined from the strength of the critical cross-section subjected to an applied compressive load at an eccentricity $(e_o + \delta)$. The calculation of secondary moments $(P\delta)$ at the ultimate limit state and the treatment of slenderness effects in design are discussed in Section 13.4. Many precast prestressed compression members, as well as some in-situ columns and piers, fall into the category of slender columns.

For very long columns, an instability or buckling failure may take place before the strength of any cross-section is reached. The resistance of a very slender member is not dependent on the cross-sectional resistance and must be determined from a non-linear stability analysis (see Reference [1]). Such an analysis is outside the scope of this book. A very slender member may buckle under a relatively small applied load, either when the load is first applied or after a period of sustained loading. The latter type of instability is caused by excessive lateral deformation due to creep and is known as creep buckling. Upper limits on the slenderness of columns are usually specified by codes of practice in order to avoid buckling failures.

13.3 CROSS-SECTION ANALYSIS: COMPRESSION AND BENDING

13.3.1 Strength interaction diagram

The design resistance of a prestressed concrete column cross-section in combined bending and uniaxial compression is calculated as for a conventionally reinforced concrete cross-section. Strength is conveniently represented by a plot of the design axial load resistance N_{Rd} versus the design moment resistance on the section M_{Rd}. This plot is called the *strength interaction curve*.

A typical strength interaction curve is shown in Figure 13.2 and represents the failure line or strength line. Any combination of axial force and bending moment applied to the column cross-section that falls inside the interaction curve is safe and can be carried by the cross-section. Any point outside the curve represents a combination of axial force and moment that exceeds

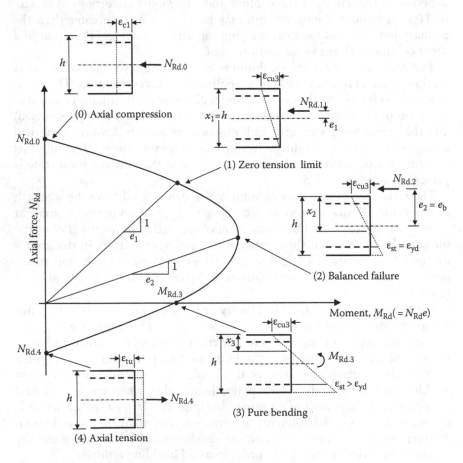

Figure 13.2 A typical strength interaction diagram.

the resistance of the cross-section. Depending on the properties of the cross-section and the relative magnitudes of the axial force and bending moment, the type of failure can range from compressive, when the moment is small, to tensile or flexural, when the axial force is small and bending predominates.

Several critical points are identified on the strength interaction curve in Figure 13.2. Point 0 on the vertical axis is the point of axial compression (zero bending) and the design resistance is $N_{Rd.0}$ (often called the *design squash load*). The cross-section is subjected to a uniform compressive strain, as shown. Point 1 represents the zero tension point. The combination of axial force $N_{Rd.1}$ and moment $N_{Rd.1}e_1$ at point 1 (when combined with pre-strain caused by prestress) produces zero strain in the extreme concrete fibre (i.e. $x/h = 1$). The extreme fibre-compressive strain at failure is ε_{cu3}. Between points 0 and 1 on the curve, the entire cross-section is in compression.

When the eccentricity of the applied load is greater than e_1, bending causes tension over part of the cross-section. Point 2 is known as the balanced failure point. The strain in the extreme compressive fibre is ε_{cu3} and the strain in the tensile steel is the design yield strain ε_{yd}. The eccentricity of the applied load at the balanced failure point is e_2 ($=e_b$). When a cross-section contains both non-prestressed and prestressed tensile steel with different yield strains and located at different positions on the cross-section, the balanced failure point is not well defined. Point 2 is usually taken as the point corresponding to a strain of ε_{yd} in the steel closest to the tensile face of the cross-section and is usually at or near the point of maximum moment capacity. At any point on the interaction curve between points 0 and 2, the tensile steel has not yielded at the ultimate limit state and failure is essentially compressive. Failures that occur between points 0 and 2 (when the eccentricity is less than e_b) are sensibly known as *primary compressive failures*.

Point 3 is the pure bending point, where the axial force is zero. The moment resistance at this point $M_{Rd.3}$ is calculated as described in Chapter 6. Point 4 is the point corresponding to direct axial tension. At any point on the interaction curve between points 2 and 4, the capacity of the tensile steel (or part of the tensile steel) is exhausted, with strains exceeding the yield strain, and the section suffers a *primary tensile failure*.

Any straight line through the origin represents a line of constant eccentricity called a loading line. Two such lines, corresponding to points 1 and 2, are drawn on Figure 13.2. The slope of each loading line is $1/e$. When a monotonically increasing compressive force N is applied to the cross-section at a particular eccentricity e_i, the plot of N versus M ($=Ne_i$) follows the loading line of slope $1/e_i$ until the strength of the cross-section is reached at the point where the loading line and the interaction curve intersect. If the eccentricity of the applied load is increased, the loading line becomes flatter and the design resistance of the cross-section N_{Rd} is reduced.

The general shape of the interaction curve shown in Figure 13.2 is typical for any cross-section that is under-reinforced in pure bending (i.e. where the tensile steel strain at point 3 exceeds the yield strain). At the pure bending point, a small increase in axial compression increases the internal compressive stress resultant on the section but does not appreciably reduce the internal tension, thus increasing the design moment resistance, as is indicated by the part of the interaction curve between points 3 and 2. For short columns, where the axial force N_{Ed} is less than about $0.1f_{cd}A_g$, it is usually acceptable to design the cross-section for bending only.

13.3.2 Strength analysis

Individual points on the strength interaction curve can be calculated using ultimate strength theory, similar to that outlined for pure bending in Section 6.3. The analysis described in the following is based on the assumptions listed in Section 6.3.1 and the idealised rectangular stress block specified in EN 1992-1-1 [2] and presented in Section 6.3.2. At any point on the interaction curve between points 1 and 3, the extreme fibre concrete compressive strain at failure is taken to be ε_{cu3} (see Table 4.2). For axial compression at point 0, the extreme fibre concrete compressive strain at failure is taken to be ε_{c1} (but not less than the design yield strain of the conventional non-prestressed reinforcement ε_{yd}).

Calculation of the design moment resistance in pure bending M_{Rd} (point 3 on the interaction curve where $M_{Rd} = M_{Rd.3}$) was discussed in Chapter 6. Other points on the strength interaction curve (between points 3 and 1) may be obtained by successively increasing the depth to the neutral axis and analysing the cross-section. With the extreme fibre strain equal to ε_{cu3}, each neutral axis position defines a particular strain distribution that corresponds to a point on the strength interaction diagram. The strain diagrams associated with points 1, 2 and 3 are also shown in Figure 13.2.

To define the interaction curve accurately, relatively few points are needed. In fact, if only points 0, 1, 2 and 3 are determined, a close approximation can be made by passing a smooth curve through each point, or even by linking successive points together by straight lines. Such an approximation is often all that is required in design.

Consider the rectangular cross-section shown in Figure 13.3a, with overall dimensions h and b. The section contains two layers of non-prestressed reinforcement $A_{s(1)}$ and $A_{s(2)}$, and two layers of bonded prestressing steel $A_{p(1)}$ and $A_{p(2)}$, as shown. A typical strain diagram in the ultimate limit state condition and the corresponding idealised stresses and stress resultants are illustrated in Figures 13.3b through d, respectively. These strains and stresses correspond to a resultant axial force N_{Rd} at an eccentricity e measured from the plastic centroid of the cross-section (as shown in Figure 13.3d). Assuming that $A_{s(1)}$ and $A_{p(1)}$ are above the neutral axis, and $A_{s(2)}$ and $A_{p(2)}$ are below, as shown, longitudinal equilibrium requires that:

$$N_{Rd} = F_{cd} + F_{sd(1)} - F_{ptd(1)} - F_{ptd(2)} - F_{sd(2)} \tag{13.1}$$

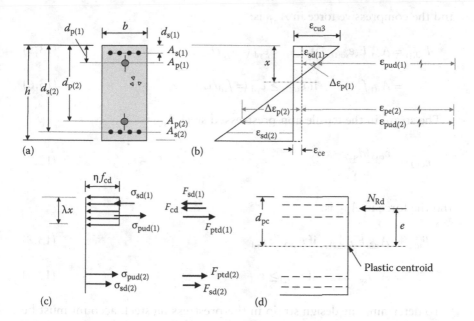

Figure 13.3 Design stresses and strains on a rectangular cross-section in compression and uniaxial bending. (a) Cross-section. (b) Strain. (c) Stresses and forces. (d) Stress resultant.

and moment equilibrium gives:

$$M_{Rd} = N_{Rd}e = F_{cd}\left(d_{pc} - \frac{\lambda x}{2}\right) + F_{sd(1)}\left(d_{pc} - d_{s(1)}\right) + F_{ptd(1)}\left(d_{p(1)} - d_{pc}\right)$$

$$+ F_{ptd(2)}\left(d_{p(2)} - d_{pc}\right) + F_{sd(2)}\left(d_{s(2)} - d_{pc}\right) \qquad (13.2)$$

where d_{pc} is the depth of the plastic centroid of the cross-section from the compressive face.

Each of the internal forces can be calculated readily from the strain diagram. The magnitude of the compressive force in the concrete F_{cd} is the volume of the rectangular stress block given by:

$$F_{cd} = \mu f_{cd}\, A'_c \qquad (13.3)$$

where A'_c is the area of concrete under the stress block and equals the gross area $\lambda x b$ minus the areas of any bonded steel and hollow ducts within this area.

The magnitude of the strain in the compressive non-prestressed steel $A_{s(1)}$ is:

$$\varepsilon_{sd(1)} = \frac{\varepsilon_{cu3}\left(x - d_{s(1)}\right)}{x} \qquad (13.4)$$

and the compressive force in $A_{s(1)}$ is:

$$F_{sd(1)} = A_{s(1)}E_s\varepsilon_{sd(1)} \quad \text{if } \varepsilon_{sd(1)} < \varepsilon_{yd} \ (= f_{yd}/E_s) \tag{13.5}$$

$$= A_{s(1)}f_{yd} \quad \text{if } \varepsilon_{sd(1)} \geq \varepsilon_{yd} \ (= f_{yd}/E_s) \tag{13.6}$$

The strain in the tensile non-prestressed steel $A_{s(2)}$ is:

$$\varepsilon_{sd(2)} = \frac{\varepsilon_{cu3}(d_{s(2)} - x)}{x} \tag{13.7}$$

and the force in $A_{s(2)}$ is:

$$F_{sd(2)} = A_{s(2)}E_s\varepsilon_{sd(2)} \quad \text{if } \varepsilon_{sd(2)} < \varepsilon_{yd} \tag{13.8}$$

$$= A_{s2}f_{sy} \quad \text{if } \varepsilon_{sd(2)} \geq \varepsilon_{yd} \tag{13.9}$$

To determine the design strain in the prestressing steel, account must be taken of the large initial tensile strain in the steel ε_{pe} caused by the effective prestress. For each area of prestressing steel:

$$\varepsilon_{pe(1)} = \frac{P_{m,t(1)}}{A_{p(1)}E_p} \quad \text{and} \quad \varepsilon_{pe(2)} = \frac{P_{m,t(2)}}{A_{p(2)}E_p} \tag{13.10}$$

The strains in the concrete at the different levels of the prestressing steel caused by the effective prestress ($\varepsilon_{ce(1)}$ and $\varepsilon_{ce(2)}$) are readily calculated. If the prestressing forces in $A_{p(1)}$ and $A_{p(2)}$ are such that the effective prestress is axial, producing uniform compressive strain ε_{ce}, i.e. $\varepsilon_{ce} = \varepsilon_{ce(1)} = \varepsilon_{ce(2)}$, the resultant effective prestressing force $P_{m,t}$ ($= P_{m,t(1)} + P_{m,t(2)}$) acts at the centroidal axis and the magnitude of ε_{ce} is:

$$\varepsilon_{ce} = \varepsilon_{ce(1)} = \varepsilon_{ce(2)} = \frac{P_{m,t}}{(\alpha_s A_{s(1)} + \alpha_s A_{s(2)} + A_c)E_{cm}}$$

$$= \frac{P_{m,t}}{[(\alpha_s - 1)(A_{s(1)} + A_{s(2)}) + A_g]E_{cm}} \tag{13.11}$$

where α_s is the modular ratio E_s/E_{cm} and A_c and A_g are, respectively, the concrete area and the gross cross-sectional area (minus the cross-sectional area of any unbonded ducts).

The changes in strain in the bonded prestressing tendons due to the application of N_{Rd} at an eccentricity e may be obtained from the strain diagram of Figure 13.3b and are given by:

$$\Delta\varepsilon_{p(1)} = -\frac{\varepsilon_{cu3}\left(x - d_{p(1)}\right)}{x} + \varepsilon_{ce(1)} \tag{13.12}$$

and

$$\Delta\varepsilon_{p(2)} = \frac{\varepsilon_{cu3}\left(d_{p(2)} - x\right)}{x} + \varepsilon_{ce(2)} \tag{13.13}$$

The final strain in each prestressing tendon is therefore:

$$\varepsilon_{pud(1)} = \varepsilon_{pe(1)} + \Delta\varepsilon_{p(1)} \tag{13.14}$$

and

$$\varepsilon_{pud(2)} = \varepsilon_{pe(2)} + \Delta\varepsilon_{p(2)} \tag{13.15}$$

The final stress in the prestressing tendons ($\sigma_{pud(1)}$ and $\sigma_{pud(2)}$) may be obtained from a stress–strain curve for the prestressing steel, such as one of the curves shown in Figure 4.10, or from one of the idealised lines in Figure 4.12. If the strain in the prestressing steel remains in the elastic range (and on the compressive side of the cross-section it does), then $\sigma_{pud} = \varepsilon_{pud}E_p$.

The design forces in the tendons are:

$$F_{ptd(1)} = \sigma_{pud(1)}A_{p(1)} \tag{13.16}$$

and

$$F_{ptd(2)} = \sigma_{pud(2)}A_{p(2)} \tag{13.17}$$

With the internal forces determined from Equations 13.3, 13.5, 13.6, 13.8, 13.9, 13.16 and 13.17, the compressive resistance N_{Rd} is obtained from Equation 13.1 and the eccentricity e is calculated using Equation 13.2. The resulting point (N_{Rd}, M_{Rd} (= $N_{Rd}e$)) represents the point on the design strength interaction curve corresponding to the assumed strain distribution.

When the cross section is subjected to pure compression (point 0 on the interaction curve), the eccentricity is zero and the strength (known as the *design squash load*) is given by:

$$N_{Rd.0} = F_{cd.0} + (A_{s(1)} + A_{s(2)})f_{yd} - \sigma_{pud(1)}A_{p(1)} - \sigma_{pud(2)}A_{p(2)} \tag{13.18}$$

where $F_{cd.0}$ is the resultant compressive force carried by the concrete part of the cross-section (A_c) in uniform compression and may be taken as $F_{cd.0} = f_{cd}A_c$.

EXAMPLE 13.1

Calculate the critical points on the strength interaction curve of the post-tensioned concrete column cross-section shown in Figure 13.4a. The cross-section is 600 mm wide and 800 mm deep. Steel quantities, prestressing details and material properties are as follows:

$A_{s(1)} = A_{s(2)} = 2250$ mm^2, $A_{p(1)} = A_{p(2)} = 1000$ mm^2, $E_s = 200 \times 10^3$ MPa,

$f_{yd} = 435$ MPa, $f_{ck} = 40$ MPa, $E_{cm} = 35,000$ MPa,   and $\alpha_s = E_s/E_{cm} = 5.71$.

From Equations 6.2 and 6.4, $\lambda = 0.8$ and $\eta = 1.0$. Assume that the ducts have been grouted.

The properties of the prestressing steel are taken from the idealised stress–strain relationship shown in Figure 13.4b, where the initial elastic modulus $E_p = 195 \times 10^3$ MPa and the effective prestress in each bonded tendon is $P_{m,t(1)} = P_{m,t(2)} = 1200$ kN.

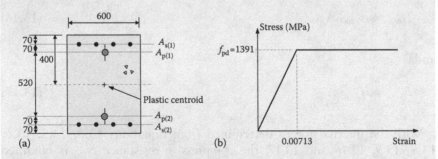

(a)   (b)

Figure 13.4 Cross-sectional details and idealised design stress–strain relationship for tendons (Example 13.1). (a) Cross-section. (b) Design stress versus strain for tendons.

The effective strain in each tendon is obtained from Equation 13.10:

$$\varepsilon_{pe(1)} = \varepsilon_{pe(2)} = \frac{1,200 \times 10^3}{1,000 \times 195,000} = 0.00615$$

Because of the symmetry of the cross-section and the fact that the resultant effective prestress passes through the centroid of the section, the strain in the concrete caused by the effective prestress is uniform over the section and obtained from Equation 13.11:

$$\varepsilon_{ce} = \varepsilon_{ce(1)} = \varepsilon_{ce(2)} = \frac{2 \times 1,200 \times 10^3}{[(5.71-1) \times (2,250+2,250)+(800 \times 600)] \times 35,000}$$

$$= 0.000137$$

Point 0: *Pure compression (e = 0)*
At the design squash load, the strain on the cross-section is uniform with a magnitude of $\varepsilon_{c1} = 0.0023$ [2]. The compressive force carried by the concrete in uniform compression is:

$$F_{cd.0} = f_{cd}A_c = 26.67 \times [600 \times 800 - (2 \times 2,250) - (2 \times 1,000)] \times 10^{-3} = 12,627 \text{ kN}$$

The magnitude of the compressive stress in the non-prestressed steel is $f_{yd} = 435$ MPa and the tensile stress in the prestressing steel is in the elastic range and given by:

$$\sigma_{pud(1)} (= \sigma_{pud(2)}) = E_p(\varepsilon_{pe(1)} - \varepsilon_{c1} + \varepsilon_{ce(1)})$$

$$= 195 \times 10^3 \times (0.00615 - 0.0023 + 0.000137) = 777.5 \text{ MPa}$$

The design resistance of the cross-section in axial compression is given by Equation 13.18:

$$N_{Rd.0} = 12,627 + (2,250+2,250) \times 435 \times 10^{-3} - (1,000+1,000) \times 777.5 \times 10^{-3}$$

$$= 13,030 \text{ kN}$$

Point 1: *Zero tension*
For the case of zero tension, $x = h = 800$ mm and the magnitude of ε_{cu3} is 0.0035 [2]. With $\lambda x = 640$ mm, Equation 13.3 gives the compressive force in the concrete:

$$F_{cd} = \mu f_{cd}A'_c = \mu f_{cd}(\lambda x b - A_{s(1)} - A_{p(1)})$$

$$= 1.0 \times 26.67 \times (640 \times 600 - 2,250 - 1,000) \times 10^{-3} = 10,153 \text{ kN}$$

From Equation 13.4, the magnitude of the compressive strain in $A_{s(1)}$ is:

$$\varepsilon_{sd(1)} = \frac{0.0035 \times (800 - 70)}{800} = 0.00319 > \varepsilon_{yd} \left(= 0.00218 \right)$$

The top layer of non-prestressed steel has yielded and the compressive force in $A_{s(1)}$ is given by Equation 13.6:

$$F_{sd(1)} = 2250 \times 435 \times 10^{-3} = 979 \text{ kN}$$

From Equation 13.7, the strain in $A_{s(2)}$ is:

$$\varepsilon_{sd(2)} = \frac{0.0035 \times (730 - 800)}{800} = -0.000306$$

Since the bottom steel $A_{s(2)}$ is above the neutral axis at Point 1 and in compression (and not above the neutral axis and in tension as shown in Figure 13.3), the force $F_{sd(2)}$ will in fact be a negative quantity. From Equation 13.8:

$$F_{sd(2)} = 2,250 \times 200,000 \times (-0.000306) \times 10^{-3}$$
$$= -138 \text{ kN (i.e. compressive)}$$

The change in strain at each level of prestressing steel is compressive and given by Equations 13.12 and 13.13:

$$\Delta\varepsilon_{p(1)} = -\frac{0.0035 \times (800 - 140)}{800} + 0.000137 = -0.00275$$

$$\Delta\varepsilon_{p(2)} = \frac{0.0035 \times (660 - 800)}{800} + 0.000137 = -0.000476$$

and the final strains in the prestressing tendons are obtained from Equations 13.14 and 13.15:

$$\varepsilon_{pud(1)} = 0.00615 - 0.00275 = 0.00340$$

$$\varepsilon_{pud(2)} = 0.00615 - 0.000476 = 0.00567$$

Both final strains are in the elastic range and the forces in the tendons are:

$$F_{ptd(1)} = \sigma_{pud(1)} A_{p(1)} = \varepsilon_{pud(1)} E_p A_{p(1)} = 0.00340 \times 195,000 \times 1,000 \times 10^{-3} = 663 \text{ kN}$$

$$F_{ptd(2)} = \sigma_{pud(2)} A_{p(2)} = \varepsilon_{pud(2)} E_p A_{p(2)} = 0.00567 \times 195,000 \times 1,000 \times 10^{-3} = 1,106 \text{ kN}$$

The resultant compressive force is obtained from Equation 13.1:

$$N_{Rd.1} = 10,153 + 979 - 663 - 1,106 + 138 = 9,501 \text{ kN}$$

and the design moment resistance for the case of zero tension is calculated using Equation 13.2:

$$M_{Rd.1} = \left[10,153 \times \left(400 - \frac{0.8 \times 800}{2}\right)1 + 979 \times (400 - 70) - 663 \times (400 - 140)\right.$$
$$\left. + 1106 \times (660 - 400) - 138 \times (730 - 400)\right] \times 10^{-3}$$
$$= 1205 \text{ kNm}$$

Therefore, the eccentricity corresponding to point 1 is:

$$e_1 = \frac{M_{Rd.1}}{N_{Rd.1}} = 126.8 \text{ mm}$$

Point 2: *The balanced failure point*
Point 2 corresponds to first yielding in the non-prestressed tensile steel, i.e. $\varepsilon_{sd(2)} = \varepsilon_{yd} = 0.00218$, and therefore the design force in the tensile non-prestressed steel is:

$$F_{sd(2)} = 2250 \times 435 \times 10^{-3} = 979 \text{ kN}$$

The depth to the neutral axis at point 2 is therefore:

$$x = \frac{\varepsilon_{cu3}}{\varepsilon_{cu3} + \varepsilon_{yd}} d_{s(2)} = \frac{0.0035}{0.0035 + 0.00218} \times 730 = 450 \text{ mm}$$

The corresponding curvature is $\kappa_{ud2} = \varepsilon_{cu3}/x = 7.78 \times 10^{-6} \text{ mm}^{-1}$. The compressive force in the concrete is (Equation 13.3):

$$F_{cd} = 1.0 \times 26.67 \times (0.8 \times 450 \times 600 - 2250 - 1000) = 5673 \text{ kN}$$

With $x = 450$ mm, the strain in the non-prestressed compressive reinforcement is:

$$\varepsilon_{sd(1)} = \frac{0.0035 \times (450 - 70)}{450} = 0.00296 > \varepsilon_{yd}$$

and from Equation 13.6:

$$F_{sd(1)} = 2250 \times 435 \times 10^{-3} = 979 \text{ kN}$$

With the strain in $A_{p(1)}$ still in the elastic range, the force $F_{ptd(1)}$ is obtained from Equations 13.12, 13.14 and 13.16:

$$F_{ptd(1)} = A_{p(1)} \left[\varepsilon_{pe(1)} - \frac{0.0035(x - d_{p(1)})}{x} + \varepsilon_{ce(1)} \right] E_p$$

$$= 1,000 \times \left[0.00615 - \frac{0.0035 \times (450 - 140)}{450} + 0.000137 \right] \times 195,000 \times 10^{-3}$$

$$= 756 \text{ kN}$$

Equations 13.13 and 13.15 give:

$$\varepsilon_{pud(2)} = \varepsilon_{pe(2)} + \frac{0.0035(d_{p(2)} - x)}{x} + \varepsilon_{ce(2)} = 0.00615 + 0.00163 + 0.000137$$

$$= 0.00792$$

which is greater than the design yield strain in Figure 13.4b and, from that figure, we get $\sigma_{pud(2)}$ = 1391 MPa. From Equation 13.17:

$$F_{ptd(2)} = 1391 \times 1000 \times 10^{-3} = 1391 \text{ kN}$$

With the strain in bottom steel $A_{s(1)}$ much greater than f_{yd}, $F_{sd(2)}$ = 979 kN. The design resistance corresponding to the balanced point (point 2) is obtained from Equations 13.1 and 13.2:

$$N_{Rd.2} = 5673 + 979 - 756 - 1391 - 979 = 3526 \text{ kN}$$

and the design moment resistance at the balanced load point is calculated using Equation 13.2:

$$M_{Rd.2} = \left[5673 \times \left(400 - \frac{0.8 \times 450}{2} \right) + 979 \times (400 - 70) - 756 \times (400 - 140) \right.$$

$$\left. + 1391 \times (660 - 400) + 979 \times (730 - 400) \right] \times 10^{-3}$$

$$= 2059 \text{ kNm}$$

The eccentricity corresponding to point 2 is:

$$e_2 = \frac{M_{Rd.2}}{N_{Rd.2}} = 584 \text{ mm}$$

Point 3: *Pure bending*

For equilibrium of the section in pure bending (i.e. bending without axial force), the magnitude of the resultant compression is equal to the magnitude of the resultant tension and $N_{Rd} = 0$. A trial-and-error approach to determine the depth to the neutral axis indicates that:

$$x = 196.0 \text{ mm}$$

and the compressive force in the concrete is (Equation 13.3):

$$F_{cd} = 1.0 \times 26.67 \times (0.8 \times 196.0 \times 600 - 2250 - 1000) \times 10^{-3} = 2422 \text{ kN}$$

From Equation 13.4:

$$\varepsilon_{sd(1)} = \frac{0.0035 \times (196 - 70)}{196} = 0.00225 > \varepsilon_{yd}$$

and therefore $F_{sd(1)} = 2250 \times 435 \times 10^{-3} = 979$ kN.
  Equations 13.12, 13.14 and 13.16 give:

$$F_{ptd(1)} = 1,000 \times \left[0.00615 - \frac{0.0035 \times (196 - 140)}{196} + 0.000137 \right] \times 195,000 \times 10^{-3}$$

$$= 1,031 \text{ kN}$$

Equations 13.13 and 13.15 give:

$$\varepsilon_{pud(2)} = \varepsilon_{pe(2)} + \frac{0.0035 \times (d_{p(2)} - x)}{x} + \varepsilon_{ce(2)} = 0.01457$$

which is greater than the design yield strain (0.00713) in Figure 13.4b and therefore $\sigma_{pud(2)} = 1391$ MPa and $F_{ptd(2)} = 1391 \times 1000 \times 10^{-3} = 1391$ kN.
  The design resistance corresponding to point 3 is obtained from Equations 13.1 and 13.2:

$$N_{Rd.3} = 2422 + 979 - 1031 - 1391 - 979 = 0$$

and the design moment resistance is calculated using Equation 13.2:

$$M_{Rd.3} = \left[2422 \times \left(400 - \frac{0.8 \times 196}{2} \right) + 979 \times \left(400 - 70 \right) - 1031 \times \left(400 - 140 \right) \right.$$

$$\left. + 1391 \times \left(660 - 400 \right) + 979 \times \left(730 - 400 \right) \right] \times 10^{-3}$$

$$= 1519 \text{ kNm}$$

Point 4: *Axial tension*

The capacity of the section in tension is dependent only on the steel strength. Therefore, taking f_{yd} = 435 MPa and f_{pd} = 1391 MPa, as indicated in Figure 13.4b, the axial tensile strength is:

$$N_{Rd.4} = (A_{s1} + A_{s2})f_{yd} + (A_{p1} + A_{p2})f_{pd} = 4740 \text{ kN}$$

Figure 13.5 shows the strength interaction curve for the cross-section (solid line).

In addition, Figure 13.5 also illustrates the interaction curve for a cross-section with the same dimensions, material properties and steel quantities

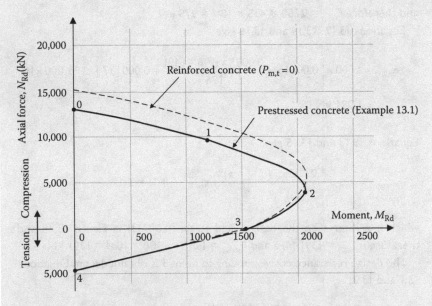

Figure 13.5 Strength interaction curve (Example 13.1).

(both non-prestressed and prestressed), but without any effective prestress, i.e. $P_{m,t(1)} = P_{m,t(2)} = 0$. A comparison between the two curves indicates the effect of prestress. In this example, the prestressing steel induced an effective prestress of 5 MPa over the column cross-section. Evidently, the prestress reduces the axial load carrying capacity by about 16% (at point 0) but slightly increases the bending strength of the cross-section in the primary tension region (i.e. between points 2 and 3).

13.3.3 Biaxial bending and compression

When a cross-section is subjected to axial compression and bending about both principal axes, such as the section shown in Figure 13.6a, the strength interaction diagram can be represented by the three-dimensional surface shown in Figure 13.6b. The shape of this surface may be defined by a set of contours obtained by taking horizontal slices through the surface. A typical contour is shown in Figure 13.6b. Each contour is associated with a particular axial force N. The equation of the contour represents the relationship between M_x and M_y at that particular value of axial force. In EN 1992-1-1 [2], the design expression given in Equation 13.19 is specified to model the shape of these contours. The form of Equation 13.19 was originally proposed by Bresler [3].

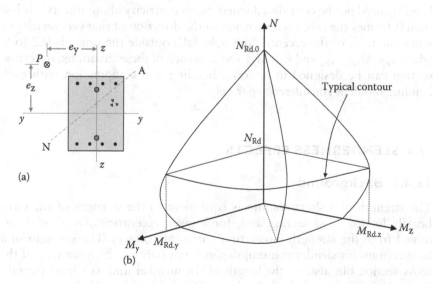

Figure 13.6 Biaxial bending and compression. (a) Cross-section. (b) Strength interaction diagram.

Table 13.1 Index a in Equation 13.19

$N_{Ed}/N_{Rd.0}$	0.1	0.7	1.0
a	1.0	1.5	2.0

If the factored design actions N_{Ed}, $M_{Ed.x}$ and $M_{Ed.y}$ fall inside the *design interaction surface*, then the cross-section is adequate. According to EN 1992-1-1 [2], a cross-section subjected to biaxial bending should satisfy the following equation:

$$\left(\frac{M_{Ed.z}}{M_{Rd.z}}\right)^a + \left(\frac{M_{Ed.y}}{M_{Rd.y}}\right)^a \leq 1.0 \tag{13.19}$$

where $M_{Ed.z}(M_{Ed.y})$ is the design moment about z-axis (y-axis), including second-order effects, and $M_{Rd.z}(M_{Rd.y})$ is the design moment resistance calculated separately about z-axis (y-axis) under the design axial force N_{Ed}. The factored design moments $M_{Ed.z}$ and $M_{Ed.y}$ are magnified to account for slenderness, if applicable (see Section 13.4). The index a is a factor that depends on the axial force and the shape of the cross-section. For a circular or elliptical cross-section, $a = 2.0$. For a rectangular cross-section, a is given in Table 13.1, where $N_{Rd.0}$ is the design axial resistance given in Equation 13.18.

Biaxial bending is not a rare phenomenon. Most columns are subjected to simultaneous bending about both principal axes. In general, biaxial bending need not be considered where the eccentricity about one axis is less than 0.1 times the column dimension in the direction of that eccentricity or when the ratio of the eccentricities e_z/e_y falls outside the range is 0.2 to 5, where $e_z = M_{Ed.y}/N_{Ed}$ and $e_y = M_{Ed.z}/N_{Ed}$. In each of these situations, the cross-section can be designed in uniaxial bending and compression with each bending moment considered separately.

13.4 SLENDERNESS EFFECTS

13.4.1 Background

The strength of a short column is equivalent to the strength of the most heavily loaded cross-section and, for a given eccentricity, may be determined from the strength interaction curve (or surface). The strength of a long column (or slender column) depends not only on the resistance of the cross-section but also on the length of the member and its support conditions. A discussion of the behaviour of a slender pin-ended column was presented in Section 13.2 and the increase in secondary moments due to slenderness effects was illustrated in Figure 13.1. In general, as the length of a compression member increases, strength decreases.

To predict accurately the second-order effects in structures as they deform under load requires an iterative non-linear computer analysis, and this generally involves considerable computational effort. For the design of concrete compression members, simplified procedures are available to account for slenderness effects and one such procedure is presented here. A more detailed study of geometric non-linearity and instability in structures is outside the scope of this book.

The critical buckling load N_B of an axially loaded, perfectly straight, pin-ended elastic column was determined by Euler and is given by:

$$N_B = \frac{\pi^2 EI}{l^2} \tag{13.20}$$

where l is the length of the Euler column between its pinned ends. In practice, concrete columns are rarely, if ever, pinned at their ends. A degree of rotational restraint is usually provided at each end of a column by the supporting beams and slabs, or by a footing. In some columns, translation of one end of the column with respect to the other may also occur in addition to rotation. Some columns are completely unsupported at one end, such as a cantilevered column. The buckling load of these columns may differ considerably from that given by Equation 13.20.

In general, for design purposes, the critical buckling load of real columns is expressed in terms of the *effective length* l_0. EN 1992-1-1 [2] defines l as the clear height of the column between end restraints and may be taken as the unsupported length of the column, i.e. the distance between the faces of members providing lateral support to the column. Where column capitals or haunches are present, l is usually measured to the lowest extremity of the capital or haunch. The ratio l_0/l is an *effective length factor* that depends on the support conditions of the column and equals 1.0 for a pin-ended Euler column. The buckling load of a concrete column is therefore:

$$N_B = \frac{\pi^2 EI}{l_0^2} \tag{13.21}$$

For the determination of N_B for cracked reinforced concrete columns, it is reasonable to take EI as the ratio of moment to curvature at the strength limit state condition of the column.

In structures that are laterally braced, the ends of columns are not able to translate appreciably relative to each other, i.e. sidesway is prevented. Many concrete structures are braced, with stiff vertical elements such as shear walls, elevator shafts and stairwalls providing bracing for more flexible columns and ensuring that the ends of columns are not able to translate appreciably relative to each other. If the attached elements at each end of a braced column provide some form of rotational restraint, the critical buckling load will be greater than that of a pin-ended column (given in Equation 13.20) and the effective length l_0 in Equation 13.21 is less than l. The effective lengths specified in EN 1992-1-1 [1] for braced columns are shown in Figure 13.7a. The effective length of any column is the length associated with single curvature buckling, i.e. the distance between the points of inflection in the deflected column, as shown in Figures 13.7a and 13.8a. For the column shown in Figure 13.7a(iv), the supports are neither pinned nor fixed. The effective length depends on the relative flexural stiffness of the column and the supporting elements at each end of the column (such as floor beams and slabs).

For columns in unbraced structures, where one end of the column can translate relative to the other (i.e. sidesway is not prevented), the effective length l_0 is greater than l, sometimes much greater, as shown in Figure 13.8b. The buckling load of an unbraced column is therefore significantly less than that of a braced column. Values of l_0 specified in EN 1992-1-1 [2] for unbraced columns under various support conditions are shown in Figure 13.7b.

A braced column is a compression member located within a storey of a building in which horizontal displacements do not significantly affect the moments in the column. EN 1992-1-1 [2] defines a braced column as

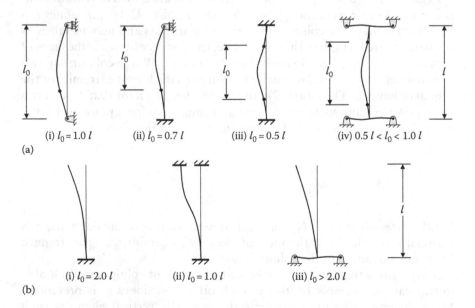

Figure 13.7 Effective lengths l_0 [2]. (a) Braced columns. (b) Unbraced columns.

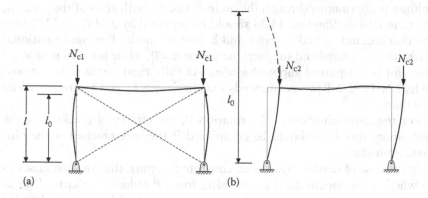

Figure 13.8 Effective lengths in a braced and an unbraced portal frame. (a) Braced frame. (b) Unbraced frame.

a column in a structure which is assumed not to contribute to the overall horizontal stability of the structure. The lateral actions applied at the ends of such a column are resisted by masonry infill panels, shear walls, lateral bracing or the like.

A compression member may be assumed to be braced, if it is located in a storey where the bracing elements (shear walls, masonry infills, bracing trusses and other types of bracing) have a total stiffness, resisting lateral movement of the storey, of at least six times the sum of the stiffnesses of all the columns within the storey.

For compression members in regular frames, the effective length for the braced column shown in Figure 13.7a(iv) is given in EN 1992-1-1 [2] as:

$$l_0 = 0.5l\sqrt{\left(1+\frac{k_1}{0.45+k_1}\right)\cdot\left(1+\frac{k_2}{0.45+k_2}\right)} \tag{13.22}$$

and the effective length for the unbraced column in Figure 13.7b(iii) may be calculated using [2]:

$$l_0 = l.\max\left\{\sqrt{1+10\frac{k_1 k_2}{k_1+k_2}};\ \left(1+\frac{k_1}{1+k_1}\right);\ \left(1+\frac{k_2}{1+k_2}\right)\right\} \tag{13.23}$$

The terms k_1 and k_2 are the relative flexibilities of the rotational restraints at ends 1 and 2, respectively, given by:

$$k = \left(\frac{\theta}{M}\right).\left(\frac{EI}{l}\right) \tag{13.24}$$

and θ is the rotation of the restraining members under end column moment M and EI is the bending stiffness of the compression member. Where the column is continuous through the node at one or both ends of the column, the term EI/l in Equation 13.24 should be replaced by $[(EI/l)_a + (EI/l)_b]$ for the two columns a and b above and below the node. For rigid rotational restraint (i.e. a fixed end support) the factor $k = 0$, while for zero rotational restraint (i.e. a pinned support) $k = \infty$. As fully rigid restraint is not possible, EN 1992-1-1 [2] recommends that in design k_1 and k_2 are not taken less than 0.1.

For the determination of the rotation θ, the effects of cracking of the restraining members should be considered if they are cracked at the ultimate limit state.

In the case of slender prestressed concrete columns, the question arises as to whether the longitudinal prestressing force P reduces the critical buckling load. In general, a concrete column prestressed with internally bonded strands or post-tensioned with tendons inside ducts within the member is no more prone to buckling than a reinforced concrete column of the same size and stiffness and under the same support conditions. As a slender, prestressed concrete column displaces laterally, the tendons do not change position within the cross-section and the eccentricity of the line of action of the prestressing force does not change. The prestressing force cannot, therefore, generate secondary moments. However, if a member is externally prestressed so that the line of action of the prestressing force remains constant, then prestress can induce secondary moments and hence affect the buckling load.

13.4.2 Slenderness criteria

The slenderness ratio λ of a column is the ratio of the effective length l_0 and the radius of gyration of the uncracked cross-section i:

$$\lambda = \frac{l_0}{i} \tag{13.25}$$

EN 1992-1-1 [2] permits secondary moments in a column to be ignored if the slenderness ratio of an isolated member is less than λ_{lim}, where:

$$\lambda_{lim} = \frac{20.A.B.C}{\sqrt{n}} \tag{13.26}$$

where $A = 1/(1 + 0.2\varphi_{ef})$; $B = \sqrt{1 + 2\omega}$; $C = 1.7 - r_m$; $n = N_{Ed}/(A_c f_{cd})$; φ_{ef} is the effective creep ratio given by $\varphi_{ef} = \varphi(\infty, t_0)M_{0Eqp}/M_{0Ed}$; $\varphi(\infty, t_0)$ is the final creep coefficient; M_{0Eqp} is the first-order moment caused by the quasi-permanent service load combination; M_{0Ed} is the first-order moment caused by the ultimate design load combination; ω is the mechanical reinforcement ratio given by $\omega = A_s f_{yd}/(A_c f_{cd})$; A_s is the area of longitudinal reinforcement in the column; r_m is the ratio of first-order end moments ($r_m = M_{01}/M_{02}$) and $|M_{02}| \geq |M_{01}|$. If the column is bent in single curvature (i.e. M_{01} and M_{02} produce tension on the same side of the column), r_m is taken to be positive and $C \leq 1.7$. If the column is bent in double curvature, r_m is negative and $C > 1.7$. In Equation 13.26, if φ_{ef} is not known, $A = 0.7$ may be assumed; if ω is unknown, B may be taken as 1.1; and if r_m is unknown, $C = 0.7$ is a conservative assumption. For unbraced members generally, $C = 0.7$. For braced members, where the primary moments arise predominantly from transverse loading or initial imperfections, $C = 0.7$.

In addition, if Equation 13.27 is satisfied, global second-order effects may be ignored in buildings where: (1) the plan layout is reasonably symmetrical; (2) global shear deformations are negligible; (3) the columns are braced by bracing members with negligible rotation at the base and whose stiffness is reasonably constant over the height of the structures; and (4) the gravity load increases by about the same amount per storey.

$$F_{V,Ed} \leq 0.31 \cdot \frac{n_s}{n_s + 1.6} \cdot \frac{\sum E_{cd} I_c}{L^2} \tag{13.27}$$

where $F_{V,Ed}$ is the total vertical load on both braced and unbraced members; n_s is the number of storeys; L is the total height of the building above the base where moment is restrained; $E_{cd} = E_{cm}/1.2$ is the design value of the elastic modulus of concrete and I_c is the second moment of areas of the bracing member(s) based on the uncracked concrete section(s). Where the bracing elements are uncracked at the ultimate limit state, the limiting value of $F_{V,Ed}$ given on the right hand side of Equation 13.27 may be doubled.

Notwithstanding the requirements mentioned earlier, EN 1992-1-1 [2] permits second-order effects to be ignored if they are less than 10% of the corresponding first-order effects. The code also permits the effects of creep to be ignored in the determination of secondary moments if the following three conditions are satisfied:

$$\varphi(\infty, t_0) \leq 2.0; \quad \lambda \leq 75 \quad \text{and} \quad M_{0Ed}/N_{Ed} \geq h \tag{13.28}$$

where M_{0Ed} is the first-order moment and h is the depth of the column cross-section at right angles to the axis of bending.

For columns subjected to biaxial bending, the slenderness criteria should be checked separately for each direction.

13.4.3 Moment magnification method

In lieu of a detailed second-order analysis to determine the effects of short-term and time-dependent deformation on the magnitude of moment and forces in slender structures, EN 1992-1-1 [2] specifies a *moment magnifier method* to account for slenderness effects in columns. The idea behind the moment magnification method is based on the concept of using a factor to magnify the column moments to account for the change in geometry of the structure and the resulting secondary actions. The axial load and the magnified moment are then used in the design of the column cross-section. The effect of second-order (secondary) moments on the strength of a slender column is shown on the strength interaction curve in Figure 13.9. Line OA is the loading line corresponding to an initial eccentricity e on a particular cross-section. If the column is short, the second-order moments are insignificant, the loading line is straight and the design strength or resistance of the column corresponds to the design axial force at A ($N_{Rd,short}$). If the column is slender, the second-order moments increase at a faster rate than the applied axial force and the loading line becomes curved, as shown. The strength of the slender column is the axial force corresponding to point B ($N_{Rd,slender}$), where the curved loading line meets the strength interaction curve. The loss of strength due to second-order moments is indicated in Figure 13.9.

The total moment at failure M_{Ed} is the sum of the first-order moment M_{0Ed} ($=Ne$) obtained from a linear analysis and the second-order moment M_2 and may be expressed as:

$$M_{Ed} = M_{0Ed}\left[1+\frac{\beta}{(N_B/N_{Ed})-1}\right] \tag{13.29}$$

where the terms inside the square brackets are used to *magnify* the primary moment to account for slenderness effects. In Equation 13.29, N_B is the buckling load (see Equation 13.21) based on nominal stiffness EI which may be taken to be M_{Rd}/κ_{Rd} at the balanced failure point on the strength

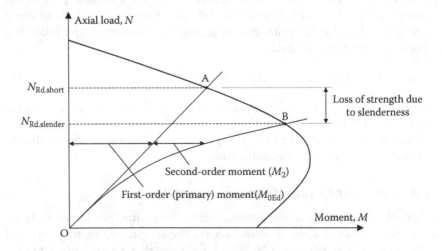

Figure 13.9 Strength interaction curve for a cross-section in a slender column.

interaction diagram multiplied by $1/(1 + A)$ to account for the effects of creep; A is defined under Equation 13.26; N_{Ed} is the design axial force on the column and β depends on the distribution of first- and second-order moments along the column, the magnitude and direction of the first-order design moments at each end of the columns M_{01} and M_{02}, the deflected shape of the column, which in turn depends on whether the column is braced or unbraced, and the rotational restraint at each end of the column.

For isolated members with constant cross-section and axial load, the second-order moment may be assumed to have a sine-shaped distribution. In this case:

$$\beta = \pi^2/c_0 \qquad\qquad (13.30)$$

and c_0 is a constant that depends on the distribution of first-order moments along the column. For constant first-order moments, $c_0 = 8$; for a parabolic distribution of moments, $c_0 = 9.6$ and for a symmetric triangular distribution, $c_0 = 12$.

For columns not subjected to transverse loads, where M_{01} and M_{02} are not the same, the first-order moment diagram may be approximated by an equivalent constant first-order moment M_{0e} given by:

$$M_{0e} = 0.6M_{02} + 0.4M_{01} \geq 0.4M_{02} \qquad\qquad (13.31)$$

where $|M_{02}| \geq |M_{01}|$. The end moments M_{01} and M_{02} both have the same sign, if they both produce tension on the same side of the column; otherwise, they have opposite signs. With this equivalent moment, $c_0 = 8$ in Equation 13.30. For columns bent in double curvature, i.e. M_{01} and M_{02} have opposite signs, the value of $c_0 = 8$ also applies.

For columns other than isolated members with constant cross-section and axial load and for columns subjected to transverse loads, it is reasonable to assume that $\beta = 1.0$ and, in this case, Equation 13.29 reduces to:

$$M_{Ed} = \frac{M_{0Ed}}{1 - (N_{Ed}/N_B)} \qquad\qquad (13.32)$$

EXAMPLE 13.2

Consider a 10 m long pin-ended column in a braced structure. The column cross-section is shown in Figure 13.4a and the material properties and steel quantities are as outlined in Example 13.1. The strength interaction curve for the cross-section was calculated in Example 13.1 and illustrated in Figure 13.5. The column is laterally supported at close centres to prevent displacement perpendicular to the weak axis of the section but is unsupported between its ends in the direction perpendicular to the strong axis. The column is loaded by a design compressive force N_{Ed} = 3500 kN at a constant eccentricity e to produce compression and uniaxial bending about the strong axis. Establish two loading lines for the column corresponding to initial eccentricities of e = 100 and 400 mm and determine the strength of the slender column N_{Rd} in each case. Assume that the ratio of the quasi-permanent service load to the design load is 0.6 and that $\varphi(\infty, t_0)$ = 2.5.

Since the column is braced and pinned at each end, the effective length associated with buckling about the strong axis is l_0 = 10 m (see Figure 13.7a). The effective length about the weak axis is small due to the specified closely spaced lateral supports. With:

$$n = \frac{N_{Ed}}{A_c f_{cd}} = \frac{3500 \times 10^3}{800 \times 600 \times 26.67} = 0.273$$

$$\varphi_{ef} = \varphi(\infty, t_0) \times 0.6 = 1.5$$

$$\omega = \frac{(A_s f_{yd} + A_p f_{pd})}{A_c f_{cd}} = \frac{(4500 \times 435 + 2000 \times 1391)}{800 \times 600 \times 26.67} = 0.37$$

$$r_m = 1.0$$

the constants in Equation 13.26 are $A = 1/(1 + 0.2\varphi_{ef}) = 0.77$, $B = \sqrt{1 + 2\omega} = 1.32$, $C = 1.7 - r_m = 0.7$ and therefore from Equation 13.26:

$$\lambda_{lim} = \frac{20 \times 0.77 \times 1.32 \times 0.7}{\sqrt{0.273}} = 27.2$$

The radius of gyration of the uncracked cross section is $i = \sqrt{I/A} = 231 \text{mm}$, and for bending about the strong axis, the slenderness ratio is (Equation 13.25):

$$\lambda = \frac{10,000}{231} = 43.3$$

This is greater than λ_{lim} and therefore the column is slender.
For this column, c_0 = 8 and from Equation 13.30:

$$\beta = \pi^2/8 = 1.234$$

Using the results from Example 13.1 at the balanced failure point, we get:

$$EI = \frac{M_{Rd.2}}{\kappa_{Rd.2}(1 + A)} = \frac{2059 \times 10^6}{7.78 \times 10^{-6} \times (1 + 0.77)} = 1.50 \times 10^{14} \text{ Nmm}^2$$

The bucking load is given by Equation 13.21:

$$N_B = \frac{\pi^2 \times 1.50 \times 10^{14}}{10,000^2} \times 10^{-3} = 14,800 \text{ kN}$$

The loading line for each initial eccentricity is obtained by calculating the total moment M_{ED} from Equation 13.29 for a series of values of axial force N_{ED} and plotting the points (N_{ED}, M_{ED}) on a graph of axial force and moment.

Sample calculations are provided for the points on the loading lines corresponding to an axial force $N_{Ed} = 3500$ kN.

The first-order moment when e = 100 mm is:

$$M_{0Ed} = 3500 \times 10^3 \times 100 \times 10^{-6} = 350 \text{ kNm}$$

From Equation 13.29:

$$M_{Ed} = 350 \times \left[1 + \frac{1.234}{(14,800 - 3,500) - 1} \right] = 484 \text{ kNm}$$

The first-order moment when e = 400 mm is:

$$M_{0Ed} = 3500 \times 10^3 \times 400 \times 10^{-6} = 1400 \text{ kNm}$$

From Equation 13.29:

$$M_{Ed} = 1,400 \times \left[1 + \frac{1.234}{(14,800/3,500) - 1} \right] = 1,935 \text{ kNm}$$

Other points on the loading line are tabulated below.

N_{Ed} (kN)	Moment magnifier $\left[1 + \dfrac{\beta}{(N_B/N_{Ed}) - 1} \right]$	e = 100 mm M_{Ed} (kNm)	e = 400 mm M_{Ed} (kNm)
2000	1.193	239	954
3000	1.314	394	1577
4000	1.457	583	2331
5000	1.630	815	3260
6000	1.841	1105	—
7000	2.107	1475	—
8000	2.452	1962	—

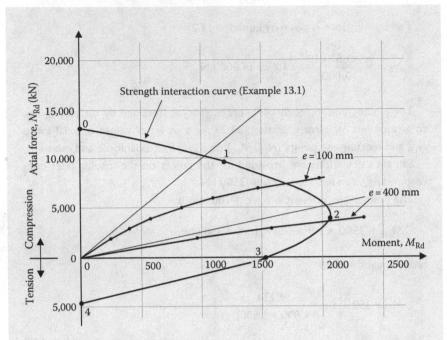

Figure 13.10 Loading lines and strength of the slender columns (Example 13.2).

The loading lines are plotted on Figure 13.10, together with the strength interaction curve reproduced from Figure 13.5. The strength of each column is the axial load corresponding to the intersection of the loading line and the strength interaction curve. The design resistance N_{Rd}, and hence the maximum factored design load N_{Ed} that can be applied to the slender column (and meet the strength design requirements of EN 1992-1-1 [2]), is obtained from the point where the loading line crosses the strength interaction curve. Note the significant reduction of strength in both columns due to slenderness. For the initial eccentricity $e = 100$ mm, the design resistance of the slender column is $N_{Rd} = 7480$ kN (compared to the cross-sectional strength of 10,050 kN). For the initial eccentricity $e = 400$ mm, the design resistance of the slender column is $N_{Rd} = 3450$ kN (compared to the cross-sectional strength of 5020 kN).

Slenderness causes a far greater reduction in strength in the primary compression region when the initial eccentricity is small than when eccentricity is large and bending predominates. Note also that for very slender columns with large eccentricities, the curved loading line crosses the strength interaction curve in the primary tension region and this is the same region in which prestress provides some additional strength (see Figure 13.5). There is some advantage in prestressing slender columns.

13.5 REINFORCEMENT REQUIREMENTS FOR COMPRESSION MEMBERS

EN 1992-1-1 [2] specifies design requirements related to the quantity and distribution of reinforcement in columns. The area of longitudinal reinforcement in a column A_s should not be less than $A_{s,min}$, where:

$$A_{s,min} = \frac{0.1N_{Ed}}{f_{yd}} \quad \text{or} \quad 0.002A_c \text{ whichever is greater} \tag{13.33}$$

and should not be greater than $0.04A_c$ in regions outside lap locations. This upper limit may be exceeded if the amount and disposition of the reinforcement will not prevent the proper placement and compaction of the concrete at splice locations and at the junctions with other members and if the full design resistance is achieved at the ultimate limit state. It is good practice to ensure that at least one longitudinal bar is placed in each corner of the column of a rectangular cross-section and at least four longitudinal bars are required in columns of circular cross-section.

In addition to longitudinal reinforcement, transverse reinforcement in the form of closed ties or spirals (helical reinforcement) is required in columns. The behaviour of a column loaded to failure depends on the nature of the transverse reinforcement. If the column contains no transverse reinforcement, when the strength of the cross-section is reached, failure will be brittle and sudden. Transverse reinforcement imparts a measure of ductility to reinforced and prestressed concrete columns by providing restraint to the highly stressed longitudinal steel and by confining the inner core of compressive concrete. Ductility is a critical design requirement for the columns in buildings located in earthquake-prone regions, where the ability to absorb large amounts of energy without failure is needed. Spirally reinforced columns, in particular, exhibit considerable ductility at failure. In addition to imparting ductility, the transverse reinforcement also carries any diagonal tensile forces associated with shear and torsion, if these actions are carried in addition to axial compression and bending.

The detailing requirements for both the longitudinal and transverse reinforcements in columns are outlined in Section 14.6.

13.6 TRANSMISSION OF AXIAL FORCE THROUGH A FLOOR SYSTEM

With the increasing use of high-strength concrete in the columns in buildings, problems may arise when a high column load is required to pass through a lower-strength concrete floor slab. Ospina and Alexander [4] showed that the strength and behaviour of column–slab joints are significantly affected by the load on the slab. The negative bending in a slab at a column location produces tension at the top of the slab. With compressive confinement only provided to the slab–column joint at the bottom of the slab where compressive struts enter the joint.

Satisfactory transmission of the column forces through the joint will usually occur, and the full column strength may be assumed in design provided the strength of the concrete in the slab $(f_{ck})_{slab}$ is greater than or equal to 75% of the strength of the concrete in the column $(f_{ck})_{col}$ and provided the longitudinal reinforcement in the column is continuous through the joint. For greater differences in column and slab strengths, calculations are required to determine whether additional longitudinal reinforcement is required through the joint. The effective strength of the concrete within the joint region at the column–slab intersection $(f_{ck})_{eff}$ is a function of the strengths of the column and slab concrete, the geometry of the connection and its location in the structure (i.e. a corner, edge or interior column).

For interior columns, restrained on four sides by beams of approximately equal depth or by a slab, the effective concrete strength within the joint may be taken as [4]:

$$(f_{ck})_{eff} = \left(1.33 - \frac{0.33}{h/b_c}\right)(f_{ck})_{slab} + \frac{0.25}{h/b_c}(f_{ck})_{col} \qquad (13.34)$$

except that $(f_{ck})_{eff}$ need not be less than $1.33(f_{ck})_{slab}$ and should not be greater than the smaller of $(f_{ck})_{col}$ and $2.5(f_{ck})_{slab}$. In Equation 13.34, h is the overall thickness of the slab at the joint and b_c is the smaller column cross-sectional dimension.

For edge columns restrained on two opposing sides by beams of approximately equal depth or by a slab:

$$(f_{ck})_{eff} = \left(1.1 - \frac{0.3}{h/b_c}\right)(f_{ck})_{slab} + \frac{0.2}{h/b_c}(f_{ck})_{col} \qquad (13.35)$$

except that $(f_{ck})_{eff}$ need not be less than $1.33(f_{ck})_{slab}$ and should not be greater than the smaller of $(f_{ck})_{col}$ and $2.0(f_{ck})_{slab}$.

For corner columns restrained on two adjacent sides by beams of approximately equal depth or by a slab:

$$(f_{ck})_{eff} = 1.33(f_{ck})_{slab} \leq (f_{ck})_{col} \qquad (13.36)$$

When $(f_{ck})_{eff} < (f_{ck})_{col}$, an increased area of longitudinal reinforcement may be placed through the joint to effectively replace the reduction in capacity from the influence of the weaker concrete. In this case, additional tie reinforcement should also be placed through the connection to confine the weaker concrete and provide ductility to the joint that is under a high stress relative to its cylinder strength.

Another procedure sometimes used in construction is that of *puddling*, in which column strength concrete is placed within the slab–column connection and the surrounding region. The *puddled* column concrete should occupy the full slab thickness and extend beyond the face of the column by a distance greater than 600 mm and twice the depth of the slab or beam at the column face, whichever is greater. Special care has to be taken to avoid placement of the weaker slab concrete within the joint region. Proper vibration of the column and slab concretes is needed for optimal melding at the faces of the two materials. Since both column and slab concrete are to be cast simultaneously, with two different concrete strength grades on site during one pour, a high level of on-site supervision and quality control is necessary.

13.7 TENSION MEMBERS

13.7.1 Advantages and applications

Prestressed concrete tension members are simple elements used in a wide variety of situations. They are frequently used as tie-backs in cantilevered construction, anchors for walls and footings, tie and chord members in trusses, hangers and stays in suspension bridges, walls of tanks and containment vessels and many other applications.

The use of reinforced concrete members in direct tension has obvious drawbacks. Cracking causes a large and sudden loss of stiffness and crack control may be difficult. Cracks occur over the entire cross-section and corrosion protection of the steel must be carefully considered in addition to aesthetic difficulties. By prestressing the concrete, however, a tension member is given strength and rigidity otherwise unobtainable from either the concrete or the steel acting alone. Provided that cracking does not occur in the concrete, the prestressing steel is protected from the environment and the tension member is suitable for its many uses. Compared with compression members, tension members usually have a high initial level of prestress.

The deformation of a prestressed concrete tension member can be carefully controlled. In situations where excessive elongation of a tension member may cause strength or serviceability problems, prestressed concrete is a design solution worth considering.

13.7.2 Behaviour

The analysis of a prestressed concrete direct tension member is straight-forward. Both the prestressing force and the external tensile loads are generally concentric with the longitudinal axis of the member, and hence bending stresses are minimised.

Prior to cracking of the concrete, the prestressing steel and the concrete act in a composite manner and behaviour may be determined by consider-ing a transformed cross-section. If required, a transformed section obtained using the effective modulus for concrete (Equation 4.23) may be used to include the time-dependent effects of creep and shrinkage.

Consider a tension member concentrically prestressed with an effective prestressing force $P_{m,t}$. The cross-section is symmetrically reinforced with an area of bonded prestressing steel A_p and non-prestressed steel A_s. The transformed area of the tie is therefore:

$$A = A_c + \alpha_p A_p + \alpha_s A_s = A_g + (\alpha_p - 1)A_p + (\alpha_s - 1)A_s \tag{13.37}$$

where α_p and α_s are the modular ratios E_p/E_{cm} and E_s/E_{cm}, respectively. The uniform stress in the concrete σ_c due to the prestressing force and the applied external load N is:

$$\sigma_c = -\frac{P_{m,t}}{A_c} + \frac{N}{A} \tag{13.38}$$

and the stress in the bonded prestressing steel is:

$$\sigma_p = \frac{P_{m,t}}{A_p} + \frac{\alpha_p N}{A} \tag{13.39}$$

For most applications, it is necessary to ensure that cracking does not occur at service loads. To provide a suitable margin against crack-ing under day-to-day loads and to ensure that cracks resulting from an unexpected overload close completely when the overload is removed, it is common in design to insist that the concrete stress remains compressive under normal in-service conditions. By setting $\sigma = 0$ in Equation 13.38 and rearranging, an upper limit to the external tensile force is established and is given by:

$$N \le P_{m,t} \frac{A}{A_c} = P_{m,t}\left(1 + \alpha_p \rho_p + \alpha_s \rho_s\right) \tag{13.40}$$

where $\rho_p = A_p/A_c$ and $\rho_s = A_s/A_c$.

If a tensile member is stressed beyond the service load range, cracking should be deemed to occur when the concrete stress reaches the lower characteristic tensile strength $f_{ctk,0.05}$ in Table 4.2. If the tensile force at first cracking is N_{cr}, then from Equation 13.38:

$$N_{cr} = \left(\frac{P_{m,t}}{A_c} + f_{ctk,0.05} \right) A \tag{13.41}$$

At a cracked section, when $N = N_{cr}$, the concrete carries no stress and the tension is carried only by the reinforcement and tendons (A_s and A_p), and the stress in non-prestressed steel is:

$$\sigma_s = \frac{N_{cr}}{A_s + (A_p E_p / E_s)} \tag{13.42}$$

and the stress in the prestressing steel is:

$$\sigma_p = \frac{N_{cr} - \sigma_s A_s}{A_p} = \frac{N_{cr}}{A_p} \left(\frac{1}{1 + (A_s E_s / A_p E_p)} \right) \tag{13.43}$$

Substituting Equation 13.41 into Equation 13.43 gives:

$$\sigma_p = \left(\frac{P_{m,t}}{A_c} + f_{ctk,0.05} \right) \frac{A}{A_p} \left(\frac{1}{1 + (A_s E_s / A_p E_p)} \right) \tag{13.44}$$

This is usually limited to a maximum of about $f_{p0.1k}/1.2$ in order to obtain a minimum acceptable margin of safety between first cracking and ultimate strength. A conservative estimate of the minimum area of prestressing steel in a tension member can be obtained from Equation 13.44 by ignoring the area of non-prestressed reinforcement as follows:

$$A_p \geq \left(\frac{P_{m,t}}{A_c} + f_{ctk,0.05} \right) \frac{1.2A}{f_{p0.1k}} \tag{13.45}$$

The design axial tensile resistance of the member is equal to the tensile strength of the steel and is given by:

$$N_{Rd.t} = A_p f_{pd} + A_s f_{yd} \tag{13.46}$$

and, in design, this must exceed the design axial tension N_{Ed}.

The axial deformation of a prestressed tension member at service loads depends on the load history (i.e. the times at which the prestressing force(s) and the external loads are applied) and the deformation characteristics of the concrete. *Stage stressing* can be used to carefully control longitudinal deformation. The axial deformation (shortening) of a tension member at any time t caused by an initial prestress P_{m0} applied to the concrete at a particular time t_0 and by shrinkage of the concrete (ε_{cs} assumed to also begin at t_0) may be approximated by:

$$e_{P.cs} = \frac{P_{m0}l}{A_c E_{c,eff.0}(1 + \alpha_{ep.0}\rho_p + \alpha_{es.0}\rho_s)} + \frac{\varepsilon_{cs}l}{(1 + \alpha_{ep.0}\rho_p + \alpha_{es.0}\rho_s)} \tag{13.47}$$

where l is the length of the member, $\rho_p = A_p/A_c$, $\rho_s = A_s/A_c$, $E_{c,eff.0}$ is the effective modulus of the concrete obtained from Equation 4.23 using the creep coefficient associated with the age of the concrete at first loading (t_0), $\alpha_{ep.0} = E_p/E_{c,eff.0}$ and $\alpha_{es.0} = E_s/E_{c,eff.0}$.

The axial elongation at any time caused by an external tensile force N applied at time t_1 may be estimated by:

$$e_N = \frac{Nl}{A_c E_{c,eff.1}(1 + \alpha_{ep.1}\rho_p + \alpha_{es.1}\rho_s)} \tag{13.48}$$

where $E_{c,eff.1}$ is the effective modulus of the concrete using the creep coefficient associated with the age of the concrete when the force N is first applied (t_1), $\alpha_{ep.1} = E_p/E_{c,eff.1}$ and $\alpha_{es.1} = E_s/E_{c,eff.1}$.

A detailed analysis of the time-dependent deformation of a tension member subjected to any load history can be made using the procedures outlined by Gilbert and Ranzi [5].

A satisfactory preliminary design of a tension member usually results if the prestressing force is initially selected so that, after losses, the effective prestress is between 10% and 20% higher than the maximum in-service tension. If the compressive stress in the concrete at transfer is limited to about $0.4f_{ck}(t_0)$, the minimum gross area of the cross-section A_g can be determined. The area of steel required to impart the necessary prestress is next calculated. The resulting member can then be checked for strength and serviceability, and details modified, if necessary.

EXAMPLE 13.3

Consider the vertical post-tensioned tension member acting as a tie-back for the cantilevered roof of a grandstand, shown in Figure 13.11. In the critical loading case, the tension member must transfer a design working dead load of 800 kN and a live load of 200 kN to the footing which is anchored to rock.

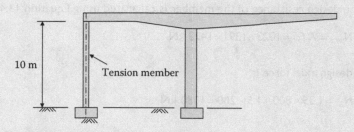

Figure 13.11 Tie-back member to be designed (Example 13.3).

The material properties are:

f_{ck} = 40 MPa; E_{cm} = 35,000 MPa; $f_{ck}(t_0)$ = 30 MPa; $E_{cm}(t_0)$ = 33,000 MPa;

f_{pk} = 1,860 MPa; E_p = 195,000 MPa and therefore α_p = 5.57 and $\alpha_{p,0}$ = 5.91.

In accordance with the preceding discussion, an effective prestress that is 10% higher than the maximum applied tension is assumed:

$P_{m,t}$ = 1.1 × (800 + 200) = 1100 kN

Owing to the small residual compression existing under sustained loads, the time-dependent loss of prestress is usually relatively small. In this short member, draw-in losses at transfer are likely to be significant. For the purposes of this example, the time-dependent losses are assumed to be 12% and the short-term losses are taken to be 15%.

The force immediately after transfer and the required jacking force are therefore:

$$P_{m0} = \frac{1100}{0.88} = 1250 \text{ kN} \quad \text{and} \quad P_{max} = P_j = \frac{1250}{0.85} = 1470 \text{ kN}$$

If the maximum steel stress at jacking is 0.9 $f_{p0.1k}$ = 1440 MPa (see Equation 5.1), then the area of prestressing steel is:

$$A_p \geq \frac{1470 \times 10^3}{1440} = 1021 \text{ mm}^2$$

Try eleven 12.5 mm diameter strands (A_p = 1023 mm²) post-tensioned within a 60 mm diameter duct located at the centroid of the cross-section.

The design resistance of the member is calculated using Equation 13.46:

$$N_{Rd.t} = A_p f_{pd} = 1023 \times 1391 = 1423 \text{ kN}$$

The design axial force is:

$$N_{Ed} = 1.35 \times 800 + 1.5 \times 200 = 1380 \text{ kN}$$

which is less than the design strength and is therefore satisfactory. If additional strength had been necessary, non-prestressed steel could be included to increase $N_{Rd.t}$ to the required level.

If the concrete stress at transfer is limited to $0.4 f_{ck}(t_0) = 12$ MPa, the area of concrete on the cross-section A_c must satisfy:

$$A_c \geq \frac{P_{m0}}{0.4 f_{ck}(t_0)} = \frac{1,250 \times 10^3}{12} = 104,200 \text{ mm}^2$$

Try a 350 mm by 350 mm square cross-section with a centrally located 60 mm duct. Therefore, before the duct is grouted:

$$A_c = 350 \times 350 - (0.25 \times \pi \times 60^2) = 119,700 \text{ mm}^2$$

Under the effective prestress, after all losses and after the duct is fully grouted, the area of the transformed section is obtained using Equation 13.37:

$$A = 350 \times 350 + (5.57 - 1) \times 1,023 = 127,180 \text{ mm}^2$$

and

$$A_c = A_g - A_p = 121,480 \text{ mm}^2$$

The uniform stress in the concrete under the full service load is given by Equation 13.38:

$$\sigma_c = -\frac{1,100 \times 10^3}{121,480} + \frac{1,000 \times 10^3}{127,180} = -1.19 \text{ MPa}$$

and the stress in the tendon is given by Equation 13.39:

$$\sigma_p = \frac{1,100 \times 10^3}{1,023} + \frac{5.57 \times 1,000 \times 10^3}{127,180} = 1,119 \text{ MPa}$$

Both stresses are satisfactory and cracking will not occur at service loads, even if the losses of prestress have been slightly underestimated.

The minimum area of steel required to ensure a factor of safety of 1.2 at cracking is checked using Equation 13.45:

$$A_p \geq \left(\frac{1,100 \times 10^3}{121,480} + 2.5\right)\frac{1.2 \times 127,180}{1,600} = 1,102 \text{ mm}^2$$

and the area of steel $A_p = 1023$ mm^2 adopted here is insufficient.

If we include four 16 mm diameter reinforcing bars ($A_s = 804$ mm^2, $E_s = 200,000$ MPa, $\alpha_s = 5.71$ and $\alpha_p = 5.57$), the revised transformed area (Equation 13.37) is $A = 130,960$ mm^2, the revised concrete area is $A_c = 120,670$ mm^2 and Equation 13.44 gives:

$$\sigma_p = \left(\frac{1,100 \times 10^3}{120,670} + 2.5\right)\frac{130,960}{1,023}\left[\frac{1}{1 + (804 \times 200,000/(1,023 \times 195,000))}\right]$$

$$= 825 \text{ MPa}$$

which is less than $f_{p0.1k}/1.2$ and therefore the margin of safety between first cracking and ultimate strength is satisfactory. The prestressed and non-prestressed reinforcement ratios are $\rho_p = A_p/A_c = 0.00848$ and $\rho_s = A_s/A_c = 0.00666$, respectively.

If the final creep coefficients associated with the age at transfer t_0 and the age when the external load is first applied t_1 are $\varphi(\infty, t_0) = 2.5$ and $\varphi(\infty, t_1) = 2.0$, then Equation 4.23 gives the appropriate effective moduli:

$$E_{c,eff.0} = \frac{33,000}{1+2.5} = 9,430 \text{ MPa} \quad \text{and} \quad E_{c,eff.1} = \frac{35,000}{1+2.0} = 11,670 \text{ MPa}$$

and therefore $\alpha_{ep.0} = 20.68$, $\alpha_{es.0} = 21.21$, $\alpha_{ep.1} = 16.71$ and $\alpha_{es.1} = 17.14$.

If the final shrinkage strain is $\varepsilon_{cs} = 650 \times 10^{-6}$, the shortening of the member caused by prestress and shrinkage is obtained using Equation 13.47:

$$e_{P.cs} = \frac{1,250 \times 10^3 \times 10,000}{120,670 \times 9,430 \times (1 + 20.68 \times 0.00848 + 21.21 \times 0.00666)}$$

$$+ \frac{650 \times 10^{-6} \times 10,000}{(1 + 20.68 \times 0.00848 + 21.21 \times 0.00666)} = 13.28 \text{ mm}$$

and the elongation caused by N is given by Equation 13.48:

$$e_N = \frac{1,000 \times 10^3 \times 10,000}{120,670 \times 11,670 \times (1 + 16.71 \times 0.00848 + 17.14 \times 0.00666)} = 5.65 \text{ mm}$$

The net effect is a shortening of the member by:

$$e = e_{P.cs} - e_N = 13.28 - 5.65 = 7.63 \text{ mm}$$

REFERENCES

1. Ranzi, G. and Gilbert, R.I. 2015. *Structural Analysis: Principles, Methods and Modelling*. Boca Raton, FL: CRC Press/Taylor & Francis Group, 562pp.
2. EN 1992-1-1. 2004. Eurocode 2: Design of concrete structures – Part 1-1: General rules and rules for buildings. British Standards Institution, London, UK.
3. Bresler, B. 1960. Design criteria for reinforced concrete columns under axial load and biaxial bending. *ACI Journal*, 57, 481–490.
4. Ospina, C.E. and Alexander, S.D.B. 1997. Transmission of high strength concrete column loads through concrete slabs, structural engineering, Report No. 214, Department of Civil Engineering, University of Alberta, Edmonton, Alberta, Canada.
5. Gilbert, R.I. and Ranzi, G. 2011. *Time-Dependent Behaviour of Concrete Structures*. London, UK: Spon Press.

Chapter 14

Detailing
Members and connections

14.1 INTRODUCTION

Detailing of a concrete structure is more than simply the preparation of working drawings that show the structural dimensions and the size and location of the reinforcing bars and tendons. It involves the communication of the engineer's ideas and specifications from the design office to the construction site and encompasses each aspect of the design process from preliminary analysis to final design. It involves the translation of a good structural design from the computer or calculation pad into the final structure. The most sophisticated or up-to-date methods of analysis and design are of little value if they remain in the calculations and do not find their way into the structure.

Detailing of the structural elements and the connections between them, perhaps more than any other single factor, decides the success or failure of a concrete structure. Good detailing ensures that the reinforcement and the concrete interact efficiently to provide satisfactory performance throughout the complete range of loading. Successful detailing requires experience, as well as a sound understanding of structural and material behaviour and an appreciation of appropriate construction practices and methodologies.

Too often detailing is the last thing considered by the engineer, or worse, not seriously considered at all. Yet it is critical if full strength and adequate ductility are to be achieved and if in-service performance is to be satisfactory. It should be remembered that reinforcement details that ensure satisfactory behaviour under service load conditions may not provide good collapse characteristics and, conversely, details that provide adequate strength and ductility do not necessarily ensure serviceability. Detailing must be considered when designing for both the ultimate and service limit state conditions.

In this chapter, guidelines for successful detailing of the structural elements and connections in prestressed concrete structures are outlined and, where necessary, the requirements of EN 1992-1-1 [1] that have been introduced in previous chapters are revisited. The reasons for providing reinforcement and the sources of tension in concrete structures (some of which may not be immediately obvious) are also discussed. The importance of adequate anchorage for reinforcement is stressed and appropriate details are recommended. Many of the principles for successful detailing were presented by Park and Paulay [2] who, in turn, gained their inspiration from the pioneering work of Leonhardt and his colleagues [3–7].

14.2 PRINCIPLES OF DETAILING

14.2.1 When is steel reinforcement required?

The detailing requirements of a reinforcement bar or tendon depend on the reasons for its inclusion in the structure. It is only after the purpose of the steel bar or tendon is identified that the detailing requirements can be specified. Reinforcement and tendons are provided in concrete structures for a variety of reasons, including:

1. to carry the internal tensile forces that are determined from analysis, i.e. to provide tensile strength in regions of the structure where tensile strength is required and to impart ductility where ductility is required;
2. to control cracking (and in the case of prestressing tendons, to control or eliminate cracking), i.e. to ensure that the cracks caused by bending, axial tension, bursting and spalling stresses and shrinkage or temperature changes are serviceable;
3. to carry compressive forces in regions where the concrete alone is inadequate (e.g. in columns and in the compressive zone of heavily reinforced beams);
4. to provide restraint to bars in compression, i.e. to prevent lateral buckling of compressive reinforcing bars prior to reaching their full strength;
5. to provide confinement to the compressive concrete in the core of columns, in beams and within connections and other disturbed regions, thereby increasing both the strength and deformability of the concrete;
6. to provide protection against spalling, for example, the *fire mesh* sometimes used in the protective concrete cover over fabricated steel sections;
7. to limit long-term deformation by providing restraint to creep and shrinkage of the concrete; and
8. to provide temporary support for other reinforcement during construction prior to and during the concreting operation.

A single reinforcing bar or tendon may be required for one or more of the reasons mentioned earlier. The ability of the bar or tendon to accomplish its task or tasks depends on how it is detailed and that is very much associated with the quality of its anchorage. When the location of a bar or tendon is determined, the question of how best to detail the steel can only be answered after its anchorage requirements are assessed. Anchorage and stress development in conventional reinforcement are discussed in more detail in Section 14.3 and the anchorages of tendons were discussed in Chapter 8.

14.2.2 Objectives of detailing

When detailing reinforcement and tendons, the objectives are the same as the broad objectives in structural design and aim at:

1. achieving the required design resistance at each cross-section in each member and at all connections, i.e. making sure the structure is strong enough;
2. preventing problems under the day-to-day in-service conditions that may impair the serviceability of the structure;
3. allowing ease of construction; and
4. maintaining economy.

To achieve these objectives, it is worthwhile to remember a number of general principles. When detailing for strength, the location and direction of all internal tensile forces should be determined by establishing the load path for the structure. Adequately anchored reinforcement or tendons should be used wherever a tensile force is required for equilibrium. The concrete should be assumed to carry no tension. The possible collapse mechanisms of the structure should be identified and adequate reinforcement should be specified to prevent premature collapse.

When detailing for serviceability, it is again essential that the location and direction of all internal tensile forces are determined. Often, this is not easy. The location of shrinkage- and temperature-induced tension may not be easily recognised or intuitively obvious. By providing restraint, the bonded reinforcement itself can create tensile forces which may crack the surrounding concrete. When the sources of tension are identified and the magnitudes and locations of the tensile forces are determined, prestressing tendons can be used to reduce or eliminate the tension in the concrete or, alternatively, cracking can be permitted and adequate quantities of appropriately anchored reinforcement, with suitable bond characteristics, must be provided to maintain stiffness after cracking and to provide crack control.

Complex reinforcement details should be drawn to a suitably large scale to ensure that they are practical, i.e. to ensure that the reinforcement can be fixed and that the concrete can be placed and compacted adequately.

Some of the sources of tension in concrete structures, some obvious and some not so obvious, are discussed in the following section.

14.2.3 Sources of tension

The sources of tension in concrete structures are numerous and the following list is by no means exhaustive. However, the list does illustrate the wide variety of reasons why concrete structures crack and why successful detailing requires a sound understanding of structural behaviour. A significant amount of on-site experience also helps.

14.2.3.1 Tension caused by bending (and axial tension)

The primary roles of the longitudinal tensile reinforcement in a flexural member are to provide strength and ductility and to ensure crack control under service conditions.

14.2.3.2 Tension caused by load reversals

Temporary propping or bad handling during construction frequently causes tension in regions where tension may not normally be expected. In Figures 14.1a and b, two common instances of temporary propping are illustrated and these should be anticipated in design. Impact and rebound loading also causes tension. For example, considerable prestress is required in driven concrete piles to overcome the tension in the concrete caused by *rebound*.

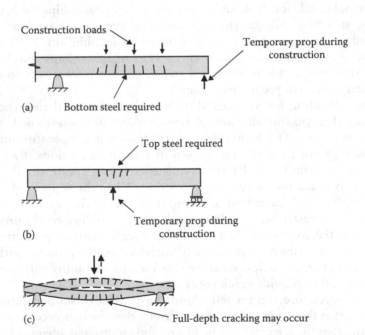

Figure 14.1 Examples where reversals of loads and internal actions may occur. (a) Cantilever. (b) Simple span. (c) Impact and rebound loading.

In slender members subjected to dynamic loads, such as precast stair treads, impact and rebound may cause tension on both sides of the member and result in full-depth cracking, as illustrated in Figure 14.1c.

14.2.3.3 Tension caused by shear and torsion

We have seen in Chapter 7 that diagonal tension caused by shear stresses causes inclined cracking in the webs of reinforced and prestressed concrete beams. Adequately anchored transverse reinforcement is required to carry this tension after inclined cracking has occurred. Some commonly used details for the anchorage of stirrups are examples of poor detailing and these are discussed in more detail in Section 14.5.4.

14.2.3.4 Tension near the supports of beams

The longitudinal tensile reinforcement required near the supports of beams is greater than indicated by the bending moment diagram. Consider the beam support shown in Figure 14.2a and the analogous truss of Figure 14.2b showing the flow of internal forces that transmit the applied loads to the support, i.e. the internal load path. The tensile force required to be carried by the longitudinal steel at the bottom of an inclined crack is equal to the compressive force at the top of the crack. This is clearly shown by the truss analogy which is a useful and convenient idealisation in any study of detailing.

Shrinkage and thermal shortening also causes tension in beams with restrained ends. Considerable reinforcement may therefore be required throughout the beam length specifically to control the resulting direct tension cracking.

For the anchorage of positive moment reinforcement at a simple support of a beam, EN 1992-1-1 [1] requires that at least 25% of the area of the positive moment reinforcement required in the span should be anchored for a length (l_{bd}) past the line of contact between the beam and the support (as shown in Figure 14.2a). See Section 14.3.2 for the determination of the

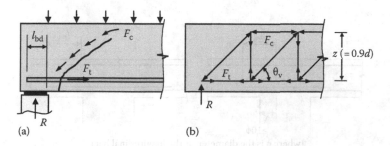

Figure 14.2 Tension at the discontinuous support of a beam. (a) Elevation at simple support. (b) Internal flow of forces – truss analogy.

anchorage length l_{bd}. For a beam with shear reinforcement at right angles to the member axis (i.e. $\alpha = 90°$), the anchored tensile reinforcement must be able to develop the tensile force F_E resulting from the transfer of the design shear force plus any axial force that develops due to any other source (including torsion and restraint to shrinkage or temperature change):

$$F_E = |V_{Ed}| \cot \theta_v + N_{Ed} \qquad (14.1)$$

where V_{Ed} is the design shear force at the face of the support, θ_v is the truss angle (as shown in Figure 14.2b) and N_{Ed} is the axial force to be added to or subtracted from the tensile force $|V_{Ed}| \cot \theta_v$.

For the anchorage of positive moment reinforcement at a continuous or flexurally restrained support of a beam, at least 25% of the area of the positive moment reinforcement required in the span should be anchored for a length past the near face of the support of at least 10 bar diameters (for straight bars). When additional reinforcement is provided to resist positive moment at an intermediate support (due to considerations of robustness, resistance to progressive collapse if the support is removed due to impact or blast, support settlement and more), it should be anchored in accordance with Figure 14.3.

14.2.3.5 Tension within the supports of beams or slabs

Shortening of beams and slabs occurs due to prestressing and due to shrinkage and drops in temperature, and this can cause tension and subsequent cracking in the supports if the longitudinal movement is restrained. For example, if adequate sliding joints are not introduced between a concrete slab and supporting masonry walls, shrinkage of the slab may cause considerable distress to the brickwork. Some illustrative examples are shown in Figure 14.4. This type of problem also frequently occurs in the supports of post-tensioned beams and slabs during the stressing operation if provision is not made for the elastic shortening of the beam or slab to be accommodated at the support. These sorts of problems are best avoided by the introduction of suitable movement joints.

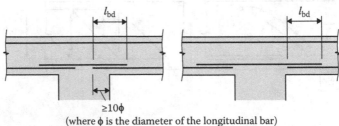

(where ϕ is the diameter of the longitudinal bar)

Figure 14.3 Anchorage of additional bottom steel at a support [1].

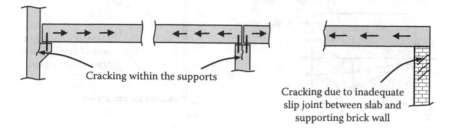

Figure 14.4 Cracking caused by tension within supports.

14.2.3.6 Tension within connections

Tension occurs within connections where there is a sudden change in direction of the internal forces. The design of connections is covered in more detail in Section 14.7. One connection, in which significant tension is often overlooked by designers, is where a secondary beam is supported by a primary girder, as shown in Figure 14.5. Most of the reaction from the secondary beam flows into the bottom region of the primary girder via diagonal compression. It is essential that this force finds an effective support. Additional stirrups are required within the connection in the primary girder to transfer this *hanging* load from the bottom to the top of the girder (see also Section 14.5.5).

14.2.3.7 Tension at concentrated loads

The transverse tension caused by the dispersion of a concentrated load, such as exists behind the bearing plate in a post-tensioning anchorage (see Figure 14.6a), was discussed in Chapter 8. A similar situation exists at the

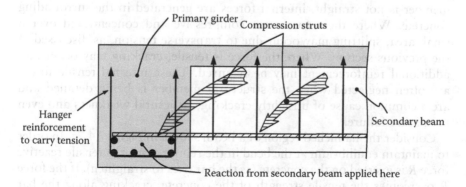

Figure 14.5 Primary girder supporting secondary beam.

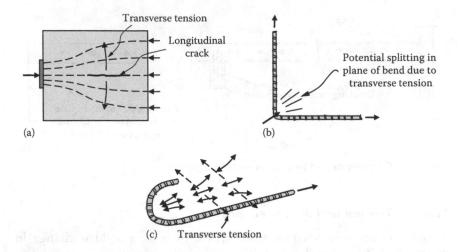

Figure 14.6 Tension due to concentrated loads. (a) Post-tensioned anchorage zone. (b) Bent bar. (c) Hooked anchorage.

anchorage of a reinforcing bar, such as the bend shown in Figure 14.6b or the hook shown in Figure 14.6c. A concentrated force is applied to the concrete where the bar changes direction, as shown. As the concentrated force disperses, a transverse tension exists in the concrete that may cause splitting of the concrete in the plane of the hook or bend, particularly if the radius of curvature of the bend is small.

14.2.3.8 Tension caused by directional changes of internal forces

Wherever a reinforcement bar changes direction or a loaded concrete member is not straight, internal forces are generated in the surrounding concrete. Where these forces are compressive and concentrated over a small area, splitting may occur due to transverse tension, as discussed in the previous section. Where the force is tensile, cracking may occur and additional reinforcement may be required. These internal tensile forces are often neglected when the structural member is being detailed and are a common cause of unsightly cracking, structural weakness and even premature failure.

Consider the haunched region of a beam shown in Figure 14.7a. In order to maintain equilibrium at the bend in the reinforcement, a tensile reactive force R is exerted on the concrete (as the bar tries to straighten). If the force R overcomes the tensile strength of the concrete, cracking along the bar

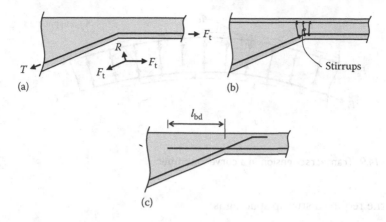

Figure 14.7 Haunched region of a beam. (a) Without transverse reinforcement. (b) With transverse reinforcement. (c) Lapped longitudinal reinforcement.

will occur. The steel will straighten resulting in the loss of concrete cover, structural integrity and strength.

Transverse reinforcement in the form of stirrups may be provided to carry the force R, thereby overcoming the problem, as shown in Figure 14.7b. However, a better solution may be to eliminate the problem altogether by anchoring each reinforcement bar with straight extensions so that no transverse force is generated (see Figure 14.7c).

The same principle applies when the internal compressive force changes direction, as illustrated in Figure 14.8. Adequately anchored transverse reinforcement must be provided in the web to carry the resultant force R.

In a curved member, such as that shown in Figure 14.9, the continuously changing direction of the internal compressive and tensile forces (caused by bending) creates a distributed transverse tensile force in the web. Stirrups at regular centres are required to carry these tensile forces. The transverse tension per unit length q_t produced by the tensile force in the longitudinal reinforcement is:

$$q_t = \frac{F_t}{r_m} = \frac{A_s f_{yd}}{r_m}$$
(14.2)

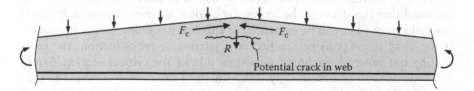

Figure 14.8 Transverse tension due to direction change of internal compression.

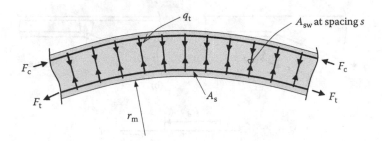

Figure 14.9 Transverse tension in a curved member.

and the required stirrup spacing is:

$$s = \frac{A_{sw} f_{ywd}}{q_t} = \frac{A_{sw}}{A_s} \frac{f_{ywd}}{f_{yd}} r_m \qquad (14.3)$$

where f_{ywd} and f_{yd} are the design yield stresses of the stirrups and the longitudinal reinforcement, respectively; A_{sw} is the total area of stirrup legs within the length s and A_s and r_m are the cross-sectional area and the radius of curvature of the curved longitudinal bars or tendons, respectively.

14.2.3.9 *Other common sources of tension*

Other sources of tension include, among others, the internal restraint to shrinkage of the concrete by the bonded reinforcement or the external restraint to shrinkage by the supports of a structural member; restraint of deformation caused by temperature changes, including heat of hydration; restraint to load-independent movement caused by formwork or more permanent cladding and settlement of supports.

14.3 ANCHORAGE OF DEFORMED BARS

14.3.1 Introductory remarks

When a non-prestressed reinforcement bar is required for strength, it is assumed that the stress in the bar at the critical section can not only reach the yield stress, but can be sustained at this level as deformation increases. If the yield stress is to be reached at a particular cross-section, the reinforcing bar must be *anchored* on either side of the critical section. Stress development can be obtained by embedment of the steel in concrete so that stress is transferred past the critical section by bond, or by some form of mechanical anchorage. Codes of practice specify a minimum length, called the *development length* or *anchorage length*, over which a straight bar in tension must be embedded in the concrete in order to develop its yield stress. The provision of anchorage lengths in excess of the specified development length for every bar at a critical section or peak stress location ensures that anchorage or bond failures do not occur before the design strength at the critical section is achieved.

At an anchorage of a deformed bar, the deformations bear on the surrounding concrete and the bearing forces F_b are inclined at an angle β to the bar axis as shown in Figure 14.10a [8]. The perpendicular components of the bearing forces exert a radial force on the surrounding concrete. Tepfers [9,10] described the concrete in the vicinity around the bar as acting like a thick-walled pipe as shown in Figure 14.10b and the radial forces exerted by the bar cause tensile stresses that may lead to splitting cracks radiating from the bar if the tensile strength of the concrete is exceeded. Bond failure may be initiated by these splitting cracks within the anchorage length of an anchored bar (Figures 14.10c and d) or within a lapped tension splice (Figure 14.10e).

Transverse reinforcement across the splitting planes (A_{st} in Figures 14.10c and e) delays the propagation of splitting cracks and improves bond strength. Compressive pressure transverse to the plane of splitting delays the onset of cracking in the anchorage region thereby improving bond strength.

For a reinforcing bar of diameter ϕ, if the design value of the bond stress is f_{bd}, the bond force that can develop over the required anchorage length $l_{b,rqd}$ is $\pi \phi l_{b,rqd} f_{bd}$ and this force must not be less than the design force in the bar $\sigma_{sd} A_s = \sigma_{sd} \pi \phi^2 / 4$. That is $\pi \phi l_{b,rqd} f_{bd} \geq \sigma_{sd} \pi \phi^2 / 4$ and therefore:

$$l_{b,rqd} \geq \frac{\phi}{4} \frac{\sigma_{sd}}{f_{bd}} \tag{14.4}$$

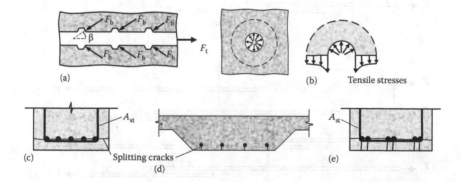

Figure 14.10 Splitting failures around developing bars. (a) Forces exerted on concrete by a deformed bar in tension [8]. (b) Tensile stresses in concrete [9]. (c) Horizontal splitting due to insufficient bar spacing. (d) Vertical splitting due to insufficient cover. (e) Splitting (bond) failure at a lapped splice.

The design value of the average ultimate bond stress f_{bd} in Equation 14.4 is proportional to the tensile strength of concrete and, in EN 1992-1-1, is specified as:

$$f_{bd} = 2.25\eta_1\eta_2 f_{ctd} \tag{14.5}$$

where f_{ctd} is design value of the concrete tensile strength (given by Equation 4.12), except that $f_{ctk,0.05}$ should not be taken higher than the value specified for C60/75 concrete (i.e. 3.1 MPa) because of the brittle nature of high-strength concretes and because of the limited experimental data available for the anchorage of deformed bars in high-strength concrete. The coefficient η_1 in Equation 14.5 depends on the position of the bar during concreting and consequently the quality of the bond: $\eta_1 = 1.0$ when the bond quality is 'good'; and $\eta_1 = 0.7$ in all other cases and for bars in elements built with slip forms. Bond quality is deemed to be 'good' for a horizontal bar within 250 mm from the soffit of the member and, for members greater than 600 mm in depth, any horizontal bar located outside the top 300 mm of concrete. The coefficient η_2 in Equation 14.5 depends on the bar diameter: $\eta_2 = 1.0$ when $\phi \leq 32$ mm and $\eta_2 = (132 - \phi)/100$ when $\phi > 32$ mm.

Reinforcing bars may be spliced together by welding or by a mechanical anchorage or by overlapping the bars by a specified length l_0 as shown in Figure 14.11. In this latter anchorage, known as a lapped splice, each bar must be able to develop the yield stress within the lap length, and the design force in the bar on either side of the splice ($A_s\sigma_{sd}$) must be safely carried

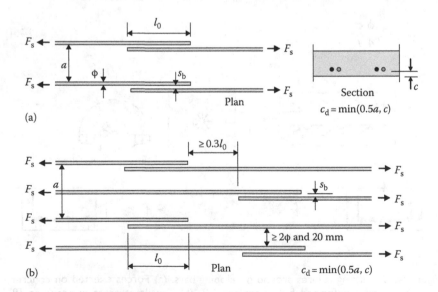

Figure 14.11 Contact and non-contact lapped splices. (a) 100% of bars spliced at the same location. (b) 50% of bars spliced at the same location (staggered splices).

across the splice without bond failure. Both contact splices ($s_b = 0$) and non-contact lapped splices ($s_b > 0$) are frequently used.

The mechanism of bond transfer at a lapped splice is different from that at a single anchored bar with no adjacent bar developing stress in close proximity, so in general where the bars at a lapped splice are required to develop the yield stress, the specified lap length is greater than the development length.

14.3.2 Design anchorage length

According to EN 1992-1-1 [1], the design anchorage length for anchoring the force $F_s = \sigma_{sd} A_s$ in a bar assuming a constant bond stress f_{bd} may be calculated from:

$$l_{bd} = \alpha_1 \alpha_2 \alpha_3 \alpha_4 \alpha_5 l_{b,rqd} \geq l_{b,min} \tag{14.6}$$

The coefficients α_1, α_2, α_3, α_4 and α_5 are given in Table 14.1 and the basic anchorage length $l_{b,rqd}$ is given by Equation 14.4. For straight anchorages and for the hooked, cogged or mechanically welded bars shown in Figure 14.12 when $c_d \leq 3\phi$, $\alpha_1 = 1.0$. For the anchorages shown in Figure 14.12, $\alpha_1 = 0.7$ when $c_d > 3\phi$. The coefficient α_2 depends on the concrete minimum cover c_d as given in Figure 14.13.

The beneficial effects of confinement by transverse steel within the anchorage length are accounted for by α_3 which is determined as $1.0 - K\lambda$ (but $0.7 \leq \alpha_3 \leq 1.0$). The term λ is given by $\lambda = (\Sigma A_{st} - \Sigma A_{st.min})/A_s$, where ΣA_{st}

Table 14.1 Anchorage coefficients $\alpha_1, \alpha_2, \alpha_3, \alpha_4$ and α_5

	Reinforcement bar	
Type of anchorage	In tension	In compression
Straight	$\alpha_1 = 1.0$	$\alpha_1 = 1.0$
Other than straight (see Figure 14.12b through d)	$\alpha_1 = 0.7$ if $c_d > 3\phi$ otherwise, $\alpha_1 = 1.0$	$\alpha_1 = 1.0$
Straight	$\alpha_2 = 1 - 0.15(c_d - \phi)/\phi$ but $0.7 \leq \alpha_2 \leq 1.0$	$\alpha_2 = 1.0$
Other than straight (see Figure 14.12b through d)	$\alpha_2 = 1 - 0.15(c_d - 3\phi)/\phi$ but $0.7 \leq \alpha_2 \leq 1.0$	$\alpha_2 = 1.0$
All types	$\alpha_3 = 1 - K\lambda$ but $0.7 \leq \alpha_3 \leq 1.0$	$\alpha_3 = 1.0$
All types (see Figure 14.12d)	$\alpha_4 = 0.7$	$\alpha_4 = 0.7$
All types	$\alpha_5 = 1.0 - 0.04p$ but $0.7 \leq \alpha_5 \leq 1.0$	—

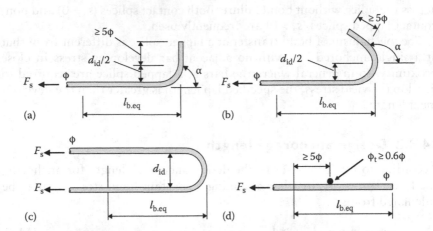

Figure 14.12 Bent, hooked, looped or welded anchorages [1]. (a) Equivalent anchorage length for standard bend ($90° \leq \alpha < 150°$). (b) Equivalent anchorage length for standard hook ($\alpha \geq 150°$). (c) Equivalent anchorage length for standard loop. (d) Equivalent anchorage length for welded transverse bar. *Notes:* $d_{id} \geq 4\phi$ when $\phi \leq 16$ mm; $d_{id} \geq 7\phi$ when $\phi > 16$ mm.

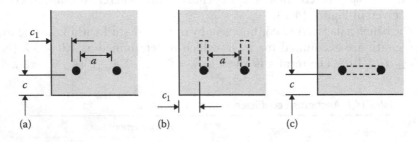

Figure 14.13 Concrete confinement dimension, c_d [1]. (a) Straight bars, $c_d = \min(a/2, c_1, c)$. (b) Bent or hooked bars, $c_d = \min(a/2, c_1)$. (c) Looped bars, $c_d = c$.

is the cross-sectional area of the transverse reinforcement along the design anchorage length l_{bd}; $\Sigma A_{st.min}$ is the cross-sectional area of the minimum transverse reinforcement, which may be taken as $0.25A_s$ for beams and 0 for slabs; A_s is the cross-sectional area of a single bar of diameter ϕ being anchored; and K is a factor that accounts for the position of the bars being anchored with respect to the transverse reinforcement, with values given in Figure 14.14.

The coefficient α_4 accounts for the influence of one or more transverse bars of diameter ϕ_t ($\geq 0.6\phi$) welded to the bar being anchored and $\alpha_5 = 1.0 - 0.04p$ (but $0.7 \leq \alpha_5 \leq 1.0$), where p is transverse compressive pressure (in MPa) at the ultimate limit state along l_{bd} perpendicular to the plane of splitting. The average design ultimate bond stress reduces when transverse

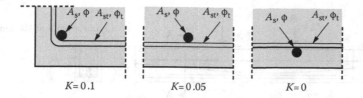

Figure 14.14 Values of K for anchored bars in beams and slabs [1].

tensile stress exists at the anchorage (i.e. when p is negative). Although this is not accounted for in EN 1992-1-1 [1], it is recommended here that, if transverse tensile stress exists, it should be considered in the determination of the development length by the inclusion of α_5 (= $1.0 - 0.04p$) greater than 1.0 in Equation 14.6.

The product $(\alpha_2\alpha_3\alpha_5)$ must not be less than 0.7.

The minimum anchorage length in Equation 14.6 is specified as

$l_{b,min} > $ max $\{0.3l_{b,rqd}, 10\phi, 100$ mm$\}$ for anchorages in tension
$l_{b,min} > $ max $\{0.6l_{b,rqd}, 10\phi, 100$ mm$\}$ for anchorages in compression

As a simple alternative to Equation 14.6 for the anchorages shown in Figure 14.12, the equivalent anchorage lengths $l_{b,eq}$ may be taken as $\alpha_1 l_{b,reqd}$ for the bent, hooked or looped anchorages shown in Figure 14.12a, b and c, and the equivalent anchorage lengths may be taken as $\alpha_4 l_{b,reqd}$ for the welded anchorage of Figure 14.12d.

The anchorage of a bar in compression is provided by end bearing of the bar, as well as bond between the concrete and the steel bar along the development length. However, bends and hooks do not contribute to compression anchorages.

EXAMPLE 14.1

Calculate the anchorage length required for the two terminated 28 mm diameter bars centrally located in the bottom of the beam shown in Figure 14.15. Take f_{yk} = 500 MPa; f_{ck} = 35 MPa, cover to the 28 mm bars c = 40 mm and the clear spacing between the bottom bars is a = 60 mm. The beam contains 12 mm diameter single stirrups at 150 mm centres throughout. The cross-sectional area of one 28 mm diameter bar is A_s = 616 mm² and one 12 mm bar is A_{st} = 113 mm².

It is assumed that the two terminated bars are required to develop the design yield stress at a distance d from the point of maximum moment in

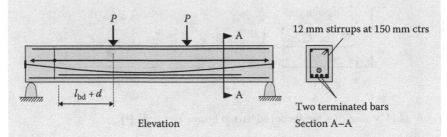

Figure 14.15 Development length of 28 mm diameter bottom bars.

the direction of increasing shear, i.e. $\sigma_{sd} = f_{yd} = 435$ MPa (tension), and may therefore be terminated at a distance $(l_{bd} + d)$ from the point load P as shown in Figure 14.15.

For these bottom bars in tension, Equation 14.5 gives:

$$f_{bd} = 2.25\eta_1\eta_2 f_{ctd} = 2.25 \times 1.0 \times 1.0 \times 1.47 = 3.30 \text{ MPa}$$

and from Equation 14.4:

$$l_{b,rqd} \geq \frac{28}{4}\frac{435}{3.30} = 923 \text{ mm}$$

From Table 14.1, for these straight 28 mm diameter bars:

$$\alpha_1 = 1.0$$

With the concrete confinement dimension $c_d = a/2 = 30$ mm (see Figure 14.13a), we have:

$$\alpha_2 = 1.0 - 0.15(30 - 28)/28 = 0.99$$

The minimum number of stirrups that can be located within length $l_{b,rqd}$ is 7. Therefore:

$$\Sigma A_{st} = 7 \times 113 = 791 \text{ mm}^2$$

Taking $\Sigma A_{st.min} = 0.25 A_s = 154$ mm^2, the parameter λ is:

$$\lambda = (791 - 154)/616 = 1.03$$

From Figure 14.14, $K = 0.05$ (for each of the two interior bars) and therefore:

$$\alpha_3 = 1.0 - 0.05 \times 1.03 = 0.95$$

It is assumed that there are no welded cross-bars within the anchorage length and that in this location the transverse pressure perpendicular to the anchored bar (p) is zero. Hence, $\alpha_4 = 1.0$ and $\alpha_5 = 1.0$.

From Equation 14.6:

$$l_{bd} = 1.0 \times 0.99 \times 0.95 \times 1.0 \times 1.0 \times 923 = 868 \text{ mm}$$

Of course, the strength of the beam must be checked at the point where the two bars are terminated (at $l_{bd} + d$ from the constant peak moment region, as shown in Figure 14.15).

14.3.3 Lapped splices

Because of the relatively poor anchorage conditions that prevail at a lapped splice (at least two adjacent bars are anchored in close proximity), the lap length l_0 specified in EN 1992-1-1 [1] is larger than the anchorage length (l_{bd}) for a single bar. Laps should normally be staggered and should not be placed in critical regions unless absolutely necessary (e.g. not at plastic hinge locations). At the ends of the splice where the bars terminate in a tension zone, a discontinuity exists and transverse cracks are usually initiated. These cracks may trigger the splitting cracks shown in Figure 14.10e. Lapped splices should therefore be staggered where possible so that no free ends line up at the same section, unless the bars are further apart than 12ϕ [2]. In addition, laps at any section should be arranged symmetrically. Notwithstanding, EN 1992-1-1 [1] does permit up to 100% of bars in tension to be lapped at the same location provided the bars are all in one layer and provided the requirements outlined in the following paragraph are satisfied. Where the bars are in several layers, the maximum percentage should be reduced to 50%. All secondary (or distribution) reinforcement and all bars in compression may be lapped at one section.

Referring to Figure 14.11, according to EN 1992-1-1 [1], the clear distance between lapped bars s_b should not exceed 4ϕ or 50 mm, whichever is the smaller, except that if s_b is larger than 4ϕ or 50 mm, the lap length should be increased to $l_0 + s_b$, where l_0 is given by Equation 14.7. In addition, EN 1992-1-1 [1] requires that the longitudinal distance between two adjacent laps should not be less than $0.3l_0$ and the clear distance between adjacent bars should not be less than 2ϕ or 20 mm.

The design lap length l_0 is:

$$l_0 = \alpha_1 \alpha_2 \alpha_3 \alpha_5 \alpha_6 l_{b,rqd} \geq l_{0,min} \tag{14.7}$$

Table 14.2 Lap length coefficient α_6

Percentage of lapped bars ρ_1	<25%	33%	50%	>50%
α_6	1.0	1.15	1.4	1.5

where the basic anchorage length $l_{b,rqd}$ is given by Equation 14.4 and:

$$l_{0,min} = \max\{0.3\alpha_6 l_{b,rqd}, 15\phi, 200 \text{ mm}\} \tag{14.8}$$

The coefficients $\alpha_1, \alpha_2, \alpha_3$ and α_5 are given in Table 14.1, but for the calculation of α_3, $\Sigma A_{st.min}$ is taken as $A_s(\sigma_{sd}/f_{yd})$, where A_s is the area of one lapped bar. The coefficient α_6 depends on ρ_1 the percentage of bars lapped within $0.65l_0$ from the centre of the lap length under consideration and is given by $\alpha_6 = (\rho_1/25)^{0.5}$, but not exceeding 1.5 and not less than 1.0. Values of α_6 are given in Table 14.2.

When the diameter of the lapped bars is greater than or equal to 20 mm, the total area of transverse reinforcement required to carry the transverse tension forces that develop in the lap zone (ΣA_{st}) should not be less than the area of one lapped bar (i.e. $\Sigma A_{st} \geq A_s$). The transverse reinforcement should be placed perpendicular to the direction of the lapped bars and between the lapped bars and the concrete surface. The transverse reinforcement should be located towards the ends of the lap as shown in Figure 14.16. When more than 50% of the longitudinal reinforcement is lapped at one point and the distance between laps is less than 10ϕ, the transverse reinforcement should be in the form of links (stirrups) or U bars anchored into the body (the web) of the cross-section.

When the diameter of the lapped bars is less than 20 mm, or the percentage of lapped bars at a section is less than 25%, no additional transverse reinforcement is required other than that which may be present for other reasons, such as distribution reinforcement, shear reinforcement and so on.

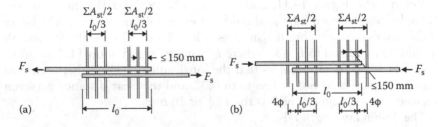

Figure 14.16 Location of transverse reinforcement at lapped splices. (a) Bars in tension. (b) Bars in compression.

14.4 STRESS DEVELOPMENT AND COUPLING OF TENDONS

A discussion of the anchorage of the tendons in a pretensioned member was presented in Section 8.2 and the minimum anchorage length specified in EN 1992-1-1 [1] of a pretensioned tendon from its end to the critical cross-section where the design stress is required was given in Equation 8.5.

When tendons are coupled, the coupler should be capable of developing the characteristic breaking force of the tendon and should be positioned so that no more than 50% of the tendons are coupled at any one cross-section. In general, couplers should not be located near intermediate supports. For post-tensioned tendons, where bonded construction is required couplers should be enclosed in grout-tight housings to facilitate grouting the duct (see Figure 3.11).

14.5 DETAILING OF BEAMS

14.5.1 Anchorage of longitudinal reinforcement: General

As mentioned earlier, a prerequisite for good detailing is favourable bond and anchorage conditions for each reinforcing bar and each tendon. The stress conditions surrounding a bar anchorage have considerable effect on the quality of bond. Where possible, bars should be anchored in regions where compressive stresses act in a transverse or normal direction to the bar. Bond strength increases considerably when normal pressure is present [11]. This increase is more pronounced for larger diameter bars. In Figure 14.17, it can be seen that the anchorage conditions for the bottom reinforcement at the support are more favourable than for the top reinforcement and consequently a shorter anchorage length is required.

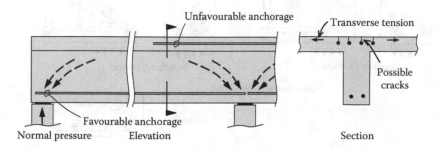

Figure 14.17 Anchorage of longitudinal reinforcement in a continuous beam.

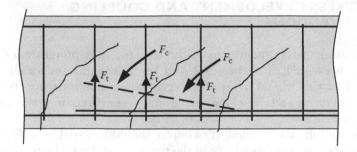

Figure 14.18 Anchorage of terminating bottom reinforcement.

When bottom reinforcement is terminated away from the support, the diagonal compression in the web improves the anchorage, provided of course that there is sufficient web reinforcement to carry the diagonal force back to the top of the beam. The anchorage conditions of the terminating bars may be further improved by bending them into the web, as shown in Figure 14.18. This also reduces the possibility of premature shear failure at the discontinuity caused by the terminating flexural reinforcement.

The transverse tension that may cause splitting in the plane of a hooked anchorage (as illustrated in Figure 14.6) can be overcome at a beam support simply by tilting the hook (or better still, laying the hook in a near horizontal plane) to expose it to the normal reaction pressure at the support, as shown in Figure 14.19.

If the bearing length at a support is small and close to the free end of a member, a *sliding shear* failure may occur along a steep inclined crack, as shown in Figure 14.20. In such a case, additional small diameter bars may be required at right angles to the potential failure plane to provide a

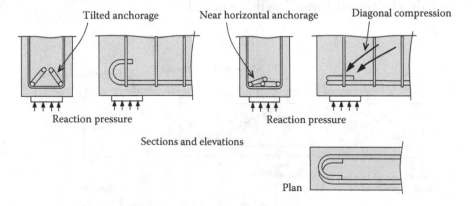

Figure 14.19 Hooked anchorages – preferred positions.

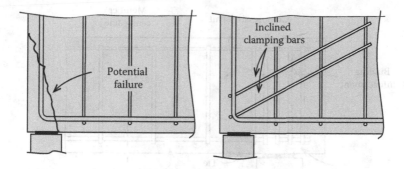

Figure 14.20 Detail when support length is short.

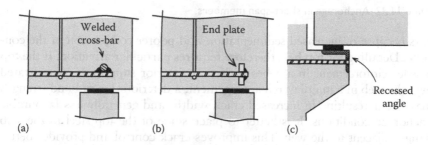

(a)   (b)   (c)

Figure 14.21 Mechanical anchorages. (a) Welded cross-bar. (b) Welded end plate (headed bar). (c) Bar welded to recessed angle.

clamping action between the crack surfaces and thereby prevent sliding and failure. These additional bars must be fully developed on both sides of the failure surface. In some instances, where the length available for anchorage is small, cross-bars may be welded to the terminating bar to ensure that it can develop its required strength. Such mechanical anchorages are commonly used in precast elements and in regions of high concentrated loads such as corbels, brackets and other support points. Typical examples are illustrated in Figure 14.21.

In short-span members, where load is carried to the support by arch action, it is essential that all bottom reinforcing bars (the tie of the arch) are fully developed at each support. To avoid bond failure in situations where the development length of each bottom bar is restricted, closely spaced transverse reinforcement in the form of stirrups can be used to bind the surfaces of a potential horizontal splitting crack in the plane of the reinforcement and vertical splitting cracks developing from the anchored bar through the cover concrete (see Figure 14.22).

Apart from the reasons discussed earlier and illustrated in Figure 14.17, the anchorage conditions for top bars are always less favourable than for bottom

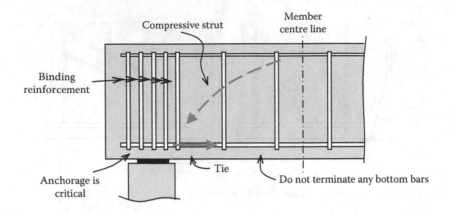

Figure 14.22 Anchorage in short-span members.

bars because of increased sedimentation and poorer compaction of the concrete. Detailing for top bars therefore requires particular attention. If the top tensile reinforcement in a T-beam over an interior support is concentrated over the web in a multilayered arrangement, a deterioration of bond strength may occur resulting in increased crack widths and generally less favourable anchorage conditions. It is better to place some of the top steel in the slab flange adjacent to the web. This improves crack control and provides better access to concrete vibrators within the beam web. The measured crack widths in two beams tested by Leonhardt et al. [7] are compared in Figure 14.23. EN 1992-1-1 [1] requires that, at interior supports, the total area of tensile reinforcement in the top of the cross-section be spread over the effective width of the flange, with part of the tensile steel concentrated over the web width and inside the stirrup cage in order to develop an efficient truss action.

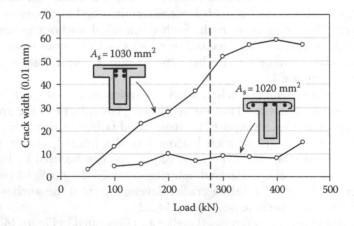

Figure 14.23 Comparison of crack widths in T-beams [7].

In monolithic construction, a beam in the region of an exterior support will be subject to an end moment even if a simple support has been assumed in design. This is as a result of the stiffness of the support and the inevitable partial fixity that exists at the connection. EN 1992-1-1 [1] requires that the section at the support be designed for a negative bending moment of magnitude equal to at least 15% of the maximum bending moment in the span.

14.5.2 Maximum and minimum requirements for longitudinal steel

In Section 5.12.1, the minimum amount of reinforcement required for crack control in each part of a cross-section (according to EN 1992-1-1) was given in Equation 5.192. In addition, EN 1992-1-1 [1] requires that the area of longitudinal reinforcement in a beam must be greater than or equal to $A_{s,min}$ given by:

$$A_{s,min} = 0.26 \frac{f_{ctm}}{f_{yk}} b_t d \tag{14.9}$$

but not less than $0.0013 b_t d$, where b_t is the mean width of the tensile zone and f_{ctm} is the mean tensile strength of the concrete specified in Table 4.2.

The area of tensile or compressive reinforcement should not exceed $A_{s,max} = 0.04 A_c$ outside lap locations.

For post-tensioned members with permanently unbonded tendons or external prestressing cables, the design bending resistance should be greater than 1.15 times the cracking moment.

If reinforcing bars in concrete structures are too closely spaced, the concrete may not be able to be placed and compacted satisfactorily and this may compromise the bond between the bars and the concrete and the appearance of the finished concrete. If the bars are spaced too widely apart, cracks may become unsightly and cause serviceability problems. EN 1992-1-1 [1] requires that the clear distance between individual parallel bars (both horizontally and vertically) should be not less than the maximum of the bar diameter ϕ, $(d_g + 5 \text{ mm})$ and 20 mm, where d_g is the maximum size of the aggregate. Lapped bars, however, are permitted to be in contact with each other within the lap.

Where bars in a beam are placed in separate horizontal layers, the bars in each layer should be located directly above or below the bars in the adjacent layers. Sufficient space is required between the resulting columns of bars to permit access for concrete vibrators so that the concrete can be well compacted around all bars.

For crack control in reinforced concrete, it is good practice to ensure that the spacing between the bonded reinforcement bars does not exceed about 300 mm. In the maximum moment regions of slabs, EN 1992-1-1 [1] states that the spacing of the principal reinforcement should not exceed 2h or 250 mm, whichever is smaller and the spacing of the secondary reinforcement should not exceed 3h or 400 mm, whichever is smaller. In one-way slabs, the area of secondary (or distribution) reinforcement placed at right angles to the principal reinforcement should not be less than 20% of the area of the principal reinforcement.

14.5.3 Curtailment of longitudinal reinforcement

The longitudinal reinforcement required in a beam at a particular cross-section must be designed to carry the tensile force arising from the maximum design bending moment at that section ($F_{td} = M_{Ed}/z$), plus the additional force $\Delta F_{td} = V(\cot \theta_v - \cot \alpha)$ arising from the inclined cracks (as discussed in Section 7.2.5 and illustrated in Figure 7.7). If the member is also subject to a design axial tensile force N_{Ed} this should also be included in the determination of F_{td}. This force ($F_s = M_{Ed}/z + N_{Ed} + \Delta F_{td}$) may be estimated by *shifting* the moment envelope by a distance a_1 given by [1]:

$$a_1 = z(\cot \theta_v - \cot \alpha)/2 \tag{14.10}$$

where the symbols have been defined in Figure 7.7. These forces are plotted on the beam axis in Figure 14.24.

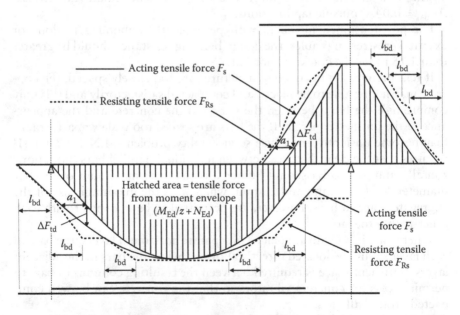

Figure 14.24 Shift rule for the determination of tensile forces in longitudinal steel and curtailment of reinforcement [1].

Figure 14.24 illustrates the so-called shift rule and also shows the appropriate cut-off points for the longitudinal bars. Within the anchorage lengths at each end of a terminating bar, a linear variation of force may be assumed within the bar varying from zero at the end of the bar to σ_{sd} at l_{bd} from the end of the bar.

14.5.4 Anchorage of stirrups

The flow of internal forces in a beam can be idealised as a parallel chord truss. This *truss analogy* was discussed in Section 7.2.3 and is illustrated in Figure 14.25. The compressive top chord and the diagonal web strut are the concrete portions of the truss, while the tensile bottom chord and vertical web ties must of course be steel reinforcement. The diagonal compression (in the concrete web strut) can only be resisted at the bottom of the beam at the intersection of the horizontal and vertical reinforcement, i.e. at the pin joints of the analogous truss. It is evident that the tension in the vertical tie is constant over its entire height (i.e. from the pin joint at the bottom chord to the pin joint at the top chord). Therefore, adequate anchorage of the stirrups must be provided at every point along the vertical leg of the stirrup. After all, when calculating the shear strength provided by the stirrups, it is assumed that every vertical stirrup leg crossed by an inclined crack is at yield, irrespective of whether the inclined crack crosses the stirrup at its mid-depth or close to its top or bottom.

EN 1992-1-1 [1] requires that the shear reinforcement should form an angle α of between 45° and 90° to the longitudinal axis of the member. The anchorage of the vertical or inclined legs of a stirrup may be achieved by a standard hook or cog (see Figure 14.12) or by welding of the fitment to the longitudinal bar or by a welded splice. The requirements for stirrup anchorages in EN 1992-1-1 [1] are given in Figure 14.26.

Stirrups hooks and cogs should be located in the compression zone where anchorage conditions are most favourable. At overloads, when diagonal cracks may have formed, the compression zone may be relatively small. Stirrup hooks should therefore be as close to the compression edge as cover

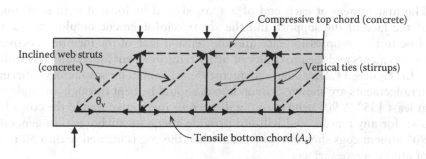

Figure 14.25 The truss analogy.

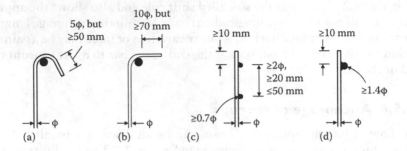

Figure 14.26 Permissible stirrup anchorages [1]. (a) Hooked anchorage. (b) Cogged anchorage. (c) Welded anchorage-two cross-bars. (d) Welded anchorage-single cross-bar.

requirements allow. Stirrups depend on this transverse pressure for anchorage of the hooks or cogs. It is common practice to locate the stirrup hooks near the top surface of a beam even in negative moment regions. When the top surface is in tension, the discontinuity created by a stirrup and its anchorage may act as a crack initiator. A primary crack therefore frequently occurs in the plane of the stirrup hook and anchorage is lost. As a consequence, in these regions, the beam may possess less than its required shear strength.

It is good practice to show the location of the stirrup hooks on the structural drawings and not to locate the hooks in regions where transverse cracking might compromise the anchorage of the stirrup. Stirrup hooks should always be located around a larger diameter longitudinal bar that disperses the concentrated force at the anchorage and reduces the likelihood of splitting in the plane of the anchorage. Longitudinal bars are in fact required in each corner of a closed stirrup to distribute the concentrated force applied to the concrete at the corner. It is essential that the stirrup and stirrup hook fit snugly and are in contact with the longitudinal bars in each corner of the stirrup.

The shear reinforcement calculated as being necessary at any cross-section should be provided for a distance h from that cross-section in the direction of decreasing shear, where h is the depth of the cross-section. The first fitment at each end of a span should be located within 50 mm of the face of the support and the shear reinforcement should extend as close to the compression face and the tension face of the member as cover requirements and the proximity of other reinforcement and tendons permit.

In Figure 14.27, some satisfactory and some unsatisfactory stirrup arrangements are shown. Stirrup hooks should be bent through an angle of at least 135°. A 90° bend (a cog) will become ineffective should the cover be lost, for any reason, and will not provide adequate anchorage. In general, 90° fitment cogs should not be used when the cog is located within 50 mm of any concrete surface.

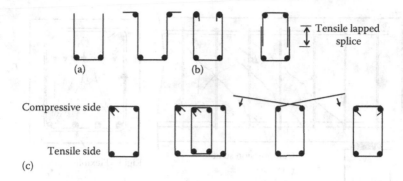

Figure 14.27 Stirrup shapes. (a) Incorrect. (b) Satisfactory in some situations. (c) Desirable.

In addition to carrying diagonal tension produced by shear, and controlling inclined web cracks, closed stirrups also provide increased ductility by confining the compressive concrete. The *open stirrups* shown in Figure 14.27b are commonly used, particularly in post-tensioned beams where the opening at the top of the stirrup facilitates the placement and positioning of the post-tensioning duct along the member. This form of stirrup does not provide confinement for the concrete in the compression zone and is undesirable in heavily reinforced beams where confinement of the compressive concrete may be required to improve the ductility of the member. It is good practice to use adequately anchored closed stirrups (Figure 14.27c) even in areas of low shear, particularly when the longitudinal tensile steel quantities are relatively high and cross-section ductility is an issue.

EN 1992-1-1 [1] permits lapped joints on the vertical legs of a stirrup near the surface of the web of a beam, such as shown in Figure 14.27b, provided the stirrup is not required to carry torsion. When a beam is subject to torsion, diagonal cracks exist on each face and the cracks spiral around the beam. Open stirrups of the type shown in Figure 14.27b are unsuitable. There is no point at which all stirrups can be effectively anchored since the spiral cracking may occur in the plane of a stirrup hook. It is likely therefore that a number of stirrup anchorages are lost as the ultimate load is approached if conventional hooks are used. This can be accounted for in design by closing up the stirrup spacing somewhat, thereby allowing for some lost anchorages. Ideally closed stirrups with welded anchorages should be used where torsion is significant.

In regions of high shear, it is desirable to use multi-leg stirrups when more than two longitudinal tensile bars are used. Park and Paulay [2] suggest that a *truss joint* should be formed at each longitudinal bar, i.e. the number of vertical stirrup legs should ideally equal the number of longitudinal bars and tendons. This is often not practical, but multi-leg stirrups

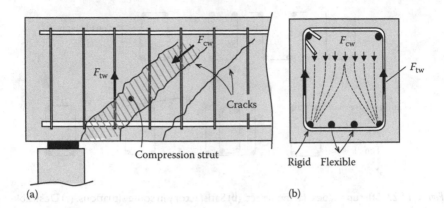

Figure 14.28 Undesirable distribution of diagonal compression due to wide stirrups [2]. (a) Elevation. (b) Cross-section.

should be used in members with wide webs to avoid the undesirable distribution of diagonal compression shown in Figure 14.28, where the untied interior longitudinal bars with no nearby vertical stirrup leg cannot effectively resist diagonal compression and are therefore relatively inefficient in receiving bond forces.

Multi-leg stirrups are also far better for controlling the longitudinal splitting cracks (known as 'dowel cracks') that precipitate bond failure of the longitudinal bars in the shear span and are illustrated in Figure 14.29. The formation and propagation of this dowel crack usually triggers the sudden catastrophic shear failure that may occur in a heavily loaded shear span in a prestressed or reinforced concrete beam.

Closely spaced inclined stirrups (although in some instances not practical) are the most efficient form of shear reinforcement both in terms of strength and crack control. Vertical stirrups also perform well. Bent-up longitudinal bars

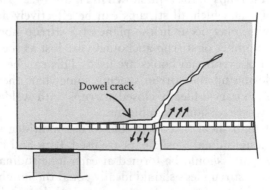

Figure 14.29 Shear failure triggered by bond failure of longitudinal bars.

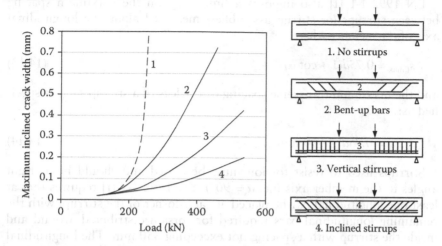

Figure 14.30 Crack control provided by various types of transverse reinforcement [2].

(once commonplace, now rarely used) are relatively inefficient. Figure 14.30 shows the effect of various stirrup types on the control of inclined cracks in the web of beams measured by Leonhardt et al. [7].

Wherever longitudinal bars are terminated in a tension zone, primary cracks are likely to occur at the discontinuity. These cracks tend to be wider than adjacent primary cracks and, if they become inclined due to the presence of shear, may lead to premature shear failure (probably due to a reduction in aggregate interlock). In the vicinity of terminating tensile reinforcement in a beam in bending and shear, it is good practice not to terminate more than 25% of the tensile reinforcement at any location. It is also good practice to ensure that the stirrup spacing determined as necessary at a termination point be reduced by at least 25% for a distance equal to the overall depth of the cross-section (h) along the terminated bar.

The shear reinforcement ratio ρ_w is defined in EN 1992-1-1 [1] as:

$$\rho_w = \frac{A_{sw}}{(sb_w\sin\alpha)} \geq \rho_{w,min} \tag{14.11}$$

where A_{sw} is the area of shear reinforcement at each stirrup location (i.e. the sum of the areas of each leg of the stirrup assembly), s is the spacing along the longitudinal axis of the member between each stirrup assembly, b_w is the width of the web of the cross-section, α is the angle between the shear reinforcement and the longitudinal axis and $\rho_{w,min}$ is the minimum shear reinforcement ratio given by:

$$\rho_{w,min} = \frac{\left(0.08\sqrt{f_{ck}}\right)}{f_{yk}} \tag{14.12}$$

EN 1992-1-1 [1] also imposes a limit $s_{l,max}$ on the maximum spacing between stirrups (or stirrup assemblies) measured along the longitudinal axis of the member where:

$$s_{l,max} = 0.75d(1 + \cot \alpha) \tag{14.13}$$

and the maximum transverse spacing of the legs of a stirrup $s_{t,max}$ is specified as:

$$s_{t,max} = 0.75d \le 600 \text{ mm} \tag{14.14}$$

Stirrups used to resist torsion must be closed and should be at right angles to the member axis (i.e. $\alpha = 90°$). EN 1992-1-1 [1] requires that at least one longitudinal bar be placed at each corner of the stirrup, with the remaining longitudinal bars required for torsion distributed around and inside the stirrup with a spacing not exceeding 350 mm. The longitudinal spacing of the torsion stirrups should not exceed the smaller of $u/8$ (where u is the circumference of the cross-section), $s_{l,max}$ (Equation 14.13) and the smallest dimension of the cross-section (b_w or h).

14.5.5 Detailing of support and loading points

When the support is at the soffit of a beam or slab, as shown in Figure 14.31a, the diagonal compression passes directly into the support as shown. However, when the support is at the top of the beam, as shown in Figure 14.31b, the diagonal compression must be carried back up to the support via an internal tie as shown. It is essential that adequately anchored reinforcement be included to act as the tension tie and the reinforcement must pass into and be anchored within the support.

Consider the suspended slab supported from above by the upturned beam shown in Figure 14.32a. The horizontal component of the diagonal compression being delivered at the support of the slab must be resisted by

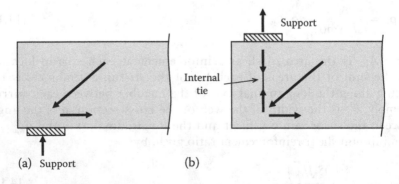

Figure 14.31 Support points. (a) Support at soffit. (b) Support at top of beam.

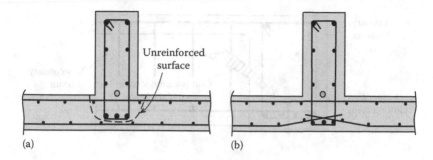

Figure 14.32 Suspended slab supported from above by upturned beam. (a) Incorrect detail. (b) Correct detail.

the bottom slab steel. The vertical component of the diagonal compression (i.e. the reaction from the slab) must be carried in tension up to the top of the upturned beam. This tension force must be carried across the unreinforced surface indicated in Figure 14.32a. The concrete on this surface may not be able to carry this tension and, if cracking occurs, premature and catastrophic failure could occur. The detail shown in Figure 14.32b overcomes the problem. The diagonal compression from the slab is now resisted by the stirrups in the upturned beam. No longer is there an unreinforced section of concrete required to carry tension. The vertical and horizontal members of the analogous truss have been effectively connected.

Consider the beam-to-beam connection shown in Figure 14.33. The reaction from the secondary beam F_{Ed} is delivered to the primary girder at the level of the bottom steel. This reaction should be carried by stirrups in the primary beam (*hanger* or *suspension reinforcement*) surrounding the principal reinforcement of the primary beam up to the top of the girder where it can be resolved into diagonal compression in a similar way to that of any other load applied to the top of the girder. EN 1992-1-1 [1] permits some of these stirrups to be located outside the volume of concrete that is common to both beams. The *suspension* reinforcement is additional to the transverse reinforcement required for shear in the primary girder. For the reasons discussed in the previous paragraph, the bottom reinforcement in the secondary beam should always pass over the bottom reinforcement in the primary girder. The reinforcement details of the primary girder together with its truss analogy are shown in Figures 14.33b and c.

The area of additional suspension reinforcement A_{sr} required to carry the factored reaction F_{Ed} may be obtained from:

$$A_{sr} = \frac{F_{Ed}}{f_{yd}} \qquad (14.15)$$

where f_{yd} is the design yield stress of the hanger reinforcement.

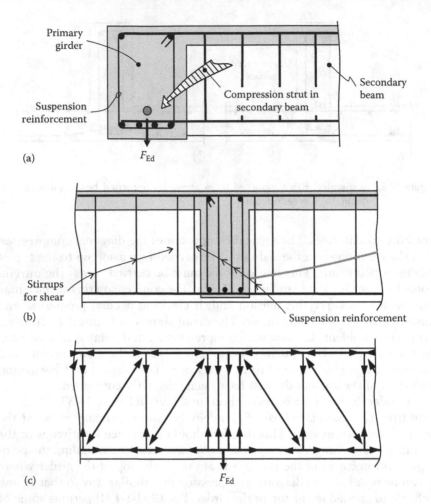

Figure 14.33 Beam-to-beam connection. (a) Section through primary girder. (b) Primary girder – elevation. (c) Primary girder – truss analogy.

When a load is applied to the underside of a concrete beam, some mechanical device must be used to transfer the *hanging* load to the top of the beam. Some typical devices are illustrated in Figure 14.34. When internal rods are used, plain round bars or bolts are suitable since bond is not required to transfer the load to the top of the girder.

To form an internal hinge, particularly in precast construction, a *half joint* or *dapped-end joint*, as shown in Figure 14.35a, is frequently used. At such a connection, careful detailing is essential. Only half the beam depth is available and the internal forces are generally relatively large. The two alternative strut-and-tie models shown in Figures 14.35b and c are

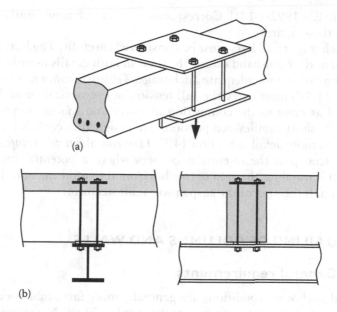

Figure 14.34 Mechanical means to support *hanging* loads. (a) External yoke. (b) Internal rods.

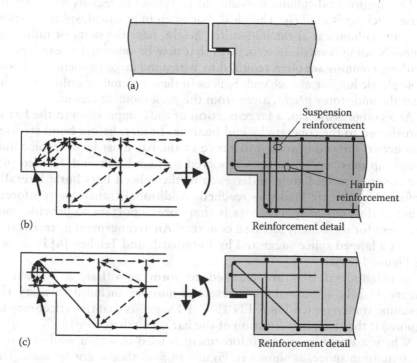

Figure 14.35 Half joint details. (a) Half joint. (b) Strut-and-tie model (No. 1). (c) Strut-and-tie model (No. 2).

outlined in EN 1992-1-1 [1]. Corresponding reinforcement details are also shown in these figures.

The anchorage of all bars must be considered carefully. The bottom reinforcement at the right-hand side of the joint in both details usually requires a hook or a 90° cog for adequate anchorage. The *suspension reinforcement* in Figure 14.35b must carry the full tension in the vertical tie and should be located as close to the connection as cover and spacing requirements permit. The short cantilevered portions are designed as corbels (which are discussed in more detail in Section 14.8). Horizontal *hairpin reinforcement* should extend past the re-entrant corners where a potential crack may develop. It is usual for the area of this horizontal reinforcement to be taken as at least half the area of the suspension reinforcement.

14.6 DETAILING OF COLUMNS AND WALLS

14.6.1 General requirements

Bond and anchorage conditions are generally more favourable in columns than in beams because transverse cracking is less likely. Nevertheless, several points need to be considered when detailing columns.

The longitudinal column bars should be spliced in regions where transverse cracking is unlikely. The ideal, but often impractical, splice location in many columns is at the mid-storey height, near the point of inflection where bending is small. In structures that may be subjected to earthquake loading, columns are often required to withstand large moments and possible plastic hinging at each end. Splices in these columns should always be near the mid-storey height, away from the peak moment region.

At a compressive lap, a large portion of the compression in the bar is transferred to the concrete by end bearing. In fact, before bond stresses can occur, the end bearing resistance of the bar must be overcome and some slip must occur. In tests reported by Leonhardt and Teichen [6], the concrete immediately under each of the spliced bars burst laterally before the ultimate load was reached. Additional transverse reinforcement at the ends of spliced bars is therefore important to provide confinement for the heavily stressed concrete. An arrangement of transverse ties at a lapped splice suggested by Leonhardt and Teichen [6] is shown in Figure 14.36a.

If longitudinal bars are cranked to form an offset, as shown in Figure 14.36b, additional transverse ties must be included to carry the resulting transverse tension F. EN 1992-1-1 suggests that this effect may be ignored if the change in direction of the bar is less than 1 in 12.

Where a single layer of reinforcement is used in a thin wall, a transverse tension splice, as shown in Figure 14.36c, should not be used. The internal couple resulting from the offset may lead to the cracking shown

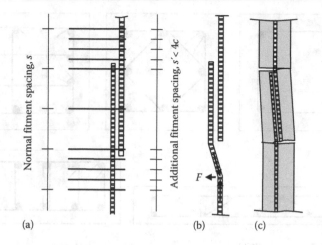

Figure 14.36 Lapped splices in columns and walls. (a) Additional fitments at compressive splice. (b) Tension at cranked bars. (c) Unsatisfactory tension splice in a thin wall.

and this could precipitate premature failure, particularly if the wall is subjected to lateral loading. Lapped splices in thin walls should be in the plane of the wall.

14.6.2 Transverse reinforcement in columns

Transverse reinforcement (or fitments) in the form of closed ties or helices are required in columns for three reasons:

1. To provide restraint to the heavily stressed longitudinal reinforcement and thereby prevent outward buckling before the full strength of the bar is reached.
2. To provide confinement to the concrete core and thereby improve both the strength and ductility of the column. This confinement occurs at the points where the ties change direction around the longitudinal bars and results from the tension induced in the ties as the concrete core dilates under axial compression.
3. To act as shear reinforcement when diagonal tension cracks are possible.

While helical reinforcement is often used in piles and circular columns, closed ties are the most common form of lateral reinforcement used in rectangular columns. Typical tie arrangements are shown in Figure 14.37. Each main longitudinal bar is tied in two directions so as to effectively prevent outward buckling.

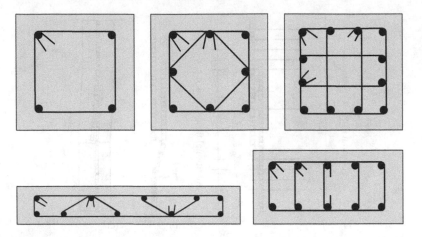

Figure 14.37 Typical tie (link) arrangements in rectangular columns.

Lateral restraint should be provided for:

1. each corner bar or bundle of bars in a cross-section;
2. every other bar when bars are spaced at centres exceeding 150 mm; and
3. at least every alternate bar, where bars are spaced at 150 mm or less.

For columns containing bundled bars, each bundle should be restrained.

Lateral restraint may be considered to be provided if the longitudinal reinforcement is placed within and in contact with a non-circular fitment and located:

1. at a bend in the fitment, where the bend has an included angle of 135° or less, e.g. position A in Figure 14.38;
2. between two 135° fitment hooks, e.g. position B in Figure 14.38; and
3. inside a single 135° hook at the end of a fitment that is approximately perpendicular to the column face, e.g. position C in Figure 14.38.

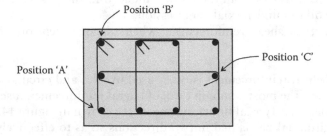

Figure 14.38 Lateral restraint to longitudinal bars.

Although used in some circumstances, 90° fitment cogs are not recommended here. In the case of circular fitments or helical reinforcement, the longitudinal reinforcement inside the helix may be deemed to be laterally restrained if the bars are equally spaced around the perimeter.

EN 1992-1-1 [1] specifies the minimum bar diameter for fitments and helical reinforcement tabulated below.

Longitudinal bar diameter (mm)	Minimum bar diameter for fitment and helix (mm)
Single bars up to 24	6
Single bars 24 to 32	8
Single bars 32 to 40	10

For bundles of bars, 10 mm is recommended as the minimum bar diameter for fitments and helical reinforcement.

The spacing of fitments, or the pitch of a helix, should not exceed $s_{cl,tmax}$, which may be taken as the smaller of 10ϕ, the lesser dimension of the column cross-section and 400 mm, where ϕ is the diameter of the longitudinal bar being restrained [1]. If the column has a circular cross-section, its diameter is taken as the lesser dimension of the column cross-section. In regions of a column within a distance equal to the larger column dimension above or below a beam or slab, the maximum spacing requirement should be reduced by a factor of 0.6. In the case of a wall, the larger column dimension here is taken to be 4 times the wall thickness. Similarly, at lapped splices of longitudinal bars of diameter greater than 14 mm in columns, the maximum spacing requirement should be reduced by a factor of 0.6, with a minimum of 3 bars evenly spaced in the lap length [1].

One fitment, or the first turn of a helix, should be located not more than 50 mm vertically above the top of a footing, or the top of a slab. Another fitment, or the final turn of a helix, should be located not more than 50 mm vertically below the soffit of a slab, except that in a column with a capital, the fitment or turn of the helical reinforcement shall be located at a level at which the area of the cross-section of the capital is not less than twice that of the column.

In situations where beams or brackets frame into a column from four directions and the column is restrained in all directions, the fitments or helical reinforcement may be terminated 50 mm below the highest soffit of the beams or brackets.

Where a splice is always in compression, the force in the longitudinal bar may be transmitted by end bearing. The mating ends of the bar must be square-cut and held in concentric contact by a sleeve. For such an end-bearing splice, an additional fitment should be placed above and below each sleeve. The bars should be rotated to achieve the maximum possible area of contact between the ends of the bars.

14.6.3 Longitudinal reinforcement in columns

EN 1992-1-1 [1] specifies that the minimum diameter of a longitudinal bar in a column is 8 mm and the minimum area of longitudinal reinforcement in a column should not be less than $A_{s,min}$, where:

$$A_{s,min} = \frac{0.10 N_{Ed}}{f_{yd}} \tag{14.16}$$

and N_{Ed} is the design axial compressive force. The maximum area of longitudinal reinforcement $A_{s,max}$ is $0.04\,A_c$ outside the lap locations and 0.08 A_c at laps. At least one longitudinal bar should be placed in each corner of a polygon-shaped cross-section, and a column of circular cross-section should have at least four, evenly spaced, longitudinal bars.

14.6.4 Requirements for walls

Walls are defined in EN 1992-1-1 [1] as members with a length to thickness ratio of 4 or more. The maximum and minimum requirements for vertical reinforcement are the same as the corresponding requirements for longitudinal reinforcement in columns given in the preceding section. Vertical reinforcement should be located at each face of the wall and the spacing of the vertical bars should not exceed 3 times the wall thickness or 400 mm, whichever is less. For walls subjected to lateral loads (out-of-plane bending), the maximum spacing requirements for crack control in slabs apply and the upper limit on bar spacing should be reduced to 300 mm.

Horizontal reinforcement is also required running parallel to the faces of the wall on each surface of area not less than 25% of the area of vertical reinforcement or $0.001 A_c$ whichever is greater. The spacing between horizontal bars should not exceed 400 mm.

Where the area of vertical reinforcement in any part of a wall exceeds $0.02 A_c$, transverse reinforcement in the form of ties (links) should be provided in accordance with the requirements for columns (see Section 14.6.2).

14.7 DETAILING OF BEAM–COLUMN CONNECTIONS

14.7.1 Introduction

Connections between structural members are often the weakest points in a structural system. Within connections, internal forces change direction abruptly and adequately anchored reinforcement must be inserted to carry all tension. Often the space available for anchorage of reinforcement is small and restricted, and due to extensive cracking within the connection at overloads, the anchorage conditions are particularly unfavourable.

Ideally, the strength of a connection should not govern the strength of the structure. Connections should therefore possess strength at least as great as the members they join. Connections should also perform satisfactorily at service loads (controlled cracks and small rotations) and be easy to construct. As will be seen, these requirements are often not easy to satisfy and detailing the reinforcement in connections requires careful attention.

The moments, shears and axial forces in a concrete structure are usually determined using elastic analyses. In such analyses, connections are usually assumed to be rigid, i.e. the ends of all members meeting at the connection are assumed to rotate by the same amount. In addition to being designed to carry the internal actions, the connections should respond with deformations at least similar to those assumed in the analysis. If the connection is too flexible, span moments will exceed those given by the analysis causing increased deformations and, perhaps, premature failure in heavily reinforced members at overload where the ductility required for redistribution of internal actions may not be available.

The difficulty in predicting the load-deformation characteristics of connections is a major obstacle to the use of collapse load analysis for reinforced and prestressed concrete frames. Of course, in any connection, ductility is essential and the use of low ductility steel should not be contemplated.

14.7.2 Knee connections (or two-member connections)

Two-member connections, such as those shown in Figure 14.39, are commonly used in concrete structures. It is often difficult to achieve 100% efficiency in such a connection (i.e. the strength of the connection is equal to the strength of the adjoining members), particularly when subjected to opening moments.

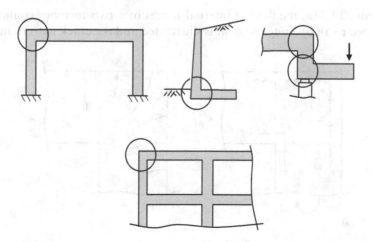

Figure 14.39 Two-member connections.

14.7.2.1 Closing moments

The load path in two-member connections subjected to *closing moments* is shown in Figure 14.40a and the crack pattern caused by increasing the applied load is illustrated in Figure 14.40b. Strut-and-tie models outlined in EN 1992-1-1 for beam–column connections subject to closing moments are shown in Figure 14.41. The strut-and-tie model for a closing moment connection where the beam and columns depths are within the range $0.67 < h_2/h_1 < 1.5$ is shown in Figure 14.41a, together with the corresponding reinforcement layout. For such a load path, the code does not require any checks on transverse link reinforcement or anchorage lengths within the beam–column joint, provided that the tension reinforcement on top of the beam is bent around the corner and continuous.

The strut-and-tie model and reinforcement layout shown in Figure 14.41b is for a closing moment connection when $h_2/h_1 < 0.67$. According to EN 1992-1-1 [1], for this model, the angle θ should be limited by $0.4 \le \tan\theta \le 1.0$ [1]. Transverse (horizontal) links are required to carry the internal tensile forces F_{td1} in the horizontal ties within the connection. The anchorage length l_{bd} for the tensile reinforcement carrying F_{td2} at the top of the column should be determined for the force $\Delta F_{td} = F_{td2} - F_{td1}$.

For the reinforcement details shown in Figure 14.41, the top tensile bars in the beam are easily developed provided the radius of bend is large enough to avoid splitting failure. If necessary, the inclusion of a larger diameter transverse bar tied inside the 90% bend in the top reinforcement could be used to distribute the concentrated force applied to the concrete at the bend and avoid the development of splitting cracks. The main tensile bars should be continuous around the corner.

14.7.2.2 Opening moments

In Figure 14.42a, the flow of internal forces in a two-member connection subjected to an *opening moment* is illustrated and the crack pattern in such

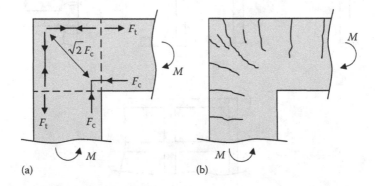

Figure 14.40 Knee connection in closing bending. (a) Internal forces. (b) Crack pattern.

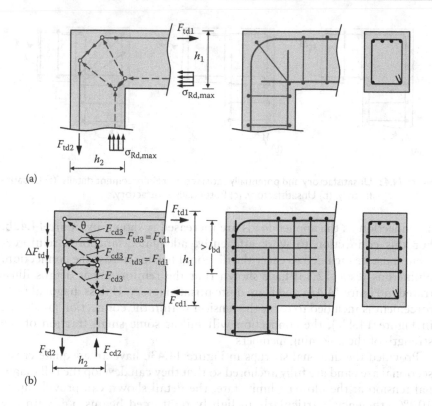

(a)

(b)

Figure 14.41 Two-member connections with closing moments – strut-and-tie models and reinforcement details. (a) Similar depths of beam and column. (b) Very different depths of beam and column $h_1 \gg h_2$.

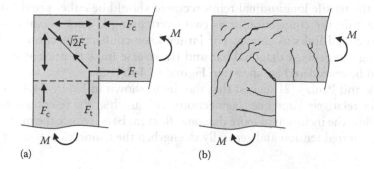

(a)   (b)

Figure 14.42 Knee connection in opening bending. (a) Internal forces. (b) Crack pattern.

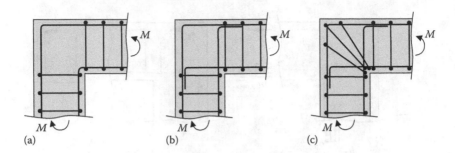

Figure 14.43 Unsatisfactory and potentially satisfactory reinforcement details. (a) Unsatisfactory. (b) Unsatisfactory. (c) Potentially satisfactory.

a connection as the applied loads are increased is shown in Figure 14.42b. For this connection to work efficiently, adequately anchored reinforcement must be included to carry the diagonal tension across the connection. Numerous tests [12,13] have shown that the reinforcement details illustrated in Figure 14.43a and b are quite unsatisfactory. Unless diagonal reinforcement is included to carry the tension within the connection (as shown in Figure 14.43c), the connection will fail at some small fraction of the strength of the adjoining members.

Provided the diagonal stirrups in Figure 14.43c have an adequate cross-sectional area and are fully anchored so that they can develop the full diagonal tension at the ultimate limit state, the detail shown can provide up to 100% efficiency, particularly in lightly reinforced beams (i.e. with $\rho = A_s/bd < 0.01$). In general, the more lightly reinforced the members, the more efficient is the connection. The diagonal stirrups must fit snuggly around the longitudinal steel to effectively control the growth of cracks.

For frame corners subjected to opening moments, when the beam and column are of similar depths, EN 1992-1-1 [1] outlines the strut-and-tie models and reinforcement details shown in Figure 14.44. According to the code, the tensile longitudinal reinforcement should be either provided as a loop within the connection or as two overlapping U bars in combination with inclined links as shown. For larger knee connections carrying large opening moments, a diagonal bar and transverse links to prevent splitting should be considered (as shown in Figure 14.44b).

Park and Paulay [2] suggest that the detail shown in Figure 14.45 is suitable for relatively large knee connections. A haunch at the re-entrant corner will allow the inclusion of more diagonal flexural bars, reduce the magnitude of the internal tension and generally strengthen the connection.

14.7.3 Exterior three-member connections

The flow of internal forces in a typical exterior beam–column connection is illustrated in Figure 14.46a and the crack pattern under increasing load

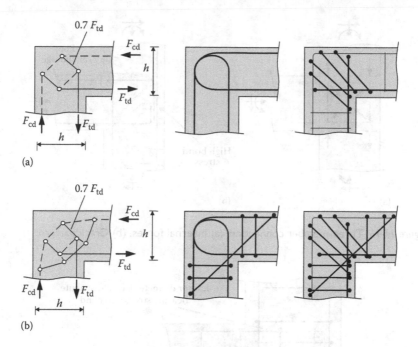

Figure 14.44 Strut-and-tie models and reinforcement details – opening bending [1]. (a) Moderate opening moment, $A_s/bh \leq 0.02$. (b) Large opening moment, $A_s/bh > 0.02$.

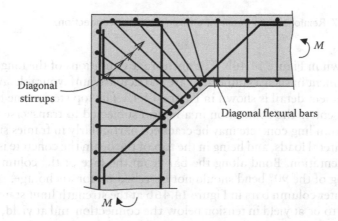

Figure 14.45 Suggested detail for large opening knee connections [2].

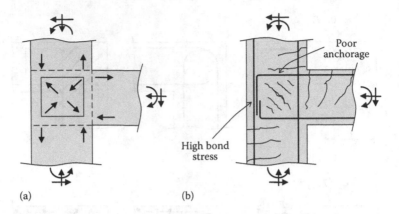

Figure 14.46 Three-member connection. (a) Internal forces. (b) Crack pattern.

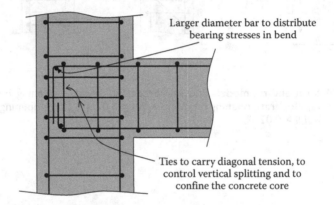

Figure 14.47 Reinforcement detail for a three-member connection.

is as shown in Figure 14.46b. The anchorage conditions of the longitudinal reinforcement in such a connection are particularly unfavourable. A typical reinforcement detail is shown in Figure 14.47. The top steel in the horizontal beam enters the connection in a region subjected to transverse tension. The surrounding concrete may be cracked, particularly in frames subjected to high lateral loads, and being in the top of the beam the concrete is subject to sedimentation. Bond along the bar from the face of the column to the beginning of the 90° bend should not be relied on for anchorage.

The outer column bars in Figure 14.46b at the strength limit state may be required to be at yield in tension below the connection and at yield, or close to it, in compression above the connection. Code provisions for anchorage may not be able to be met here. Tightly fitting and closely spaced column ties within the connection are required to control the splitting cracks caused by the usually high bond stresses around these bars.

When an exterior three-member connection is subjected to reversals of load, such as may occur under seismic loading, the anchorage conditions in such a connection are at their worst. Often mechanical anchorages are required for the longitudinal reinforcement in the beams and a large amount of transverse steel is required within the connection. Park and Paulay [2] point out that under conditions of alternating plasticity, it is unwise to assume that the concrete within the connection will contribute to the shear strength. Confinement of this highly stressed concrete block by closely fitting vertical and horizontal ties is a prerequisite for adequate reinforcement anchorage. Care should be taken to ensure that the reinforcement layout permits the placement and proper compaction of the concrete within the connection.

14.7.4 Interior four-member connections

Similar remarks apply to the interior connections in frames. If the beam moments equilibrate each other (or nearly so), i.e. the column moments are small compared to the beam moments (which is usually the case under gravity loads), no particular problems should arise. However, for frames carrying large lateral loads, the beam moments may be of opposite sign, as shown in Figure 14.48, and the anchorage conditions of the longitudinal bars are critical. Once again, closely spaced horizontal and vertical ties are required to carry diagonal tension across the connection, to control splitting cracks caused by high bond stresses, to control diagonal cracking, to confine the concrete core and, therefore, to improve the anchorage conditions generally within the connection.

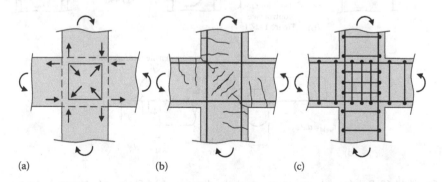

Figure 14.48 Interior four-member connection. (a) Internal forces. (b) Crack pattern. (c) Reinforcement detail.

14.8 DETAILING OF CORBELS

Corbels are short cantilevers that tend to act as non-flexural deep beams, rather than flexural members. Corbels carry reactions or concentrated loads into columns or walls as shown in Figure 14.49a and have shear span-to-depth ratios generally not greater than unity. From studies of stress trajectories in photoelastic models, Franz and Niedenhoff [14] verified the flow of internal forces shown. A typical reinforcement detail for a corbel is shown in Figure 14.49b. The horizontal main primary reinforcement $A_{s,main}$ is required to carry the tension from under the bearing pad back into the column and must therefore be fully anchored at both ends. It should be anchored in the supporting element on the far face, with the anchorage length measured from the position of the vertical column reinforcement at the near face. In the corbel, the main tensile reinforcement must be anchored under the bearing plate with the anchorage length measured from the inner face of the plate. The main tensile reinforcement is sometimes looped horizontally to provide anchorage, rather than bent vertically as shown in Figure 14.49b. Often the main tensile reinforcement is welded to a cross-bar of at least the same diameter (as shown in Figure 14.49c and d) or directly to a steel end ring plate or another anchoring device to ensure

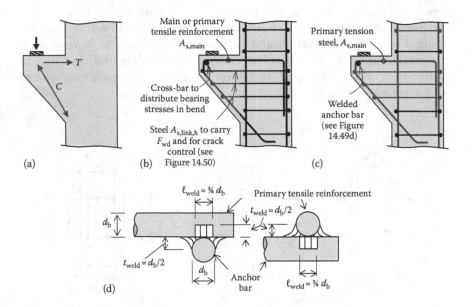

Figure 14.49 Reinforced concrete corbel details. (a) Flow of internal forces. (b) Reinforcement detail. (c) Welded primary tensile steel. (d) Satisfactory weld details [15].

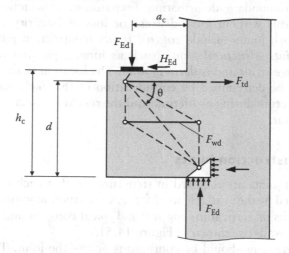

Figure 14.50 Strut-and-tie model of a corbel [1].

adequate anchorage. The welds should be designed to develop the yield strength of the primary tension reinforcement. A weld detail that has been used successfully in corbel tests is shown in Figure 14.49d [15]. Horizontal links (of cross-sectional area $A_{s,link,h}$) are usually included as shown in Figure 14.49b to control the inclined cracking which occurs from the top surface parallel to the compressive strut and to carry the web tension F_{wd} in the intermediate tie shown in the strut-and-tie model of Figure 14.50.

EN 1992-1-1 [1] suggests that corbels may be designed using strut-and-tie models, such as that shown in Figure 14.50, where the inclination θ of the main compressive strut is limited by $1.0 \leq \tan\theta \leq 2.5$. In no case, should $A_{s,link,h} < 0.25 A_{s,main}$. In addition, for heavily loaded corbels, where $F_{Ed} > V_{Rd,c}$ and $a_c > 0.5h_c$, closed vertical links of total area $A_{s,link,v} < 0.5$ F_{Ed}/f_{yd} should be provided over the full depth of the corbel and evenly spaced between the inside edge of the bearing plate and the face of the column.

14.9 JOINTS IN STRUCTURES

14.9.1 Introduction

Joints are introduced into concrete structures for two main reasons:

1. As stopping places in the concreting operation. The location of these *construction joints* depends on the size and production capacity of the construction site and work force.

2. To accommodate deformation (expansion, contraction, rotation, settlement) without local distress or loss of integrity of the structure. Such joints include *control joints* (contraction joints), *expansion joints*, *structural joints* (such as hinges, pin and roller joints), *shrinkage strips* and *isolation joints*. The location of these joints can usually be determined by consideration of the likely movements of the structure during its lifetime and the resulting effects on structural behaviour.

14.9.2 Construction joints

Construction joints are required in structures so that each concrete pour can be handled by the available workforce. Construction joints may be horizontal in slabs or vertical in long walls. Typical construction joint details for slabs and walls are shown in Figure 14.51.

All reinforcement should be continuous across the joint. The keyed or dowelled joints of Figures 14.51b and c are of questionable value, particularly in slabs with top and bottom reinforcement ratios exceeding about 0.0035. If the surface of the hardened concrete at the joint is properly prepared, friction and aggregate interlock on the concrete surface, together with the dowel action of the reinforcement, can provide shear strength at the joint as high as that of adjacent sections placed monolithically.

For a sound joint, after the first concrete pour, the exposed reinforcement should be cleaned and the aggregate of the hardened concrete exposed by wire brushing or water or sand blasting. The hardened concrete should be thoroughly wetted before the new concrete is poured. For a waterproof construction joint, a continuous plastic or rubber waterstop or waterbar is essential. Compaction of the concrete around the waterstop should be thorough and careful.

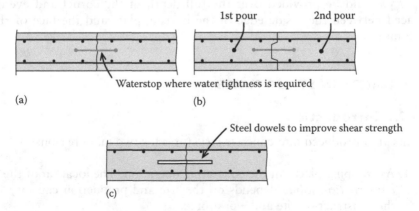

Figure 14.51 Construction joint details. (a) Butt joint. (b) Keyed joint. (c) Dowelled joint.

If possible, construction joints should coincide with other joints (e.g. expansion or contraction joints) so as to minimise the total number of joints in the structure. Construction joints should be disguised (or otherwise hidden) or incorporated as architectural features. It is therefore necessary that the number and location of joints be decided well before the concrete trucks arrive onto the construction site. Ideally, construction joints should not be placed in regions of high moment or shear and should not occur where they could create stress concentrations.

14.9.3 Control joints (contraction joints)

When concrete is not free to contract, restraint to shrinkage and temperature changes produces tension that can, and frequently does, cause excessive cracking. Restraint to shrinkage and temperature changes can be handled in two different ways. Sufficient reinforcement can be inserted to control cracking and cause a large number of very fine, serviceable cracks. Alternatively, control joints can be used in walls and slabs to concentrate the cracking into preformed grooves. The contraction of the concrete is taken up at the control joint and the restraint to shrinkage between cracks is largely removed.

A control joint is an intentionally introduced plane of weakness in a slab or wall. Control joints are spaced and positioned so that cracking and contraction take place only on these preselected straight lines. The joints should be close enough together so that shrinkage and temperature-induced tension in the concrete between the joints remains small. Typical control joint details in reinforced concrete slabs and walls are shown in Figure 14.52. The joint must allow contraction of the concrete (i.e. it must

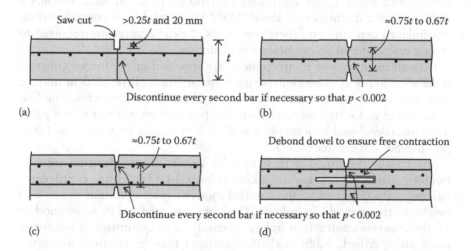

Figure 14.52 Control joint details. (a) Saw-cut joint in slab on ground. (b) Wall ($t < 200$ mm). (c) Wall ($t \geq 200$ mm). (d) Dowelled joint.

be able to open longitudinally), but must resist relative transverse movement (i.e. it must allow for shear transfer, if required).

Control joints are formed by locally reducing the cross-sectional area of the slab or wall by about 25%–33%. Form inserts can be used to form the joint (as illustrated in Figures 14.52b and c) or a saw cut in the fresh concrete on the surface of a slab can be used to create the weakened plane. To ensure relatively free contraction at the joint after a crack has formed, the reinforcement crossing the joint should be no more than $0.002A_c$, where A_c is the cross-sectional area of the concrete slab or wall at the joint.

If significant shear is to be transferred across the joint, dowels can be used to improve shear strength, as shown in Figure 14.52d. The dowels should be debonded on one side of the joint to allow free contraction.

Control joints should be located in regions of low moment since the flexural strength of a slab or wall is usually quite low at a joint. Water tightness is always a potential problem at a control joint and a continuous flexible (expandable) waterstop should be used if a waterproof joint is required.

The position and spacing of control joints depends on many factors including the shrinkage characteristics of the concrete, the curing and exposure conditions, the external restraint (due to supports, adjacent parts of the structure or friction with the ground) and the structural layout. No hard and fast rules can be made. However, as a general rule, a joint should be located wherever an abrupt change in dimension of the structure occurs (thickness, width or height).

In the design of walls and slabs, the designer in general has a choice between joints at close spacings (3–5 m – see Reference [16]) with small quantities of reinforcement (about $0.002A_c$) or joints at much wider centres and much larger quantities of reinforcement (at least $0.006A_c$). Reinforcement quantities of about $0.002A_c$ are unable to control shrinkage-induced cracking and therefore joints at close centres are required to reduce restraint and accommodate contraction.

If small quantities of reinforcement are specified in reinforced concrete slabs with joints at close centres, joint spacing of as little as 3 m may be necessary in dry environments or with high-shrinkage concretes. The first joint should be located no more than 3 m from a corner. The ratio of panel dimensions enclosed by joints in a wall or slab should be as close to 1.0 as possible but not greater than 1.5.

Short walls restrained at their base by a more massive footing are particularly prone to shrinkage cracking (Figure 14.53a), as are cantilevered balcony slabs (Figure 14.53b). Control joints at spacing similar to the wall height (or the span of a cantilevered balcony) are required to accommodate all the concrete contraction unless relatively large quantities of reinforcement are specified. Additional reinforcement may be required along the balcony edge to prevent unsightly cracking at each joint location.

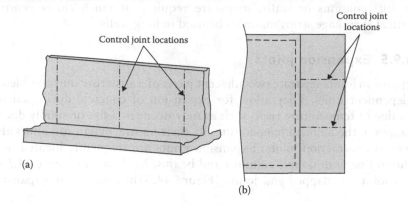

Figure 14.53 Control joint locations. (a) Wall elevation. (b) Balcony plan.

14.9.4 Shrinkage strips

Shrinkage strips serve the same purpose as control joints. A strip of about 1 m width across a building is left during concreting, thus allowing the concrete on either side of the strip to shrink freely. After several weeks, when a significant amount of the drying shrinkage has occurred, the strips are poured and continuity is established.

The reinforcement crossing a shrinkage strip is usually continuous but is lapped or bent horizontally as shown in Figure 14.54 to allow unrestrained contraction on either side of the strip.

In long multistorey framed structures without stiff columns or walls, shrinkage strips are often placed in slabs at about 40 m centre. When there

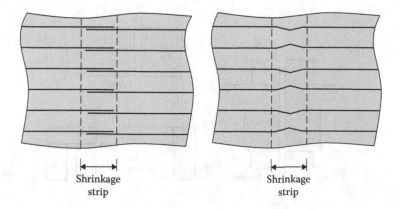

Figure 14.54 Alternative reinforcement details at shrinkage strips.

are stiff columns or walls, strips are required at much closer centres. Vertical shrinkage strips may also be used in long walls.

14.9.5 Expansion joints

Expansion joints separate two adjacent parts of a structure into completely independent units. They allow for expansion of concrete during curing and due to temperature rises, such as may occur in a fire or simply due to changes in the ambient temperature. By their nature, expansion joints also serve as contraction joints. Expansion joints are frequently located on a column line with double columns and beams, as shown in Figure 14.55a. Half joints or dapped-end joints (Figure 14.55b) also act as expansion joints.

The use of expansion joints in buildings is somewhat controversial. The contraction caused by shrinkage is usually several times greater than the expansion caused by ambient temperature rises. Indeed, many large buildings have been built successfully without expansion joints. Notwithstanding this, it is good practice to include expansion joints at abrupt changes in the plan dimensions of a building, as shown in Figure 14.55c, to avoid the stress concentrations and cracking that would otherwise occur at these locations.

It is important that movement joints in the concrete structure be accompanied by and be compatible with movement joints in the finishes, partitions and cladding attached to it. Movement in the concrete structure should not impose loads on the attached non-structural elements.

14.9.6 Structural joints

Structural joints allow free movement (translation and/or rotation) between two parts of a structure. The half joint of Figure 14.55b allows unrestrained

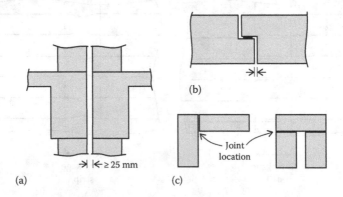

Figure 14.55 Expansion joint details. (a) Double column and beams. (b) Half joint. (c) Building plans – joint locations.

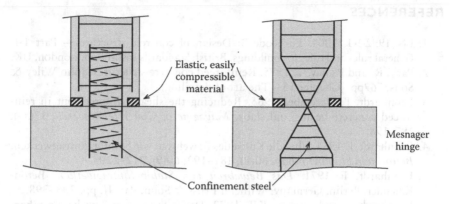

Figure 14.56 Structural hinge joints [16].

translation along the axis of the member and unrestrained rotation and will serve the dual function of a structural hinge and an expansion/contraction joint.

When a hinge is required at the base of a column or when moment is not to be transferred from one element to another, the hinge joints shown in Figure 14.56 may be used. The joints must be able to transmit the imposed axial force and shear. It is essential that the strength and strain capacity of the concrete within the hinge is increased as much as possible using closely spaced confinement reinforcement, usually in the form of helical reinforcement.

When a concrete beam or slab is supported by a masonry wall, it is prudent to ensure that the concrete movements do not cause distress to the masonry. A sliding joint should always be used to break the bond between the concrete and the wall. Two layers of galvanised steel flashing (or equivalent) usually provide a satisfactory sliding joint between a concrete slab and a load-bearing masonry wall.

Isolation joints separate different parts of a structure and ensure that deformation in one part of the structure does not impose loads on the other. Differential shortening of adjacent columns in framed structures can induce significant moments and shears into floor slabs and beams. Isolation joints are frequently used to separate portions of a structure with significantly different sustained stress levels. For example, tower columns in a multistorey building may need to be separated from adjacent columns with lower stress levels due to differences in their long-term axial shortening. Structural members supporting vibrating machinery or pumps are usually isolated from the remainder of the structure. To avoid differential settlement problems, slabs on ground are isolated from the walls and columns passing through them. In structures subjected to seismic loads, parts with dissimilar mass and stiffness are separated so that they are able to oscillate without hammering against each other.

REFERENCES

1. EN 1992-1-1. 2004. Eurocode 2: Design of concrete structures – Part 1-1: General rules and rules for buildings. British Standards Institution, London, UK.
2. Park, R. and Paulay, T. 1975. Reinforced concrete structures. John Wiley & Sons, 769pp. (Chapter 13 – The art of detailing).
3. Leonhardt, F. December 1965. Reducing the shear reinforcement in reinforced concrete beams and slabs. *Magazine of Concrete Research*, 17(53), 187–198.
4. Leonhardt, F. 1965. Über die Kunst des Bewehrens von Stahlbetontragwerken. *Beton- und Stahlbetonbau*, 60(8), 181–192; 60(9), 212–220.
5. Leonhardt, F. 1971. *Das Bewehren von Stahlbetontragwerken*. Beton-Kalender. Berlin, Germany: Wilhelm Ernst & Sohn, Part II, pp. 303–398.
6. Leonhardt, F. and Teichen, K.T. 1972. Druck-Stösse von Bewehrungstäben. *Deutscher Ausschuss für Stahlbeton*, Bulletin No. 222, Wilhelm Ernst & Sohn, Berlin, Germany, pp. 1–53.
7. Leonhardt, F., Walther, R. and Dilger, W. 1964. *Schubversuche an Durchlaufträgern*. Deutscher Ausschuss für Stahlbeton. Heft 163. Berlin, Germany: Ernst & Sohn.
8. Goto, Y. 1971. Cracks formed in concrete around deformed tension bars. *ACI Journal*, 68(4), 244–251.
9. Tepfers, R. 1979. Cracking of concrete cover along anchored deformed reinforcing bars. *Magazine of Concrete Research*, 31(106), 3–12.
10. Tepfers, R. 1982. Lapped tensile reinforcement splices. *Journal of the Structural Division, ASCE*, 108(1), 283–301.
11. Untrauer, R.E. and Henry, R.L. 1965. Influence of normal pressure on bond strength. *ACI Journal*, 62(5), 577–586.
12. Mayfield, B., Kong, F.-K., Bennison A., and Triston Davies, J.C.D. 1971. Corner joint details in structural lightweight concrete. *ACI Journal*, 68(5), 366–372.
13. Swann, R.A. 1969. Flexural strength of corners of reinforced concrete portal frames. Technical Report TRA 434. Cement and Concrete Association, London, UK.
14. Franz, G. and Niedenhoff, H. 1963. *The Reinforcement of Brackets and Short Deep Beams*. Library Translation No. 114. London, UK: Cement and Concrete Association.
15. ACI 318-14M. 2014. *Building Code Requirements for Reinforced Concrete*. Detroit, MI: American Concrete Institute.
16. Fintel, M. 1974. *Handbook of Concrete Engineering*. New York: Van Nostrand Reinhold Company.

Index

Printed in the United States
by Baker & Taylor Publisher Services